S.Chand's IIT Foundation Series

A Compact and Comprehensive Book of
IIT Foundation
Mathematics

CLASS – VIII

S. Chand's IIT Foundation Series

A Compact and Comprehensive Book of

IIT Foundation Mathematics

CLASS – VIII

S.K. GUPTA
Principal (Retd.)
Birla Vidya Mandir,
Nainital

ANUBHUTI GANGAL
M.A. (Gold Medallist), M.Ed.
Formerly, Senior Faculty Member
The Daly College, Indore
Birla Vidya Mandir, Nainital

S. CHAND & COMPANY LTD.
(AN ISO 9001: 2008 COMPANY)
RAM NAGAR, NEW DELHI-110 055

S. CHAND & COMPANY LTD.
(An ISO 9001 : 2008 Company)
Head Office: 7361, RAM NAGAR, NEW DELHI - 110 055
Phone: 23672080-81-82, 9899107446, 9911310888
Fax: 91-11-23677446
Shop at: **schandgroup.com**; *e-mail:* **info@schandgroup.com**

Branches :

AHMEDABAD : 1st Floor, Heritage, Near Gujarat Vidhyapeeth, Ashram Road, **Ahmedabad** - 380 014, Ph: 27541965, 27542369, ahmedabad@schandgroup.com

BENGALURU : No. 6, Ahuja Chambers, 1st Cross, Kumara Krupa Road, **Bengaluru** - 560 001, Ph: 22268048, 22354008, bangalore@schandgroup.com

BHOPAL : Bajaj Tower, Plot No. 243, Lala Lajpat Rai Colony, Raisen Road, **Bhopal** - 462 011, Ph: 4274723. bhopal@schandgroup.com

CHANDIGARH : S.C.O. 2419-20, First Floor, Sector - 22-C (Near Aroma Hotel), **Chandigarh** -160 022, Ph: 2725443, 2725446, chandigarh@schandgroup.com

CHENNAI : 152, Anna Salai, **Chennai** - 600 002, Ph: 28460026, 28460027, chennai@schandgroup.com

COIMBATORE : No. 5, 30 Feet Road, Krishnasamy Nagar, Ramanathapuram, **Coimbatore** -641045, Ph: 0422-2323620 coimbatore@schandgroup.com **(Marketing Office)**

CUTTACK : 1st Floor, Bhartia Tower, Badambadi, **Cuttack** - 753 009, Ph: 2332580; 2332581, cuttack@schandgroup.com

DEHRADUN : 1st Floor, 20, New Road, Near Dwarka Store, **Dehradun** - 248 001, Ph: 2711101, 2710861, dehradun@schandgroup.com

GUWAHATI : Pan Bazar, **Guwahati** - 781 001, Ph: 2738811, 2735640 guwahati@schandgroup.com

HYDERABAD : Padma Plaza, H.No. 3-4-630, Opp. Ratna College, Narayanaguda, **Hyderabad** - 500 029, Ph: 24651135, 24744815, hyderabad@schandgroup.com

JAIPUR : A-14, Janta Store Shopping Complex, University Marg, Bapu Nagar, **Jaipur** - 302 015, Ph: 2719126, jaipur@schandgroup.com

JALANDHAR : Mai Hiran Gate, **Jalandhar** - 144 008, Ph: 2401630, 5000630, jalandhar@schandgroup.com

JAMMU : 67/B, B-Block, Gandhi Nagar, **Jammu** - 180 004, (M) 09878651464 **(Marketing Office)**

KOCHI : Kachapilly Square, Mullassery Canal Road, Ernakulam, **Kochi** - 682 011, Ph: 2378207, cochin@schandgroup.com

KOLKATA : 285/J, Bipin Bihari Ganguli Street, **Kolkata** - 700 012, Ph: 22367459, 22373914, kolkata@schandgroup.com

LUCKNOW : Mahabeer Market, 25 Gwynne Road, Aminabad, **Lucknow** - 226 018, Ph: 2626801, 2284815, lucknow@schandgroup.com

MUMBAI : Blackie House, 103/5, Walchand Hirachand Marg, Opp. G.P.O., **Mumbai** - 400 001, Ph: 22690881, 22610885, mumbai@schandgroup.com

NAGPUR : Karnal Bag, Model Mill Chowk, Umrer Road, **Nagpur** - 440 032, Ph: 2723901, 2777666 nagpur@schandgroup.com

PATNA : 104, Citicentre Ashok, Govind Mitra Road, **Patna** - 800 004, Ph: 2300489, 2302100, patna@schandgroup.com

PUNE : 291/1, Ganesh Gayatri Complex, 1st Floor, Somwarpeth, Near Jain Mandir, **Pune** - 411 011, Ph: 64017298, pune@schandgroup.com **(Marketing Office)**

RAIPUR : Kailash Residency, Plot No. 4B, Bottle House Road, Shankar Nagar, **Raipur -** 492 007, Ph: 09981200834, raipur@schandgroup.com **(Marketing Office)**

RANCHI : Flat No. 104, Sri Draupadi Smriti Apartments, East of Jaipal Singh Stadium, Neel Ratan Street, Upper Bazar, **Ranchi** - 834 001, Ph: 2208761, ranchi@schandgroup.com **(Marketing Office)**

SILIGURI : 122, Raja Ram Mohan Roy Road, East Vivekanandapally, P.O., **Siliguri**-734001, Dist., Jalpaiguri, (W.B.) Ph. 0353-2520750 **(Marketing Office)**

VISAKHAPATNAM : Plot No. 7, 1st Floor, Allipuram Extension, Opp. Radhakrishna Towers, Seethammadhara North Extn., **Visakhapatnam** - 530 013, (M) 09347580841, visakhapatnam@schandgroup.com **(Marketing Office)**

First Edition 2012

ISBN: 81-219-3961-5   **Code :** 14 639

PRINTED IN INDIA
By Rajendra Ravindra Printers Pvt. Ltd., 7361, Ram Nagar, New Delhi -110 055
and published by S. Chand & Company Ltd., 7361, Ram Nagar, New Delhi -110 055.

PREFACE AND A NOTE FOR THE STUDENTS

ARE YOU ASPIRING TO BECOME AN ENGINEER AND BECOME AN IIT SCHOLAR ?

Here is the book especially designed to motivate you, to sharpen your intellect and develop the right attitude and aptitude and lay solid foundation for your success in various entrance examinations like **IIT, AIEEE, EAMCET, WBJEE, MPPET, SCRA, Kerala PET, OJEE, Raj PET, AMU,** etc.

SALIENT FEATURES

1. Content based on the curriculum of the classes for *CBSE, ICSE, Andhra Pradesh* and *Boards of School Education of Other States.*

2. Full and comprehensive coverage of all the topics.

3. Detailed synopsis of each chapter at the beginning in the form of *'Key Concepts'*. This will not only facilitate thorough *'Revision'* and *'Recall'* of every topic but also greatly help the students in understanding and mastering the concepts besides providing a *back-up* to classroom teaching.

4. The books are enriched with an exhaustive range of hundreds of thought provoking objective questions in the form of solved examples and practice questions in Question Banks which not only offer a great variety and reflect the modern trends but also invite, explore, develop and put to test the *thinking, analysing* and *problem solving skills of the students*.

5. **Answers, Hints** and **Solutions** have been provided to boost up the morale and increase the confidence level.

6. **Self Assessment Sheets** have been given at the end of each chapter to help the students to assess and evaluate their understanding of the concepts and learn to attack the problems independently.

We hope the series will be able to fulfil its aims and objectives and will be found immensely useful by the students aspiring to become topclass engineers.

Suggestions for improvement and also the feedback received from various sources would be most welcome and gratefully acknowledged.

AUTHORS

CONTENTS

NUMBER SYSTEM

- *Numbers*
- *Fractions and Decimals*
- *HCF and LCM*
- *Square Roots and Cube Roots*
- *Exponents*
- *Surds*

Chapter 1

NUMBERS

1. Numbers of the form $\dfrac{p}{q}$, $q \neq 0$, where p and q are integers and which can be expressed in the form of terminating or repeating decimals are called **rational numbers**.

   e.g., $\dfrac{7}{32} = 0.21875$, $\dfrac{8}{15} = 0.5\overline{3}$ are rational numbers.

2. **Properties of operations of rational numbers**

   For any rational numbers a, b, c.

   (i) Rational numbers are closed under addition, multiplication and subtraction,
   i.e., $(a + b)$, $(a \times b)$ and $(a - b)$ are also rational numbers.

   (ii) Rational numbers follow the **commutative law** of addition and multiplication,
   i.e., $a + b = b + a$ and $a \times b = b \times a$.

   (iii) Rational numbers follow the **associative law** of addition and multiplication,
   i.e., $(a + b) + c = a + (b + c)$ and $(a \times b) \times c = a \times (b \times c)$.

   (iv) **Additive identity** : 0 is the additive identity for rational numbers as $a + 0 = 0 + a = 0$.

   (v) **Multiplicative identity** : 1 is the multiplicative identity for rational numbers as $a \times 1 = 1 \times a = a$.

   (vi) **Additive inverse** : For every rational number 'a', there is a rational number '$-a$' such that $a + (-a) = 0$.

   (vii) **Multiplicative inverse** : For every rational number 'a' except 0, there is a rational number $\dfrac{1}{a}$ such that $a \times \dfrac{1}{a} = 1$.

   (viii) **Distributive property** : Multiplication distributes over addition in rational numbers,
   i.e., $a(b + c) = a \times b + a \times c$.

   (ix) Between any two different rational numbers, there are infinitely many rational numbers. Rational numbers between any two given rational numbers a and b are $q_1 = \dfrac{1}{2}(a + b)$, $q_2 = \dfrac{1}{2}(q_1 + b)$, $q_3 = \dfrac{1}{2}(q_2 + b)$ and so on.

3. Numbers which when converted into decimals are expressible neither as terminating nor as repeating decimals are called **irrational numers**.

   e.g., $\sqrt{2} = 1.41421356237.......$ is an irrational number.

   An irrational number cannot be expressed in the form $\dfrac{p}{q}$ $(q \neq 0)$ and its exact root cannot be found.

4. The totality of all rational and all irrational numbers is called real numbers which is denoted by R.

   The natural numbers, whole numbers, integers non-integral rationals $\left(\dfrac{4}{7}, -\dfrac{7}{8}\right)$ etc., are all contained in real numbers.

5. **Unit's digit of any number that can be expressed as a power of any natural number between 1 and 9.**

   (i) The units' digit of any number expressed as a power of 2 is any of the digits 2, 4, 8, 6 as $2^1 = 2$, $2^2 = 4$, $2^3 = 8$, $2^4 = 16$, $2^5 = 32$, ... etc.

   (ii) The units' digit of any number expressed as a power of 3 is any of the digits 3, 9, 7, 1 as $3^1 = 3$, $3^2 = 9$, $3^3 = 27$, $3^4 = 81$, $3^5 = 243$, ... etc.

   (iii) The units' digit of any number expressed as a power of 4 is 4 if the power is odd and 6 if the power is even as $4^1 = 4$, $4^2 = 16$, $4^3 = 64$, $4^4 = 256$, ...etc.

   (iv) The units' digit of any number expressed as a power of 5 is always 5 as $5^1 = 5$, $5^2 = 25$, $5^3 = 125$, ... etc.

   (v) The units' digit of any number expressed as a power of 6 is always 6 as $6^1 = 6$, $6^2 = 36$, $6^3 = 216$, ... etc.

   (vi) The units' digit of any number expressed as a power of 7 is any of the digits 7, 9, 3, 1 as $7^1 = 7$, $7^2 = 49$, $7^3 = 343$, $7^4 = 2401$, ... etc.

   (vii) The units' digit of any number expressed as a power of 8 is any of the digits 8, 4, 2, 6 as $8^1 = 8$, $8^2 = 64$, $8^3 = 512$, $8^4 = 4096$, ... etc.

   (viii) The units' digit of any number expressed as a power of 9 is 9 if the power is odd and 1 if the power is even as $9^1 = 9$, $9^2 = 81$, $9^3 = 729$, $9^4 = 6561$, ... etc.

6. **Total number of factors of any number.**

   The number has to be resolved into prime factors. If the prime factorisation of the given number $= 2^m \times 3^n \times 5^p \times 11^q$, then the total number of factors $= (m + 1) \times (n + 1) \times (p + 1) \times (q + 1)$.

   For example, if $N = 2^4 \times 5^6 \times 7^4 \times 13^1$, then

   Number of factors of $N = (4 + 1) \times (6 + 1) \times (4 + 1) \times (1 + 1) = 5 \times 7 \times 5 \times 2 = 350$.

   $2^4 = 16$ ∴ factors of 16 are 1, 2, 4, 8, 16, i.e., 5 in number, i.e., $(4 + 1)$ in numbers.

   Similarly $7^4 = 2401$ ∴ factors of 2401 and 1, 7, 49, 343 and 2401, i.e., 5 in number and so on.

Solved Examples

Ex. 1. *If a number 573 xy is divisible by 90, then what is the value of x + y ?*

Sol. 573 xy is divisible by 90, i.e., (9×10)

   ⇒ 573 xy is divisible by both 9 and 10.

   ⇒ $y = 0$ as a number is divisible by 10 if its ones' digit $= 0$.

   Also, sum of digits $= 5 + 7 + 3 + x + 0 = 15 + x$

   For divisibility by 9, $15 + x = 18 \Rightarrow x = 3$

   ∴ $x + y = 3 + 0 = 3$.

Ex. 2. *The product of two whole numbers is 13. What is the sum of the squares of their reciprocals ?*

Sol. The two whole numbers whose product is 13 are 1 and 13.

   ∴ Required sum $= 1 + \dfrac{1}{13^2} = 1 + \dfrac{1}{169} = \dfrac{170}{169}$.

Ex. 3. *What per cent is the least rational number of the greatest rational number, if $\dfrac{1}{2}, \dfrac{2}{5}, \dfrac{1}{3}$ and $\dfrac{5}{9}$ are arranged in ascending order ?*

Sol. Since LCM of 2, 5, 3, 9 = 270, $\dfrac{1}{2} = \dfrac{135}{270}, \dfrac{2}{5} = \dfrac{108}{270}, \dfrac{1}{3} = \dfrac{90}{270}, \dfrac{5}{9} = \dfrac{150}{270}$

   ∴ Arranged in ascending order the numbers are $\dfrac{90}{270}, \dfrac{108}{270}, \dfrac{135}{270}, \dfrac{150}{270}$, i.e., $\dfrac{1}{3}, \dfrac{2}{5}, \dfrac{1}{2}$ and $\dfrac{5}{9}$.

   ∴ Required per cent $= \left(\dfrac{1}{3} \div \dfrac{5}{9}\right) \times 100\% = \left(\dfrac{1}{3} \times \dfrac{9}{5} \times 100\right)\% = \mathbf{60\%}$.

Ex. 4. *Find the units' digit in the expression $11^1 . 12^2 . 13^3 . 14^4 . 15^5 . 16^6$?*

Sol. Units' digit in the given expression

= Units' digit of 11^1 × Units' digit of 2^2 × Units' digit of 3^3 × Units' digit of 4^4 × Units' digit of 5^5 × Units' digit of 6^6

= Units' digit of $(1 \times 4 \times 7 \times 6 \times 5 \times 6)$

= Units' digit of 5040 = **0**.

Ex. 5. *Find the units' digit in the expression $(515)^{31} + (525)^{90}$?*

Sol. Since the units' digit of any number written as a power of 5 is always 5,

Units' digit in the expression $(515)^{31} + (525)^{90}$

= Units' digit of 5^{31} + Units digit of 5^{90}

= Units' digit of $(5 + 5) = 10 = $ **0**.

Ex. 6. *What is the total number of factors of the number $N = 4^{11} \times 14^5 \times 11^2$?*

Sol. $N = 4^{11} \times 14^5 \times 11^2 = (2^2)^{11} \times (2 \times 7)^5 \times 11^2 = 2^{22} \times 2^5 \times 7^5 \times 11^2 = 2^{27} \times 7^5 \times 11^2$

∴ Total number of prime factors of

$N = (27 + 1) \times (5 + 1) \times (2 + 1) = 28 \times 6 \times 3 = $ **504**.

Ex. 7. *The numbers 1, 3, 5, ..., 25 are multiplied together. What is the number of zeros at the right end of the product ?*

Sol. Since all the numbers to be multiplied are odd number, the digit at the units' place will be 5. Hence the number of zeros at the right end of the product is zero.

Ex. 8. *In a division sum, the remainder is 6 and the divisor is 5 times the quotient and is obtained by adding 2 to thrice of remainder. Find the dividend.*

Sol. Divisor = $3 \times 6 + 2 = 20$ (∵ Remainder = 6)

Quotient = $\dfrac{20}{5} = 4$

∴ Dividend = Divisor × Quotient + Remainder

= $20 \times 4 + 6 = $ **86**.

Ex. 9. *A three digit number 3a5 is added to another 3-digit number 933 to give a 4-digit number 12b8, which is divisible by 11. Then, find the value of a + b ?*

Sol. Given, $a + 3 = b$ ... (i)

$$\begin{array}{r} 3\ a\ 5 \\ +\ 9\ 3\ 3 \\ \hline 12\ b\ 8 \end{array}$$

For 12b8 to be divisible by 11,

$(8 + 2) - (b + 1) = 0 \Rightarrow 10 - b - 1 = 0 \Rightarrow 9 - b = 0 \Rightarrow b = 9$

∴ From (i) $a = 6$

∴ $a + b = 6 + 9 = $ **15**.

Ex. 10. *If $\dfrac{1}{891} = 0.00112233445566778899........$ Then what is the value of $\dfrac{198}{891}$?*

Sol. $\dfrac{198}{891} = \dfrac{1}{891} \times 198 = \dfrac{2}{9} = 0.2222.........$

Ex. 11. *What is the value of $\dfrac{1}{5} \div \dfrac{1}{5} \div \dfrac{1}{5} \div \dfrac{1}{5} \div \dfrac{1}{5} \div 5 \div 5 \div 5?$*

Sol. $\dfrac{1}{5} \div \dfrac{1}{5} \div \dfrac{1}{5} \div \dfrac{1}{5} \div \dfrac{1}{5} \div 5 \div 5 \div 5$

$= \dfrac{1}{5} \div \dfrac{1}{5} \div \dfrac{1}{5} \div \dfrac{1}{5} \div \dfrac{1}{5} \div 5 \div 1 = \dfrac{1}{5} \div \dfrac{1}{5} \div \dfrac{1}{5} \div \dfrac{1}{5} \div \dfrac{1}{5} \div 5$

$= \dfrac{1}{5} \div \dfrac{1}{5} \div \dfrac{1}{5} \div \dfrac{1}{5} \div \dfrac{1}{25} = \dfrac{1}{5} \div \dfrac{1}{5} \div \dfrac{1}{5} \div \dfrac{1}{5} \times 25 = \dfrac{1}{5} \div \dfrac{1}{5} \div \dfrac{1}{5} \div 5$

$= \dfrac{1}{5} \div \dfrac{1}{5} \div \left(\dfrac{1}{5} \times \dfrac{1}{5} \right) = \dfrac{1}{5} \div \dfrac{1}{5} \div \dfrac{1}{25} = \dfrac{1}{5} \div \dfrac{1}{5} \times 25 = \dfrac{1}{5} \div 5 = \dfrac{1}{25}$.

Question Bank–1

1. If the remainder obtained by subtracting a number from its own square is 4 times the number, what is the number ?

   (a) 4
   (b) 3
   (c) 6
   (d) 5

2. The difference between the squares of two consecutive odd integers is always divisible by

   (a) 8
   (b) 7
   (c) 6
   (d) 3

3. A number when divided by 296 leaves 75 as remainder. If the same number is divided by 37, the remainder obtained is

   (a) 2
   (b) 1
   (c) 11
   (d) 8

4. A number when divided by 5 leaves 3 as remainder. If the square of the same number is divided by 5, the remainder obtained is

   (a) 9
   (b) 4
   (c) 1
   (d) 3

5. When N is divided by 4, the remainder is 3. What is the remainder when $2N$ is divided by 4.

   (a) 2
   (b) 3
   (c) 4
   (d) 8

6. The units' digit in the expression $(11^1 + 12^2 + 13^2 + 14^4 + 15^5 + 16^6)$ is

   (a) 1
   (b) 9
   (c) 7
   (d) 0

7. The digit in the units' place of $[(251)^{98} + (21)^{29} - (106)^{100} + (705)^{35} - (16)^4 + 259]$ is

   (a) 1
   (b) 4
   (c) 5
   (d) 6

8. The number of digits in $(48^4 \times 5^{12})$ is

   (a) 18
   (b) 16

   (c) 14
   (d) 12

9. Which one of the following is a rational number ?

   (a) $(\sqrt{2})^2$
   (b) $2\sqrt{2}$
   (c) $2 + \sqrt{2}$
   (d) $\dfrac{\sqrt{2}}{2}$

10. If x is a rational number and y is an irrational number, then

    (a) Both $x + y$ and xy are necessarily irrational.
    (b) Both $x + y$ and xy are necessarily rational.
    (c) xy is necessarily irrational, but $x + y$ can be either rational or irrational.
    (d) $x + y$ is necessarily irrational, but xy can be either rational or irrational.

11. Consider the following statements :

    A number $a_1\, a_2\, a_3\, a_4\, a_5\, a_6$ is divisible by 11 if

    1. $(a_1 + a_3 + a_5) - (a_2 + a_4 + a_6) = 0$

    2. $(a_1 + a_3 + a_5) - (a_2 + a_4 + a_6)$ is divisible by 11

    Which of these statements is/are correct ?

    (a) 1 alone
    (b) 2 alone
    (c) Both 1 and 2
    (d) Neither 1 nor 2

12. The number 111,111,111,111 is divisible by

    (a) 9 and 11
    (b) 5 and 11
    (c) 3 and 9
    (d) 3 and 11

13. If $1^2 + 2^2 + 3^2 + \ldots\ldots + 512^2 = m$, then $2^2 + 4^2 + 6^2 + \ldots\ldots + 1024^2$ is equal to

    (a) $3m$
    (b) $4m$
    (c) m^2
    (d) m^3

14. If m and n are positive integers, then the digit in the units' place of $5^n + 6^m$ is always

    (a) 1
    (b) 5
    (c) 6
    (d) $n + m$

15. What number should replaced M in this multiplication problem ?

$$\begin{array}{r} 3\,M\,4 \\ \times\quad 4 \\ \hline 1\,2\,1\,6 \end{array}$$

(a) 0 (b) 5
(c) 7 (d) 8

16. m and n are integers and $\sqrt{mn} = 10$. Which of the following cannot be a value of $m + n$?

(a) 25 (b) 52
(c) 101 (d) 50

17. If a six digit number $93p25q$ is divisible by 88, then the values of p and q are respectively

(a) 2 and 8 (b) 8 and 2
(c) 8 and 6 (d) 6 and 8

18. What is the 25th digit to the right of the decimal point in the decimal of $\dfrac{6}{11}$?

(a) 5 (b) 3
(c) 4 (d) 6

19. A number when divided by the sum of 555 and 445 gives two times their difference as quotient and 30 as remainder. The number is

(a) 220030 (b) 22030
(c) 1220 (d) 1250

20. There are four prime numbers written is ascending order. The product of the first three is 385 and that of last three is 1001. Find the first number.

(a) 5 (b) 7
(c) 11 (d) 17

21. What is the highest power of 5 that divides $90 \times 80 \times 70 \times 60 \times 50 \times 40 \times 30 \times 20 \times 10$

(a) 10 (b) 12
(c) 14 (d) 15

22. Five digit numbers are formed such that either all the digits are even or all the digits are odd. If no digit is allowed to be repeated in one number find the difference between the maximum possible number with odd digits and the minimum possible number with even digits

(a) 77063 (b) 79999
(c) 72841 (d) 86420

23. Which one of the following numbers will completely divide $(3^{25} + 3^{26} + 3^{27} + 3^{28})$?

(a) 11 (b) 16
(c) 25 (d) 30

24. A number when divided successively by 4 and 5 leaves remainders 1 and 4 respectively. When it is successively divided by 5 and 4, then the respective remainders will be

(a) 1, 2 (b) 2, 3
(c) 3, 2 (d) 4, 1

25. If $2.5252525........ = \dfrac{p}{q}$ (in the lowest form) then what is the value of $\dfrac{q}{p}$?

(a) 0.4 (b) 0.42525
(c) 0.0396 (d) 0.396

26. What is the sum of two numbers whose difference is 45, and the quotient of the greater number by the lesser number is 4 ?

(a) 100 (b) 90
(c) 80 (d) 75

27. The number of factors of a number $N = 2^3 \times 3^2 \times 5^3$ is

(a) 18 (b) 45
(c) 48 (d) 9

28. $P = 441 \times 484 \times 529 \times 576 \times 625$. The total number of factors are

(a) 607 (b) 5706
(c) 1024 (d) 6075

29. If $N = 12^3 \times 3^4 \times 5^2$, then the total number of even factors of N is

(a) 25 (b) 121
(c) 144 (d) 84

30. The units' digit of the sum $1 + 9 + 9^2 + ... 9^{1006}$ is

(a) 2 (b) 1
(c) 9 (d) 0

Answers

1. (d)	2. (a)	3. (b)	4. (b)	5. (a)	6. (b)	7. (b)	8. (b)	9. (a)	10. (d)
11. (c)	12. (d)	13. (b)	14. (a)	15. (a)	16. (d)	17. (c)	18. (a)	19. (a)	20. (a)
21. (a)	22. (a)	23. (d)	24. (b)	25. (d)	26. (d)	27. (c)	28. (d)	29. (c)	30. (d)

Hints and Solutions

1. (d) Let the number be x. Then,

$$x^2 - x = 4x \Rightarrow x^2 = 5x \Rightarrow x = 5$$

2. (a) Let the two consecutive odd integers be $(2x + 1)$ and $(2x + 3)$.

$$\therefore (2x+3)^2 - (2x+1)^2 = 4x^2 + 12x + 9 - 4x^2 - 4x - 1$$
$$= 8x + 8 = 8(x + 1)$$

Hence the difference is divisible by 8.

3. (b) Given number $= 296n + 75$

$$= 37 \times 8 \times n + (37 \times 2 + 1)$$
$$= 37(8n + 2) + 1$$

$\therefore$ Required remainder $= 1$

4. (b) Let the given number $= 5n + 3$

$\therefore$ Square of the number

$$= (5n + 3)^2$$
$$= 25n^2 + 30n + 9$$
$$= 5 \times 5n^2 + 5 \times 6n + 5 + 4$$
$$= 5(5n^2 + 6n + 1) + 4$$

$\therefore$ Required remainder $= 4$.

5. (a) $N = 4Q + 3$, where Q is the quotient.

$$\therefore 2N = 8Q + 6 = 4 \times 2Q + 4 + 2 = 4(2Q + 1) + 2$$

$\Rightarrow$ Required remainder $= 2$.

6. (b) Units' digit of the given expression

$=$ Units' digit of $1^1 +$ Units' digit of $2^2 +$ Units' digit of $3^3 +$ Units' digit of $4^4 +$ Units' digit of $5^5 +$ Units' digit of 6^6

$=$ Units' digit of $(1 + 4 + 7 + 6 + 5 + 6)$

$=$ Units' digit of $29 = \mathbf{9}$.

[**Note:** $3^3 = 27$, $4^4 = 256$; units' digit for any power of 5 and 6 = 5 and 6 respectively].

7. (b) Units' place digit

$=$ Units' digit of $1^{98} +$ Units' digit of 1^{29} $-$ Units' digit of $6^{100} +$ Units' digit of 5^{35} $-$ Units' digit of $6^4 + 9$

$= 1 + 1 - 6 + 5 - 6 + 9 = \mathbf{4}$

[**Note:** The units' digit of any positive integral power of 1 is always 1, of any positive integral power of 5 is always 5 and of any positive integral power of 6 is always 6.]

8. (b) $48^4 \times 5^{12} = (3 \times 16)^4 \times 5^{12} = (3 \times 2^4)^4 \times 5^{12}$

$$= 3^4 \times 2^{16} \times 5^{12} = 3^4 \times 2^4 \times 2^{12} \times 5^{12}$$
$$= (3 \times 2)^4 \times (2 \times 5)^{12} = 6^4 \times 10^{12}$$
$$= 1296 \times 10^{12}$$

$\therefore$ Number of digits $= 4 + 12 = 16$.

9. (a) $(\sqrt{2})^2 = 2$, a rational number.

12. (d) Sum of digits $= 12$, hence divisible by 3

Sum of digits at odd place – Sum of digits at even places

$= 6 - 6 = 0$, hence divisible by 11.

13. (b) $2^2 + 4^2 + 6^2 + \dots + (1024)^2$

$$= 2^2(1^2 + 2^2 + 3^2 + \dots + 512^2) = 4m.$$

14. (a) The units' digit of any positive integer power of 5 is always 5 and that of any positive integral power of 6 is always 6.

$\therefore$ Units' digit of $5^m + 6^n =$ Units' digit of $(5 + 6)$
$$= \text{Units' digit of } 11 = \mathbf{1}.$$

16. (d) $\sqrt{mn} = 10 \Rightarrow mn = 100$

$\therefore$ The possible pairs of m and n are

$(m, n) = (1, 100), (2, 50), (4, 25), (5, 20), (10, 10)$

$\Rightarrow m + n$ can be 101, 52, 29, 25, 20

So 50 cannot be a value of $m + n$.

17. (c) A number is divisible by 88 if it is divisible by (8×11), i.e., by both 8 and 11.

For a number to be divisible by 8 the last three digits should be divisible by 8, i.e., from the given options we can see that $q = 6$ as 256 is div. by 8.

$\therefore$ Sum of digits at odd places of given number
$= 6 + 2 + 3 = 11$

Sum of digits at even places $= 5 + p + 9$
$= 14 + p$

For divisibility by 11, $14 + p - 11 = 22 \Rightarrow p = 8$.

18. (a) $\dfrac{6}{11} = 0.54545454\dots$

Thus we can see that odd places are occupied by the digit 5 and even places by the digit 4. Therefore, the 25th digits to the right of the decimal point in $\dfrac{6}{11}$ is 5.

19. (a) Divisor = 555 + 445 = 1000

Quotient = 2 × (555 − 445) = 220

Remainder = 30

∴ Dividend = Divisor × Quotient + Remainder

= 1000 × 220 + 30 = 220030.

20. (a) Let the numbers be a, b, c, d. Then,

$abc = 385$ and $bcd = 1001$

$\Rightarrow \dfrac{abc}{bcd} = \dfrac{385}{1001} = \dfrac{a}{d} = \dfrac{5}{13} \Rightarrow a = 5, d = 13$

21. (a) All the numbers that are multiplied have 0 as units' digit, so all of them are divisible by 5 once. Also 50 can be divided by 5^2. So the highest power of 5 that divides the given product = (9 + 1) = 10.

22. (a) Maximum possible number with odd digits = 97531 and the minimum possible number with even digits = 20468.

∴ Required difference = 97531 − 20468 = 77063

23. (d) $3^{25} + 3^{26} + 3^{27} + 3^{28}$

$= 3^{25}(1 + 3 + 3^2 + 3^3)$

$= 3^{25} \times (1 + 3 + 9 + 27) = 3^{25} \times 40$

$= 3^{24} \times 3 \times 4 \times 10 = 3^{24} \times 4 \times 30$,

which is divisible by 30.

24. (b) Let the number be x. Then,

∴ $y = 5 \times 1 + 4 = 9$

$\Rightarrow x = 4 \times y + 1 = 4 \times 9 + 1 = 37$

```
4 | x
5 | y  1 (Remainder)
  | 1  4 (Remainder)
```

∴ When 37 is successively divided by 5 and 4, we get the remainder as 2 and 3 respectively.

```
5 | 37
4 | 7 − 2
  | 1 − 3
```

25. (d) $\dfrac{p}{q} = 2.525252........$

$2.\overline{52} = 2\dfrac{52}{99} = \dfrac{250}{99}$

∴ $\dfrac{q}{p} = \dfrac{99}{250} = 0.396$.

26. (d) Let the lesser number be x. Then,

Greater number = $x + 45$

Given, $\dfrac{x+45}{x} = 4 \Rightarrow x + 45 = 4x$

$\Rightarrow 3x = 45 \Rightarrow x = 15$

Then, required sum = $x + x + 45 = 30 + 45 = 75$

27. (c) Number of factors of $N(2^3 \times 3^2 \times 5^3)$

$= (3+1) \times (2+1) \times (3+1)$

$= 4 \times 3 \times 4 = 48$

28. (d) $P = (21)^2 \times (22)^2 \times (23)^2 \times (24)^2 \times (25)^2$

$= (3 \times 7)^2 \times (2 \times 11)^2 \times (23)^2 \times (2^3 \times 3)^2 \times (5^2)^2$

$= 3^2 \times 7^2 \times 2^2 \times 11^2 \times 23^2 \times 2^6 \times 3^2 \times 5^4$

$= 2^8 \times 3^4 \times 5^4 \times 7^2 \times 11^2 \times 23^2$

Total number of factors

$= (8+1) \times (4+1) \times (4+1) \times (2 \times 1) \times (2+1) \times (2+1)$

$= 9 \times 5 \times 5 \times 3 \times 3 \times 3 = 6075$

29. (c) $N = (2^2 \times 3)^3 \times 3^4 \times 5^2$

$= 2^6 \times 3^7 \times 5^2$

∴ Total number of factors of N

$= (6+1) \times (7+1) \times (2+1)$

$= 7 \times 8 \times 3 = 168$

Some of these are odd factors and some of these are even factors. The odd factors are formed with the combination of 3s and 5s.

∴ Total number of odd factors

$= (7+1) \times (2+1) = 8 + 3 = 24$

∴ Number of even factors = 168 − 24 = 144.

30. (d) The units' digit of each term successively is 1, 9, 1, 9, ... as the power of 9 is odd in the second term and even in the third term and so on. First term = 1

The units' digit of first two term = 0 as 1 + 9 = 10,

The units' digit of first three terms = 1 as 1 + 9 + 1 = 11

The units' digit of first four terms = 0 as 1 + 9 + 1 + 9 = 20

$\Rightarrow$ Units' digit is 0 or 1 depending upon the number of terms whether even or odd.

∴ Units' digit of the sum of 1006 terms, i.e., even number of terms = 0.

Self Assessment Sheet–1

1. $16^5 + 2^{15}$ is divisible by
   (a) 31   (b) 13
   (c) 27   (d) 33

2. Find the number of divisors of 10800.
   (a) 57   (b) 60
   (c) 72   (d) none of these

3. The units digit of the expression $125^{813} \times 553^{3703} \times 4532^{828}$ is
   (a) 4   (b) 2
   (c) 0   (d) 5

4. If $M = 2^2 \times 3^5$, $N = 2^3 \times 3^4$, then the number of factors of N that are common with factors of M is
   (a) 8   (b) 5
   (c) 18   (d) 15

5. Find the greatest number by which the expression $7^{2n} - 3^{2n}$ is always exactly divisible.
   (a) 4   (b) 10
   (c) 20   (d) 40

6. Given that $1^2 + 2^2 + 3^2 + \ldots\ldots\ldots + 10^2 = 385$, then the value of $(2^2 + 4^2 + 6^2 + \ldots\ldots\ldots + 20^2)$ is equal to
   (a) 770   (b) 1540
   (c) 1155   (d) $(385)^2$

7. There is one number which is formed by writing one digit 6 times. Such number is always divisible by:
   (*e.g.*, 0.111111, 0.444444 etc.)
   (a) 7   (b) 11
   (c) 13   (d) all of these

8. The number of integers x for which the number $\sqrt{x^2 + x + 1}$ is rational is:
   (a) infinite   (b) one
   (c) two   (d) three

9. A number divided by 296 leaves 75 as remainder. If the same number is divided by 37, the remainder obtained is
   (a) 2   (b) 1
   (c) 11   (d) 8

10. How many prime factors are there in the expression $(12)^{43} \times (34)^{48} \times (2)^{57}$?
    (a) 282   (b) 237
    (c) 142   (d) 61

Answers

1. (d)	2. (b)	3. (c)	4. (d)	5. (d)	6. (b)	7. (d)	8. (c)	9. (b)	10. (a)

Chapter

2

FRACTIONS AND DECIMALS

KEY FACTS

1. Number of the form $\dfrac{a}{b}$ where $b \neq 0$ are called fractions.

2. **Types of fractions**:

   (a) **Proper fraction**: A fraction $\dfrac{a}{b}$ in which $b \neq 0$ and $a < b$.

   (b) **Improper fraction** : A fraction $\dfrac{a}{b}$ in which $b \neq 0$ and $a = b$ or $a > b$.

   (c) **Mixed fraction** : A fraction which can be expressed as the sum of a natural number and a fraction.

   (d) **Decimal fraction** : A fraction whose denominator is a multiple of 10 is called a **decimal fraction**.

   (e) **Vulgar fractions** : Fractions having denominators as whole numbers other than a power of 10, i.e., 10, 100, 1000 .... etc.

3. Fractions obtained on multiplying or dividing both numerator and denominator of a given fraction by the same non-zero number, and the given fraction are called **equivalent fractions**.

4. A fraction of the form $\dfrac{a}{b}, b \neq 0$, where a and b are whole numbers is called a **simple fraction**, whereas a fraction of the form $\dfrac{a}{b}$ where a and b are fractions is called a **complex fractions.**

5. A fraction $\dfrac{a}{b}$ is said to be in its lowest form if the HCF of a and b is 1.

6. **Comparing fractions** $\dfrac{a}{b}$ **and** $\dfrac{c}{d}$ **:**

   (i) If $ad > bc$, then $\dfrac{a}{b} > \dfrac{c}{d}$ (ii) If $ad = bc$, then $\dfrac{a}{b} = \dfrac{c}{d}$ (iii) If $ad < bc$, then $\dfrac{a}{b} < \dfrac{c}{d}$ $\boxed{\dfrac{a}{b} \times \dfrac{c}{d}}$

Note. Before applying the rule of cross-multiplication, put the given fractions in standard from. Thus, $\dfrac{a}{-b}$ should be written as $\dfrac{-a}{b}$.

---

7. (i) For addition or subtraction of like fractions, numerators are added (or subtracted), denominator remaining the same.

   (ii) For addition or subtraction of unlike fractions, first change them to equivalent like fractions (by finding the LCM of denominators) and then do as in (i).

> **Note.** To add or subtract mixed fractions, you may add or subtract the whole numbers parts separately and fractional parts separately.

8. If $\dfrac{a}{b}$ and $\dfrac{c}{d}$ are two fractions, then., $\dfrac{a}{b} \times \dfrac{c}{d} = \dfrac{a \times c}{b \times d}$

9. If $\dfrac{a}{b}$ and $\dfrac{c}{d}$ are two fractions, then $\dfrac{a}{b} \div \dfrac{c}{d} = \dfrac{a}{b} \times \dfrac{d}{c}$. Where $\dfrac{d}{c}$ is the **reciprocal** of $\dfrac{c}{d}$.

10. (a) The sequence of operations followed while simplifying a numerical expression is **BODMAS**.

    (b) In case of brackets the sequence followed is '—', ( ), { }, [ ], i.e., from the innermost to the outer most.

11. A decimal number has two parts, the **whole number part** on the left of the decimal point and the **decimal part** on the right of the decimal point. E.g. 12.56.

    The **number of digits in the decimal part** is the **number of decimal places**. E.g., the number 12.56 has two decimal places.

12. To *add* or *subtract* decimal numbers write the numbers one below the other so that the decimal points are in one vertical line. Then add or subtract.

13. **To multiply :**

    (*i*) **a decimal by 10 or powers of 10 :** Shift the decimal point to the right by as many places as there are number of zeros, e.g., $4.952 \times 10 = 45.92$, $4.952 \times 100 = 495.2$ etc.

    (*ii*) **a decimal by a decimal :** Ignore the decimal point and multiply them as whole numbers. The number of decimal places in the product is the sum of the number of decimal places both in the multiplicand and multiplier.

14. **To divide :**

    (*i*) **a decimal by 10 or powers of 10 :** Shift the decimal point to the left by as many places as there are numbers of zeros, e.g., $65.958 \div 10 = 6.5958$, $65.958 \div 1000 = 0.065958$ etc.

    (*ii*) **a decimal by a decimal :** Convert the divisor to a whole number by shifting the decimal point to the right by the same number of decimal places as there are in the divisor. Then shift the decimal point of the dividend also by the same number of places to the right, e.g., $16.403 \div 6.98 = 1640.3 \div 698$.

15. A fraction $\dfrac{p}{q}$ is a terminating decimal if in its lowest form the denominator has factors as 2 or 5 or both 2 and 5, otherwise $\dfrac{p}{q}$ is expressible as a non-terminating repeating or recurring decimal. For example : $\dfrac{7}{20} = 0.35$, $\dfrac{5}{9} = 0.5555 \ldots\ldots$

16. **To change decimals to fractions :**

    (*i*) **When the decimals are terminating :** Remove the decimal point and write the resulting number in the numerator. In the denominator write 1 followed by as many zeros as the number of decimal places in the given decimal number $45.93 = \dfrac{4593}{100}$.

    (*ii*) **When the decimals are repeating :**

        (a) In a **pure recurring decimal** (a decimal number in which all the digits after the decimal point are repeated). Write the repeated figure only once in the numerator and write as many nines in the denominator as is the number of repeated digits, e.g., $0.\overline{7} = \dfrac{7}{9}$, $0.\overline{59} = \dfrac{59}{99}$, $2.\overline{63} = 2 + \dfrac{63}{99} = 2\dfrac{7}{11}$.

        (b) In a **mixed recurring decimal** (a decimal in which at least one digit after the decimal point is not repeated and then some digits are repeated), consider the decimal point. Subtract the non-repeating part from the whole and divide by (as many nines as there are repeating digits × as many tens as are non-repeating digits), e.g., $5.1\overline{925} = 5 + \dfrac{1925 - 1}{9990} = 5 + \dfrac{1924}{9990} = 5 + \dfrac{26}{135} = \dfrac{701}{135}$.

        (Three repeating digits (925) and one non-repeating digit (1))

Solved Examples

Ex. 1. *Evaluate* $8 - 8 \times \dfrac{2\frac{1}{5} - 1\frac{2}{7}}{2 - \dfrac{1}{6 - \frac{1}{6}}}$.

Sol. $8 - 8 \times \dfrac{\frac{11}{5} - \frac{9}{7}}{2 - \dfrac{1}{\frac{35}{6}}} = 8 - 8 \times \dfrac{\frac{77-45}{35}}{2 - \frac{6}{35}} = 8 - 8 + \dfrac{\frac{32}{35}}{\frac{64}{35}} = 8 - 8 \times \dfrac{1}{2} = 8 - 4 = \textbf{4.}$

Ex. 2. *What is the value of* $\left(1 + \dfrac{1}{2}\right)\left(1 + \dfrac{1}{3}\right)\left(1 + \dfrac{1}{4}\right) \dots\dots \left(1 + \dfrac{1}{120}\right)$?

Sol. Given exp. $= \dfrac{3}{2} \times \dfrac{4}{3} \times \dfrac{5}{4} \times \dots\dots \times \dfrac{120}{119} \times \dfrac{121}{120} = \dfrac{121}{2} = \textbf{60.5}$

Ex. 3. *If* $\dfrac{2x}{1 + \dfrac{1}{1 + \frac{x}{1-x}}} = 1$ *, then find the value of x.*

Sol. Given, $\dfrac{2x}{1 + \dfrac{1}{\frac{1-x+x}{1-x}}} = 1 \Rightarrow \dfrac{2x}{1 + \dfrac{1}{\frac{1}{1-x}}} = 1 \Rightarrow \dfrac{2x}{1 + (1-x)} = 1 \Rightarrow 2x = 2 - x \Rightarrow 3x = 2 \Rightarrow x = \dfrac{2}{3}.$

Ex. 4. *If we multiply a fraction by itself and divide the product by its reciprocal, the fraction thus obtained is* $18\dfrac{26}{27}$ *.*
What is the original fraction?

Sol. Let the fraction be x. Then,

$\dfrac{x \times x}{\frac{1}{x}} = 18\dfrac{26}{27} \quad \Rightarrow \quad x^3 = \dfrac{512}{27} \quad \Rightarrow \quad x = \sqrt[3]{\dfrac{512}{27}} = \dfrac{8}{3} = \mathbf{2\dfrac{2}{3}}.$

Ex. 5. *The fluid contained in a bucket can fill four large bottles or seven small bottles. A full large bottle is used to fill an empty small bottle. What fraction of the fluid is left over in the large bottle when the small one is full ?*

(a) $\dfrac{2}{7}$ (b) $\dfrac{3}{7}$ (c) $\dfrac{4}{7}$ (d) $\dfrac{5}{7}$

Sol. Let the capacity of the bucket be x litres. Then,

Capacity of 1 large bottle $= \dfrac{x}{4}$

Capacity of 1 small bottle $= \dfrac{x}{7}$

Fluid left in large bottle $= \dfrac{x}{4} - \dfrac{x}{7} = \dfrac{3x}{28}$

$\therefore$ Required fraction $= \dfrac{3x/28}{x/4} = \dfrac{3x}{28} \times \dfrac{4}{x} = \dfrac{3}{7}$.

Ex. 6. *A man has divided his total money in his will in such a way that half of it goes to his wife, $\dfrac{2}{3}$rd of the remaining among his three sons equally and the rest among his four daughters equally. If each daughter gets Rs 20,000, how much money will each son get ?*

Sol. Let the total money, the man had $=$ Rs x

Money gone to wife $=$ Rs $\dfrac{x}{2}$

Remaining money $= \dfrac{x}{2}$

$\therefore$ Sons' share $= \dfrac{2}{3}$ of $\dfrac{x}{2} =$ Rs $\dfrac{x}{3}$

Each son's share $= \dfrac{1}{3}$ of Rs $\dfrac{x}{3} =$ Rs $\dfrac{x}{9}$

Daughters' share $= \dfrac{x}{2} - \dfrac{x}{3} = \dfrac{x}{6}$

Each daughter's share $= \dfrac{1}{4} \times$ Rs $\dfrac{x}{6} =$ Rs $\dfrac{x}{24}$

Given, $\dfrac{x}{24} = 20000 \Rightarrow x =$ Rs 480000

$\therefore$ Each son's share $= \dfrac{480000}{9} =$ **Rs 53,333.33**

Ex. 7. *Simplify* : $\dfrac{\left(3\frac{2}{3}\right)^2 - \left(2\frac{1}{2}\right)^2}{\left(4\frac{3}{4}\right)^2 - \left(3\frac{1}{3}\right)^2} \div \dfrac{3\frac{2}{3} - 2\frac{1}{2}}{4\frac{3}{4} - 3\frac{1}{3}}$.

Sol. Given exp. $= \dfrac{\left(\frac{11}{3}\right)^2 - \left(\frac{5}{2}\right)^2}{\left(\frac{19}{4}\right)^2 - \left(\frac{10}{3}\right)^2} \times \dfrac{\left(\frac{19}{4} - \frac{10}{3}\right)}{\left(\frac{11}{3} - \frac{5}{2}\right)} = \dfrac{\left(\frac{11}{3} + \frac{5}{2}\right)\left(\frac{11}{3} - \frac{5}{2}\right)}{\left(\frac{19}{4} + \frac{10}{3}\right)\left(\frac{19}{4} - \frac{10}{3}\right)} \times \dfrac{\left(\frac{19}{4} - \frac{10}{3}\right)}{\left(\frac{11}{3} - \frac{5}{2}\right)}$

$= \dfrac{\dfrac{22+15}{6}}{\dfrac{57+40}{12}} = \dfrac{\dfrac{37}{6}}{\dfrac{97}{12}} = \dfrac{37 \times 2}{97} = \dfrac{\mathbf{74}}{\mathbf{97}}$.

Ex. 8. *If $1.5x = 0.04y$ then what is the value of $\dfrac{y-x}{y+x}$?*

Sol. $1.5x = 0.04y \implies \dfrac{y}{x} = \dfrac{1.5}{0.04} = \dfrac{150}{4} = 37.5$

$\therefore \dfrac{y-x}{y+x} = \dfrac{\dfrac{y}{x}-1}{\dfrac{y}{x}+1} = \dfrac{37.5-1}{37.5+1} = \dfrac{36.5}{38.5} = \dfrac{73}{77}.$

Ex. 9. *Evaluate :* $\dfrac{(6.4)^2 - (5.4)^2}{(8.9)^2 + (8.9 \times 2.2) + (1.1)^2}$

Sol. Given exp. $= \dfrac{(6.4+5.4)(6.4-5.4)}{(8.9+1.1)^2} = \dfrac{11.8 \times 1}{100} = \mathbf{0.118}.$

Ex. 10. *Simplify :* $\dfrac{(0.06)^2 + (0.47)^2 + (0.079)^2}{(0.006)^2 + (0.047)^2 + (0.0079)^2}$

Sol. Given exp. $= \dfrac{(0.06)^2 + (0.47)^2 + (0.079)^2}{\left(\dfrac{0.06}{10}\right)^2 + \left(\dfrac{0.47}{10}\right)^2 + \left(\dfrac{0.079}{10}\right)^2} = 100\left[\dfrac{(0.06)^2 + (0.47)^2 + (0.079)^2}{(0.06)^2 + (0.47)^2 + (0.079)^2}\right] = \mathbf{100}.$

Ex. 11. *The value of* $\dfrac{3.157 \times 4126 \times 3.198}{63.972 \times 2835.121}$ *is closest to*

(a) 0.002   *(b) 0.02*   *(c) 0.2*   *(d) 2*

Sol. (c) The expression approximately $= \dfrac{3.2 \times 4126 \times 3.2}{64 \times 2835}$

$= 0.232$

$= \mathbf{0.2}$ (approx)

Ex. 12. *Which among the following numbers is the greatest ?*

$0.07 + \sqrt{0.16}$, $\sqrt{1.44}$, 1.2×0.83 , $1.02 - \dfrac{0.6}{24}$

Sol. $0.07 + \sqrt{0.16} = 0.07 + 0.4 = 0.47$

$\sqrt{1.44} = 1.2$

$1.2 \times 0.83 = 0.996$

$1.02 - \dfrac{0.6}{24} = 1.02 - \dfrac{\overset{1}{\cancel{6}}}{\underset{40}{\cancel{240}}} = 1.02 - 0.025 = 0.995$

$\therefore$ The greatest number is 1.2, i.e., $\sqrt{\mathbf{1.44}}$.

Question Bank–2

1. Find the value of $\dfrac{2}{3} \times \dfrac{3}{\dfrac{5}{6} \div \dfrac{2}{3} \text{ of } 1\dfrac{1}{4}}$

    (a) $\dfrac{1}{2}$ (b) $\dfrac{2}{3}$

    (c) 1   (d) 2

2. $\dfrac{3\dfrac{1}{4} - \dfrac{4}{5} \text{ of } \dfrac{5}{6}}{4\dfrac{1}{3} \div \dfrac{1}{5} - \left(\dfrac{3}{10} + 21\dfrac{1}{5}\right)} - \left(1\dfrac{2}{3} \text{ of } 1\dfrac{1}{2}\right)$ is equal to

    (a) 9   (b) $11\dfrac{1}{2}$

    (c) 13   (d) $15\dfrac{1}{2}$

3. Evaluate $\dfrac{1\dfrac{1}{7} - \dfrac{2}{3} + \dfrac{\dfrac{2}{5}}{1 - \dfrac{1}{25}}}{1 - \dfrac{1}{7}\left(\dfrac{1}{3} + \dfrac{\dfrac{2}{5}}{1 - \dfrac{2}{5}}\right)}$

    (a) $\dfrac{3}{4}$ (b) $\dfrac{24}{25}$

    (c) 1   (d) $1\dfrac{1}{24}$

4. Simplify: $\dfrac{1}{1 + \dfrac{\dfrac{2}{3}}{1 + \dfrac{2}{3} + \dfrac{\dfrac{8}{9}}{1 - \dfrac{2}{3}}}}$

    (a) $\dfrac{11}{13}$ (b) $\dfrac{13}{15}$

    (c) $\dfrac{13}{11}$ (d) $\dfrac{15}{13}$

5. If $\dfrac{37}{13} = 2 + \dfrac{1}{x + \dfrac{1}{y + \dfrac{1}{z}}}$, where x, y, z are natural numbers, then x, y, z are

    (a) 1, 2, 5   (b) 1, 5, 2

    (c) 5, 2, 11   (d) 11, 2, 5

6. The sum of the first 25 terms of the series

$$\dfrac{1}{3} + \dfrac{1}{4} - \dfrac{1}{2} - \dfrac{1}{3} + \dfrac{1}{2} - \dfrac{1}{4} + \dfrac{1}{3} + \dfrac{1}{4} - \dfrac{1}{2} - \dfrac{1}{3} + \dots \text{ is}$$

    (a) $\dfrac{1}{2}$ (b) $-\dfrac{1}{4}$

    (c) $\dfrac{1}{3}$ (d) $\dfrac{1}{4}$

7. $\dfrac{17}{15} \times \dfrac{17}{15} + \dfrac{2}{15} \times \dfrac{2}{15} - \dfrac{17}{15} \times \dfrac{4}{15}$ is equal to

    (a) 0   (b) 1

    (c) 10   (d) 11

8. $\left(4\dfrac{11}{15} + \dfrac{15}{71}\right)^2 - \left(4\dfrac{11}{15} - \dfrac{15}{71}\right)^2$ is equal to

    (a) 1   (b) 2

    (c) 3   (d) 4

9. The square root of $\dfrac{\left(3\dfrac{1}{4}\right)^4 - \left(4\dfrac{1}{3}\right)^4}{\left(3\dfrac{1}{4}\right)^2 - \left(4\dfrac{1}{3}\right)^2}$ is

    (a) $7\dfrac{1}{2}$ (b) $5\dfrac{5}{12}$

    (c) $1\dfrac{1}{12}$ (d) $1\dfrac{7}{12}$

10. What is the value of

$$\dfrac{3}{4} + \dfrac{5}{36} + \dfrac{7}{144} + \dots\dots + \dfrac{17}{5184} + \dfrac{19}{8100}?$$

    (a) 0.95   (b) 1

    (c) 0.99   (d) 0.98

11. $\left[\left(1 + \dfrac{1}{10 + \dfrac{1}{10}}\right) \times \left(1 + \dfrac{1}{10 + \dfrac{1}{10}}\right) - \left(1 - \dfrac{1}{10 + \dfrac{1}{10}}\right)\right.$

$\left. \times \left(1 - \dfrac{1}{10 + \dfrac{1}{10}}\right)\right] \div \left[\left(1 + \dfrac{1}{10 + \dfrac{1}{10}}\right) + \left(1 - \dfrac{1}{10 + \dfrac{1}{10}}\right)\right]$

simplifies to

    (a) $\dfrac{100}{101}$ (b) $\dfrac{90}{101}$

    (c) $\dfrac{20}{101}$ (d) $\dfrac{101}{100}$

12. A student was asked to simplify the following :

$$\frac{7}{5-2\frac{2}{3}}\div\frac{3-\dfrac{2}{3-1\frac{1}{2}}}{4-1\frac{1}{2}}-\frac{5}{7}\times\left[\frac{7}{10}+1\frac{1}{5}\times\frac{3\frac{1}{3}-2\frac{1}{2}}{2\frac{5}{21}-2}\right]+\frac{\dfrac{3}{1.6}+\dfrac{5}{3.2}}{\dfrac{5}{4.8}+\dfrac{1}{9.6}}$$

His answer was $3\frac{1}{5}$. Find the per cent error in the answer.

(a) 10%   (b) 20%

(c) 25%   (d) 30%

13. In a school, $\frac{3}{7}$ of the students are girls and the rest are boys. $\frac{1}{4}$ of the boys are below ten years of age and $\frac{5}{6}$ of the girls are also below ten years of age.

If the number of students above ten years of age is 500, then find the total number of students in the school.

(a) 600   (b) 1000

(c) 900   (d) 1100

14. A man first sold $\frac{2}{3}$rd of his total quantity of rice and 100 kg. Again he sold $\frac{1}{2}$ of the remaining quantity and 100 kg. If the total remaining quantity of the stock is 150 kg. Then, what was the original stock of rice ?

(a) 2100 kg   (b) 1800 kg

(c) 2400 kg   (d) 2000 kg

15. The value of $1-\frac{1}{20}+\frac{1}{20^2}-\frac{1}{20^3}+........$ correct to 5 places of decimal is

(a) 1.05   (b) 0.95238

(c) 0.95239   (d) 10.5

16. $\left(\dfrac{1}{1\times4}+\dfrac{1}{4\times7}+\dfrac{1}{7\times10}+\dfrac{1}{10\times13}+\dfrac{1}{13\times16}\right)$ is equal to

(a) $\dfrac{1}{3}$ (b) $\dfrac{5}{16}$

(c) $\dfrac{3}{8}$ (d) $\dfrac{41}{7280}$

17. If $\dfrac{a}{b}=\dfrac{1}{3}, \dfrac{b}{c}=2, \dfrac{c}{d}=\dfrac{1}{2}, \dfrac{d}{e}=3$ and $\dfrac{e}{f}=\dfrac{1}{4}$, then what is the value of $\dfrac{abc}{def}$?

(a) $\dfrac{3}{8}$ (b) $\dfrac{27}{8}$

(c) $\dfrac{3}{4}$ (d) $\dfrac{27}{4}$

18. Evaluate : 515.15 – 15.51 – 1.51 – 5.11 – 1.11.

(a) 491.91   (b) 419.91

(c) 499.19   (d) 411.19

19. Simplify : $12.28 \times 1.5 - 36 \div 2.4$

(a) 3.24   (b) 3.42

(c) 4.32   (d) 4.23

20. Which of the following is equal to 1 ?

(a) $\dfrac{(0.11)^2}{(1.1)^2\times0.1}$ (b) $\dfrac{(1.1)^2}{11^2\times(0.01)^2}$

(c) $\dfrac{(0.011)^2}{1.1^2\times0.012}$ (d) $\dfrac{(0.11)^2}{11^2\times0.01}$

21. When the number $N = 0.73\overline{545}$ is written as a fraction in its lowest terms, the denominator exceeds the numerator by

(a) 199   (b) 299

(c) 109   (d) 219

22. What is the value of $(7.5 \times 7.5 + 37.5 + 2.5 \times 2.5)$?

(a) 30   (b) 60

(c) 80   (d) 100

23. Evaluate $\overline{0.142857}\div\overline{0.285714}$

(a) $\dfrac{1}{2}$ (b) $\dfrac{1}{3}$

(c) 2   (d) 10

24. If $1^3 + 2^3 + ... + 9^3 = 2025$, then the value of $(0.11)^3 + (0.22)^3 + ... + (0.99)^3$ is close to

(a) 0.2695   (b) 0.3695

(c) 2.695   (d) 3.695

25. $\dfrac{5.42\times6+5.42\times24}{32.71\times32.71-27.29\times27.29}\div\dfrac{6.54\times6.54-3.46\times3.46}{3.08\times5+3.08\times45}$

is equal to

(a) 0.3   (b) 0.4

(c) 0.7   (d) 2.5

26. The value of $\dfrac{(67.542)^2 - (32.458)^2}{75.458 - 40.374}$ is

    (a) 1   (b) 10
    (c) 100   (d) 0.1

27. $\dfrac{0.1 \times 0.1 \times 0.1 + 0.02 \times 0.02 \times 0.02}{0.2 \times 0.2 \times 0.2 + 0.04 \times 0.04 \times 0.04}$ is equal to

    (a) 0.0125   (b) 0.125
    (c) 0.25   (d) 0.5

28. $\sqrt{(0.798)^2 + 0.404 \times 0.798 + (0.202)^2} + 1$ is equal to

    (a) 0   (b) 2

    (c) 1.596   (d) 0.404

29. Which of the following numbers is the least ?

    $(0.5)^2$, $\sqrt{0.49}$, $\sqrt[3]{0.008}$, 0.23

    (a) $(0.5)^2$ (b) $\sqrt{0.49}$
    (c) $\sqrt[3]{0.008}$ (d) 0.23

30. The value of $\dfrac{489.1375 \times 0.0483 \times 1.956}{0.0873 \times 92.581 \times 99.749}$ is closest to

    (a) 0.006   (b) 0.06
    (c) 0.6   (d) 6

Answers

1. (d)	2. (c)	3. (d)	4. (b)	5. (b)	6. (c)	7. (b)	8. (d)	9. (b)	10. (c)
11. (c)	12. (c)	13. (b)	14. (b)	15. (b)	16. (b)	17. (c)	18. (a)	19. (b)	20. (c)
21. (b)	22. (d)	23. (a)	24. (c)	25. (d)	26. (c)	27. (b)	28. (b)	29. (c)	30. (b)

Hints and Solutions

1. (d) $\dfrac{2}{3} \times \dfrac{3}{\dfrac{5}{6} - \dfrac{2}{3} \div \text{of } 1\dfrac{1}{4}}$

$= \dfrac{2}{3} \times \dfrac{3}{\dfrac{5}{6} - \dfrac{2}{3} \div \dfrac{5}{4}} = \dfrac{2}{3} \times \dfrac{3}{\dfrac{5}{6} - \dfrac{5}{6}}$.

$= \dfrac{2}{3} \times \dfrac{3}{1} = \mathbf{2}.$

2. (c) $\dfrac{3\dfrac{1}{4} - \dfrac{4}{5} \text{ of } \dfrac{5}{6}}{4\dfrac{1}{3} \div \dfrac{1}{5} - \left(\dfrac{3}{10} + 21\dfrac{1}{5}\right)} - \left(1\dfrac{2}{3} \text{ of } 1\dfrac{1}{2}\right)$

$= \dfrac{\dfrac{13}{4} - \dfrac{4}{5} \times \dfrac{5}{6}}{\dfrac{13}{3} \div \dfrac{1}{5} - \left(\dfrac{3}{10} + \dfrac{106}{5}\right)} - \left(\dfrac{5}{3} \times \dfrac{3}{2}\right)$

$= \dfrac{\dfrac{13}{4} - \dfrac{2}{3}}{\dfrac{13}{3} \times \dfrac{5}{1} - \left(\dfrac{3 + 212}{10}\right)} - \dfrac{5}{2}$

$= \dfrac{\dfrac{39 - 8}{12}}{\dfrac{65}{3} - \dfrac{215}{10}} - \dfrac{5}{2} = \dfrac{\dfrac{31}{12}}{\dfrac{650 - 645}{30}} - \dfrac{5}{2}$

$= \dfrac{31}{12} \div \dfrac{5}{30} - \dfrac{5}{2} = \dfrac{31}{12} \times \dfrac{30}{5} - \dfrac{5}{2}$

$= \dfrac{31}{2} - \dfrac{5}{2} = \dfrac{26}{2} = \mathbf{13}.$

3. (d) $\dfrac{1\dfrac{1}{7} - \dfrac{2}{3} + \dfrac{\dfrac{2}{5}}{1 - \dfrac{1}{25}}}{1 - \dfrac{1}{7}\left(\dfrac{1}{3} + \dfrac{\dfrac{2}{5}}{1 - \dfrac{2}{5}}\right)} = \dfrac{\dfrac{8}{7} - \dfrac{2}{3} + \dfrac{\dfrac{2}{5}}{\dfrac{24}{25}}}{1 - \dfrac{1}{7}\left(\dfrac{1}{3} + \dfrac{\dfrac{2}{5}}{\dfrac{3}{5}}\right)}$

$= \dfrac{\dfrac{8}{7} - \dfrac{2}{3} + \dfrac{2}{5} \times \dfrac{25}{24}}{1 - \dfrac{1}{7}\left(\dfrac{1}{3} + \dfrac{2}{3}\right)}$

$= \dfrac{\dfrac{8}{7} - \dfrac{2}{3} + \dfrac{5}{12}}{1 - \dfrac{1}{7}} = \dfrac{\dfrac{96 - 56 + 35}{84}}{\dfrac{6}{7}}$

$= \dfrac{75}{84} \times \dfrac{7}{6} = \dfrac{25}{24} = \mathbf{1\dfrac{1}{24}}$

4. (b) $\dfrac{1}{1 + \dfrac{\dfrac{2}{3}}{1 + \dfrac{2}{3} + \dfrac{\dfrac{8}{9}}{1 - \dfrac{2}{3}}}} = \dfrac{1}{1 + \dfrac{\dfrac{2}{3}}{1 + \dfrac{2}{3} + \dfrac{\dfrac{8}{9}}{\dfrac{1}{3}}}}$

$$= \cfrac{1}{1+\cfrac{2}{3}{1+\cfrac{2}{3}+\cfrac{8}{3}}} = \cfrac{1}{1+\cfrac{\frac{2}{3}}{\frac{13}{3}}}$$

$$= \cfrac{1}{1+\cfrac{2}{13}} = \cfrac{1}{\frac{15}{13}} = \mathbf{\frac{13}{15}}.$$

5. (b) $2 + \cfrac{1}{x+\cfrac{1}{y+\cfrac{1}{z}}} = \frac{37}{13} = 2\frac{11}{13} = 2 + \frac{11}{13}$

$$\Rightarrow \cfrac{1}{x+\cfrac{1}{y+\cfrac{1}{z}}} = \frac{11}{13} \Rightarrow x + \cfrac{1}{y+\cfrac{1}{z}} = \frac{13}{11}$$

$$\Rightarrow x + \cfrac{1}{y+\cfrac{1}{z}} = 1 + \frac{2}{11}$$

$$\Rightarrow x = 1,\ y + \frac{1}{z} = \frac{11}{2} = 5\frac{1}{2} = 5 + \frac{1}{2}$$

$$\Rightarrow x = \mathbf{1},\ y = \mathbf{5},\ z = \mathbf{2}$$

6. (c) Sum of the first 6 terms

$$= \frac{1}{3} + \frac{1}{4} - \frac{1}{2} - \frac{1}{3} + \frac{1}{2} - \frac{1}{4} = 0$$

$\therefore$ Sum of first 24 terms = 0

$\therefore$ Required sum = 25th term = $\mathbf{\frac{1}{3}}$.

7. (b) Given exp. $= \frac{17}{15} \times \frac{17}{15} + \frac{2}{15} \times \frac{2}{15} - 2 \times \frac{17}{15} \times \frac{2}{15}$

$$= \left(\frac{17}{15}\right)^2 + \left(\frac{2}{15}\right)^2 - 2 \times \frac{17}{15} \times \frac{2}{15}$$

$$= \left(\frac{17}{15} - \frac{2}{15}\right)^2 \text{ [Using } a^2 + b^2 - 2ab = (a-b)^2]$$

$$= \left(\frac{15}{15}\right)^2 = 1^2 = \mathbf{1}$$

8. (d) $\left(4\frac{11}{15} + \frac{15}{71}\right)^2 - \left(4\frac{11}{15} - \frac{15}{71}\right)^2$

$$= \left(4\frac{11}{15} + \frac{15}{71} + 4\frac{11}{15} - \frac{15}{71}\right)$$

$$\left(4\frac{11}{15} + \frac{15}{71} - 4\frac{11}{15} + \frac{15}{71}\right)$$

[Using $(a^2 - b^2) = (a+b)(a-b)$]

$$= \left(8\frac{22}{15}\right) \times \left(\frac{30}{71}\right) = \frac{142}{15} \times \frac{30}{71} = \mathbf{4}.$$

9. (b) $\cfrac{\left(3\frac{1}{4}\right)^4 - \left(4\frac{1}{3}\right)^4}{\left(3\frac{1}{4}\right)^2 - \left(4\frac{1}{3}\right)^2}$

$$= \cfrac{\left(\left(3\frac{1}{4}\right)^2 + \left(4\frac{1}{3}\right)^2\right)\left(\left(3\frac{1}{4}\right)^2 - \left(4\frac{1}{3}\right)^2\right)}{\left(3\frac{1}{4}\right)^2 - \left(4\frac{1}{3}\right)^2}$$

$$= \left(\frac{13}{4}\right)^2 + \left(\frac{13}{3}\right)^2 = \frac{169}{16} + \frac{169}{9}$$

[$a^4 - b^4 = (a^2 + b^2)(a^2 - b^2)$]

$$= 169\left(\frac{1}{16} + \frac{1}{9}\right) = 169 \times \left(\frac{9+16}{144}\right)$$

$$= \frac{169 \times 25}{144}$$

$\therefore$ Square root of given exp. $= \sqrt{\dfrac{169 \times 25}{144}}$

$$= \frac{13 \times 5}{12} = \frac{65}{12} = \mathbf{5\frac{5}{12}}.$$

10. (c) $\frac{3}{4} + \frac{5}{6} + \frac{7}{144} + \dots + \frac{17}{5184} + \frac{19}{8100}$

$$= \frac{3}{1^2 . 2^2} + \frac{5}{2^2 . 3^2} + \frac{7}{3^2 . 4^2} + \dots + \frac{17}{8^2 . 9^2} + \frac{19}{9^2 . 10^2}$$

$$= \left(1 - \frac{1}{2^2}\right) + \left(\frac{1}{2^2} - \frac{1}{3^2}\right) + \left(\frac{1}{3^2} - \frac{1}{4^2}\right) + \dots$$

$$+ \left(\frac{1}{8^2} - \frac{1}{9^2}\right) + \left(\frac{1}{9^2} - \frac{1}{10^2}\right)$$

$$= 1 - \frac{1}{10^2} = 1 - \frac{1}{100} = \frac{99}{100} = \mathbf{0.99}.$$

11. (c) Given exp. $= \dfrac{\left(1+\dfrac{1}{10+\dfrac{1}{10}}\right)^2 - \left(1-\dfrac{1}{10+\dfrac{1}{10}}\right)^2}{\left(1+\dfrac{1}{10+\dfrac{1}{10}}\right) + \left(1-\dfrac{1}{10+\dfrac{1}{10}}\right)}$

$= \dfrac{\left[\left(1+\dfrac{1}{10+\dfrac{1}{10}}\right) + \left(1-\dfrac{1}{10+\dfrac{1}{10}}\right)\right]\left[\left(1+\dfrac{1}{10+\dfrac{1}{10}}\right) - \left(1-\dfrac{1}{10+\dfrac{1}{10}}\right)\right]}{\left[\left(1+\dfrac{1}{10+\dfrac{1}{10}}\right) + \left(1-\dfrac{1}{10+\dfrac{1}{10}}\right)\right]}$

$= \left(1+\dfrac{1}{10+\dfrac{1}{10}}\right) - \left(1-\dfrac{1}{10+\dfrac{1}{10}}\right)$

$= \left(1+\dfrac{1}{\dfrac{101}{10}}\right) - \left(1-\dfrac{1}{\dfrac{101}{10}}\right)$

$= \left(1+\dfrac{10}{101}\right) - \left(1-\dfrac{10}{101}\right)$

$= \left(\dfrac{101+10}{101}\right) - \left(\dfrac{101-10}{101}\right)$

$= \left(\dfrac{111}{101}\right) - \left(\dfrac{91}{101}\right) = \dfrac{\mathbf{20}}{\mathbf{101}}.$

12. (c) Given exp. $= \dfrac{3-\dfrac{2}{3-\dfrac{3}{2}}}{5-\dfrac{8}{3}} \div \dfrac{3-\dfrac{3}{2}}{4-\dfrac{3}{2}}$

$-\dfrac{5}{7}\left[\dfrac{7}{10} + \dfrac{6}{5} \times \dfrac{\dfrac{10}{3}-\dfrac{5}{2}}{\dfrac{47}{21}-2}\right] + \dfrac{\dfrac{30}{16}+\dfrac{50}{32}}{\dfrac{50}{48}+\dfrac{10}{96}}$

$= \dfrac{3-\dfrac{2}{\dfrac{3}{2}}}{\dfrac{7}{\dfrac{7}{3}}} \div \dfrac{\dfrac{3}{2}}{\dfrac{5}{2}} - \dfrac{5}{7}\left[\dfrac{7}{10} + \dfrac{6}{5} \times \dfrac{\dfrac{20-15}{6}}{\dfrac{47-42}{21}}\right] + \dfrac{\dfrac{15}{8}+\dfrac{25}{16}}{\dfrac{25}{24}+\dfrac{5}{48}}$

$= 3 \div \left[\left(3-\dfrac{4}{3}\right) \times \dfrac{2}{5}\right] - \dfrac{5}{7}\left[\dfrac{7}{10} + \dfrac{6}{5} \times \dfrac{\dfrac{5}{6}}{\dfrac{5}{21}}\right] + \dfrac{\dfrac{55}{16}}{\dfrac{55}{48}}$

$= 3 \div \left[\dfrac{5}{3} \times \dfrac{2}{5}\right] - \dfrac{5}{7}\left[\dfrac{7}{10} + \dfrac{6}{5} \times \dfrac{21}{6}\right] + \dfrac{48}{16}$

$= 3 \div \dfrac{2}{3} - \dfrac{5}{7}\left[\dfrac{7}{10} + \dfrac{21}{5}\right] + 3$

$= \dfrac{9}{2} - \dfrac{5}{7} \times \dfrac{7}{10} - \dfrac{5}{7} \times \dfrac{21}{5} + 3$

$= \dfrac{9}{2} - \dfrac{1}{2} - 3 + 3 = \dfrac{8}{2} = 4$

$\therefore$ Percentage error $= \left(\dfrac{4-3\dfrac{1}{5}}{3\dfrac{1}{5}} \times 100\right)\%$

$= \left(\dfrac{4-\dfrac{16}{5}}{\dfrac{16}{5}} \times 100\right)\% = \left(\dfrac{\dfrac{4}{5}}{\dfrac{16}{5}} \times 100\right)\% = \mathbf{25\%}.$

13. (b) Let the total number of students in the school be x.

Then, Number of girls $= \dfrac{3x}{7}$

Number of boys $= \dfrac{4x}{7}$

Number of boys below ten years of age

$= \dfrac{1}{4} \times \dfrac{4x}{7} = \dfrac{x}{7}$

Number of girls below ten years of age

$= \dfrac{5}{6} \times \dfrac{3x}{7} = \dfrac{5x}{14}$

$\therefore$ Total number of students below 10 years of age

$= \dfrac{x}{7} + \dfrac{5x}{14} = \dfrac{7x}{14} = \dfrac{x}{2}$

$\therefore$ Total number of students above 10 years of age

$= x - \dfrac{x}{2} = \dfrac{x}{2}$

Given, $\dfrac{x}{2} = 500 \Rightarrow x = \mathbf{1000}.$

14. (b) Let the original stock of rice be x kg.

Parts of the stock sold first time $= \left(\dfrac{2x}{3} + 100\right)$ kg

$\therefore$ Remaining stock $= \left[x - \left(\dfrac{2x}{3} + 100\right)\right]$ kg

$= \left(\dfrac{x}{3} - 100\right)$ kg

Part of the stock sold second time

$= \left[\dfrac{1}{2}\left(\dfrac{x}{3} - 100\right) + 100\right]$

$= \left(\dfrac{x}{6} - 50 + 100\right)$ kg $= \left(\dfrac{x}{6} + 50\right)$ kg

$\therefore$ Remaining stock $= \left(\dfrac{x}{3} - 100\right) - \left(\dfrac{x}{6} + 50\right)$

$= \left(\dfrac{x}{3} - \dfrac{x}{6} - 100 - 50\right)$ kg

$= \left(\dfrac{x}{6} - 50\right)$ kg

Given, $\dfrac{x}{6} - 150 = 150 \Rightarrow \dfrac{x}{6} = 300 \Rightarrow x = \mathbf{1800\ kg}$.

15. (b) Given exp. $= 1 - \dfrac{1}{20} + \dfrac{1}{20^2} - \dfrac{1}{20^3} + \ldots$

$= 1 - \dfrac{1}{20} + \dfrac{1}{400} - \dfrac{1}{8000} + \ldots$

$= 1 - 0.05 + 0.0025 - 0.000125 + \ldots$

$= 1.0025 - 0.050125 = 0.952375 = \mathbf{0.95238}.$

16. (b) $\dfrac{1}{1 \times 4} + \dfrac{1}{4 \times 7} + \dfrac{1}{7 \times 10} + \dfrac{1}{10 \times 13} + \dfrac{1}{13 \times 16}$

$= \dfrac{1}{3}\left[\left(1 - \dfrac{1}{4}\right) + \left(\dfrac{1}{4} - \dfrac{1}{7}\right) + \left(\dfrac{1}{7} - \dfrac{1}{10}\right)\right.$

$\left. + \left(\dfrac{1}{10} - \dfrac{1}{13}\right) + \left(\dfrac{1}{13} - \dfrac{1}{16}\right)\right]$

$= \dfrac{1}{3}\left(1 - \dfrac{1}{16}\right) = \dfrac{1}{3} \times \dfrac{15}{16} = \dfrac{\mathbf{5}}{\mathbf{16}}$

17. (a) $\dfrac{a}{b} = \dfrac{1}{3} \Rightarrow a = \dfrac{b}{3}$

$\dfrac{b}{c} = 2 \Rightarrow c = \dfrac{b}{2}$

$\dfrac{c}{d} = \dfrac{1}{2} \Rightarrow d = 2c \Rightarrow d = 2 \times \dfrac{b}{2} = b$

$\dfrac{d}{e} = 3 \Rightarrow e = \dfrac{d}{3} \qquad \Rightarrow e = \dfrac{b}{3}$

$\dfrac{e}{f} = \dfrac{1}{4} \Rightarrow f = 4e \qquad \Rightarrow f = \dfrac{4b}{3}$

$\therefore \dfrac{abc}{def} = \dfrac{b/3 \times b \times b/2}{b \times b/3 \times 4b/3} = \dfrac{\mathbf{3}}{\mathbf{8}}.$

18. (a) Given exp.
$= 515.15 - (15.51 + 1.51 + 5.11 + 1.11)$
$= 515.15 - 23.24$
$= \mathbf{491.91.}$

19. (b) $12.28 \times 1.5 - \dfrac{36}{2.4} = 18.42 - 15 = \mathbf{3.42}$

20. (c) $\dfrac{(0.11)^2}{(1.1)^2 \times 0.1} = \dfrac{0.0121}{1.21 \times 0.1} = \dfrac{0.0121}{0.121} = 0.1;$

$\dfrac{(1.1)^2}{11^2 \times (0.01)^2} = \dfrac{1.21}{121 \times 0.0001} = \dfrac{0.01}{0.0001} = 100;$

$\dfrac{(0.011)^2}{(1.1)^2 \times (0.01)^2} = \dfrac{0.000121}{1.21 \times 0.0001} = 1;$

$\dfrac{(0.11)^2}{11^2 \times 0.01} = \dfrac{0.0121}{121 \times 0.01} = \dfrac{0.0121}{1.21} = \mathbf{0.01}.$

Hence, option (c) is the correct answer.

21. (b) $N = 0.73\overline{545} = \dfrac{73545 - 735}{99000} = \dfrac{72810}{99000} = \dfrac{809}{1100}$

$\therefore$ Required difference $= 1100 - 809 = \mathbf{299.}$

22. (d) Given exp. $= (7.5)^2 + 2 \times 7.5 \times 2.5 + (2.5)^2$
$= (7.5 + 2.5)^2 = 10^2 = \mathbf{100.}$

23. (a) $0.\overline{142857} \div 0.\overline{285714}$

$= \dfrac{142857}{999999} \div \dfrac{285714}{999999}$

$= \dfrac{\overset{1}{\cancel{142857}}}{\cancel{999999}} \div \dfrac{\cancel{999999}}{\underset{2}{\cancel{2857124}}} = \dfrac{\mathbf{1}}{\mathbf{2}}.$

24. (c) $(0.11)^3 + (0.22)^3 + \ldots + (0.99)^3$
$= (0.11)^3 \left[1^3 + 2^3 + 3^3 + \ldots + 9^3\right]$
$= 0.001331 \times 2025 = 2.695275 = \mathbf{2.695}$ (approx.)

25. (d) Given exp.

$= \dfrac{5.42 \times (6 + 24)}{(32.71)^2 - (27.29)^2} \div \dfrac{(6.54)^2 - (3.46)^2}{3.08 \times (5 + 45)}$

$= \dfrac{5.42 \times 30}{(32.71 + 27.29)(32.71 - 27.29)}$

$\div \dfrac{(6.54 + 3.46)(6.54 - 3.46)}{3.08 \times 50}$

$= \dfrac{5.42 \times 30}{60 \times 5.42} \div \dfrac{10 \times 3.08}{3.08 \times 50}$

$= \dfrac{1}{2} \div \dfrac{1}{5} = \dfrac{5}{2} = \mathbf{2.5.}$

26. (c) Given exp.

$$= \frac{(67.542 + 32.458)(67.542 - 32.458)}{35.084}$$

$$= \frac{100 \times 35.084}{35.084} = \mathbf{100}.$$

27. (b) Given exp.

$$= \frac{0.1 \times 0.1 \times 0.1 + 0.02 \times 0.02 \times 0.02}{8(0.1 \times 0.1 \times 0.1) + 8(0.02 \times 0.02 \times 0.02)}$$

$$= \frac{1}{8}\left(\frac{0.1 \times 0.1 \times 0.1 + 0.02 \times 0.02 \times 0.02}{0.1 \times 0.1 \times 0.1 + 0.02 \times 0.02 \times 0.02}\right)$$

$$= \frac{1}{8} = \mathbf{0.125}.$$

28. (b) $\sqrt{(0.798)^2 + 0.404 \times 0.798 + (0.202)^2} + 1$

$$= \sqrt{(0.798)^2 + 2 \times 0.202 \times 0.798 + (0.202)^2} + 1$$

$$= \sqrt{(0.798 + 0.202)^2} + 1 = \sqrt{1} + 1 = 1 + 1 = \mathbf{2}.$$

29. (c) $(0.5)^2 = 0.25$; $\sqrt{0.49} = 0.7$;

$\sqrt[3]{0.008} = 0.2$; 0.23

Arranging in ascending order the numbers are 0.2, 0.23, 0.25, 0.7.

$\therefore \sqrt[3]{0.008} = 0.2$ is the least.

30. (b) $\dfrac{489.1375 \times 0.0483 \times 1.956}{0.0873 \times 92.581 \times 99.749} = \dfrac{489 \times 0.05 \times 2}{0.09 \times 93 \times 100}$

$$= \frac{489}{9 \times 93 \times 10} = \frac{163}{279} \times \frac{1}{10} = \frac{0.58}{10} = 0.058 = \mathbf{0.06}.$$

Self Assessment Sheet–2

1. If $4\dfrac{1}{a} \times b\dfrac{2}{3} = 7$, find the values of a and b.

(a) 1, 5   (b) 2, 3

(c) 3, 2   (d) 5, 1

2. Simplify : $\dfrac{4\frac{1}{7} - 2\frac{1}{4}}{3\frac{1}{2} + 1\frac{1}{7}} \div \cfrac{1}{2 + \cfrac{1}{2 + \cfrac{1}{5 - \frac{1}{5}}}}$

(a) 0   (b) –1

(c) $3\dfrac{1}{24}$ (d) 1

3. Find the number which when multiplied by

$\dfrac{0.0016 \times 0.025}{0.325 \times 0.05} \div \dfrac{0.1216 \times 0.105 \times 0.002}{0.08512 \times 0.625 \times 0.039}$ yields the product 20?

(a) 25   (b) 100

(c) 80   (d) 200

4. The value of $0.\overline{2} + 0.\overline{3} + 0.\overline{4} + 0.\overline{9} + 0.\overline{39}$ is

(a) $0.\overline{57}$ (b) $1\dfrac{20}{33}$

(c) $2\dfrac{13}{39}$ (d) $2\dfrac{13}{33}$

5. If $8.5 - \{5\dfrac{1}{2} - (7\dfrac{1}{2} + 2.8 \div x)\} \times 4.25 \div (0.2)^2 = 306$, the value of x is

(a) 1.75   (b) 3.5

(c) 7   (d) 1.4

6. A student was asked to simplify

$\dfrac{0.6 \times 0.6 \times 0.6 + 0.5 \times 0.5 \times 0.5 + 0.1 \times 0.1 \times 0.1 - 0.09}{0.6 \times 0.6 + 0.5 \times 0.5 + 0.1 \times 0.1 - 0.41}$

and his answer was 0.6. By what per cent was his answer wrong.

(a) 25%   (b) 100%

(c) 50%   (d) 120%

7. The simplest value of

$\dfrac{\left(1 + \frac{1}{2}\right)\left(1 + \frac{1}{3}\right)\left(1 + \frac{1}{4}\right) \ldots \left(1 + \frac{1}{50}\right)}{\left(1 - \frac{1}{2}\right)\left(1 - \frac{1}{3}\right)\left(1 - \frac{1}{4}\right) \ldots \left(1 - \frac{1}{50}\right)}$ is

(a) $\dfrac{7}{32}$ (b) $\dfrac{4}{7}$

(c) $1\dfrac{87}{256}$ (d) $\dfrac{256}{343}$

8. $\dfrac{(0.22)^3 + (0.11)^3 + (0.32)^3}{(0.66)^3 + (0.96)^3 + (0.33)^3} + \dfrac{(0.32)^3 + (0.45)^3 - (0.77)^3}{81(0.32)(0.45)(0.77)}$ equals

(a) 1   (b) $\dfrac{1}{11}$

(c) 0   (d) –1

9. Find the value of

$\left(1 - \dfrac{1}{3^2}\right)\left(1 - \dfrac{1}{4^2}\right)\left(1 - \dfrac{1}{5^2}\right) \ldots \ldots \left(1 - \dfrac{1}{11^2}\right)\left(1 - \dfrac{1}{12^2}\right)$

(a) $\dfrac{17}{18}$ (b) $\dfrac{13}{18}$

(c) $\dfrac{1}{144}$ (d) $\dfrac{1}{9}$

10. A student was asked to simplify the following:

$$\frac{7}{5-2\frac{2}{3}} \div \frac{3-\frac{2}{3-1\frac{1}{2}}}{4-1\frac{1}{2}} - \frac{5}{7} \times \left[\frac{7}{10} + 1\frac{1}{5} \times \frac{3\frac{1}{3}-2\frac{1}{2}}{2\frac{5}{21}-2} \right. $$

$$\left. + \frac{\frac{3}{1.6}+\frac{5}{3.2}}{\frac{5}{4.8}+\frac{1}{9.6}} \right]$$

His answer was $3\frac{1}{5}$. Find the per cent error.

(a) 10% (b) 20%

(c) 25% (d) 50%

Answers
1. (d)   **2.** (d)   **3.** (b)   **4.** (d)   **5.** (b)   **6.** (c)   **7.** (c)   **8.** (c)   **9.** (b)   **10.** (b)

Chapter

3

HCF AND LCM

1. The HCF of two or more numbers is the greatest number that divides each of them exactly.

2. **Methods of finding HCF**

   (*i*) **Prime Factorization :** Express each of the given numbers as the product of their prime factors. The HCF of the given numbers is the product of the least powers of common factors.

   $24 = 2^3 \times 3, 32 = 2^5 \Rightarrow$ HCF $(24, 32) = 2^3 = 8$.

   (*ii*) **Continued Division Method :** Divide the larger number by the smaller number. If the remainder is zero, the divisor is the HCF, otherwise divide the previous divisor by the remainder last obtained. Repeat this until the remainder becomes zero.

   To find the HCF of more than two numbers, first find the HCF of any two numbers and then find the HCF of the result and the third number and so on. The final HCF is the required HCF.

   > **Note :** The HCF of two co-prime numbers is 1, as they have no factors in common.

3. The LCM of two or more numbers is the least number that is divisible by all these numbers.

4. **Methods of finding LCM**

   (*i*) **Prime Factorization Method :** Express each number as a product of prime factors. The LCM of the given numbers is the product of the greatest powers of these prime factors.

   $64 = 2^6, 56 = 2^3 \times 7 \Rightarrow$ LCM $(64, 56) = 2^6 \times 7 = 448$

   (*ii*) **Common Division Method :** Arrange the given numbers in a row in any order. Now divide by a prime number which divides exactly at least two of the given numbers and carry forward the numbers which are not divisible. Repeat the process till no numbers have a common factor other than 1. The product of the divisors and the remaining numbers is the LCM of the given numbers.

   > **Note:** The LCM of two co-prime numbers is equal to their product.

5. The product of the HCF and LCM of two numbers is equal to the product of the numbers.

6. The HCF always completely divides the LCM for a given set of numbers.

7. HCF of given fractions $= \dfrac{\text{HCF of numerators}}{\text{LCM of denominators}}$

   LCM of given fractions $= \dfrac{\text{LCM of numerators}}{\text{HCF of denominators}}$

8. To find the HCF and LCM of decimal numbers, convert the given numbers to like decimals. Now find HCF and LCM of the numbers treating them as whole numbers (ignoring the decimal point). The number of decimal places in the answer are eaqua' to the number of decimal places in the like decimals. Accordingly, put the decimal point in the answer.

Some Important Results

9. The greatest number that will divide x, y and z leaving remainders a, b and c respectively is given by the HCF of $(x-a)$, $(y-b)$, $(z-c)$.

10. The greatest number that will divide x, y and z leaving the same remainder in each case is given by HCF of $(x-y)$, $(y-z)$, $(z-x)$.

11. The least number which when divided by x, y and z leaves the same remainder R in each case is given by LCM of $(x, y, z) + R$.

12. The least number which when divided by x, y and z leaves the remainders a, b and c respectively is given by LCM of $(x, y, z) - p$ where $p = (x-a) = (y-b) = (z-c)$.

Solved Examples

Ex. 1. *What is the least number which when divided by 15, 18 and 21 leaves remainders 2, 5 and 8 respectively ?*

Sol. Since the difference between the divisors and the respective remainders is same, i.e.,
$15 - 2 = 18 - 5 = 21 - 8 = 13$,
The required number = LCM of $(15, 18, 21) - 13 = (3 \times 5 \times 6 \times 7) - 13 = 630 - 13 = \mathbf{617}.$

3	15, 18, 21
	5, 6, 7

Ex. 2. *There are three numbers. The HCF of each pair is 15 and the LCM of all the three numbers is 1890. What is the product of the numbers ?*

Sol. Since the HCF of each pair of numbers is 15, the HCF of the three numbers is 15.
∴ Product of the numbers = HCF × LCM
$= 15 \times 1890 = \mathbf{28350}.$

Ex. 3. *What is the least number which when divided by 4, 6, 8 and 9 leaves zero remainder in each case but when divided by 13 leaves a remainder of 7 ?*

Sol. LCM of 4, 6, 8 and 9 = 72
Dividing 72 by 13,

$$\begin{array}{r} 5 \\ 13\overline{)72} \\ -65 \\ \hline 7 \end{array}$$

2	4, 6, 8, 9
2	2, 3, 4, 9
3	1, 3, 2, 9
	1, 1, 2, 3

∴ 72 is the required number.

Ex. 4. *Find the least number which when divided by 12, 16, 18, 30 leaves remainder 4 in each case but it is completely divisible by 7 ?*

Sol. LCM of $(12, 16, 18, 20) = 720$

2	12, 16, 18, 30
2	6, 8, 9, 15
3	3, 4, 9, 15
	1, 4, 3, 5

∴ The required number is of the form $720k + 4$
Checking for values of $k = 1, 2, 3, ...$ we see that the least value of k for which $720k + 4$ is divisible by 7 is $k = 4$,
∴ Required number $= 720 \times 9 + 4 = \mathbf{2884}.$

Ex. 5. *The LCM and HCF of two positive numbers are 175 and 5 respectively. If the sum of the numbers is 60, what is the difference between them ?*

Sol. Let the two numbers be $5a$ and $5b$ as HCF of the two numbers = 5

$\Rightarrow$ Product of the two number = HCF × LCM

$\Rightarrow 5a \times 5b = 5 \times 175 \quad \Rightarrow \quad ab = \dfrac{175}{5} = 35$

$\therefore$ (a, b) can be $(1, 35)$ or $(5, 7)$

Thus, the numbers can be $(1 \times 5$ and $35 \times 5)$ or $(5 \times 5$ and $5 \times 7)$, i.e., $(5$ and $175)$ or $(25$ and $35)$ The sum = 60 is satisfied by the pair $(25, 35)$. Hence, the difference of the numbers is 10.

Question Bank–3

1. Find the number of pairs of natural numbers with LCM as 56.
   - (a) 3
   - (b) 4
   - (c) 10
   - (d) Can't be determined

2. A General can draw up his soldiers in the rows of 10, 15 and 18 soldiers and he can also draw them up in the form of a solid square. Find the least number of soldiers with the general.
   - (a) 100
   - (b) 3600
   - (c) 900
   - (d) 90

3. The circumferences of the fore and hind wheels of a carriage are $6\dfrac{3}{14}$ m and $8\dfrac{1}{18}$ m respectively. At any given moment, a chalk mark is put on the point of contact of each wheel with the ground. Find the distance travelled by the carriage so that both the chalk marks are again on the ground at the same time.
   - (a) 218 m
   - (b) 217.5 m
   - (c) 218.25 m
   - (d) 217 m

4. The LCM of two numbers is 28 times of their HCF. The sum of their LCM and HCF is 1740. If one of the numbers is 240, find the other number.
   - (a) 240
   - (b) 620
   - (c) 540
   - (d) 420

5. Find the two largest numbers of four digits having 531 as their HCF.
   - (a) 9231, 9762
   - (b) 9027, 9558
   - (c) 9037, 9568
   - (d) 9127, 9658

6. Find the greatest number of five digits which become exactly divisible by 10, 12, 15 and 18 when 3769 is added to it.
   - (a) 99819
   - (b) 99911
   - (c) 99900
   - (d) 99111

7. Two numbers both greater than 29 have HCF = 29 and LCM = 4147. The sum of the numbers is
   - (a) 666
   - (b) 669
   - (c) 696
   - (d) 966

8. The HCF of two numbers each consisting of 4 digits is 103 and their LCM is 19261. The numbers are
   - (a) 1133, 1751
   - (b) 1053, 1657
   - (c) 1061, 1111
   - (d) 1591, 1377

9. Four prime numbers are written in ascending order of their magnitudes. The product of the first three is 715 and that of the last three is 2431. What is the largest given prime number ?
   - (a) 5
   - (b) 19
   - (c) 17
   - (d) 23

10. A number lying between 1000 and 2000 is such that on division by 2, 3, 4, 5, 6, 7 and 8 leaves remainders 1, 2, 3, 4, 5, 6 and 7 respectively. The number is
    - (a) 518
    - (b) 416
    - (c) 364
    - (d) 1679

11. Find the greatest number of five digits which when divided by 4, 6, 14 and 20 leaves respectively 1, 3, 11 and 17 as remainders.
    - (a) 99930
    - (b) 99960
    - (c) 99997
    - (d) 99957

12. Find the least number which when divided by 12, 24, 36 and 40 leaves a remainder 1, but when divided by 7 leaves no remainder.
    - (a) 361
    - (b) 1080
    - (c) 721
    - (d) 371

13. What is the least number which when divided by the numbers 3, 5, 6, 8, 10 and 12 leaves in each case a remainder 2, but when divided by 13 leaves no remainder.

(a) 312   (b) 962
(c) 1562   (d) 1586

14. A heap of stones can be made up into groups of 21. When made up into groups of 16, 20, 25 and 45, there are 3 stones left in each case. How many stones at least can there be in the heap ?
(a) 7203   (b) 2403
(c) 3603   (d) 4803

15. Find the least number which when divided by 2, 3, 4, 5 and 6 leaves 1, 2, 3, 4 and 5 as remainders respectively, but when divided by 7 leaves no remainder.
(a) 210   (b) 119
(c) 126   (d) 154

16. The HCF and LCM of two numbers are 12 and 72 respectively. If the sum of the two numbers is 60, then one of the numbers will be
(a) 12   (b) 24
(c) 60   (d) 72

17. The difference of two numbers is 20 and their product is 56.25 times their difference. Find the LCM of the numbers.
(a) 70   (b) 1125
(c) 225   (d) 5

18. There are 4 numbers, The HCF of each pair is 7 and the LCM of all the numbers is 1470. What is the product of the 4 numbers ?
(a) 504210   (b) 502410
(c) 504120   (d) 501420

19. Two persons A and B walk round a circle whose diameter is 1.4 km. A walks at a speed of 165 metres per munute while B walks at a speed of 110 metres per minute. If they both start at the same time from the same point and walk in the same direction at what interval of time would they both be at the same starting point again.
(a) 1 h   (b) $1\frac{1}{3}$ h
(c) $1\frac{2}{3}$ h   (d) $1\frac{1}{2}$ h

20. The LCM of two numbers is 495 and their HCF is 5. If the sum of the numbers is 100, then their difference is
(a) 10   (b) 46
(c) 70   (d) 90

Answers

1. (c)	2. (c)	3. (b)	4. (d)	5. (b)	6. (b)	7. (c)	8. (a)	9. (c)	10. (d)
11. (d)	12. (c)	13. (b)	14. (a)	15. (b)	16. (b)	17. (c)	18. (a)	19. (b)	20. (a)

Hints and Solutions

1. (c) Let a, b be the two numbers with LCM 56. Since 56 is the LCM a and b will be the divisors of 56. The divisors of 56 are 1, 2, 4, 7, 8, 14, 28 and 56. Combining two at a time, the possible pairs with LCM as 56 are (1, 56), (2, 56), (4, 56), (7, 56), (8, 56), (14, 56), (28, 56), (8, 14), (8, 28) and (7, 8). Thus, in all there are 10 pairs.

2. (c) LCM of (10, 15, 18) = 2 × 3 × 5 × 3 = 90

2	10,	15,	18
3	5,	15,	9
5	5,	5,	3
	1,	1,	3

∴ To draw the soldier in the form of a solid square the number of soldiers = 90k, where k is a natural number. The least value of k for which 90 k is a square number is $k = 10$, i.e, the required number of soldiers = 90 × 10 = **900**.

3. (b) Required distance travelled
$$= \text{LCM of } \left(6\frac{3}{14} \text{ m}, 8\frac{1}{18} \text{ m}\right)$$

$$\left.\begin{array}{l}87 = 3 \times 29 \\ 145 = 5 \times 29\end{array}\right| = \text{LCM of } \left(\frac{87}{14}, \frac{145}{18}\right)$$

$$\left.\begin{array}{l}14 = 2 \times 7 \\ 18 = 2 \times 9\end{array}\right| = \frac{\text{LCM of } (87,145)}{\text{HCF of } (14,18)}$$

$$= \frac{3 \times 5 \times 29}{2} = \frac{435}{2} = \textbf{217.5 m.}$$

4. (d) LCM = 28 HCF
Also, LCM + HCF = 1740
⇒ 28 HCF + HCF = 1740
⇒ 29 HCF = 1740 ⇒ HCF = $\frac{1740}{29}$ = 60
⇒ LCM = 28 × 60 = 1680
Since, one number = 240
∴ Other number = $\frac{\text{HCF} \times \text{LCM}}{\text{One number}}$
$$= \frac{60 \times 1680}{240} = \textbf{420}$$

5. (b) The largest number of four digits = 9999

Dividing 9999 by 531, we get

$$531 \overline{)9999} \left(18\right.$$
$$\underline{-531}$$
$$4689$$
$$\underline{-4248}$$
$$441$$

∴ Greatest number of four digits divisible by 531
= 9999 – 441 = **9558**

The other number = 9558 – 531 = **9027**.

6. (b) LCM of (10, 12, 15 and 18)
= $2 \times 3 \times 5 \times 2 \times 3 = 180$

2	10, 12, 15, 18
3	5, 6, 15, 9
5	5, 2, 5, 3
	1, 2, 1, 3

The greatest five digit number = 99999

Dividing (99999 + 3769) = 103768 by 180, we get

$$180 \overline{)103768} \left(576\right.$$
$$\underline{-900}$$
$$1376$$
$$\underline{-1260}$$
$$1168$$
$$\underline{-1080}$$
$$88$$

Remainder = 88

∴ Required number = 99999 – 88 = **99911**.

7. (c) Since the HCF of both the numbers is 29, let the numbers be $29a$ and $29b$, where a and b are co-prime to each other.

Given, $29a \times 29b = 29 \times 4147 \Rightarrow ab = 143$

The co-prime pairs (a, b) whose product = 143 are (1, 143) and (11, 13)

Since both the required numbers are greater than 29, the co-prime pair satisfying the condition is (11, 13).

Therefore the numbers are 29×11 and 29×13, i.e., 319 and 377.

∴ Sum of the numbers = 319 + 377 = **696**.

8. (a) Similar to Q. No. 7.

9. (c) Let the four prime numbers in the ascending order of their magnitudes be a, b, c and d.

Given, $a \times b \times c = 715$ and $b \times c \times d = 2431$

HCF of $(abc$ and $bcd) = bc$

∴ Now, $715 = 11 \times 13 \times 5$ and

$2431 = 11 \times 13 \times 17$

⇒ HCF of 715 and 2431 = $11 \times 13 = 143$

∴ Largest number $(d) = \dfrac{bcd}{bc} = \dfrac{2431}{143} = \mathbf{17}$.

10. (d) The common difference between the divisors and the respective remainders = $(2 – 1) (3 – 2)$
= $(4 – 3) = \ldots\ldots\ldots = (8 – 7) = 1$

LCM of 2, 3, 4, 5, 6, 7 and 8 = 840

∴ Required number = $840 k – 1$

Since the number lies between 1000 and 2000, $k = 2$.

∴ Required number = $840 \times 2 – 1$
= 1680 – 1 = **1679**.

11. (d) Common difference between divisors and respective remainders
= $(4 – 1) = (6 – 3) = (14 – 11) = (20 – 17) = 3$

LCM of $(4, 6, 14, 20) = 2 \times 2 \times 3 \times 7 \times 5$
= 420.

2	4, 6, 14, 20
2	2, 3, 7, 10
	1, 3, 7, 5

Greatest number of five digits = 99999

Dividing 99999 by 420 and subtracting the remainder 39 from 99999, we get 99999 – 39 = 99960.

$$420 \overline{)99999} \left(238\right.$$
$$\underline{-840}$$
$$1599$$
$$\underline{-1260}$$
$$3399$$
$$\underline{-3360}$$
$$39$$

∴ The required number = 99960 – 3
= **99957**.

12. (c) LCM of 12, 24, 36 and 40 = 360

Any number which when divided by 12, 24, 36 and 40 leaving a remainder 1 is of the form $360k + 1$. Now, we have to find the least value of k for which $360k + 1$ is divisible by 7.

$$7 \overline{)360\,k + 1} \left(51k\right.$$
$$\underline{-357\,k}$$
$$3k + 1$$

By inspection we find that for $k = 2$,
$3 \times 2 + 1 = 7$

∴ Required number = $360 \times 2 + 1 = 721$.

13. (b) Similar to Q. No. 12.

14. (a) Number of stones

= LCM of (16, 20, 25, 45) × k + 3

= 3600 k + 3

Since the stones can be made up into groups of 21.

$$21 \overline{) 3600\,k + 3} (171\,k$$
$$\underline{-\,3591k}$$
$$9k + 3$$

∴ $9k + 3$ is divisible by 21 when $k = 2$

∴ Least number of stones = 3600 × 2 + 3 = **7203.**

15. (b) Common difference between divisors and respective remainders

= (2 − 1) = (3 − 2) = (4 − 3)

= (5 − 4) = (6 − 5) = 1

LCM of (2, 3, 4, 5, 6) = 60

∴ Required number = 60 k − 1

Now we have to find the least value of k for which $60k − 1$ is divisible by 7.

$$7 \overline{) 60k-1} (8k$$
$$\underline{-56k}$$
$$4k-1$$

By inspection, we find that for $k = 2$, $4 × 2 − 1 = 7$

∴ Required number = 60 × 2 − 1 = 120 − 1 = 119.

16. (b) Given, HCF = 12, LCM = 72

One number = x, othe number = 60 − x

∴ Product of the two numbers = HCF × LCM

⇒ $x(60 − x) = 12 × 72$

⇒ $x^2 − 60x + 864 = 0$

⇒ $x^2 − 36x − 24x + 864 = 0$

⇒ $x(x − 36) − 24(x − 36) = 0$

⇒ $(x − 36)(x − 24) = 0 ⇒ x = $ **36** or **24.**

∴ One of the number is 24.

17. (c) Let the numbers be a and b. Then,

$a − b = 20$ .... (i)

$ab = 56.25 × 20 = 1125$

⇒ $b = \dfrac{1125}{a}$

∴ Putting the value of b in (i) we get

$a − \dfrac{1125}{a} = 20 ⇒ a^2 − 1125 = 20a$

⇒ $a^2 − 20a − 1125 = 0$

⇒ $a^2 − 45a + 25a − 1125 = 0$

⇒ $a(a − 45) + 25(a − 45) = 0$

⇒ $(a − 45)(a + 25) = 0$

⇒ $a = 45$ or $−25$

Neglecting the negative value of a,

$b = 45 − 20 = 25$

Now, $a = 45 = 3^2 × 5$, $b = 25 = 5^2$

∴ LCM of (45, 25) = $3^2 × 5^2 = 225$

18. (a) Since the HCF of each pair = 7, let the four numbers be $7a$, $7b$, $7c$, $7d$.

Also, LCM = $abcd × $ HCF

⇒ $abcd = \dfrac{1470}{7} = 210$

∴ Product of the numbers

= $7a × 7b × 7c × 7d$

= $7^4 × abcd = 7^4 × 210 = $ **504210.**

19. (b) Diameter of the circle = 1.4 cm

⇒ Circumference = $2\pi r$

$= 2 × \dfrac{22}{7} × \dfrac{1.4}{2}$ km = 4.4 km

A's speed = 165 m/min

∴ Time taken by A to travel 4.4 km (4400 m)

$= \dfrac{4400}{165}$ min $= \dfrac{80}{3}$ min

B's speed = 110 m/ min

∴ Time taken by B's to travel 4.4 km (4400 m)

$= \dfrac{4400}{110}$ min = 40 min

Required interval of time = LCM of $\left(\dfrac{80}{3}, 40\right)$

$= \dfrac{\text{LCM}(80, 40)}{\text{HCF}(3, 1)} = \dfrac{80}{1}$ min

$= \dfrac{80}{60}$ hrs $= 1\dfrac{1}{3}$ **hrs**

20. (a) Let one number = x. Then,

Other number = 100 − x

LCM = 495, HCF = 5

∴ $x(100 − x) = 495 × 5$

⇒ $100x − x^2 = 2475$

⇒ $x^2 − 100x + 2475 = 0$

⇒ $x^2 − 45x − 55x + 2475 = 0$

⇒ $x(x − 45) − 55(x − 45) = 0$

⇒ $(x − 45)(x − 55) = 0$

⇒ $x = 45$ or 55.

∴ The numbers and 45, 100 − 45 = 55 or 55, 100 − 55 = 45

Required difference = 55 − 45 = **10.**

Self Assessment Sheet–3

1. What is the greatest number of 4-digits that, which when divided by any of the numbers 6, 9, 12 and 17 leaves a remainder of 1?
   (a) 9997   (b) 9793
   (c) 9895   (d) 9487

2. Find the greatest number that will divide 55, 127 and 175, so as to leave the same remainder in each case.
   (a) 1   (b) 16
   (c) 24   (d) 15

3. The sum of two numbers is 1215 and their HCF is 81. How many pairs of such numbers can be formed?
   (a) 2   (b) 6
   (c) 4   (d) None

4. How many numbers between 200 and 600 are exactly divisible by 4, 5 and 6?
   (a) 5   (b) 16
   (c) 10   (d) 6

5. The HCF of two numbers of same number of digits is 45 and their LCM is 540. The numbers are
   (a) 270, 540   (b) 135, 270
   (c) 180, 270   (d) 135, 180

6. Find the least number which on being divided by 5, 6, 8, 9, 12 leaves in each case a remainder 1 but when divided by 13 leaves no remainder?
   (a) 3601   (b) 1469
   (c) 2091   (d) 4879

7. If $x = 103$, then the LCM of $x^2 - 4$ and $x^2 - 5x + 6$ is
   (a) 105105   (b) 1051050
   (c) 106050   (d) 1060500

8. The LCM of two numbers is 12 times their HCF. The sum of HCF and LCM is 403. If one of the numbers is 93, then the other number is
   (a) 124   (b) 128
   (c) 134   (d) 138

9. The HCF and LCM of two numbers x and y is 6 and 210 respectively. If $x + y = 72$, which of the following relation is correct?
   (a) $\dfrac{1}{x} + \dfrac{1}{y} = \dfrac{3}{35}$ (b) $\dfrac{1}{x} + \dfrac{1}{y} = \dfrac{2}{35}$
   (c) $\dfrac{1}{x} + \dfrac{1}{y} = \dfrac{35}{2}$ (d) not sufficient

10. If the HCF of $x^2 - x - 6$ and $x^2 + 9x + 14$ is $(x + m)$, then the value of m is:
    (a) 1   (b) 2
    (c) – 2   (d) – 1

Answers

1. (b)	2. (c)	3. (c)	4. (d)	5. (d)	6. (a)	7. (d)	8. (a)	9. (b)	10. (b)

Chapter 4

SQUARE ROOTS AND CUBE ROOTS

KEY FACTS

1. If a number is multiplied by itself, the product so obtained is called the square of that number. It is a number raised to the power 2.

   In the statement $13 \times 13 = 169$, 169 is the **square** of 13 and 13 is the **square root** of 169.

2. The square of a natural number is called a **perfect square**. Following are some important properties of square numbers.

   (*i*) A square number is never negative.

   (*ii*) A square number never ends in 2, 3, 7 or 8.

   (*iii*) The number of zeros at the end of a perfect square is always even.

   (*iv*) The square of an even number is even.

   (*v*) The square of an odd number is odd.

   (*vi*) For any natural number n. n^2 = Sum of first n odd natural numbers.

3. **Finding the square root :**

   (*i*) The square root of a perfect square number can be obtained by finding the prime factorization of the square number, pairing equal factors and picking out one prime factor of each pair. The product of the prime factors thus picked gives the square root of the number.

   > **Note:** We may also write the product of prime factors in exponential form and for finding the square root, we take half of the index value of each factor and then multiply.

   For example :

   $196 = 2 \times 2 \times 7 \times 7 \Rightarrow \sqrt{196} = 2 \times 7 = 14$ or

   $196 = 2^2 \times 7^2 \Rightarrow \sqrt{196} = 2^{2/2} \times 7^{2/2} = 2 \times 7 = 14$

   (*ii*) $\sqrt{p \times q} = \sqrt{p} \times \sqrt{q}$

   (*iii*) $\sqrt{\dfrac{p}{q}} = \dfrac{\sqrt{p}}{\sqrt{q}}$

   (*iv*) The square root of a number can also be found by *division method.* You will be explained this method with the help of an example.

Example: Find the square root of 17424.

Step 1. Take the first pair of digits and find the nearest perfect square. Here $1^2 = 1$.

Step 2. Twice of $1 = 2$

Step 3. 2 goes into 7 three times. Put 3 on the top and in the divisor as shown. $23 \times 3 = 69$.

Step 4. Double 13. You get 26. 26 goes into 52, 2 times. Place 2 on top and in the divisor as shown $2 \times 262 = 524$.

Step 5. Subtract. The remainder is 0. Therefore, 132 is the exact square root of 17424.

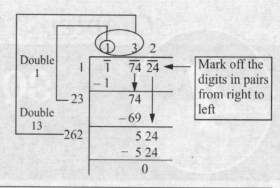

Note: In a decimal number, the pairing of numbers starts from the decimal point. For the integral part it goes from right to left ($\leftarrow$) and for the decimal part it goes from left to right, i.e., $\overset{\leftarrow}{31}\,\overset{\leftarrow}{61}.\overset{\rightarrow}{81}\,\overset{\rightarrow}{29}$. The procedure followed is the same as in integral numbers explained above.

(v) If a positive number is not a perfect square, then an approximate value of its square root may be obtained by the division method. $\sqrt{2}$ can be found as :

$$\therefore \sqrt{2} = 1.414 \text{ (approx.)}$$

Also, if n is not a perfect square as 2, then $\sqrt{n}$ is not a rational number, e.g., $\sqrt{2}, \sqrt{3}, \sqrt{7}$ are not rational numbers.

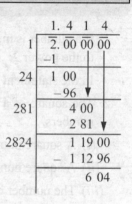

4. The **cube** of a number is the *number raised to the power* 3, e.g., cube of $8 = 8^3 = 8 \times 8 \times 8 = 512$.

A natural number n is a perfect cube if there exists a natural number m such that $n = m \times m \times m$, i.e., $m^3 = n$. 64 is a perfect cube as $64 = 4^3$.

5. (i) *The cube of an even number is even*, i.e., $6^3 = 216$.

   (ii) *The cube of an odd number is odd*, i.e., $5^3 = 125$.

6. The **cube root** of a number n is the number whose cube is n. It is denoted by $\sqrt[3]{n}$, e.g., $\sqrt[3]{8} = 2$.

7. The cube root of a number can be found by resolving the number into prime factors, making groups of 3 equal factors, picking out one of the equal factors from each group and multiplying the factors so picked.

$$1728 = \underline{2 \times 2 \times 2} \times \underline{2 \times 2 \times 2} \times \underline{3 \times 3 \times 3}$$

$$\sqrt[3]{1728} = 2 \times 2 \times 3 = 12$$

8. The cube root of a negative perfect cube is negative, e.g., $\sqrt[3]{-125} = -5$

9. For any integer a and b, we have

   (i) $\sqrt[3]{a \times b} = \sqrt[3]{a} \times \sqrt[3]{b}$ (ii) $\sqrt[3]{\dfrac{a}{b}} = \dfrac{\sqrt[3]{a}}{\sqrt[3]{b}}$

Question Bank–4

1. Find the value of $\sqrt{11.981 + 7\sqrt{1.2996}}$

   (a) 5.181   (b) 3.354
   (c) 4.467   (d) 4.924

2. What is the least number that must be added to 1901 so that the sum may be a perfect square is

   (a) 35   (b) 32
   (c) 30   (d) 29

3. The positive square root of 45.5625 is

   (a) 5.25   (b) 5.65
   (c) 6.35   (d) 6.75

4. The least perfect square number which is divisible by each of 21, 36 and 66 is

   (a) 213444   (b) 214344
   (c) 214434   (d) 231444

5. The square root of $0.09 + 2 \times 0.21 + 0.49$ is

(a) $\sqrt{0.09} + \sqrt{0.49}$ (b) $2\sqrt{0.21}$

(c) 1   (d) 0.58

6. The least integer that must be added to (9798×9792) to make it a perfect square is

(a) 9   (b) 8

(c) 7   (d) 6

7. Assume that $\sqrt{13} = 3.605$ (approx.) and $\sqrt{130} = 11.40$ (approx.).

Find the value of $\sqrt{1.3} + \sqrt{1300} + \sqrt{0.013}$

(a) 36.164   (b) 37.304

(c) 36.304   (d) 37.164

8. The digit in the units' place in the square root of 15876 is

(a) 8   (b) 6

(c) 4   (d) 2

9. The smallest number that must be added to 680621 to make the sum a perfect square is

(a) 4   (b) 5

(c) 6   (d) 8

10. $\sqrt{(0.798)^2 + 0.404 \times 0.798 + (0.202)^2} + 1$ is equal to

(a) 0   (b) 2

(c) 1.596   (d) 0.404

11. What is the least number that should be subtracted from 0.000326 to have a perfect square is

(a) 0.000004   (b) 0.000002

(c) 0.04   (d) 0.02

12. Each member of a picnic party contributed twice as many rupees as the total number of members and the total collection was Rs 3042. The number of members present in the party was

(a) 2   (b) 32

(c) 40   (d) 39

13. What is the least number which must be subtracted from 10420 to make it a perfect square ?

(a) 219   (b) 200

(c) 189   (d) 16

14. $\sqrt{86.49} + \sqrt{5 + k^2} = 12.3$. So k is equal to

(a) $\sqrt{10}$ (b) $2\sqrt{5}$

(c) $3\sqrt{5}$ (d) 2

15. The number whose square is equal to the difference of the squares of 75.15 and 60.12 is

(a) 46.09   (b) 48.09

(c) 45.09   (d) 47.09

16. If $\sqrt{1369} + \sqrt{0.0615 + x} = 37.25$, then x is equal to

(a) 10^{-1} (b) 10^{-2}

(c) 10^{-3} (d) 10

17. If $\sqrt{(x-1)(y+2)} = 7$, x and y being positive whole numbers, then the values of x and y are respectively

(a) 8, 5   (b) 15, 12

(c) 22, 19   (d) 6, 8

18. If $\sqrt{0.04 \times 0.4 \times a} = 0.004 \times 0.4 \times \sqrt{b}$, then $\dfrac{a}{b}$ is

(a) 16×10^{-3} (b) 16×10^{-4}

(c) 16×10^{-5} (d) 16×10^{-2}

19. If $a = 0.1039$, then the value of $\sqrt{4a^2 - 4a + 1} + 3a$ is

(a) 0.1039   (b) 0.2078

(c) 1.1039   (d) 2.1039

20. If $3a = 4b = 6c$ and $a + b + c = 27\sqrt{29}$, then $\sqrt{a^2 + b^2 + c^2}$ is

(a) $3\sqrt{29}$ (b) 81

(c) 87   (d) 29

21. If $3\sqrt{5} + \sqrt{125} = 17.88$, then what will be the value of $\sqrt{80} + 6\sqrt{5}$?

(a) 13.41   (b) 20.46

(c) 21.66   (d) 22.35

22. The number of trees in each row of a garden is equal to the total number of rows in the garden. After 111 trees have been uprooted in a storm, their remain 10914 trees in the garden. The number of rows of trees in the garden is

(a) 100   (b) 105

(c) 115   (d) 125

23. If the product of four consecutive natural numbers increased by a natural number p is a perfect square, then the value of p is

(a) 8   (b) 4

(c) 2   (d) 1

24. $\sqrt[3]{\sqrt{0.000064}}$ is equal to

   (a) 0.02   (b) 0.2

   (c) 2   (d) 0.4

25. $\sqrt[2]{\sqrt[3]{x \times 0.000001}} = 0.2$. The value of x is

   (a) 8   (b) 16

   (c) 32   (d) 64

26. The digit in the units' place in the cube root of 21952 is

   (a) 8   (b) 6

   (c) 4   (d) 2

27. Cube root of a number when divided by 5 results in 25, what is the number ?

(a) 5   (b) 125^3

(c) 5^3 (d) 125

28. The smallest of $\sqrt{8} + \sqrt{5}$, $\sqrt{7} + \sqrt{6}$, $\sqrt{10} + \sqrt{3}$ and $\sqrt{11} + \sqrt{2}$ is

   (a) $\sqrt{8} + \sqrt{5}$ (b) $\sqrt{7} + \sqrt{6}$

   (c) $\sqrt{10} + \sqrt{3}$ (d) $\sqrt{11} + \sqrt{2}$

29. The smallest positive integer n for which $864 \times n$ is a perfect cube is

   (a) 1   (b) 2

   (c) 3   (d) 4

30. $\sqrt[3]{\sqrt[3]{a^3}}$ is equal to

   (a) a (b) 1

   (c) $a^{1/3}$ (d) a^3

Answers

1. (c)	**2.** (a)	**3.** (d)	**4.** (b)	**5.** (c)	**6.** (a)	**7.** (b)	**8.** (b)	**9.** (a)	**10.** (b)
11. (b)	**12.** (d)	**13.** (d)	**14.** (d)	**15.** (c)	**16.** (c)	**17.** (a)	**18.** (c)	**19.** (c)	**20.** (c)
21. (d)	**22.** (b)	**23.** (d)	**24.** (b)	**25.** (d)	**26.** (a)	**27.** (b)	**28.** (d)	**29.** (b)	**30.** (c)

Hints and Solutions

1. (c) $\sqrt{11.981 + 7\sqrt{1.2996}} = \sqrt{11.981 + 7 \times 1.14}$

   $= \sqrt{11.981 + 7.98} = \sqrt{19.961} = 4.468$

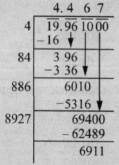

2. (a) Number to be added

   $= (44)^2 - 1901$

   $= 1936 - 1901$

   $= 35$

3. (d) $\therefore \sqrt{45.5625} = 6.75$

4. (b) LCM of 21, 36 and 66 = 2772

   $\begin{array}{r|l} 3 & 21, 36, 66 \\ \hline 2 & 7, 12, 22 \\ \hline & 7, 6, 11 \end{array}$

   $\therefore$ Least perfect square number

      $= (2^2 \times 3^2 \times 7 \times 11) \times 7 \times 11 = 213444$

5. (c) $\sqrt{0.09 + 2 \times 0.21 + 0.49}$

      $= \sqrt{(0.3)^2 + 2 \times 0.3 \times 0.7 + (0.7)^2}$

      $= \sqrt{(0.3 + 0.7)^2} = \sqrt{1} = 1.$

6. (a) $9798 \times 9792 = (9792 + 6) \times 9792$

                $= (9792)^2 + 6 \times 9792$

                $= (9792)^2 + 2 \times 3 \times 9792$

   $\therefore$ Perfect square number

                $= (9792)^2 + 2 \times 3 \times 9792 + (3)^2$

   $\therefore$ Least integer to be added to make 9798×9792 a perfect square $= 3^2 = 9$.

7. (b) $\sqrt{1.3} + \sqrt{1300} + \sqrt{0.013}$

      $= \sqrt{\dfrac{130}{100}} + \sqrt{13 \times 100} + \sqrt{\dfrac{130}{10000}}$

      $= \dfrac{\sqrt{130}}{\sqrt{100}} + \sqrt{13} \times \sqrt{100} + \dfrac{\sqrt{130}}{\sqrt{10000}}$

$$= \frac{11.40}{10} + 3.605 \times 10 + \frac{11.40}{100}$$

$$= 1.14 + 36.05 + 0.114 = 37.304$$

8. (b)

```
        1  2  6
    1 | 1  58  76
      |-1  ↓      ↓
   22 |   58
      |  -44       ↓
  246 |    14 76
      |   -14 76
      |        0
```

∴ $\sqrt{15876} = 126$

⇒   Digit in units' place in $\sqrt{15876} = 6$.

9. (a)

```
         8   2   4
    8 | 68  06  21
      |-64  ↓
  162 |   4 06
      |  -3 24     ↓
 1644 |      82 21
      |     -65 76
      |      16 45
```

∴   Least number to be added to 680621 to make
the sum a perfect square
= $(825)^2 - 680621$
= 680625 − 680621 = 4.

10. (b) Given exp.

$$= \sqrt{(0.798)^2 + 2 \times 0.202 \times 0.798 + (0.202)^2 + 1}$$

$$= \sqrt{(0.798 + 0.202)^2 + 1}$$

$$= \sqrt{1^2} + 1 = 1 + 1 = 2.$$

11. (b) $0.000326 = \frac{326}{10^6}$

```
        1  8
    1 | 3  26
      |-1
   28 | 2  26
      |-2  24
      |      2
```

∴   Least number to be subtracted = $\frac{2}{10^6}$ = 0.000002

12. (d) Let the total number of members be x. Then,
Each member's contribution = Rs $2x$
Given, $x \times 2x = 3042$

⇒ $2x^2 = 3042$

⇒ $x^2 = 1521$

⇒ $x = 39$.

```
        3  9
    3 | 15 21
      | -9
   69 |  6 21
      | -6 21
      |     0
```

13. (d)

```
         1  0  2
    1 | 1  04  20
      |-1   ↓   ↓
  202 | 0  04  20
      |   -4 04
      |       16
```

∴ Least number to be subtracted from 10420 to
make it a perfect square = 16.

14. (d) $\sqrt{86.49} + \sqrt{5 + k^2} = 12.3$

⇒ $\sqrt{5 + k^2} = 12.3 - \sqrt{86.49} = 12.3 - 9.3 = 3$

⇒ $5 + k^2 = 9$ (on squaring both the sides)

⇒ $k^2 = 9 - 5 = 4 \Rightarrow k = 2$

15. (c) Let the required number be x. Then.
$$x^2 = (75.15)^2 - (60.12)^2$$
$$= (75.15 + 60.12)(75.15 - 60.12)$$
$$= 135.27 \times 15.03 = 2033.1081$$

```
          4  5 . 0  9
    4 | 2033.10 81
      | -16 ↓
   85 |  433
      | -425   ↓    ↓
 9009 |     8 10 81
      |    -8 10 81
      |          0
```

∴ $x = \sqrt{2033.1081}$
= 45.09.

16. (c) $\sqrt{1369} + \sqrt{0.0615 + x} = 37.25$

⇒ $\sqrt{0.0615 + x} = 37.25 - \sqrt{1369}$
$= 37.25 - 37 = 0.25$

⇒ $0.0615 + x = (0.25)^2 = 0.0625$

⇒ $x = 0.0625 - 0.0615 = 0.001 = \frac{1}{10^3} = 10^{-3}$.

17. (a) $\sqrt{(x-1)(y+2)} = 7 \Rightarrow (x-1)(y+2) = 7^2$

⇒ $(x-1) = 7$ and $(y+2) = 7$

⇒ $x = 8$ and $y = 5$.

18. (c) Given exp. $\Rightarrow \sqrt{0.016 \times a} = 0.0016 \times \sqrt{b}$

$$\Rightarrow \sqrt{0.016} \times \sqrt{a} = 0.0016 \times \sqrt{b} \Rightarrow \frac{\sqrt{a}}{\sqrt{b}} = \frac{0.0016}{\sqrt{0.016}}$$

$$\Rightarrow \frac{a}{b} = \frac{0.0016 \times 0.0016}{0.016} = 0.00016$$

$$= \frac{16}{100000} = 16 \times 10^{-5}.$$

19. (c) $\sqrt{4a^2 - 4a + 1} + 3a$

$$= \sqrt{(1)^2 - 2 \times 2a \times 1 + (2a)^2} + 3a$$

$$= \sqrt{(1 - 2a)^2} + 3a = (1 - 2a) + 3a = 1 + a$$

$$= 1 + 0.1039 = 1.1039.$$

20. (c) $3a = 4b = 6c \Rightarrow 4b = 6c \Rightarrow b = \dfrac{3}{2}c$

and $3a = 4b \Rightarrow a = \dfrac{4}{3}b = \dfrac{4}{3} \times \dfrac{3}{2}c = 2c$

$\therefore a + b + c = 27\sqrt{29}$

$\Rightarrow 2c + \dfrac{3}{2}c + c = 27\sqrt{29}$

$\Rightarrow \dfrac{9}{2}c = 27\sqrt{29} \Rightarrow c = 6\sqrt{29}$

Now, $\sqrt{a^2 + b^2 + c^2}$

$$= \sqrt{(a + b + c)^2 - 2(ab + bc + ca)}$$

$$= \sqrt{(27\sqrt{29})^2 - 2\left(2c \times \dfrac{3}{2}c + \dfrac{3}{2}c \times c + c \times 2c\right)}$$

$$= \sqrt{729 \times 29 - 2\left(3c^2 + \dfrac{3}{2}c^2 + 2c^2\right)}$$

$$= \sqrt{729 \times 29 - 2 \times \dfrac{13c^2}{2}}$$

$$= \sqrt{729 \times 29 - 13 \times (6\sqrt{29})^2}$$

$$= \sqrt{29(729 - 468)} = \sqrt{29 \times 261} = \sqrt{29 \times 29 \times 9}$$

$$= 29 \times 3 = 87.$$

21. (d) $3\sqrt{5} + \sqrt{125} = 17.88$

$\Rightarrow 3\sqrt{5} + \sqrt{5^2 \times 5} = 17.88$

$\Rightarrow 3\sqrt{5} + 5\sqrt{5} = 17.88$

$\Rightarrow 8\sqrt{5} = 17.88 \Rightarrow \sqrt{5} = \dfrac{17.88}{8} = 2.235$

Now, $\sqrt{80} + 6\sqrt{5} = \sqrt{16 \times 5} + 6\sqrt{5}$

$$= 4\sqrt{5} + 6\sqrt{5} = 10\sqrt{5}$$
$$= 10 \times 2.235 = 22.35.$$

22. (b) Let the number of rows = number of trees = x

$\therefore$ Total number of trees in the garden

$= x \times x = x^2 = 10914 + 111 = 11025$

$\therefore$ No. of rows of trees $= \sqrt{11025} = 105$

23. (d) Let the four consecutive natural numbers be
x, $x + 1$, $x + 2$ and $x + 3$. Then,

A perfect square $= x(x + 1)(x + 2)(x + 3) + p$

$$= x(x + 3)(x + 1)(x + 2) + p$$
$$= (x^2 + 3x) \times (x^2 + 3x + 2) + p$$
$$= (x^2 + 3x)^2 + 2(x^2 + 3x + 2) + p$$
$$= (x^2 + 3x)^2 + 2(x^2 + 3x) + p$$

The expression on the right hand side will be a perfect square if and only $p = 1$.

Perfect square number

$$= [(x^2 + 3x)^2 + 2(x^2 + 3x) + 1]$$
$$= (x^2 + 3x + 1)^2$$

24. (b) $\sqrt[3]{\sqrt{0.000064}} = \sqrt[3]{\sqrt{\dfrac{64}{1000000}}}$

$$= \sqrt[3]{\dfrac{8}{1000}} = \dfrac{2}{10} = 0.2$$

25. (d) $\sqrt[2]{\sqrt[3]{x \times 0.000001}} = 0.2$

$\Rightarrow \sqrt[3]{x \times 0.000001} = (0.2)^2$

$= 0.04$ (Squaring both the sides)

$\Rightarrow x \times 0.000001 = (0.04)^3 = 0.000064$

(on taking the cube of both the sides)

$\Rightarrow x = \dfrac{0.000064}{0.000001} = 64.$

26. (a)

2	21952
2	10976
2	5488
2	2744
2	1372
2	686
7	343
7	49
	7

$\therefore 21952 = 2 \times 2 \times 2 \times 2 \times 2 \times 2 \times 7 \times 7 \times 7$
$= 2^3 \times 2^3 \times 7^3$

$\therefore \sqrt[3]{21952} = \sqrt[3]{2^3 \times 2^3 \times 7^3} = 2 \times 2 \times 7 = 28$

$\therefore$ Digit in the units' place of $\sqrt[3]{21952} = 8$.

27. (b) Let the number be x. Then,

$\dfrac{\sqrt[3]{x}}{5} = 25 \Rightarrow \sqrt[3]{x} = 125 \Rightarrow x = (125)^3$

28. (d) $\sqrt{8} + \sqrt{5} = 2.83 + 2.24 = 5.07$

$\sqrt{7} + \sqrt{6} = 2.65 + 2.45 = 5.09$

$\sqrt{10} + \sqrt{13} = 3.16 + 3.61 = 6.77$

$\sqrt{11} + \sqrt{2} = 3.32 + 1.41 = 4.73$

$\therefore$ Smallest is $\sqrt{11} + \sqrt{2}$

29. (b)

2	864
2	432
2	216
2	108
2	54
3	27
3	9
	3

$864 = \underline{2 \times 2 \times 2} \times 2 \times 2 \times \underline{3 \times 3 \times 3}$

Since 2×2 is the only in complete triplet, so 864 has to be multiplied by 2 to make it a perfect cube.

$\Rightarrow n = 2.$

30. (c) $\sqrt[3]{\sqrt[3]{a^3}} = \left((a^3)^{\frac{1}{3}} \right)^{\frac{1}{3}} = a^{3 \times \frac{1}{9}} = a^{\frac{1}{3}}.$

Self Assessment Sheet–4

1. If $\sqrt{(x-1)(y+2)} = 7$ and x and y are positive whole numbers, their values respectively are
   (a) 8, 5   (b) 15, 12
   (c) 22, 19   (d) None of these

2. The square root of the expression

   $\dfrac{(12.1)^2 - (8.1)^2}{(0.25)^2 + (0.25)(19.95)}$ is

   (a) 1   (b) 2
   (c) 3   (d) 4

3. Consider the following values of the three given numbers: $\sqrt{103}, \sqrt{99.35}, \sqrt{102.20}$

   1. 10.1489 (approx.)

   2. 10.109 (approx.)

   3. 9.967 (approx.)

   The correct sequence of these values matching with the above numbers is:
   (a) 1, 2, 3   (b) 1, 3, 2
   (c) 2, 3, 1   (d) 3, 1, 2

4. What value should come in place of the question mark (?) in the following equation?

   $$48\sqrt{?} + 32\sqrt{?} = 320$$

   (a) 16   (b) 2
   (c) 4   (d) 32

5. Which is greater $(\sqrt{7} + \sqrt{10})$ or $(\sqrt{3} + \sqrt{19})$?
   (a) $\sqrt{7} + \sqrt{10}$ (b) $\sqrt{3} + \sqrt{19}$
   (c) both are equal   (d) none of these

6. Find the least number which if added to 17420 will make it a perfect square?
   (a) 3   (b) 5
   (c) 9   (d) 4

7. Calculate the value of N in the given series and then find the value of x using the given equation.

   99   163 N 248   273   289

   If $\sqrt{2N + 17} = x$, then x equals
   (a) 20.5   (b) 20.0
   (c) 21.5   (d) 21.0

8. The largest number of 5-digits that is a perfect square is
   (a) 99900   (b) 99856
   (c) 99981   (d) 99801

9. If $99 \times 21 - \sqrt[3]{x} = 1968$, then x equals
   (a) 1367631   (b) 1366731
   (c) 1367   (d) 111

10. If $P = 999$, then $\sqrt[3]{P(P^2 + 3P + 3) + 1} =$
    (a) 1000   (b) 999
    (c) 1002   (d) 998

Answers

1. (a)	2. (d)	3. (b)	4. (a)	5. (b)	6. (d)	7. (d)	8. (b)	9. (a)	10. (a)

Self Assessment Sheet

Answers

Chapter 5

EXPONENTS

KEY FACTS

1. The short cut for writing product of a number by itself several times such as, $a \times a \times a \times a \ldots \ldots \times a$ (n times) $= a^n$, is known as exponential notation, where a is any real number and n is an integer.

   a^n is read as **"a to the nth power"**.

   For example : The expression 3^6 is read as **three to the sixth power**. Here, 3 is called the **base** and 6 is called the **exponent** and 3^6 is the **exponential form**.

2. **Laws of exponents**

   If $a \neq 0$, $b \neq 0$ be any real number and m, n be any rational numbers. Then,

   Law I : $a^m \times a^n = a^{m+n}$

   Law II : $(a^m)^n = a^{mn}$

   Law III : $(ab)^m = a^m \times b^m$

   Law IV : $\left(\dfrac{a}{b}\right)^m = \dfrac{a^m}{b^m}, (b \neq 0)$

   Law V : $\dfrac{a^m}{a^n} = a^{m-n}$

   Law VI : $\left(\dfrac{a}{b}\right)^{-m} = \left(\dfrac{b}{a}\right)^m$

   Law VII : $a^0 = 1$

   Law VIII : $a^{-n} = \dfrac{1}{a^n}$; $a^n = \dfrac{1}{a^{-n}}$

Solved Examples

Ex. 1. Simplify: $(0.04)^{-1.5}$

Sol. $(0.04)^{-1.5} = (0.04)^{-\frac{3}{2}} = \dfrac{1}{(0.04)^{\frac{3}{2}}} = \dfrac{1}{(\sqrt{0.04})^3} = \dfrac{1}{(0.2)^3} = \dfrac{1}{0.008} = \dfrac{1000}{8} = 125.$

Ex. 2. Simplify : $(100)^{\frac{1}{2}} \times (0.001)^{\frac{1}{3}} - (0.0016)^{\frac{1}{4}} \times 3^0 + \left(\dfrac{5}{4}\right)^{-1}$

Sol. Given exp. $= (10^2)^{\frac{1}{2}} \times \left(\dfrac{1}{1000}\right)^{\frac{1}{3}} - \left(\dfrac{16}{10000}\right)^{\frac{1}{4}} \times 3^0 + \left(\dfrac{5}{4}\right)^{-1} = 10^{2 \times \frac{1}{2}} \times \left(\dfrac{1}{10}\right)^{3 \times \frac{1}{3}} - \left(\dfrac{2}{10}\right)^{4 \times \frac{1}{4}} \times 3^0 + \dfrac{4}{5}$

$$= 10 \times \dfrac{1}{10} - \dfrac{1}{5} + \dfrac{4}{5} = 10 \times 0.1 - 0.2 + 0.8 = 1 + 0.6 = 1.6.$$

Ex. 3. *Evaluate :* $\dfrac{6^{\frac{2}{3}} \times \sqrt[3]{6^7}}{\sqrt[3]{6^6}}$.

Sol. Given exp. $= \dfrac{6^{\frac{2}{3}} \times 6^{\frac{7}{3}}}{6^{\frac{6}{3}}} = 6^{\frac{2}{3}+\frac{7}{3}-\frac{6}{3}} = 6^{\frac{3}{3}} = 6^1 = 6$.

Ex. 4. *Given that* $10^{0.48} = x$ *and* $10^{0.70} = y$ *and* $x^z = y^2$, *then find the approximate value of z ?*

Sol. Given, $x^z = y^2 \Rightarrow (10^{0.48})^z = (10^{0.70})^2$

$\Rightarrow 10^{0.48z} = 10^{1.40} \Rightarrow 0.48z = 1.40 \Rightarrow z = \dfrac{140}{48} = 2.9$ (approx)

Ex. 5. *If* $(\sqrt{3})^5 \times 9^2 = 3^n \times 3\sqrt{3}$, *then what is the value of n ?*

Sol. Given, $\left(3^{\frac{1}{2}}\right)^5 \times (3^2)^2 = 3^n \times 3 \times 3^{\frac{1}{2}} \Rightarrow 3^{\frac{5}{2}} \times 3^4 = 3^{n+1+\frac{1}{2}}$

$\Rightarrow 3^{\frac{5}{2}+4} = 3^{n+\frac{3}{2}} \Rightarrow 3^{13/2} = 3^{n+3/2} \Rightarrow n + \dfrac{3}{2} = \dfrac{13}{2} \Rightarrow n = \dfrac{13}{2} - \dfrac{3}{2} = \dfrac{10}{2} = 5.$

Ex. 6. *If* $3^{x-y} = 27$ *and* $3^{x+y} = 243$, *then what is the value of x?*

Sol. Given, $3^{x-y} = 27 = 3^3$ and $3^{x+y} = 243 = 3^5$

$\Rightarrow \qquad\qquad x - y = 3 \qquad\qquad\qquad\qquad\qquad\qquad\qquad\qquad ... (i)$

$\qquad\qquad\qquad x + y = 5 \qquad\qquad\qquad\qquad\qquad\qquad\qquad\qquad ... (ii)$

Adding equations (*i*) and (*ii*), we get

$2x = 8 \Rightarrow x = 4$

Ex. 7. *Simplify :* $\left(\dfrac{x^a}{x^b}\right)^{(a+b)} \cdot \left(\dfrac{x^b}{x^c}\right)^{(b+c)} \cdot \left(\dfrac{x^c}{x^a}\right)^{(c+a)}$

Sol. $\left(\dfrac{x^a}{x^b}\right)^{(a+b)} \cdot \left(\dfrac{x^b}{x^c}\right)^{(b+c)} \cdot \left(\dfrac{x^c}{x^a}\right)^{(c+a)} = \left(x^{(a-b)}\right)^{(a+b)} \cdot \left(x^{(b-c)}\right)^{(b+c)} \cdot \left(x^{(c-a)}\right)^{(c+a)}$

$$= x^{a^2-b^2} \cdot x^{b^2-c^2} \cdot x^{c^2-a^2} = x^{a^2-b^2+b^2-c^2+c^2-a^2} = x^0 = 1.$$

Ex. 8. *If* $2 = 10^m$ *and* $3 = 10^n$, *then find the value of 0.15.*

Sol. $0.15 = \dfrac{1.5}{10} = \dfrac{3}{2 \times 10} = \dfrac{10^n}{10^m \times 10} = \dfrac{10^n}{10^{m+1}} = 10^{n-(m+1)} = 10^{n-m-1}$

Ex. 9. *What is the value of the expression* $\dfrac{x-1}{x^{3/4}+x^{1/2}} \cdot \dfrac{x^{1/2}+x^{1/4}}{x^{1/2}+1} \cdot x^{1/4}$ *when x = 16 ?*

Sol. Required value $= \dfrac{16-1}{16^{3/4}+16^{1/2}} \cdot \dfrac{16^{1/2}+16^{1/4}}{16^{1/2}+1} \cdot 16^{1/4}$

$= \dfrac{15}{(2^4)^{3/4}+(4^2)^{1/2}} \cdot \dfrac{(4^2)^{1/2}+(2^4)^{1/4}}{(4^2)^{1/2}+1} \cdot (2^4)^{1/4} = \dfrac{15}{2^3+4} \cdot \dfrac{4+2}{4+1} \cdot 2 = \dfrac{15}{12} \times \dfrac{6}{5} \times 2 = 3$.

Ex. 10. *What is the simplified value of* $\left\{ \dfrac{4^{m+\frac{1}{4}} \times \sqrt{2 \cdot 2^m}}{2\sqrt{2^{-m}}} \right\}^{\frac{1}{m}}$ *?*

Sol. $\left\{ \dfrac{4^{m+\frac{1}{4}} \times \sqrt{2 \cdot 2^m}}{2\sqrt{2^{-m}}} \right\}^{\frac{1}{m}} = \left\{ \dfrac{(2^2)^{m+\frac{1}{4}} \times (2^{m+1})^{\frac{1}{2}}}{2 \times 2^{\frac{-m}{2}}} \right\}^{\frac{1}{m}} = \left\{ \dfrac{2^{2m+\frac{1}{2}} \times 2^{\frac{m}{2}+\frac{1}{2}}}{2^{1-\frac{m}{2}}} \right\}^{\frac{1}{m}} = \left\{ 2^{2m+\frac{1}{2}+\frac{m}{2}+\frac{1}{2}-1+\frac{m}{2}} \right\}^{\frac{1}{m}} = (2^{3m})^{\frac{1}{m}} = 2^3 = 8.$

Ex. 11. *Given, $a = 2^x$, $b = 4^y$, $c = 8^z$ and $ac = b^2$. Find the relation between x, y and z.*

Sol. $ac = b^2$

$\Rightarrow 2^x \cdot 8^z = (4^y)^2 \Rightarrow 2^x \cdot (2^3)^z = ((2^2)^y)^2 \Rightarrow 2^x \cdot 2^{3z} = 2^{4y}$

$\Rightarrow 2^{x+3z} = 2^{4y} \Rightarrow x+3z = 4y.$

Question Bank–5

1. $\dfrac{\left(\dfrac{1}{216}\right)^{-\frac{2}{3}}}{\left(\dfrac{1}{27}\right)^{-\frac{4}{3}}} = x$. The value of x is

(a) $\dfrac{3}{4}$ (b) $\dfrac{4}{9}$

(c) $\dfrac{2}{3}$ (d) $\dfrac{1}{8}$

2. If $\sqrt{3^n} = 81$. Then, n is equal to

(a) 2   (b) 4

(c) 6   (d) 8

3. $(64)^{-\frac{2}{3}} \times \left(\dfrac{1}{4}\right)^{-3}$ equals

(a) $\dfrac{1}{4}$ (b) 1

(c) 4   (d) 16

4. $\dfrac{1}{(216)^{-\frac{2}{3}}} + \dfrac{1}{(256)^{-\frac{3}{4}}} + \dfrac{1}{(243)^{-\frac{1}{5}}}$ is equal to

(a) 103   (b) 105

(c) 107   (d) 101

5. $(4)^{0.5} \times (0.5)^4$ is equal to

(a) 1   (b) 4

(c) $\dfrac{1}{8}$ (d) $\dfrac{1}{32}$

6. $\left(\dfrac{1}{64}\right)^0 + (64)^{\frac{-1}{2}} + (32)^{\frac{4}{5}} - (32)^{\frac{-4}{5}}$ is equal to

(a) $16\dfrac{1}{8}$ (b) $17\dfrac{1}{8}$

(c) $17\dfrac{1}{16}$ (d) $-17\dfrac{1}{16}$

7. **Simplify :**

$\left[\left(2\dfrac{10}{27}\right)^{-\frac{2}{3}} \div \left(11\dfrac{1}{9}\right)^{-0.5} \right] + \left[(6.25)^{0.5} \div (-4)^{-1} \right]$

(a) $-8\dfrac{1}{8}$ (b) $8\dfrac{1}{8}$

(c) $1\dfrac{7}{8}$ (d) $-1\dfrac{7}{8}$

8. Simplify : $\dfrac{(6.25)^{\frac{1}{2}} \times (0.0144)^{\frac{1}{2}} + 1}{(0.027)^{\frac{1}{3}} \times (81)^{\frac{1}{4}}}$

(a) 0.14
(b) 1.4
(c) 1
(d) $1.\overline{4}$

9. $4^{3.5} : 2^5$ is the same as

(a) 4 : 1
(b) 2 : 1
(c) 7 : 5
(d) 7 : 10

10. Simplify : $\left[\sqrt[3]{\sqrt[6]{5^9}}\right]^4 \left[\sqrt[6]{\sqrt[3]{5^9}}\right]^4$

(a) 5^2
(b) 5^4
(c) 5^8
(d) 5^{12}

11. The value of $\dfrac{(243)^{\frac{n}{5}} \cdot 3^{2n+1}}{9^n \cdot 3^{n-1}}$ is

(a) 1
(b) 9
(c) 3
(d) 3^n

12. If $x^{x\sqrt{x}} = (x\sqrt{x})^x$. then x is equal to

(a) $\dfrac{3}{2}$
(b) $\dfrac{2}{9}$
(c) $\dfrac{9}{4}$
(d) $\dfrac{4}{9}$

13. $\left[1 - \{1 - (1 - a^4)^{-1}\}^{-1}\right]^{-\frac{1}{4}}$ is equal to

(a) a^4
(b) a^2
(c) a
(d) $\dfrac{1}{a}$

14. If $64^a = \dfrac{1}{256^b}$, then $3a + 4b$ equals

(a) 2
(b) 4
(c) 8
(d) 0

15. If $a = b^{\frac{2}{3}}$ and $b = c^{-2}$, what is the value of a in terms of c ?

(a) $\dfrac{4}{c^3}$
(b) $\sqrt[3]{c^4}$
(c) $\dfrac{1}{\sqrt[3]{c^4}}$
(d) $\sqrt[4]{c^3}$

16. The value of $\dfrac{5 \cdot (25)^{n+1} + 25 \cdot (5)^{2n-1}}{25 \cdot (5)^{2n} - 105(25)^{n-1}}$ is

(a) 0
(b) 1
(c) $6\dfrac{1}{4}$
(d) $5\dfrac{1}{4}$

17. If $5\sqrt{5} \times 5^3 \div 5^{-\frac{3}{2}} = 5^{a+2}$, then the value of a is

(a) 4
(b) 5
(c) 6
(d) 8

18. What is the expression $(x + y)^{-1} (x^{-1} + y^{-1}) (xy^{-1} + x^{-1}y)^{-1}$ equal to

(a) $x + y$
(b) $(x^2 + y^2)^{-1}$
(c) xy
(d) $x^2 + y^2$

19. If $2^x - 2^{x-1} = 4$, then what is the value of $2^x + 2^{x-1}$?

(a) 8
(b) 12
(c) 10
(d) 16

20. If $x = y^z$, $y = z^x$ and $z = x^y$, then

(a) $\dfrac{xy}{z} = 1$
(b) $xyz = 1$
(c) $x + y + z = 1$
(d) $xz = y$

21. $\dfrac{1}{1 + x^{b-a} + x^{c-a}} + \dfrac{1}{1 + x^{a-b} + x^{c-b}} + \dfrac{1}{1 + x^{b-c} + x^{a-c}}$ equals.

(a) $\dfrac{1}{2}$
(b) 2
(c) 1
(d) 0

22. Find the value of x if $\left[3^{2x-2} + 10\right] \div 13 = 7$.

(a) 1
(b) 3
(c) 4
(d) 2

23. The value of $\dfrac{2^{\frac{1}{2}} \cdot 3^{\frac{1}{3}} \cdot 4^{\frac{1}{4}}}{10^{-\frac{1}{5}} \cdot 5^{\frac{3}{5}}} \div \dfrac{3^{\frac{4}{3}} \cdot 5^{\frac{7}{5}}}{4^{-\frac{3}{5}} \cdot 6}$ is

(a) 5
(b) 6
(c) 10
(d) 15

24. The value of $\left(x^{\frac{b+c}{c-a}}\right)^{\frac{1}{a-b}} \left(x^{\frac{c+a}{a-b}}\right)^{\frac{1}{b-c}} \left(x^{\frac{a+b}{b-c}}\right)^{\frac{1}{c-a}}$ is

(a) 1
(b) a
(c) b
(d) c

25. The value of $\left(\dfrac{x^q}{x^r}\right)^{\frac{1}{qr}} \times \left(\dfrac{x^r}{x^p}\right)^{\frac{1}{rp}} \times \left(\dfrac{x^p}{x^q}\right)^{\frac{1}{pq}}$ is equal to

(a) $x^{\frac{1}{p} + \frac{1}{q} + \frac{1}{r}}$
(b) 0
(c) $x^{pq+qr+rp}$
(d) 1

26. The largest number among the following is

(a) $3^{2^{2^2}}$ (b) $\left\{(3^2)^2\right\}^2$

(c) $3^2 \times 3^2 \times 3^2$ (d) 3222

27. If $6^x - 6^{x-3} = 7740$, then $x^x =$

(a) 7796   (b) 243

(c) 3125   (d) 46656

28. The value of

$$\frac{9^x(9^{x-1})^x}{9^{x+1}\cdot 3^{2x-2}}\left\{\frac{729^{\frac{x}{3}}}{81}\right\}^{-x} \div \frac{3^a - 2^3\cdot 3^{a-2}}{3^a - 3^{a-1}}$$

(a) 9   (b) 6

(c) 12   (d) 16

29. Find the value of $(2^{\frac{1}{4}}-1)(2^{\frac{3}{4}}+2^{\frac{1}{2}}+2^{\frac{1}{4}}+1)$

(a) 2   (b) 3

(c) 5   (d) 1

30. Simplify : $\dfrac{a^{\frac{1}{2}}+a^{-\frac{1}{2}}}{1-a}+\dfrac{1-a^{-\frac{1}{2}}}{1-\sqrt{a}}$

(a) 1   (b) 0

(c) $\dfrac{2}{1-a}$ (d) $1+a$

31. If $3^{x+y}=81$ and $81^{x-y}=3$, then the value of x and y are

(a) $\dfrac{17}{8},\dfrac{9}{8}$ (b) $\dfrac{17}{8},\dfrac{15}{8}$

(c) $\dfrac{17}{8},\dfrac{11}{8}$ (d) $\dfrac{15}{8},\dfrac{11}{8}$

32. Find x, if $8^{x-2}\times\left(\dfrac{1}{2}\right)^{4-3x}=(0.0625)^x$

(a) 0   (b) 4

(c) 2   (d) 1

33. Find the value of the expression :

$$\frac{(x^{a+b})^2\times(x^{b+c})^2\times(x^{c+a})^2}{(x^a x^b x^c)^4}$$

if $x=2, a=1, b=2, c=3$

(a) 16   (b) 32

(c) 24   (d) 1

34. If $(2.4)^x=(0.24)^y=10^z$ then show that $\dfrac{1}{x}-\dfrac{1}{z}=\dfrac{1}{y}$

35. If $2^x=4^y=8^z$ and $xyz=288$, then $\dfrac{1}{2x}+\dfrac{1}{4y}+\dfrac{1}{8z}$ equals

(a) $\dfrac{11}{8}$ (b) $\dfrac{11}{24}$

(c) $\dfrac{11}{48}$ (d) $\dfrac{11}{96}$

Answers

1. (b)	**2.** (d)	**3.** (c)	**4.** (a)	**5.** (c)	**6.** (c)	**7.** (a)	**8.** (d)	**9.** (a)	**10.** (b)
11. (b)	**12.** (c)	**13.** (c)	**14.** (d)	**15.** (c)	**16.** (c)	**17.** (a)	**18.** (b)	**19.** (b)	**20.** (b)
21. (c)	**22.** (b)	**23.** (c)	**24.** (a)	**25.** (d)	**26.** (a)	**27.** (c)	**28.** (b)	**29.** (d)	**30.** (c)
31. (b)	**32.** (d)	**33.** (d)	**35.** (d)						

Hints and Solutions

1. (b) $\dfrac{\left(\frac{1}{216}\right)^{-\frac{2}{3}}}{\left(\frac{1}{27}\right)^{-\frac{4}{3}}}=\dfrac{\left(\left(\frac{1}{6}\right)^3\right)^{-\frac{2}{3}}}{\left(\left(\frac{1}{3}\right)^3\right)^{-\frac{4}{3}}}=\dfrac{\frac{1}{6^{-2}}}{\frac{1}{3^{-4}}}=\dfrac{6^2}{3^4}=\dfrac{36}{81}=\dfrac{4}{9}.$

2. (d) $\sqrt{3^n}=81 \Rightarrow 3^{n/2}=3^4 \Rightarrow \dfrac{n}{2}=4 \Rightarrow n=8$

3. (c) $(64)^{\frac{-2}{3}}\times\left(\dfrac{1}{4}\right)^{-3}=(4^3)^{\frac{-2}{3}}\times(4^{-1})^{-3}$

$=4^{-2}\times 4^3 = 4^{-2+3}=4^1=4$

4. (a) Given exp. $=\dfrac{1}{(6^3)^{-\frac{2}{3}}}+\dfrac{1}{(4^4)^{-\frac{3}{4}}}+\dfrac{1}{(3^5)^{-\frac{1}{5}}}$

$=\dfrac{1}{6^{-2}}+\dfrac{1}{4^{-3}}+\dfrac{1}{3^{-1}}$

$=6^2+4^3+3=36+64+3=103$

5. (c) $(4)^{0.5} \times (0.5)^4 = (2^2)^{0.5} \times \left(\dfrac{1}{2}\right)^4$

$$= 2^1 \times 2^{-4} = 2^{-3} = \dfrac{1}{2^3} = \dfrac{1}{8}.$$

6. (c) Given exp. $= 1 + (8^2)^{-\frac{1}{2}} + (2^5)^{\frac{4}{5}} - (2^5)^{-\frac{4}{5}}$

$$= 1 + 8^{-1} + 2^4 - 2^{-4}$$

$$= 1 + \dfrac{1}{8} + 16 - \dfrac{1}{16} = \dfrac{16 + 2 + 256 - 1}{16}$$

$$= \dfrac{273}{16} = 17\dfrac{1}{16}$$

7. (a) Given exp.

$$= \left[\left(\dfrac{64}{27}\right)^{-\frac{2}{3}} \div \left(\dfrac{100}{9}\right)^{-\frac{1}{2}}\right] + \left[(6.25)^{\frac{1}{2}} \div \dfrac{1}{-4}\right]$$

$$= \left[\left(\left(\dfrac{4}{3}\right)^3\right)^{-\frac{2}{3}} \div \left(\left(\dfrac{10}{3}\right)^2\right)^{-\frac{1}{2}}\right] + \left[\left((2.5)^2\right)^{\frac{1}{2}} \div \dfrac{1}{-4}\right]$$

$$= \left[\left(\dfrac{4}{3}\right)^{-2} \div \left(\dfrac{10}{3}\right)^{-1}\right] + 2.5 \times -4$$

$$= \left[\left(\dfrac{3}{4}\right)^2 \div \dfrac{3}{10}\right] - 10 = \dfrac{9}{16} \times \dfrac{10}{3} - 10$$

$$= \dfrac{15}{8} - 10 = \dfrac{15 - 80}{8} = \dfrac{-65}{8} = -8\dfrac{1}{8}.$$

8. (d) Given exp. $= \dfrac{\left((2.5)^2\right)^{\frac{1}{2}} \times \left((0.12)^2\right)^{\frac{1}{2}} + 1}{\left((0.3)^3\right)^{\frac{1}{3}} \times (3^4)^{\frac{1}{4}}}$

$$= \dfrac{2.5 \times 0.12 + 1}{0.3 \times 3}$$

$$= \dfrac{0.3 + 1}{0.9} = \dfrac{1.3}{0.9} = \dfrac{13}{9} = 1.\overline{4}.$$

9. (a) $4^{3.5} : 2^5 = (2^2)^{3.5} : 2^5 = 2^7 : 2^5$

$$= \dfrac{2^7}{2^5} : 1 = 2^{7-5} : 1 = 2^2 : 1 = 4 : 1.$$

10. (b) $\left[\sqrt[3]{\sqrt[6]{5^9}}\right]^4 \times \left[\sqrt[6]{\sqrt[3]{5^9}}\right]^4$

$$= 5^{9 \times \frac{1}{6} \times \frac{1}{3} \times 4} \times 5^{9 \times \frac{1}{3} \times \frac{1}{6} \times 4}$$

$$= 5^2 \times 5^2 = 5^4.$$

11. (b) Given exp. $= \dfrac{(3^5)^{\frac{n}{5}} \cdot 3^{2n+1}}{(3^2)^n \cdot 3^{n-1}} = \dfrac{3^n \cdot 3^{2n+1}}{3^{2n} \cdot 3^{n-1}}$

$$= \dfrac{3^{3n+1}}{3^{3n-1}} = 3^{(3n+1)-(3n-1)} = 3^2 = 9.$$

12. (c) $x^{x\sqrt{x}} = (x\sqrt{x})^x$

$$\Rightarrow x^{x\sqrt{x}} = (x^{3/2})^x \Rightarrow x^{x\sqrt{x}} = x^{3/2x}$$

$$\Rightarrow x\sqrt{x} = \dfrac{3}{2}x \Rightarrow \sqrt{x} = \dfrac{3}{2} \Rightarrow x = \dfrac{9}{4}.$$

13. (c) Given exp. $= \left[1 - \left\{1 - \dfrac{1}{1-a^4}\right\}^{-1}\right]^{-\frac{1}{4}}$

$$= \left[1 - \left\{\dfrac{1-a^4-1}{1-a^4}\right\}^{-1}\right]^{-\frac{1}{4}}$$

$$= \left[1 - \left\{\dfrac{-a^4}{1-a^4}\right\}^{-1}\right]^{-\frac{1}{4}} = \left[1 - \left\{\dfrac{1-a^4}{-a^4}\right\}\right]^{-\frac{1}{4}}$$

$$= \left[1 - \left\{\dfrac{a^4-1}{a^4}\right\}\right]^{-\frac{1}{4}} = \left[\dfrac{a^4-a^4+1}{a^4}\right]^{-\frac{1}{4}}$$

$$= \left(\dfrac{1}{a^4}\right)^{-\frac{1}{4}} = (a^{-4})^{-\frac{1}{4}} = a^1 = a.$$

14. (d) $64^a = \dfrac{1}{256^b} \Rightarrow (2^6)^a = \dfrac{1}{(2^8)^b}$

$$\Rightarrow 2^{6a} \times 2^{8b} = 1 \Rightarrow 2^{6a+8b} = 2^0$$

$$\Rightarrow 6a + 8b = 0 \Rightarrow 3a + 4b = 0$$

15. (c) $a = b^{\frac{2}{3}}$ and $b = c^{-2}$

$$\therefore a = (c^{-2})^{\frac{2}{3}} = c^{-\frac{4}{3}} = \dfrac{1}{c^{4/3}} = \dfrac{1}{\sqrt[3]{c^4}}$$

16. (c) Given exp. $= \dfrac{5 \times (5^2)^{n+1} + 5^2 \times 5^{2n-1}}{5^2 \times 5^{2n} - 21 \times 5 \times (5^2)^{n-1}}$

$= \dfrac{5^{2n+3} + 5^{2n+1}}{5^{2n+2} - 21 \times 5^{2n-1}} = \dfrac{5^{2n+1}(5^2 + 1)}{5^{2n-1}(5^3 - 21)}$

$= \dfrac{5^{(2n+1)-(2n-1)} \times 26}{(125 - 21)} = \dfrac{5^2 \times 26}{104} = \dfrac{25}{4} = 6\dfrac{1}{4}.$

17. (a) $5\sqrt{5} \times 5^3 \div 5^{-\frac{3}{2}} = 5^{a+2}$

$\Rightarrow 5^{\frac{3}{2}} \times 5^3 \div 5^{-\frac{3}{2}} = 5^{a+2}$

$\Rightarrow 5^{\frac{3}{2} + 3 - \left(-\frac{3}{2}\right)} = 5^{a+2}$

$\Rightarrow 5^6 = 5^{a+2} \Rightarrow a + 2 = 6 \Rightarrow a = 4$

18. (b) $(x+y)^{-1}(x^{-1} + y^{-1})(xy^{-1} + x^{-1}y)^{-1}$

$= \dfrac{1}{(x+y)} \times \left(\dfrac{1}{x} + \dfrac{1}{y}\right) \times \left(\dfrac{x}{y} + \dfrac{y}{x}\right)^{-1}$

$= \dfrac{1}{(x+y)} \times \left(\dfrac{y+x}{xy}\right) \times \left(\dfrac{x^2 + y^2}{xy}\right)^{-1}$

$= \dfrac{1}{(x+y)} \times \dfrac{(y+x)}{xy} \times \dfrac{xy}{(x^2 + y^2)}$

$= (x^2 + y^2)^{-1}$

19. (b) $2^x - 2^{x-1} = 4 \Rightarrow 2^x - \dfrac{2^x}{2} = 4$

$\Rightarrow 2^x\left(1 - \dfrac{1}{2}\right) = 4 \Rightarrow 2^x \times \dfrac{1}{2} = 4$

$\Rightarrow 2^x = 8 \Rightarrow 2^x = 2^3 \Rightarrow x = 3$

$\therefore 2^x + 2^{x-1} = 2^3 + 2^2 = 8 + 4 = 12.$

20. (b) $z = x^y = (y^z)^y$ $(\because x = y^z)$

$= y^{zy} = (z^x)^{zy} = z^{xyz}$ $(\because y = z^x)$

$\therefore z^1 = z^{xyz} \Rightarrow xyz = 1$

21. (c) $\dfrac{1}{1 + x^{b-a} + x^{c-a}} + \dfrac{1}{1 + x^{a-b} + x^{c-b}} + \dfrac{1}{1 + x^{b-c} + x^{a-c}}$

$= \dfrac{1}{1 + \dfrac{x^b}{x^a} + \dfrac{x^c}{x^a}} + \dfrac{1}{1 + \dfrac{x^a}{x^b} + \dfrac{x^c}{x^b}} + \dfrac{1}{1 + \dfrac{x^b}{x^c} + \dfrac{x^a}{x^c}}$

$= \dfrac{x^a}{x^a + x^b + x^c} + \dfrac{x^b}{x^b + x^a + x^c} + \dfrac{x^c}{x^c + x^b + x^a}$

$= \dfrac{x^a + x^b + x^c}{x^a + x^b + x^c} = 1.$

22. (b) $\left[3^{2x-2} + 10\right] \div 13 = 7$

$\Rightarrow 3^{2x-2} + 10 = 7 \times 13 = 91$

$\Rightarrow 3^{2x-2} = 91 - 10 = 81 = 3^4$

$\Rightarrow 2x - 2 = 4 \Rightarrow 2x = 6 \Rightarrow x = 3.$

23. (c) Given exp. $= \dfrac{2^{\frac{1}{2}} \times 3^{\frac{1}{3}} \times (2^2)^{\frac{1}{4}}}{(2 \times 5)^{-\frac{1}{5}} \times 5^{\frac{3}{5}}} \div \dfrac{3^{\frac{4}{3}} \times 5^{\frac{7}{5}}}{(2^2)^{-\frac{3}{5}} \times 2 \times 3}$

$= \dfrac{2^{\frac{1}{2}} \times 3^{\frac{1}{3}} \times (2)^{\frac{1}{2}}}{(2)^{-\frac{1}{5}} \times 5^{-\frac{1}{5}} \times 5^{\frac{3}{5}}} \times \dfrac{(2)^{\frac{-6}{5}} \times 2 \times 3}{3^{\frac{4}{3}} \times 5^{\frac{7}{5}}}$

$= \dfrac{2^{\left(\frac{1}{2} + \frac{1}{2} - \frac{6}{5} + 1 + \frac{1}{5}\right)} \times 3^{\left(\frac{1}{3} + 1 - \frac{4}{3}\right)}}{5^{\left(-\frac{1}{5} + \frac{3}{5} - \frac{7}{5}\right)}}$

$= \dfrac{2 \times 3^0}{5^{-1}} = 2 \times 5 = 10.$

24. (a) $\left(x^{\frac{b+c}{c-a}}\right)^{\frac{1}{a-b}} \times \left(x^{\frac{c+a}{a-b}}\right)^{\frac{1}{b-c}} \times \left(x^{\frac{a+b}{b-c}}\right)^{\frac{1}{c-a}}$

$= x^{\frac{b+c}{(c-a)(a-b)}} \times x^{\frac{c+a}{(a-b)(b-c)}} \times x^{\frac{a+b}{(b-c)(c-a)}}$

$= x^{\frac{b+c}{(c-a)(a-b)} + \frac{c+a}{(a-b)(b-c)} + \frac{a+b}{(b-c)(c-a)}}$

$= x^{\frac{(b+c)(b-c)+(c+a)(c-a)+(a+b)(a-b)}{(a-b)(b-c)(c-a)}}$

$= x^{\frac{b^2-c^2+c^2-a^2+a^2-b^2}{(a-b)(b-c)(c-a)}}$

$= x^{\frac{0}{(a-b)(b-c)(c-a)}} = x^0 = 1.$

25. (d) $\left(\dfrac{x^q}{x^r}\right)^{\frac{1}{qr}} \times \left(\dfrac{x^r}{x^p}\right)^{\frac{1}{rp}} \times \left(\dfrac{x^p}{x^q}\right)^{\frac{1}{pq}}$

$= (x^{q-r})^{\frac{1}{qr}} \times (x^{r-p})^{\frac{1}{rp}} \times (x^{p-q})^{\frac{1}{pq}}$

$= x^{\frac{q-r}{qr}} \times x^{\frac{r-p}{rp}} \times x^{\frac{p-q}{pq}}$

$= x^{\frac{1}{r} - \frac{1}{q}} \times x^{\frac{1}{p} - \frac{1}{r}} \times x^{\frac{1}{q} - \frac{1}{p}}$

$= x^{\frac{1}{r} - \frac{1}{q} + \frac{1}{p} - \frac{1}{r} + \frac{1}{q} - \frac{1}{p}} = x^0 = 1.$

26. (a) $3^{2^{2^2}} = 3^{2^4} = 3^{16}$;

$\left\{ \left(3^2\right)^2 \right\}^2 = 3^8 = 6561$;

$3^2 \times 3^2 \times 3^2 = 3^{2+2+2} = 3^6 = 729$;

$\therefore \ 3^{16} > 3^8 > 3222 > 3^6$

27. (c) $6^x - 6^{x-3} = 7740 \ \Rightarrow \ 6^x - \dfrac{6^x}{6^3} = 7740$

$\Rightarrow \ 6^x\left(1 - \dfrac{1}{216}\right) = 7740 \ \Rightarrow \ 6^x \times \dfrac{215}{216} = 7740$

$\Rightarrow \ 6^x = \dfrac{7740 \times 216}{215} = 36 \times 216 = 6^5 \ \Rightarrow \ x = 5$

$\therefore \ x^x = 5^5 = 3125.$

28. (b) Given exp.

$= \dfrac{3^{2x}(3^{2x-2})^x}{3^{2x+2} \cdot 3^{2x-2}} \left\{ \dfrac{(3^6)^{\frac{x}{3}}}{3^4} \right\}^{-x} \div \dfrac{3^{a-2}(3^2 - 2^3)}{3^{a-1}(3-1)}$

$= \dfrac{3^{2x} \cdot 3^{2x^2 - 2x}}{3^{2x+2+2x-2}} \left\{ \dfrac{3^{2x}}{3^4} \right\}^{-x} \div \dfrac{3^{a-2-a+1} \times (9-8)}{2}$

$= \dfrac{3^{2x^2}}{3^{4x}} (3^{2x-4})^{-x} \div \dfrac{1}{2 \times 3}$

$= 3^{2x^2 - 4x} \times 3^{-2x^2 + 4x} \div \dfrac{1}{6} = 3^0 \times 6 = 1 \times 6 = 6.$

29. (d) $\left(2^{\frac{1}{4}} - 1 \right)\left(2^{\frac{3}{4}} + 2^{\frac{1}{2}} + 2^{\frac{1}{4}} + 1 \right)$

Let $2^{\frac{1}{4}} = a$. Then,

Given exp. $= (a-1)(a^3 + a^2 + a + 1)$

$= (a-1)(a^2(a+1) + 1(a+1))$

$= (a-1)(a+1)(a^2+1) = (a^2-1)(a^2+1)$

$= a^4 - 1$

$\therefore$ Required value $= \left(2^{\frac{1}{4}} \right)^4 - 1 = 2 - 1 = 1.$

30. (c) $\dfrac{a^{\frac{1}{2}} + a^{\frac{-1}{2}}}{1-a} + \dfrac{1 - a^{\frac{-1}{2}}}{1+\sqrt{a}} = \dfrac{\sqrt{a} + \dfrac{1}{\sqrt{a}}}{1-a} + \dfrac{1 - \dfrac{1}{\sqrt{a}}}{1+\sqrt{a}}$

$= \dfrac{\dfrac{a+1}{\sqrt{a}}}{1-a} + \dfrac{\dfrac{\sqrt{a}-1}{\sqrt{a}}}{1+\sqrt{a}} = \dfrac{a+1}{\sqrt{a}(1-a)} + \dfrac{\sqrt{a}-1}{\sqrt{a}(1+\sqrt{a})}$

$= \dfrac{(a+1)(1+\sqrt{a}) + (\sqrt{a}-1)(1-a)}{\sqrt{a}(1-a)(1+\sqrt{a})}$

$= \dfrac{(a+1 + a\sqrt{a} + \sqrt{a}) + \sqrt{a} - 1 - a\sqrt{a} + a}{\sqrt{a}(1-a)(1+\sqrt{a})}$

$= \dfrac{2a + 2\sqrt{a}}{\sqrt{a}(1-a)(1+\sqrt{a})}$

$= \dfrac{2\sqrt{a}(\sqrt{a}+1)}{\sqrt{a}(1-a)(1+\sqrt{a})} = \dfrac{2}{1-a}.$

31. (b) $3^{x+y} = 81 \ \Rightarrow \ 3^{x+y} = 3^4 = x + y = 4$ ... (i)

$81^{x-y} = 3 \ \Rightarrow \ (3^4)^{x-y} = 3^1$

$\Rightarrow \ 4x - 4y = 1$ ... (ii)

Eqn (i) $\times$ 4 + Eqn (ii) gives

$4x + 4y + 4x - 4y = 16 + 1$

$\Rightarrow \ 8x = 17 \ \Rightarrow \ x = \dfrac{17}{8}$

Putting $x = \dfrac{17}{8}$ in (i), we get $\dfrac{17}{8} + y = 4$

$\Rightarrow \ y = 4 - \dfrac{17}{8} = \dfrac{15}{8}$

$\therefore \ x = \dfrac{17}{8}, \ y = \dfrac{15}{8}.$

32. (d) $8^{x-2} \times \left(\dfrac{1}{2} \right)^{4-3x} = (0.0625)^x$

$\Rightarrow \ (2^3)^{x-2} \times (2^{-1})^{4-3x} = \left(\dfrac{625}{10000} \right)^x$

$\Rightarrow \ 2^{3x-6} \times 2^{-4+3x} = \left(\dfrac{1}{16} \right)^x = (2^{-4})^x = 2^{-4x}$

$\Rightarrow \ 2^{3x-6-4+3x} = 2^{-4x}$

$\Rightarrow \ 2^{6x-10} = 2^{-4x}$

$\Rightarrow \ 6x - 10 = -4x$

$\Rightarrow \ 10x = 10 \ \Rightarrow \ x = 1.$

33. (d) $\dfrac{(x^{a+b})^2 \times (x^{b+c})^2 \times (x^{c+a})^2}{(x^a \, x^b \, x^c)^4}$

$= \dfrac{x^{2a+2b} \times x^{2b+2c} \times x^{2c+2a}}{(x^{a+b+c})^4}$

$= \dfrac{x^{2a+2b+2b+2c+2c+2a}}{x^{4a+4b+4c}} = \dfrac{x^{4a+4b+4c}}{x^{4a+4b+4c}} = 1.$

34. $(2.4)^x = 10^z \Rightarrow 2.4 = 10^{\frac{z}{x}}$

and $(0.24)^y = 10^z \Rightarrow 0.24 = 10^{\frac{z}{y}}$

$\therefore \dfrac{2.4}{0.24} = \dfrac{10^{\frac{z}{x}}}{10^{\frac{z}{y}}} \Rightarrow 10 = 10^{\frac{z}{x}-\frac{z}{y}}$

$\Rightarrow 1 = \dfrac{z}{x} - \dfrac{z}{y} = z\left(\dfrac{1}{x} - \dfrac{1}{y}\right).$

$\Rightarrow \dfrac{1}{x} - \dfrac{1}{y} = \dfrac{1}{z}$ or $\dfrac{1}{x} - \dfrac{1}{z} = \dfrac{1}{y}.$

35. (d) $2^x = (2^2)^y = (2^3)^z \Rightarrow x = 2y = 3z$

Given $xyz = 288 \Rightarrow x \times \dfrac{x}{2} \times \dfrac{x}{3} = 288$

$\Rightarrow x^3 = 6 \times 288 \Rightarrow x^3 = 1728$

$\Rightarrow x = \sqrt[3]{1728} = 12$

$\therefore y = \dfrac{12}{2} = 6$ and $z = \dfrac{12}{3} = 4$

$\therefore \dfrac{1}{2x} + \dfrac{1}{4y} + \dfrac{1}{8z} = \dfrac{1}{24} + \dfrac{1}{24} + \dfrac{1}{32}$

$= \dfrac{4+4+3}{96} = \dfrac{11}{96}.$

Self Assessment Sheet–5

1. Find the value of $(27)^{-2/3} + ((2^{-2/3})^{-5/3})^{-9/10}$

(a) $\dfrac{1}{9}$ (b) $\dfrac{2}{9}$

(c) $\dfrac{11}{18}$ (d) 1

2. Given that: $10^{0.48} = x$ and $10^{0.7} = y$ and $x^z = y^2$, find the value of z.

(a) $2\dfrac{11}{12}$ (b) $\dfrac{4}{9}$

(c) $1\dfrac{1}{48}$ (d) $\dfrac{48}{49}$

3. The expression $(x^{-2p}\, y^{3q})^6 \div (x^3 y^{-1})^{-4p}$ after simplification becomes:

(a) independent of x but not of y.

(b) independent of y but not of x.

(c) independent of both x and y.

(d) dependent of both x and y but independent of p and q.

4. If $2^{x-1} + 2^{x+1} = 320$, then x equals

(a) 4   (b) 5

(c) 6   (d) 7

5. $\left(\dfrac{x^a}{x^b}\right)^{a^2+ab+b^2}\left(\dfrac{x^b}{x^c}\right)^{b^2+bc+c^2}\left(\dfrac{x^c}{x^a}\right)^{c^2+ca+a^2}$ equals

(a) $a^3 + b^3 + c^3$ (b) 1

(c) $(a + b + c)^3$ (d) 0

6. $\dfrac{16 \times 2^{n+1} - 4 \times 2^n}{16 \times 2^{n+2} - 2 \times 2^{n+2}}$ equals

(a) $\dfrac{1}{4}$ (b) $-\dfrac{1}{2}$

(c) $-\dfrac{1}{4}$ (d) $\dfrac{1}{2}$

7. If $3^{x+8} = 27^{2x+1}$, then the value of

$\left[\left(\dfrac{\sqrt{289}}{\sqrt[3]{216}}\right)^x \div \left(\dfrac{17}{\sqrt[4]{1296}}\right)^x\right]^{1/2}$ is

(a) 1   (b) 0

(c) $\dfrac{17}{6}$ (d) $\dfrac{6}{17}$

8. Find the value of

$$\dfrac{(0.3)^{1/3}\left(\dfrac{1}{27}\right)^{1/4}(9)^{1/6}(0.81)^{2/3}}{(0.9)^{2/3}(3)^{-1/2}(243)^{-1/4}}$$

(a) 0.9   (b) 2.7

(c) 0.27   (d) 0.09

9. $\dfrac{\sqrt[6]{2}\left[(625)^{3/5} \times (1024)^{-6/5} \div (25)^{3/5}\right]^{1/2}}{[(\sqrt[3]{128})^{-5/2}] \times (125)^{1/5}}$

$\div \dfrac{(10^3)^2 \div (10^2)^3}{(10^2)^3 \div (10^3)^2}$ equals

(a) $\dfrac{1}{5}$ (b) $\dfrac{1}{125}$

(c) 1   (d) $\dfrac{1}{10}$

10. If $a = (\sqrt{5} + \sqrt{4})^{-3}$ and $b = (\sqrt{5} - \sqrt{4})^{-3}$, then the value of

$(a + 1)^{-1} + (b + 1)^{-1}$ is

(a) $20\sqrt{5}$ (b) 4

(c) 1   (d) $16\sqrt{5}$

Answers

1. (c)	2. (a)	3. (a)	4. (d)	5. (b)	6. (d)	7. (a)	8. (b)	9. (c)	10. (c)

Chapter 6

SURDS

1. **What are surds ?** Irrational numbers of the type $\sqrt{2}, \sqrt{3}\ \sqrt{5}\ \sqrt{17},\ldots\ldots$ which are square roots of positive rational numbers that cannot be expressed as squares of any rational numbers are called surds.

   Similarly, $\sqrt[3]{2}, \sqrt[3]{4}, \sqrt[3]{19}$ etc., are numbers which are the cube roots of positive rational numbers that cannot be expressed as the cubes of any rational numbers and so are surds.

   Definition : If x is a positive rational number and n is a positive integer such that $x^{1/n}$, i.e., $\sqrt[n]{x}$ is irrational, then $\sqrt[n]{x}$ is called a **surd** or **radical** of order n.

   $\sqrt[5]{25}$ is a surd of order 5.

2. **Comparison of surds :**

   (a) If a and b are surds of the same order, say n, then $\sqrt[n]{a} > \sqrt[n]{b}$ if $a > b$.

      For example, $\sqrt[5]{27} > \sqrt[5]{20}$ as $27 > 20$.

   (b) If the given surds are not of the same order, then first convert them to surds of the same order and then compare. For example, to compare $\sqrt[3]{2}$ and $\sqrt[4]{3}$, we take the LCM of the orders, i.e., 3 and 4, i.e., 12.

      Now, $\sqrt[3]{2} = \sqrt[12]{2^4} = \sqrt[12]{16}$ $(\because \sqrt[3]{2} = 2^{\frac{1}{3}} = 2^{\frac{4}{12}})$

      $\sqrt[4]{3} = \sqrt[12]{3^3} = \sqrt[12]{27}$ $(\because \sqrt[4]{3} = 3^{\frac{1}{4}} = 3^{\frac{3}{12}})$

      Since, $27 > 16$, therefore, $\sqrt[12]{27} > \sqrt[12]{16}$, i.e., $\sqrt[4]{3} > \sqrt[3]{2}$.

3. **Rationalisation :** When surds occur in the denominator of a fraction, it is customary to rid the denominator of the radicals. The surd in the denominator is multiplied by an appropriate expression, such that the product is a rational number. The given surd and the expression by which it is multiplied are called rationalising factors of each other. For example,

   (i) $a^{1-\frac{1}{n}}$ is the rationalising factor of $a^{\frac{1}{n}}$, as, $a^{\frac{1}{n}} \cdot a^{1-\frac{1}{n}} = a^{\frac{1}{n}+1-\frac{1}{n}} = a^1 = a$, which is a rational number.

      Hence, the rationalising factor of $6^{\frac{1}{5}}$ is $6^{1-\frac{1}{5}} = 6^{\frac{1}{5}}$.

   (ii) $(a + \sqrt{b})$ is the rationalising factor of $(a - \sqrt{b})$, as, $(a + \sqrt{b})(a - \sqrt{b}) = a^2 - b$, which is a rational.

      Hence, $(3 + \sqrt{2})$ is the rationalising factor of $(3 - \sqrt{2})$.

      $(3 + \sqrt{2})(3 - \sqrt{2}) = 9 - 2 = 7$

(iii) $(\sqrt{a}+\sqrt{b})$ is the rationalising factor of $(\sqrt{a}-\sqrt{b})$ as $(\sqrt{a}+\sqrt{b})(\sqrt{a}-\sqrt{b})=(\sqrt{a})^2-(\sqrt{b})^2=a-b$, which is a rational number.

> **Note:** Such binominal surds as $(\sqrt{a}+\sqrt{b})$ and $(\sqrt{a}-\sqrt{b})$ which differ only in the sign connecting their terms are said to be **conjugate surds**. The product of conjugate surds is always a rational.

Thus, the process of multiplication of a surd by its rationalising factor is called **rationalisation**.

4. Surds also follows all the laws of exponents which you have studied in the previous section.
   Laws of radicals.

   (i) $\sqrt[n]{a^n}=a$, e.g., $\sqrt[4]{16}=\sqrt[4]{2^4}=2$

   (ii) $\sqrt[n]{a}\,\sqrt[n]{b}=\sqrt[n]{ab}$, e.g., $\sqrt{8}\times\sqrt{6}=\sqrt{48}=\sqrt{16\times3}=\sqrt{4^2}\times\sqrt{3}=4\sqrt{3}$

   (iii) $\sqrt[m]{\sqrt[n]{a}}=\sqrt[mn]{a}$, e.g., $\sqrt{\sqrt[3]{729}}=\sqrt[2\times3]{3^6}=\sqrt[6]{3^6}=3$

   (iv) $\dfrac{\sqrt[n]{a}}{\sqrt[n]{b}}=\sqrt[n]{\dfrac{a}{b}}$, e.g., $\sqrt[4]{\dfrac{81}{16}}=\dfrac{\sqrt[4]{81}}{\sqrt[4]{16}}=\dfrac{\sqrt[4]{3^4}}{\sqrt[4]{2^4}}=\dfrac{3}{2}$

   (v) $\sqrt[n]{\sqrt[m]{(a^p)^m}}=\sqrt[n]{a^p}=\sqrt[mn]{a^{pm}}$

Solved Examples

Ex. 1. Which among the following numbers is the greatest?
$$\sqrt[3]{4},\sqrt{2},\sqrt[6]{16},\sqrt[4]{5}$$

Sol. LCM of (3, 6, 4) = 12, Raising each given number to power 12,

$$\therefore \sqrt[3]{4}=(4)^{1/3}=(4^{1/3})^{12}=4^4=256$$
$$\sqrt{2}=(2)^{1/2}=(2^{1/2})^{12}=2^6=64$$
$$\sqrt[6]{13}=(13)^{1/6}=(13^{1/6})^{12}=13^2=169$$
$$\sqrt[4]{5}=(5)^{1/4}=(5^{1/4})^{12}=5^3=125$$

$\Rightarrow \sqrt[3]{4}$ is the greatest.

Ex. 2. If $x=\dfrac{\sqrt{5}+\sqrt{3}}{\sqrt{5}-\sqrt{3}}$ **and** $y=\dfrac{\sqrt{5}-\sqrt{3}}{\sqrt{5}+\sqrt{3}}$, **then find** $(x+y)$.

Sol. $x+y=\dfrac{\sqrt{5}+\sqrt{3}}{\sqrt{5}-\sqrt{3}}+\dfrac{\sqrt{5}-\sqrt{3}}{\sqrt{5}+\sqrt{3}}=\dfrac{(\sqrt{5}+\sqrt{3})^2+(\sqrt{5}-\sqrt{3})^2}{(\sqrt{5})^2-(\sqrt{3})^2}$

$$=\dfrac{(\sqrt{5})^2+(\sqrt{3})^2+2\sqrt{5}\sqrt{3}+(\sqrt{5})^2+(\sqrt{3})^2-2\sqrt{5}\sqrt{3}}{5-3}=\dfrac{5+3+5+3}{2}=\dfrac{16}{2}=\mathbf{8.}$$

Ex. 3. By how much does $\sqrt{12}+\sqrt{18}$ **exceed** $\sqrt{3}+\sqrt{2}$ **?**

Sol. Required difference $=(\sqrt{12}+\sqrt{18})-(\sqrt{3}+\sqrt{2})$

$$=(\sqrt{4\times3}+\sqrt{9\times2})-(\sqrt{3}+\sqrt{2})=2\sqrt{3}+3\sqrt{2}-\sqrt{3}-\sqrt{2}$$
$$=(2\sqrt{3}-\sqrt{3})+(3\sqrt{2}-\sqrt{2})=\mathbf{\sqrt{3}+2\sqrt{2}}$$

Ex. 4. *Evaluate :* $\dfrac{\sqrt{5}}{\sqrt{3}+\sqrt{2}}-\dfrac{3\sqrt{3}}{\sqrt{5}+\sqrt{2}}+\dfrac{2\sqrt{2}}{\sqrt{5}+\sqrt{3}}$ *Rationalising the denominator of each term of the expression.*

Sol. Given exp. $=\dfrac{\sqrt{5}(\sqrt{3}-\sqrt{2})}{(\sqrt{3}+\sqrt{2})(\sqrt{3}-\sqrt{2})}-\dfrac{3\sqrt{3}(\sqrt{5}-\sqrt{2})}{(\sqrt{5}+\sqrt{2})(\sqrt{5}-\sqrt{2})}+\dfrac{2\sqrt{2}(\sqrt{5}-\sqrt{3})}{(\sqrt{5}+\sqrt{3})(\sqrt{5}-\sqrt{3})}$

$=\dfrac{\sqrt{15}-\sqrt{10}}{(\sqrt{3})^2-(\sqrt{2})^2}-\dfrac{(3\sqrt{15}-3\sqrt{6})}{(\sqrt{5})^2-(\sqrt{2})^2}+\dfrac{2\sqrt{10}-2\sqrt{6}}{(\sqrt{5})^2-(\sqrt{3})^2}=\dfrac{\sqrt{15}-\sqrt{10}}{3-2}-\dfrac{3(\sqrt{15}-\sqrt{6})}{5-2}+\dfrac{2(\sqrt{10}-\sqrt{6})}{5-3}$

$=\sqrt{15}-\sqrt{10}-\dfrac{3(\sqrt{15}-\sqrt{6})}{3}+\dfrac{2(\sqrt{10}-\sqrt{6})}{2}=\sqrt{15}-\sqrt{10}-\sqrt{15}+\sqrt{6}+\sqrt{10}-\sqrt{6}=\mathbf{0}.$

Ex. 5. *If $A=5+2\sqrt{6}$, then find the value of $\sqrt{A}+\dfrac{1}{\sqrt{A}}$.*

Sol. $\sqrt{A}=\sqrt{5+2\sqrt{6}}=\sqrt{3+2+2\sqrt{3\times 2}}=\sqrt{(\sqrt{3})^2+(\sqrt{2})^2+2\sqrt{3}\sqrt{2}}=\sqrt{(\sqrt{3}+\sqrt{2})^2}=\sqrt{3}+\sqrt{2}$

$\therefore\ \ \sqrt{A}+\dfrac{1}{\sqrt{A}}=(\sqrt{3}+\sqrt{2})+\dfrac{1}{(\sqrt{3}+\sqrt{2})}=(\sqrt{3}+\sqrt{2})+\dfrac{\sqrt{3}-\sqrt{2}}{(\sqrt{3}+\sqrt{2})(\sqrt{3}-\sqrt{2})}$

$=(\sqrt{3}+\sqrt{2})+\dfrac{\sqrt{3}-\sqrt{2}}{3-2}=\sqrt{3}+\sqrt{2}+\sqrt{3}-\sqrt{2}=\mathbf{2\sqrt{3}}\ .$

Ex. 6. *Evaluate :* $\sqrt{6+\sqrt{6+\sqrt{6+\ldots}}}$

Sol. Let $y=\sqrt{6+\sqrt{6+\sqrt{6+\ldots}}}$

$\Rightarrow\ y=\sqrt{6+y}$

$\Rightarrow\ y^2=6+y$ (On squaring both the sides)

$\Rightarrow\ y^2-y-6=0\ \Rightarrow\ (y-3)(y+2)=0\ \Rightarrow\ y-3=0$ or $y+2=0\ \Rightarrow\ y=3\ $ or $\ -2$

Neglecting negative value $y=\mathbf{3}.$

Ex. 7. *Find the value of* $\sqrt{-\sqrt{3}+\sqrt{3+8\sqrt{7+4\sqrt{3}}}}$.

Sol. $\sqrt{-\sqrt{3}+\sqrt{3+8\sqrt{7+4\sqrt{3}}}}=\sqrt{-\sqrt{3}+\sqrt{3+8\sqrt{4+3+2\times 2\times\sqrt{3}}}}=\sqrt{-\sqrt{3}+\sqrt{3+8\sqrt{(2)^2+(\sqrt{3})^2+2\times 2\times\sqrt{3}}}}$

$=\sqrt{-\sqrt{3}+\sqrt{3+8\sqrt{(2+\sqrt{3})^2}}}=\sqrt{-\sqrt{3}+\sqrt{3+8(2+\sqrt{3})}}=\sqrt{-\sqrt{3}+\sqrt{3+16+8\sqrt{3}}}$

$=\sqrt{-\sqrt{3}+\sqrt{(\sqrt{3})^2+(4)^2+2\times 4\times\sqrt{3}}}=\sqrt{-\sqrt{3}+\sqrt{(4+\sqrt{3})^2}}$

$=\sqrt{-\sqrt{3}+4+\sqrt{3}}=\sqrt{4}=\mathbf{2}\cdot$

Ex. 8. *Choose the correct answer. The number* $\sqrt{14+6\sqrt{5}}+\sqrt{14-6\sqrt{5}}$

    (a) is not a rational number   *(b) is a rational number ≥ 14*

    (c) simplifies to 5   *(d) simplifies to 6. (Take positive root only)*

Sol. $\sqrt{14+6\sqrt{5}}=\sqrt{9+5+2\times 3\times\sqrt{5}}=\sqrt{(3)^2+(\sqrt{5})^2+2\times 3\sqrt{5}}=\sqrt{(3+\sqrt{5})^2}=3+\sqrt{5}$

Similarly, $\sqrt{14-6\sqrt{5}}=3-\sqrt{5}$

$\therefore\ \ \sqrt{14+6\sqrt{5}}+\sqrt{14-6\sqrt{5}}=3+\sqrt{5}+3-\sqrt{5}=\mathbf{6}.$

$\therefore$ (d) is the correct answer.

Ex. 9. *Simplify :* $\dfrac{\sqrt{8+\sqrt{28}}-\sqrt{8-\sqrt{28}}}{\sqrt{8+\sqrt{28}}+\sqrt{8-\sqrt{28}}}$.

Sol. Given exp. $= \dfrac{\left(\sqrt{8+\sqrt{28}}-\sqrt{8-\sqrt{28}}\right)\times\left(\sqrt{8+\sqrt{28}}-\sqrt{8-\sqrt{28}}\right)}{\left(\sqrt{8+\sqrt{28}}+\sqrt{8-\sqrt{28}}\right)\times\left(\sqrt{8+\sqrt{28}}-\sqrt{8-\sqrt{28}}\right)} = \dfrac{\left(\sqrt{8+\sqrt{28}}-\sqrt{8-\sqrt{28}}\right)^2}{\left(\sqrt{8+\sqrt{28}}\right)^2-\left(\sqrt{8-\sqrt{28}}\right)^2}$

$= \dfrac{(\sqrt{\sqrt{8}+\sqrt{28}})^2+(\sqrt{\sqrt{8}-\sqrt{28}})^2-2\sqrt{8+\sqrt{28}}\times\sqrt{8-\sqrt{28}}}{8+\sqrt{28}-\left(8-\sqrt{28}\right)}$

$= \dfrac{(8+\sqrt{28})+(8-\sqrt{28})-2\sqrt{64-28}}{2\sqrt{28}}$

$= \dfrac{16-2\sqrt{64-28}}{2\sqrt{28}} = \dfrac{4}{2\sqrt{28}} = \dfrac{4}{2\times 2\sqrt{7}} = \dfrac{1}{\sqrt{7}}$.

Ex. 10. *Find the value of the expression* $\dfrac{1}{3-\sqrt{8}}-\dfrac{1}{\sqrt{8}-\sqrt{7}}+\dfrac{1}{\sqrt{7}-\sqrt{6}}-\dfrac{1}{\sqrt{6}-\sqrt{5}}+\dfrac{1}{\sqrt{5}-\sqrt{4}}$.

Sol. $\dfrac{1}{3-\sqrt{8}} = \dfrac{1}{(3-\sqrt{8})}\times\dfrac{3+\sqrt{8}}{3+\sqrt{8}} = \dfrac{3+\sqrt{8}}{9-8} = 3+\sqrt{8}$

$\dfrac{1}{\sqrt{8}-\sqrt{7}} = \dfrac{1}{\sqrt{8}-\sqrt{7}}\times\dfrac{\sqrt{8}+\sqrt{7}}{\sqrt{8}+\sqrt{7}} = \dfrac{\sqrt{8}+\sqrt{7}}{8-7} = \sqrt{8}+\sqrt{7}$

$\dfrac{1}{\sqrt{7}-\sqrt{6}} = \dfrac{1}{\sqrt{7}-\sqrt{6}}\times\dfrac{\sqrt{7}+\sqrt{6}}{\sqrt{7}+\sqrt{6}} = \dfrac{\sqrt{7}+\sqrt{6}}{7-6} = \sqrt{7}+\sqrt{6}$

$\dfrac{1}{\sqrt{6}-\sqrt{5}} = \dfrac{1}{\sqrt{6}-\sqrt{5}}\times\dfrac{\sqrt{6}+\sqrt{5}}{\sqrt{6}+\sqrt{5}} = \dfrac{\sqrt{6}+\sqrt{5}}{6-5} = \sqrt{6}+\sqrt{5}$

$\dfrac{1}{\sqrt{5}-\sqrt{4}} = \dfrac{1}{\sqrt{5}-\sqrt{4}}\times\dfrac{\sqrt{5}+\sqrt{4}}{\sqrt{5}+\sqrt{4}} = \dfrac{\sqrt{5}+\sqrt{4}}{5-4} = \sqrt{5}+\sqrt{4}$

∴ Given exp. $= 3+\sqrt{8}-(\sqrt{8}+\sqrt{7})+\sqrt{7}+\sqrt{6}-(\sqrt{6}+\sqrt{5})+\sqrt{5}+\sqrt{4}$

$= 3+\cancel{\sqrt{8}}-\cancel{\sqrt{8}}-\cancel{\sqrt{7}}+\cancel{\sqrt{7}}+\cancel{\sqrt{6}}-\cancel{\sqrt{6}}-\cancel{\sqrt{5}}+\cancel{\sqrt{5}}+\sqrt{4} = 3+\sqrt{4} = 3+2 = \mathbf{5}$.

Question Bank–6

1. The greatest number among $\sqrt[3]{2}, \sqrt{3}, \sqrt[3]{5}$ and 1.5 is

(a) $\sqrt[3]{2}$ (b) $\sqrt{3}$

(c) $\sqrt[3]{5}$ (d) 1.5

2. $\sqrt[4]{3}, \sqrt[6]{10}, \sqrt[12]{25}$, when arranged in descending order will be

(a) $\sqrt[4]{3}, \sqrt[6]{10}, \sqrt[12]{25}$ (b) $\sqrt[6]{10}, \sqrt[4]{3}, \sqrt[12]{25}$

(c) $\sqrt[6]{10}, \sqrt[12]{25}, \sqrt[4]{3}$ (d) $\sqrt[4]{3}, \sqrt[12]{25}, \sqrt[6]{10}$

3. $\dfrac{\sqrt{2}(2+\sqrt{3})}{\sqrt{3}(\sqrt{3}+1)}\times\dfrac{\sqrt{2}(2-\sqrt{3})}{\sqrt{3}(\sqrt{3}-1)}$ is equal to

(a) $\dfrac{1}{3}$ (b) $\dfrac{2}{3}$

(c) $\dfrac{\sqrt{2}}{3}$ (d) $3\sqrt{2}$

4. $1-\left[\dfrac{1+\sqrt{3}}{2}-\dfrac{1}{\sqrt{3}+1}\right]$ is equal to

(a) $\sqrt{3}$ (b) 1

(c) $2\sqrt{3}$ (d) 0

5. Simplify : $\dfrac{2}{\sqrt{7}+\sqrt{5}}+\dfrac{7}{\sqrt{12}-\sqrt{5}}-\dfrac{5}{\sqrt{12}-\sqrt{7}}$

(a) 5   (b) 2

(c) 1   (d) 0

6. Given that $\sqrt{3} = 1.732$; $(\sqrt{6}+\sqrt{2})\div(\sqrt{6}-\sqrt{2})$ is equal to

(a) 3.713   (b) 3.721

(c) 3.732   (d) 3.752

7. If $x = 8 + 2\sqrt{15}$, the value of $\sqrt{x} + \dfrac{1}{\sqrt{x}}$ is

  (a) $2\sqrt{3}$ (b) $2\sqrt{5}$

  (c) $\dfrac{3}{2}\sqrt{5} + \dfrac{\sqrt{3}}{2}$ (d) $\dfrac{\sqrt{5}}{2} + \dfrac{3}{2}\sqrt{3}$

8. $\left(1 + \dfrac{4\sqrt{3}}{2-\sqrt{2}} - \dfrac{30}{4\sqrt{3}-\sqrt{18}} - \dfrac{\sqrt{18}}{3+2\sqrt{3}}\right)$ is simplified to

  (a) 0   (b) 1

  (c) $\sqrt{2}$ (d) $\sqrt{3}$

9. Given that $\sqrt{3} = 1.732$, then $\left(\sqrt{147} - \dfrac{1}{4}\sqrt{48} - \sqrt{75}\right)$ is equal to

  (a) 5.196   (b) 3.464

  (c) 1.732   (d) 0.866

10. If $\sqrt{6} = 2.55$, then the value of $\sqrt{\dfrac{2}{3}} + \sqrt[3]{\dfrac{3}{2}}$ is

  (a) 4.48   (b) 4.49

  (c) 4.5   (d) 4.675

11. The value of $\sqrt{5\sqrt{5\sqrt{5....}}}$ is

  (a) 1   (b) 2.5

  (c) 5   (d) 25

12. The value of $\dfrac{1}{\sqrt{3.25}+\sqrt{2.25}} + \dfrac{1}{\sqrt{4.25}+\sqrt{3.25}} + \dfrac{1}{\sqrt{5.25}+\sqrt{4.25}} + \dfrac{1}{\sqrt{6.25}+\sqrt{5.25}}$ is

  (a) 1.00   (b) 1.25

  (c) 1.50   (d) 2.25

13. If $\sqrt{2} = 1.414$, the square root of $\dfrac{\sqrt{2}-1}{\sqrt{2}+1}$ is nearest to

  (a) 0.172   (b) 0.414

  (c) 0.586   (d) 1.414

14. The value of $\dfrac{1}{\sqrt{12-\sqrt{140}}} - \dfrac{1}{\sqrt{8-\sqrt{60}}} - \dfrac{2}{\sqrt{10+\sqrt{84}}}$ is

  (a) 0   (b) 1

  (c) 2   (d) 3

15. $\sqrt{2+\sqrt{2+\sqrt{2+....}}}$ is equal to

  (a) 1   (b) 2

  (c) 1.5   (d) 2.5

16. The value of $\sqrt{32} - \sqrt{128} + \sqrt{50}$ correct to 3 places of decimal is

  (a) 1.732   (b) 1.141

  (c) 1.414   (d) 1.441

17. The value of $\sqrt{\dfrac{(\sqrt{12}-\sqrt{8})(\sqrt{3}+\sqrt{2})}{5+\sqrt{24}}}$ is

  (a) $\sqrt{6}-\sqrt{2}$ (b) $\sqrt{6}+\sqrt{2}$

  (c) $\sqrt{6}-2$ (d) $2-\sqrt{6}$

18. What is the sum of the squares of the following numbers?

  $\dfrac{\sqrt{3}}{\sqrt{2}+1}, \dfrac{\sqrt{3}}{\sqrt{2}-1}, \dfrac{\sqrt{2}}{\sqrt{3}}$

  (a) 16   (b) $16\dfrac{2}{3}$

  (c) 18   (d) $18\dfrac{2}{3}$

19. Which is the greatest out of the following?

  (a) $\sqrt[3]{1.728}$ (b) $\dfrac{\sqrt{3}-1}{\sqrt{3}+1}$

  (c) $\left(\dfrac{1}{2}\right)^{-2}$ (d) $\dfrac{17}{8}$

20. If $a = (\sqrt{3}+\sqrt{2})^{-3}$ and $b = (\sqrt{3}-\sqrt{2})^{-3}$, then the value of $(a+1)^{-1} + (b+1)^{-1}$ is

  (a) $50\sqrt{3}$ (b) $48\sqrt{2}$

  (c) 1   (d) 5

21. By how much does $5\sqrt{7}-2\sqrt{5}$ exceed $3\sqrt{7}-4\sqrt{5}$?

  (a) $5(\sqrt{7}+\sqrt{5})$ (b) $\sqrt{7}+\sqrt{5}$

  (c) $2(\sqrt{7}+\sqrt{5})$ (d) $7(\sqrt{2}+\sqrt{5})$

22. If $a = \dfrac{\sqrt{5}+1}{\sqrt{5}-1}$ and $b = \dfrac{\sqrt{5}-1}{\sqrt{5}-1}$, the value of $\dfrac{a^2+ab+b^2}{a^2-ab+b^2}$ is

  (a) $\dfrac{3}{4}$ (b) $\dfrac{4}{3}$

  (c) $\dfrac{3}{5}$ (d) $\dfrac{5}{3}$

23. $\dfrac{3+\sqrt{6}}{5\sqrt{3}-2\sqrt{12}-\sqrt{32}+\sqrt{50}}$ is equal to

  (a) $3\sqrt{2}$ (b) 3

  (c) 6   (d) $\sqrt{3}$

24. The sum

$$\frac{1}{\sqrt{2}+1}+\frac{1}{\sqrt{3}+\sqrt{2}}+\frac{1}{\sqrt{4}+\sqrt{3}}+\cdots\frac{1}{\sqrt{100}+\sqrt{99}}\text{ is}$$

equal to

  (a) 9   (b) 10

  (c) 11   (d) 0

25. If $4=\sqrt{x+\sqrt{x+\sqrt{x+\cdots}}}$, then the value of x will be

  (a) 20   (b) 16

  (c) 12   (d) 8

Answers

1. (b)	2. (b)	3. (a)	4. (d)	5. (d)	6. (c)	7. (c)	8. (b)	9. (c)	10. (d)
11. (c)	12. (a)	13. (b)	14. (a)	15. (b)	16. (c)	17. (c)	18. (d)	19. (c)	20. (c)
21. (c)	22. (b)	23. (d)	24. (a)	25. (c)					

Hints and Solutions

1. (b) LCM of 3 and 2 = 6

Given numbers are $\sqrt[3]{2},\sqrt{3},\sqrt[3]{5}$, 1.5, i.e.,

$2^{1/3},3^{1/2},5^{1/3},1.5$

∴ Raising each number to power 6, we get

$(2^{1/3})^6,(3^{1/2})^6,(5^{1/3})^6,(1.5)^6$, i.e., $2^2, 3^3, 5^2,$

$\left(\dfrac{3}{2}\right)^6$, i.e., $4, 27, 25, \dfrac{729}{64}$.

Of all these numbers 27 is the greatest $\Rightarrow \sqrt{3}$ is the greatest.

2. (b) LCM of 4, 6 and 12 = 12

∴ Raising each of the given numbers to power 12, we have

$(3^{1/4})^{12},(10^{1/6})^{12},(25^{1/12})^{12}$, i.e., 3^3, 10^2 and 25, i.e., 27, 100 and 25

Arranged in descending order, the numbers are 100, 27, 25, i.e., $\sqrt[6]{10},\sqrt[4]{3},\sqrt[12]{25}$

3. (a) Given exp. $=\dfrac{(\sqrt{2})^2\times(2+\sqrt{3})(2-\sqrt{3})}{(\sqrt{3})^2\times(\sqrt{3}+1)(\sqrt{3}-1)}$

$=\dfrac{2\times\left[(2)^2-(\sqrt{3})^2\right]}{3\times\left[(\sqrt{3})^2-1^2\right]}=\dfrac{2\times(4-3)}{3\times(3-1)}$

$=\dfrac{2\times1}{3\times2}=\dfrac{1}{3}$.

4. (d) Rationalising the denominator of third term, we have

Given exp. $=1-\left[\dfrac{1+\sqrt{3}}{2}-\dfrac{1}{(\sqrt{3}+1)}\dfrac{(\sqrt{3}-1)}{(\sqrt{3}-1)}\right]$

$=1-\left[\dfrac{1+\sqrt{3}}{2}-\dfrac{(\sqrt{3}-1)}{(\sqrt{3})^2-1)}\right]$

$=1-\left[\dfrac{1+\sqrt{3}}{2}-\dfrac{(\sqrt{3}-1)}{2}\right]$

$=1-\left[\dfrac{1+\sqrt{3}-\sqrt{3}+1}{2}\right]$

$=1-\dfrac{2}{2}=1-1=\mathbf{0}.$

5. (d) Rationalising the denominators, we have

Given exp.

$=\dfrac{2}{(\sqrt{7}+\sqrt{5})}\times\dfrac{(\sqrt{7}-\sqrt{5})}{(\sqrt{7}-\sqrt{5})}+\dfrac{7}{(\sqrt{12}-\sqrt{5})}$

$\times\dfrac{(\sqrt{12}+\sqrt{5})}{(\sqrt{12}+\sqrt{5})}-\dfrac{5}{(\sqrt{12}-\sqrt{7})}\times\dfrac{(\sqrt{12}+\sqrt{7})}{(\sqrt{12}+\sqrt{7})}$

$=\dfrac{2(\sqrt{7}-\sqrt{5})}{7-5}+\dfrac{7(\sqrt{12}+\sqrt{5})}{12-5}-\dfrac{5(\sqrt{12}+\sqrt{7})}{12-7}$

$=\sqrt{7}-\sqrt{5}+\sqrt{12}+\sqrt{5}-\sqrt{12}-\sqrt{7}=\mathbf{0}.$

6. (c) $\dfrac{\sqrt{6}+\sqrt{2}}{\sqrt{6}-\sqrt{2}}=\dfrac{\sqrt{3\times2}+\sqrt{2}}{\sqrt{3\times2}-\sqrt{2}}=\dfrac{\sqrt{3}.\sqrt{2}+\sqrt{2}}{\sqrt{3}.\sqrt{2}-\sqrt{2}}$

$=\dfrac{\sqrt{2}(\sqrt{3}+1)}{\sqrt{2}(\sqrt{3}-1)}=\dfrac{\sqrt{3}+1}{\sqrt{3}-1}=\dfrac{(\sqrt{3}+1)(\sqrt{3}+1)}{(\sqrt{3}-1)(\sqrt{3}+1)}$

$=\dfrac{(\sqrt{3})^2+2\sqrt{3}+1}{3-1}=\dfrac{3+2\sqrt{3}+1}{2}=\dfrac{4+2\sqrt{3}}{2}$

$=2+\sqrt{3}=2+1.732=\mathbf{3.732}.$

7. (c) $\sqrt{x}=\sqrt{8+2\sqrt{15}}=\sqrt{5+3+2\sqrt{5\times3}}$

$=\sqrt{(\sqrt{5})^2+(\sqrt{3})^2+2\times\sqrt{5}\times\sqrt{3}}$

$=\sqrt{(\sqrt{5}+\sqrt{3})^2}=\sqrt{5}+\sqrt{3}$

$$\therefore \quad \sqrt{x} + \frac{1}{\sqrt{x}} = (\sqrt{5} + \sqrt{3}) + \frac{1}{(\sqrt{5} + \sqrt{3})}$$

$$= (\sqrt{5} + \sqrt{3}) + \frac{(\sqrt{5} - \sqrt{3})}{(\sqrt{5} + \sqrt{3})(\sqrt{5} - \sqrt{3})}$$

$$= (\sqrt{5} + \sqrt{3}) + \frac{(\sqrt{5} - \sqrt{3})}{5 - 3}$$

$$= \sqrt{5} + \sqrt{3} + \frac{(\sqrt{5} - \sqrt{3})}{2}$$

$$= \frac{2\sqrt{5} + 2\sqrt{3} + \sqrt{5} - \sqrt{3}}{2}$$

$$= \frac{3\sqrt{5}}{2} + \frac{\sqrt{3}}{2}.$$

8. (b) $1 + \dfrac{4\sqrt{3}}{2 - \sqrt{2}} - \dfrac{30}{4\sqrt{3} - \sqrt{18}} - \dfrac{\sqrt{18}}{3 + 2\sqrt{3}}$

$$= 1 + \frac{4\sqrt{3}}{2 - \sqrt{2}} - \frac{30}{4\sqrt{3} - 3\sqrt{2}} - \frac{3\sqrt{2}}{3 + 2\sqrt{3}}$$

$$= 1 + \frac{4\sqrt{3}(2 + \sqrt{2})}{(2 - \sqrt{2})(2 + \sqrt{2})}$$

$$- \frac{30(4\sqrt{3} + 3\sqrt{2})}{(4\sqrt{3} - 3\sqrt{2})(4\sqrt{3} + 3\sqrt{2})}$$

$$- \frac{3\sqrt{2}(3 - 2\sqrt{3})}{(3 + 2\sqrt{3})(3 - 2\sqrt{3})}$$

$$= 1 + \frac{4(2\sqrt{3} + \sqrt{6})}{(2)^2 - (\sqrt{2})^2} - \frac{30(4\sqrt{3} + 3\sqrt{2})}{(4\sqrt{3})^2 - (3\sqrt{2})^2}$$

$$- \frac{3(3\sqrt{2} - 2\sqrt{6})}{3^2 - (2\sqrt{3})^2}$$

$$= 1 + \frac{4(2\sqrt{3} + \sqrt{6})}{4 - 2} - \frac{30(4\sqrt{3} + 3\sqrt{2})}{48 - 18}$$

$$- \frac{3(3\sqrt{2} - 2\sqrt{6})}{9 - 12}$$

$$= 1 + \frac{4(2\sqrt{3} + \sqrt{6})}{2} - \frac{30(4\sqrt{3} + 3\sqrt{2})}{30}$$

$$- \frac{3(3\sqrt{2} - 2\sqrt{6})}{-3}$$

$$= 1 + 2(2\sqrt{3} + \sqrt{6}) - (4\sqrt{3} + 3\sqrt{2}) + (3\sqrt{2} - 2\sqrt{6})$$

$$= 1 + 4\sqrt{3} + 2\sqrt{6} - 4\sqrt{3} - 3\sqrt{2} + 3\sqrt{2} - 2\sqrt{6}$$

$$= \mathbf{1}.$$

9. (c) $\sqrt{147} - \dfrac{1}{4}\sqrt{48} - \sqrt{75}$

$$= \sqrt{3 \times 49} - \frac{1}{4}\sqrt{3 \times 16} - \sqrt{3 \times 25}$$

$$= 7\sqrt{3} - \frac{1}{4} \times 4\sqrt{3} - 5\sqrt{3}$$

$$= 7\sqrt{3} - \sqrt{3} - 5\sqrt{3}$$

$$= 7\sqrt{3} - 6\sqrt{3} = \sqrt{3} = \mathbf{1.732}.$$

10. (d) $\sqrt{\dfrac{2}{3}} + 3 \times \sqrt{\dfrac{3}{2}} = \dfrac{\sqrt{2}}{\sqrt{3}} + 3 \times \dfrac{\sqrt{3}}{\sqrt{2}}$

$$= \frac{\sqrt{2} \times \sqrt{3}}{\sqrt{3} \times \sqrt{3}} + 3 \times \frac{\sqrt{3} \times \sqrt{2}}{\sqrt{2} \times \sqrt{2}}$$

(Rationalising the denominator)

$$= \frac{\sqrt{6}}{3} + \frac{3 \times \sqrt{6}}{2} = \frac{2\sqrt{6} + 9\sqrt{6}}{6}$$

$$= \frac{11\sqrt{6}}{6} = \frac{11 \times 2.45}{6} = \frac{26.95}{6} = \mathbf{4.492}.$$

11. (c) Let $x = \sqrt{5\sqrt{5\sqrt{5}\cdots}}$

$$\Rightarrow x = \sqrt{5x} \Rightarrow x^2 = 5x$$

$$\Rightarrow x^2 - 5x = 0 \Rightarrow x(x - 5) = 0 \Rightarrow x = 0 \text{ or } 5$$

Neglecting $x = 0$, we have $x = \mathbf{5}$.

12. (a) $\dfrac{1}{\sqrt{3.25} + \sqrt{2.25}}$

$$= \frac{(\sqrt{3.25} - \sqrt{2.25})}{(\sqrt{3.25} + \sqrt{2.25})(\sqrt{3.25} - \sqrt{2.25})}$$

$$= \frac{\sqrt{3.25} - \sqrt{2.25}}{(\sqrt{3.25})^2 - (\sqrt{2.25})^2} = \frac{\sqrt{3.25} - \sqrt{2.25}}{3.25 - 2.25}$$

$$= \sqrt{3.25} - \sqrt{2.25}$$

Similarly, $\dfrac{1}{\sqrt{4.25} + \sqrt{3.25}} = \sqrt{4.25} - \sqrt{3.25}$,

$$\frac{1}{\sqrt{5.25} + \sqrt{4.25}} = \sqrt{5.25} - \sqrt{4.25} \text{ and }$$

$$\frac{1}{\sqrt{6.25} + \sqrt{5.25}} = \sqrt{6.25} - \sqrt{5.25}$$

$\therefore$ Given exp. $= \sqrt{3.25} - \sqrt{2.25} + \sqrt{4.25} - \sqrt{3.25}$

$$+ \sqrt{5.25} - \sqrt{4.25} + \sqrt{6.25} - \sqrt{5.25}$$

$$= \sqrt{6.25} - \sqrt{2.25} = 2.5 - 1.5 = \mathbf{1}.$$

13. (b) $\dfrac{\sqrt{2} - 1}{\sqrt{2} + 1} = \dfrac{(\sqrt{2} - 1)^2}{(\sqrt{2} + 1)(\sqrt{2} - 1)} = \dfrac{2 + 1 - 2\sqrt{2}}{(\sqrt{2})^2 - 1^2}$

$$= \frac{3 - 2\sqrt{2}}{2 - 1} = 3 - 2\sqrt{2}$$

$\therefore \sqrt{\dfrac{\sqrt{2}-1}{\sqrt{2}+1}} = \sqrt{3 - 2\sqrt{2}} = \sqrt{2 + 1 - 2\sqrt{2}}$

$\qquad = \sqrt{(\sqrt{2})^2 + 1^2 - 2\sqrt{2}}$

$\qquad = \sqrt{(\sqrt{2}-1)^2} = \sqrt{2} - 1$

$\qquad = 1.414 - 1 = \mathbf{0.414}.$

14. (a) Given exp.

$= \dfrac{1}{\sqrt{7 + 5 - \sqrt{4 \times 7 \times 5}}} - \dfrac{1}{\sqrt{5 + 3 - \sqrt{4 \times 5 \times 3}}}$

$\qquad - \dfrac{2}{\sqrt{7 + 3 + \sqrt{4 \times 7 \times 3}}}$

$= \dfrac{1}{\sqrt{(\sqrt{7})^2 + (\sqrt{5})^2 - 2\sqrt{7 \times 5}}}$

$\qquad - \dfrac{1}{\sqrt{(\sqrt{5})^2 + (\sqrt{3})^2 - 2\sqrt{5 \times 3}}}$

$\qquad - \dfrac{2}{\sqrt{(\sqrt{7})^2 + (\sqrt{3})^2 + 2\sqrt{7 \times 3}}}$

$= \dfrac{1}{\sqrt{(\sqrt{7} - \sqrt{5})^2}} - \dfrac{1}{\sqrt{(\sqrt{5} - \sqrt{3})^2}} - \dfrac{2}{\sqrt{(\sqrt{7} + \sqrt{3})^2}}$

$= \dfrac{1}{\sqrt{7} - \sqrt{5}} - \dfrac{1}{\sqrt{5} - \sqrt{3}} - \dfrac{2}{\sqrt{7} + \sqrt{3}}$

Rationalising the denominator, we get

Given exp. $= \dfrac{\sqrt{7} + \sqrt{5}}{7 - 5} - \dfrac{(\sqrt{5} + \sqrt{3})}{5 - 3} - \dfrac{2(\sqrt{7} - \sqrt{3})}{7 - 3}$

$= \dfrac{\sqrt{7} + \sqrt{5}}{2} - \dfrac{(\sqrt{5} + \sqrt{3})}{2} - \dfrac{(\sqrt{7} - \sqrt{3})}{2}$

$= \dfrac{\sqrt{7} + \sqrt{5} - \sqrt{5} - \sqrt{3} - \sqrt{7} + \sqrt{3}}{2} = \dfrac{0}{2} = \mathbf{0}.$

15. (b) Let $y = \sqrt{2 + \sqrt{2 + \sqrt{2 + \dots}}}$

$\Rightarrow y = \sqrt{2 + y} \Rightarrow y^2 = 2 + y$

$\Rightarrow y^2 - y - 2 = 0 \Rightarrow y^2 - 2y + y - 2 = 0$

$\Rightarrow y(y - 2) + 1(y - 2) = 0$

$\Rightarrow (y + 1)(y - 2) = 0$

$\Rightarrow y = -1$ or $2 \Rightarrow y = \mathbf{2}$ (Neglecting $-$ve value)

16. (c) $\sqrt{32} - \sqrt{128} + \sqrt{50} = \sqrt{2^5} - \sqrt{2^7} + \sqrt{25 \times 2}$

$\qquad = \sqrt{2^4 . 2} - \sqrt{2^6 . 2} + 5\sqrt{2}$

$\qquad = 4\sqrt{2} - 8\sqrt{2} + 5\sqrt{2}$

$\qquad = \sqrt{2} = \mathbf{1.414}.$

17. (c) $\sqrt{\dfrac{(\sqrt{12} - \sqrt{8})(\sqrt{3} + \sqrt{2})}{5 + \sqrt{24}}}$

$= \sqrt{\dfrac{(2\sqrt{3} - 2\sqrt{2})(\sqrt{3} + \sqrt{2})}{5 + 2\sqrt{6}}}$

$= \sqrt{\dfrac{2(\sqrt{3})^2 - 2\sqrt{6} + 2\sqrt{6} - 2(\sqrt{2})^2}{5 + 2\sqrt{6}}}$

$= \sqrt{\dfrac{6 - 4}{5 + 2\sqrt{6}}} = \sqrt{\dfrac{2(5 - 2\sqrt{6})}{(5 + 2\sqrt{6})(5 - 2\sqrt{6})}}$

$= \sqrt{\dfrac{2(5 - 2\sqrt{6})}{25 - 24}} = \sqrt{10 - 4\sqrt{6}}$

$= \sqrt{6 + 4 - 4\sqrt{6}} = \sqrt{(\sqrt{6})^2 + (2)^2 - 2 \times 2 \times \sqrt{6}}$

$= \sqrt{(\sqrt{6} - 2)^2} = \mathbf{\sqrt{6} - 2}.$

18. (d) Required sum

$= \left(\dfrac{\sqrt{3}}{\sqrt{2} + 1}\right)^2 + \left(\dfrac{\sqrt{3}}{\sqrt{2} - 1}\right)^2 + \left(\dfrac{\sqrt{2}}{\sqrt{3}}\right)^2$

$= \dfrac{3}{(\sqrt{2} + 1)^2} + \dfrac{3}{(\sqrt{2} - 1)^2} + \dfrac{2}{3}$

$= \dfrac{3}{2 + 1 + 2\sqrt{2}} + \dfrac{3}{2 + 1 - 2\sqrt{2}} + \dfrac{2}{3}$

$= \dfrac{3}{3 + 2\sqrt{2}} + \dfrac{3}{3 - 2\sqrt{2}} + \dfrac{2}{3}$

$= \dfrac{\begin{array}{c}9(3 - 2\sqrt{2}) + 9(3 + 2\sqrt{2}) \\ + 2(3 + 2\sqrt{2})(3 - 2\sqrt{2})\end{array}}{3(3 + 2\sqrt{2})(3 - 2\sqrt{2})}$

$= \dfrac{27 - 18\sqrt{2} + 27 + 18\sqrt{2} + 2(9 - 8)}{3(9 - 8)}$

$= \dfrac{54 + 2}{3} = \dfrac{56}{3} = \mathbf{18\dfrac{2}{3}}.$

19. (c) $\sqrt[3]{1.728} = 1.2$

$\dfrac{\sqrt{3} - 1}{\sqrt{3} + 1} = \dfrac{(\sqrt{3} - 1)^2}{(\sqrt{3} + 1)(\sqrt{3} - 1)} = \dfrac{3 + 1 - 2\sqrt{3}}{3 - 1}$

$\qquad = \dfrac{4 - 2\sqrt{3}}{2} = 2 - \sqrt{3}$

$\qquad = 2 - 1.732 = 0.268$

$\left(\dfrac{1}{2}\right)^{-2} = 2^2 = 4$

$$\frac{17}{8} = 2.2125$$

$\Rightarrow \left(\frac{1}{2}\right)^{-2}$ is the greatest.

20. (c) $a = \dfrac{1}{(\sqrt{3}+\sqrt{2})^3}, b = \dfrac{1}{(\sqrt{3}-\sqrt{2})^3}$

$(a+1)^{-1} + (b+1)^{-1} = \dfrac{1}{a+1} + \dfrac{1}{b+1}$

$= \dfrac{b+1+a+1}{(a+1)(b+1)} = \dfrac{a+b+2}{ab+a+b+1}$

$ab = \dfrac{1}{(\sqrt{3}+\sqrt{2})^3} \times \dfrac{1}{(\sqrt{3}-\sqrt{2})^3}$

$= \dfrac{1}{[(\sqrt{3}+\sqrt{2})(\sqrt{3}-\sqrt{2})]^3} = \dfrac{1}{(3-2)^2} = \dfrac{1}{1^3} = 1$

$\therefore (a+1)^{-1}+(b+1)^{-1} = \dfrac{a+b+2}{1+a+b+1} = \dfrac{a+b+2}{a+b+2} = \mathbf{1}.$

21. (c) $(5\sqrt{7}-2\sqrt{5}) - (3\sqrt{7}-4\sqrt{5})$

$= 5\sqrt{7}-2\sqrt{5}-3\sqrt{7}+4\sqrt{5}$

$= 2\sqrt{7}+2\sqrt{5} = \mathbf{2(\sqrt{7}+\sqrt{5})}.$

22. (b) $a = \dfrac{(\sqrt{5}+1)}{(\sqrt{5}-1)} \times \dfrac{(\sqrt{5}+1)}{(\sqrt{5}+1)} = \dfrac{(\sqrt{5}+1)^2}{5-1}$

$= \dfrac{5+1+2\sqrt{5}}{4} = \dfrac{3+\sqrt{5}}{2}$

$b = \dfrac{\sqrt{5}-1}{\sqrt{5}+1} \times \dfrac{\sqrt{5}-1}{\sqrt{5}-1} = \dfrac{(\sqrt{5}-1)^2}{5-1}$

$= \dfrac{5+1-2\sqrt{5}}{4} = \dfrac{3-\sqrt{5}}{2}$

$\therefore a^2+b^2 = \left(\dfrac{3+\sqrt{5}}{2}\right)^2 + \left(\dfrac{3-\sqrt{5}}{2}\right)^2$

$= \dfrac{9+5+6\sqrt{5}+9+5-6\sqrt{5}}{4}$

$= \dfrac{28}{4} = 7$

$ab = \left(\dfrac{3+\sqrt{5}}{2}\right) \times \left(\dfrac{3-\sqrt{5}}{2}\right) = \dfrac{9-5}{4} = \dfrac{4}{4} = 1$

$\therefore \dfrac{a^2+ab+b^2}{a^2-ab+b^2} = \dfrac{7+1}{7-1} = \dfrac{8}{6} = \dfrac{4}{3}.$

23. (d) Given exp. $= \dfrac{3+\sqrt{6}}{5\sqrt{3}-4\sqrt{3}-4\sqrt{2}+5\sqrt{2}}$

$= \dfrac{3+\sqrt{6}}{\sqrt{3}+\sqrt{2}} = \dfrac{(3+\sqrt{6})(\sqrt{3}-\sqrt{2})}{(\sqrt{3}+\sqrt{2})(\sqrt{3}-\sqrt{2})}$

$= \dfrac{3\sqrt{3}-3\sqrt{2}+3\sqrt{2}-2\sqrt{3}}{3-2} = \sqrt{3}.$

24. (a) $\dfrac{1}{\sqrt{2}+1} = \dfrac{\sqrt{2}-1}{(\sqrt{2}+1)(\sqrt{2}-1)} = \dfrac{\sqrt{2}-1}{2-1} = \sqrt{2}-1$

Similarly, $\dfrac{1}{\sqrt{3}+\sqrt{2}} = \sqrt{3}-\sqrt{2}$,

$\dfrac{1}{\sqrt{4}+\sqrt{3}} = \sqrt{4}-\sqrt{3}$,......,

$\dfrac{1}{\sqrt{99}+\sqrt{98}} = \sqrt{99}-\sqrt{98}$,

$\dfrac{1}{\sqrt{100}+\sqrt{99}} = \sqrt{100}-\sqrt{99}$

$\therefore$ Given exp. $= \sqrt{2}-1+\sqrt{3}-\sqrt{2}+\sqrt{4}-\sqrt{3}+$ ..... $+\sqrt{99}-\sqrt{98}+\sqrt{100}-\sqrt{99}$

$= \sqrt{100}-1 = 10-1 = \mathbf{9}.$

25. (c) $4 = \sqrt{x+\sqrt{x+\sqrt{x+....}}}$

$\Rightarrow 4 = \sqrt{x+4}$
$\Rightarrow 16 = x+4 \Rightarrow x = \mathbf{12}.$

Self Assessment Sheet–6

1. Which one of the following set of surds is incorrect sequence of ascending order of their values?

(a) $\sqrt[4]{10}, \sqrt[3]{6}, \sqrt{3}$ (b) $\sqrt{3}, \sqrt[4]{10}, \sqrt[3]{6}$

(c) $\sqrt{3}, \sqrt[3]{6}, \sqrt[4]{10}$ (d) $\sqrt[4]{10}, \sqrt{3}, \sqrt[3]{6}$

2. $\dfrac{\sqrt{31}-\sqrt{29}}{\sqrt{31}+\sqrt{29}}$ equals

(a) $60-2\sqrt{899}$ (b) $30-\sqrt{899}$

(c) $30+\sqrt{899}$ (d) $\dfrac{1}{30-\sqrt{899}}$

3. The expression $\dfrac{(5\sqrt{3}+\sqrt{50})(5-\sqrt{24})}{\sqrt{75}-5\sqrt{2}}$ simplifies to

(a) 1   (b) $\sqrt{3}-\sqrt{2}$

(c) $\sqrt{6}-\sqrt{5}$ (d) $\sqrt{2}$

4. $\sqrt{6+\sqrt{6+\sqrt{6+...}}}$ equals

(a) $6^{\frac{2}{3}}$　　　(b) 6

(c) $3^{\frac{1}{3}}$　　　(d) 3

5. If $x = 3+2\sqrt{2}$, then the value of $\sqrt{x}-\dfrac{1}{\sqrt{x}}$ is

(a) 1　　　(b) $2\sqrt{2}$

(c) 2　　　(d) $3\sqrt{3}$

6. $\dfrac{\dfrac{1}{\sqrt{9}}-\dfrac{1}{\sqrt{11}}}{\dfrac{1}{\sqrt{9}}+\dfrac{1}{\sqrt{11}}} \times \dfrac{10+\sqrt{99}}{x} = \dfrac{1}{2}$. Then x equals

(a) 2　　　(b) 3

(c) 10　　　(d) 1/10

7. If $\sqrt{\dfrac{19+8\sqrt{3}}{7-4\sqrt{3}}} = a+b\sqrt{3}$, then a equals

(a) 6　　　(b) 4

(c) 11　　　(d) 7

8. The value of $\sqrt{2\sqrt{2\sqrt{2....\infty}}}$ is

(a) 0　　　(b) 1

(c) $2\sqrt{2}$　　　(d) 2

9. $\dfrac{4\sqrt{3}}{2-\sqrt{2}} - \dfrac{30}{4\sqrt{3}-\sqrt{18}} - \dfrac{\sqrt{18}}{3-2\sqrt{3}}$ equals

(a) 1　　　(b) –1

(c) –2　　　(d) $4\sqrt{6}$

10. $\dfrac{\dfrac{\sqrt{2}-1}{\sqrt{2}+1}+\dfrac{\sqrt{2}+1}{\sqrt{2}-1}}{\dfrac{\sqrt{3}-1}{\sqrt{3}+1}+\dfrac{\sqrt{3}+1}{\sqrt{3}-1}}$ equals

(a) $\dfrac{2}{3}$　　　(b) 4

(c) $\dfrac{3}{2}$　　　(d) 6

Answers

| 1. (b) | 2. (b) | 3. (a) | 4. (d) | 5. (c) | 6. (d) | 7. (c) | 8. (d) | 9. (d) | 10. (c) |

Unit Test–1

1. In a problem involving division, the divisor is eight times the quotient and four times the remainder. If the remainder is 12, then the dividend is

(a) 300　　　(b) 288

(c) 512　　　(d) 524

2. If x is any natural number, then $x^5 - x$ is divisible by

(a) 6 but not by 10　　(b) 10 but not by 6

(c) Both 6 and 10　　(d) Neither 6 nor 10

3. The units' digit in the product $7^{35} \times 3^{71} \times 11^{55}$ is

(a) 1　　　(b) 3

(c) 7　　　(d) 9

4. The number of prime factors of $(6)^{10} \times (7)^{17} \times (55)^{27}$ is

(a) 54　　　(b) 64

(c) 81　　　(d) 91

5. xy is a number that is divided by ab where $xy < ab$ and gives a result 0. $xyxy.......$, then ab equals

(a) 11　　　(b) 33

(c) 99　　　(d) 66

6. The square root of

$\dfrac{0.2\times0.2\times0.2+0.02\times0.02\times0.02}{0.6\times0.6\times0.6+0.06\times0.06\times0.06} \div \dfrac{2\frac{1}{3}-1\frac{1}{6}}{2\frac{1}{3}+1\frac{1}{6}}$ is

(a) $\dfrac{1}{9}$　　　(b) $\dfrac{1}{3}$

(c) $\dfrac{1}{3\sqrt{3}}$　　　(d) 3

7. Simplify: $\dfrac{0.\overline{3}\times1.\overline{06}}{0.\overline{5}\times0.\overline{4}}$

(a) $\dfrac{31}{44}$　　　(b) $\dfrac{63}{44}$

(c) $\dfrac{32}{63}$　　　(d) $\dfrac{44}{111}$

8. Simplify:

$\sqrt{\dfrac{(12.12)^2-(8.12)^2}{(0.25)^2+(0.25)(19.99)}} + \dfrac{\left[(8^{-3/4})^{5/2}\right]^{8/15}\times16^{3/4}}{\sqrt[3]{\left[\{(128)^{-5}\}^{3/7}\right]^{-1/5}}}$

(a) $\dfrac{3}{2}$　　　(b) $\dfrac{9}{2}$

(c) 4　　　(d) 1

9. Find the value of $(28 + 10\sqrt{3})^{1/2} - (7 - 4\sqrt{3})^{-1/2}$

(a) 3 (b) $\dfrac{1}{3}$

(c) 1 (d) 0

10. The value of $\sqrt{\sqrt[3]{0.046656}}$ is

(a) 6 (b) 0.6
(c) 0.06 (d) 0.006

11. The value of $\sqrt{20 + \sqrt{20 + \sqrt{20 + ...}}}$ is

(a) 4 (b) 5
(c) 6 (d) Greater than 6

12. The value of $\sqrt{a\sqrt{b\sqrt{c\sqrt{d}}}}$ is

(a) $a^{1/2}b^{1/2}c^{1/2}d^{1/2}$ (b) $a^{1/2}b^{1/4}c^{1/8}d^{1/16}$
(c) $(abcd)^{1/12}$ (d) $(abcd)^{1/8}$

13. Simplified value of $\left\{ \dfrac{4^{m+\frac{1}{4}} \times \sqrt{2.2^m}}{2\sqrt{2^{-m}}} \right\}^{1/m}$ is

(a) 8 (b) 4
(c) 16 (d) 2

14. The sum of two numbers is 684 and their HCF is 57. The number of possible pairs of such numbers is

(a) 2 (b) 3
(c) 4 (d) None of these

15. Find the greatest number of six digits which on being divided by 6, 7, 8, 9 and 10 leaves 4, 5, 6, 7 and 8 as remainder respectively.

(a) 997920 (b) 997918
(c) 997922 (d) 997930

Answers

1. (a)	2. (c)	3. (a)	4. (d)	5. (c)	6. (b)	7. (b)	8. (b)	9. (a)	10. (b)
11. (b)	12. (b)	13. (a)	14. (a)	15. (b)					

UNIT-2

ALGEBRA

- *Algebraic Expressions and Identities*
- *Factorization of Algebraic Expressions*
- *HCF and LCM of Polynomials and Rational Expressions*
- *Linear Equations in One Variable (Revision)*
- *Simultaneous Linear Equations*
- *Quadratic Equations*
- *Linear Inequalities*
- *Matrices*

Chapter
7
ALGEBRAIC EXPRESSIONS AND IDENTITIES

KEY FACTS

1. Algebraic expressions in which the variables involved have only non- negative integral exponents are called **polynomials.**

   For example, $by^2 - y + 2$ is a polynomial while $a^{1/2} + 8a^3 - 3$ and $y^{-2} + y$ are not polynomials.

2. A polynomial that involves only one variable is called a **polynomial in one variable.**

3. (i) Polynomials having only one term are called **monomials.**

   (ii) Polynomials having only two terms are called **binomials.**

   (iii) Polynomials having only three terms are called **trinomials.**

4. (i) The highest exponent of the variable in a polynomial in one variable is called the degree of the polynomial, e.g., degree of the polynomial $7x^4 - 4x^2 + 9$ is 4.

   (ii) A constant is a polynomial of degree zero.

   (iii) In a polynomial in more than one variable, the highest sum of the powers of the variables is called the degree of the polynomial, e.g., degree of the polynomial $6 - 4x^6 + 2x^3 y^4 + 3xy^3$ is $3 + 4 = 7$.

5. The standard form of a polynomial in one variable is that in which the terms of the polynomial are written in the decreasing or descending order of the exponents of the variable.

6. A polynomial is said to be **linear, quadratic, cubic** or **biquadratic** if its degree is 1, 2, 3 or 4 respectively.

7. (i) The **product of two monomials** is the product of the numerical coefficients and literals in the two monomials, the exponent of each literal being the sum of its exponents in the given monomials.

   $$e.g., (4x^4 y^3 z^4 x) \times (-7 x^5 y^2 z) \times a = (4 \times -7) \times x^{4+5} \times y^{3+2} \times z^{4+1} \times a$$
   $$= -28 x^9 y^5 z^5 a$$

   (ii) The **product of a monomial and a binomial** is obtained on multiplying each term of the binomial by the monomial and adding up the two products.

   $$e.g., -4 a^2 b (2x + 4y^2) = (-4 a^2 b)(2x) + (-4 a^2 b)(4y^2)$$
   $$= -8 a^2 bx - 16 a^2 by^2$$

   (iii) The **product of two binomials is a polynomial** obtained on multiplying each term of one binomial by each term of the other binomial and adding up the products.

   $$e.g., (6x + 4y)(2x^2 - 6y) = 6x(2x^2 - 6y) + 4y(2x^2 - 6y)$$

   $$= 12 x^3 - 36 xy + 8 x^2 y - 24 y^2$$

(iv) The **product of a binomial and a polynomial** is obtained on multiplying each term of the binomial with each term of the polynomial and adding up the products.

e.g., $(4a - b)(3a^2 + 4ab - 2b^2) = 4a(3a^2 + 4ab - 2b^2) - b(3a^2 + 4ab - 2b^2)$

$$= 12a^3 + 16a^2b - 8ab^2 - 3a^2b - 4ab^2 + 2b^3$$

$$= 12a^3 + 13a^2b - 12ab^2 + 2b^3$$

8. (i) If a polynomial is divided by another polynomial then we have,

   dividend = (divisor × quotient + remainder)

   (ii) If on dividing a polynomial (called dividend), a zero remainder is obtained, then the dividing polynomial (called divisor) is a factor of the dividend. In such cases, dividend = divisor × quotient.

   (iii) Before performing long division, the divisor and dividend must be written in the standard form.

   (iv) While performing long division, like terms are written one below the other, leaving gaps whenever necessary.

   (v) The degree of the remainder is always less than that of the divisor.

9. **Special products :**

(i) $(x + a)(x + b) = x^2 + (a + b)x + ab$

(ii) $(a + b)^2 = a^2 + 2ab + b^2$ ⎫ **Perfect squares**

(iii) $(a - b)^2 = a^2 - 2ab + b^2$ (iv) $a^2 + b^2 = (a + b)^2 - 2ab = (a - b)^2 + 2ab$

(v) $x^2 + \dfrac{1}{x^2} = \left(x + \dfrac{1}{x}\right)^2 - 2 = \left(x - \dfrac{1}{x}\right)^2 + 2$ (vi) $(a + b)^2 - (a - b)^2 = 4ab$

(vii) $(a + b)^2 + (a - b)^2 = 2(a^2 + b^2)$ (viii) $(a + b)(a - b) = a^2 - b^2$

(ix) $(a + b + c)^2 = a^2 + b^2 + c^2 + 2(ab + bc + ca)$ (x) $(x + y)^3 = x^3 + y^3 + 3x^2y + 3xy^2$

(xi) $(x - y)^3 = x^3 - y^3 - 3x^2y + 3xy^2 = x^3 - y^3 - 3xy(x - y)$ $= x^3 + y^3 + 3xy(x + y)$

(xii) $(x + y)(x^2 - xy + y^2) = x^3 + y^3$ (xiii) $(x - y)(x^2 + xy + y^2) = x^3 - y^3$

(xiv) $(x + y + z)(x^2 + y^2 + z^2 - xy - yz - zx) = x^3 + y^3 + z^3 - 3xyz$

Solved Examples

Ex. 1. *What should be added to the product of $(x^2 + xy - y^2)$ and $(x^2 - xy + y^2)$ to get x^2y^2?*

Sol. $(x^2 + xy - y^2)(x^2 - xy + y^2) = x^2(x^2 - xy + y^2) + xy(x^2 - xy + y^2) - y^2(x^2 - xy + y^2)$

$$= x^4 - x^3y + x^2y^2 + x^3y - x^2y^2 + xy^3 - y^2x^2 + xy^3 - y^4$$

$$= x^4 + 2xy^3 - x^2y^2 - y^4$$

Let the algebraic expression to be added be A. Then,

$$A = x^2y^2 - (x^4 + 2xy^3 - x^2y^2 - y^4)$$

$$= x^2y^2 - x^4 - 2xy^3 + x^2y^2 + y^4$$

$$= 2x^2y^2 - x^4 - 2xy^3 + y^4.$$

Ex. 2. *What is the quotient when ($a^4 - b^4$) is divided by $a - b$?*

Sol.

$$
a - b \overline{\big)\begin{array}{l} a^3 + a^2b + ab^2 + b^3 \\ a^4 \qquad\qquad\qquad - b^4 \\ \underline{\overset{-}{a^4} \overset{+}{-} a^3b} \\ a^3b \\ \underline{\overset{-}{a^3b} \overset{+}{-} a^2b^2} \\ a^2b^2 \\ \underline{\overset{-}{a^2b^2} \overset{+}{-} ab^3} \\ ab^3 - b^4 \\ \underline{\overset{-}{ab^3} \overset{+}{-} b^4} \\ 0 \end{array}}
$$

$\therefore$ Reqd. quotient $= a^3 + a^2b + ab^2 + b^3$.

Ex. 3. *If $2x + 3y = 6\sqrt{3}$ and $2x - 3y = 6$, find the value of xy ?*

Sol. $(2x + 3y)^2 - (2x - 3y)^2 = (6\sqrt{3})^2 - 6^2 \Rightarrow (4x^2 + 12xy + 9y^2) - (4x^2 - 12xy + 9y^2) = 108 - 36$

$\Rightarrow \cancel{4x^2} + 12xy + \cancel{9y^2} - \cancel{4x^2} + 12xy - \cancel{9y^2} = 72 \Rightarrow 24\,xy = 72 \Rightarrow xy = \mathbf{3}.$

Ex. 4. *If $x = 3 + 2\sqrt{2}$, find the value $\sqrt{x} - \dfrac{1}{\sqrt{x}}$.*

Sol. $x = 3 + 2\sqrt{2} \Rightarrow \dfrac{1}{x} = \dfrac{1}{3 + 2\sqrt{2}}$

$\Rightarrow \dfrac{1}{x} = \dfrac{1}{(3 + 2\sqrt{2})} \times \dfrac{(3 - 2\sqrt{2})}{(3 - 2\sqrt{2})} = \dfrac{3 - 2\sqrt{2}}{(3^2 - (2\sqrt{2})^2)} = \dfrac{3 - 2\sqrt{2}}{9 - 8} = 3 - 2\sqrt{2}$

$\therefore x + \dfrac{1}{x} = 3 + 2\sqrt{2} + 3 - 2\sqrt{2}$

$\Rightarrow x + \dfrac{1}{x} = 6 \Rightarrow x + \dfrac{1}{x} - 2 = 6 - 2 \Rightarrow \left((\sqrt{x})^2 + \left(\dfrac{1}{\sqrt{x}} \right)^2 - 2 \times \sqrt{x} \times \dfrac{1}{\sqrt{x}} \right) = 4$

$\Rightarrow \left(\sqrt{x} - \dfrac{1}{\sqrt{x}} \right)^2 = 4 \Rightarrow \sqrt{x} - \dfrac{1}{\sqrt{x}} = \pm\sqrt{4} = \pm\mathbf{2}.$

Ex. 5. *If $a^4 + \dfrac{1}{a^4} = 322$, then find the value of $a^3 - \dfrac{1}{a^3}$.*

Sol. $a^4 + \dfrac{1}{a^4} = 322 \Rightarrow a^4 + \dfrac{1}{a^4} + 2 = 322 + 2$

$\Rightarrow \left(a^2 + \dfrac{1}{a^2} \right)^2 = 324 \Rightarrow a^2 + \dfrac{1}{a^2} = \sqrt{324} = 18$

Now, $a^2 + \dfrac{1}{a^2} - 2 = 18 - 2 \Rightarrow \left(a - \dfrac{1}{a} \right)^2 = 16 \Rightarrow a - \dfrac{1}{a} = \sqrt{16} = 4$

$\therefore \left(a - \dfrac{1}{a} \right)^3 = a^3 - \dfrac{1}{a^3} - 3 \times a \times \dfrac{1}{a} \left(a - \dfrac{1}{a} \right)$

$\Rightarrow 4^3 = a^3 - \dfrac{1}{a^3} - 3 \times 4 \Rightarrow a^3 - \dfrac{1}{a^3} = 64 + 12 = \mathbf{76}.$

Ex. 6. *Find the value of $(25a^2 + 16b^2 + 9 + 40ab - 24b - 30a)$ at $a = -1$ and $b = 2$.*

Sol. Using the identity

$(a + b + c)^2 = a^2 + b^2 + c^2 + 2ab + 2bc + 2ca,$

$25a^2 + 16b^2 + 9 + 40ab - 24b - 30a$

$\qquad = (5a)^2 + (4b)^2 + (-3)^2 + 2 \times 5a \times 4b + 2 \times 4b \times (-3) + 2 \times 5a \times (-3)$

$\qquad = (5a + 4b - 3)^2$

$\therefore$ Required value $= (5 \times (-1) + 4 \times 2 - 3)^2$

$\qquad\qquad\qquad\quad = (-5 + 8 - 3)^2 = (-8 + 8)^2 = \mathbf{0}.$

Ex. 7. *Simplify: $(x + y)^3 + (x - y)^3 + 6x(x^2 - y^2)$*

Sol. Let $(x + y) = a$ and $(x - y) = b$

Then, $(x + y)^3 + (x - y)^3 + 6x(x^2 - y^2) = (x + y)^3 + (x - y)^3 + 3 \times 2x \times (x + y)(x - y)$

$\qquad\qquad = (x + y)^3 + (x - y)^3 - 3 \times [(x + y) + (x - y)] \times (x + y)(x - y)$

$\qquad\qquad = a^3 + b^3 + 3ab(a + b)$

$\qquad\qquad = (a + b)^3 = (x + y + x - y)^3 = (2x)^3 = \mathbf{8x^3}.$

Ex. 8. *Find the product $(2a + 3b - 4c) (4a^2 + 9b^2 + 16c^2 - 6ab + 12bc + 8ca)$.*

Sol. Use the identity

$(x + y + z) (x^2 + y^2 + z^2 - xy - yz - zx) = x^3 + y^3 + z^3 - 3xyz$

Here, $x = 2a, y = 3b$ and $z = -4c$

$\therefore (2a + 3b - 4c) (4a^2 + 9b^2 + 16c^2 - 6ab + 12bc + 8ca)$

$\qquad\qquad = (2a)^3 + (3b)^3 + (-4c)^3 - 3 \times (2a) \times (3b) \times (-4c)$

$\qquad\qquad = \mathbf{8a^3 + 27b^3 - 64c^3 + 72abc.}$

Ex. 9. *If $a + b + c = 0$, then find $\dfrac{a^2}{bc} + \dfrac{b^2}{ca} + \dfrac{c^2}{ab}$.*

Sol. If $a + b + c = 0$, then $a^3 + b^3 + c^3 = 3abc$

$\therefore \dfrac{a^2}{bc} + \dfrac{b^2}{ca} + \dfrac{c^2}{ab} = \dfrac{a^3}{abc} + \dfrac{b^3}{abc} + \dfrac{c^3}{abc} = \dfrac{a^3 + b^3 + c^3}{abc} = \dfrac{0}{abc} = \mathbf{0.}$

Ex. 10. *If $x + y + z = 1$, $xy + yz + zx = -1$ and $xyz = -1$, find the value of $x^3 + y^3 + z^3$.*

Sol. $(x + y + z) (x^2 + y^2 + z^2 - xy - yz - zx) = x^3 + y^3 + z^3 - 3xyz$ $\qquad\qquad$ ... (i)

Given, $x + y + z = 1$, $xy + yz + zx = -1$ and $xyz = -1$.

To find $x^2 + y^2 + z^2$, we use the identity :

$(x + y + z)^2 = x^2 + y^2 + z^2 + 2(xy + yz + zx)$

$\Rightarrow 1 = x^2 + y^2 + z^2 + (2 \times -1)$

$\Rightarrow x^2 + y^2 + z^2 = 1 + 2 = 3$

$\therefore$ Substituting all the values in eqn (i), we get

$1 \times (3 - (-1)) = x^3 + y^3 + z^3 - (3 \times -1)$

$\Rightarrow 1 \times (3 + 1) = x^3 + y^3 + z^3 + 3$

$\Rightarrow 4 = x^3 + y^3 + z^3 + 3 \Rightarrow x^3 + y^3 + z^3 = \mathbf{1.}$

Ex. 11. *If $ab + bc + ca = 0$, then prove that* $\dfrac{1}{a^2 - bc} + \dfrac{1}{b^2 - ac} + \dfrac{1}{c^2 - ab} = 0.$

Sol. $ab + bc + ca = 0 \Rightarrow ab + ca = -bc \Rightarrow a(b+c) = -bc$

(Adding a^2 to both the sides)

$\Rightarrow a^2 + a(b+c) = a^2 - bc \Rightarrow a^2 - bc = a(a+b+c)$

Similarly, $b^2 - ac = b(a+b+c)$ and $c^2 - ab = c(a+b+c)$

$\therefore \dfrac{1}{a^2 - bc} + \dfrac{1}{b^2 - ca} + \dfrac{1}{c^2 - ab} = \dfrac{1}{a(a+b+c)} + \dfrac{1}{b(a+b+c)} + \dfrac{1}{c(a+b+c)}$

$= \dfrac{bc + ca + ab}{abc\,(a+b+c)} = \dfrac{0}{abc(a+b+c)} = \mathbf{0}.$

Ex. 12. *If $(x+y)^2 - z^2 = 4$, $(y+z)^2 - x^2 = 9$ and $(z+x)^2 - y^2 = 36$, then find the value of $x+y+z$.*

Sol. $(x+y)^2 - z^2 = 4 \Rightarrow (x+y+z)(x+y-z) = 4$

$\Rightarrow (x+y-z) = \dfrac{4}{x+y+z}$ ... (i)

$(y+z)^2 - x^2 = 9 \Rightarrow (y+z+x)(y+z-x) = 9$

$\Rightarrow (y+z-x) = \dfrac{9}{x+y+z}$ ... (ii)

$(z+x)^2 - y^2 = 36 \Rightarrow (z+x+y)(z+x-y) = 36$

$\Rightarrow (z+x-y) = \dfrac{36}{x+y+z}$ ... (iii)

Adding (i), (ii) and (iii) we get

$(x+y-z)+(y+z-x)+(z+x-y) = \dfrac{4}{x+y+z} + \dfrac{9}{x+y+z} + \dfrac{36}{x+y+z}$

$\Rightarrow (x+y+z) = \dfrac{49}{(x+y+z)} \Rightarrow (x+y+z)^2 = 49$

$\Rightarrow x+y+z = 7.$

Question Bank–7

1. Which of the following expressions are exactly equal in value ?

1. $(3x-y)^2 - (5x^2 - 2xy)$

2. $(2x-y)^2$

3. $(2x+y)^2 - 2xy$

4. $(2x+3y)^2 - 8y(2x+y)$

(a) 1 and 2 only (b) 1, 2 and 3 only
(c) 2 and 4 only (d) 1, 2 and 4 only

2. What will be the value of

$n^4 - 10n^3 + 36n^2 - 49n + 24$ if $n = 1$?

(a) 21 (b) 2
(c) 1 (d) 22

3. If $x+y+z = 0$, then $x^2 + xy + y^2$ equals :

(a) $y^2 + yz + z^2$ (b) $y^2 - yz + z^2$
(c) $z^2 - xy$ (d) $z^2 + zx + x^2$

4. If $a^3 = 117 + b^3$ and $a = 3 + b$, then the value of $a + b$ is :

(a) ± 7 (b) 49
(c) 0 (d) ± 13

5. If $a + b + c = 0$, then what is the value of $(a+b-c)^3 + (c+a-b)^3 + (b+c-a)^3$?

(a) $-8(a^3 + b^3 + c^3)$ (b) $a^3 + b^3 + c^3$
(c) $24\,abc$ (d) $-24\,abc$

6. If $a^{1/3} + b^{1/3} + c^{1/3} = 0$, then the value of $(a+b+c)^3$ will be :

(a) $9a^2b^2c^2$ (b) $3abc$

(c) $6abc$ (d) $27abc$

7. What is the value of the following expression ?

$(1+x)(1+x^2)(1+x^4)(1+x^8)(1-x)$

(a) $1+x^{16}$ (b) $1-x^{16}$

(c) $x^{16}-1$ (d) x^8+1

8. An expression, which when divided by $(x+1)$ gives $(x^2 - x + 1)$ as quotient and 3 as remainder is equal to :

(a) $x^3 - 2$ (b) $x^3 - 1$

(c) $x^3 + 2$ (d) $(x^3 + 4)$

9. The remainder obtained on dividing the polynomial $3x^4 - 4x^3 - 3x - 1$ by $(x-1)$ is :

(a) 0   (b) 5

(c) –5   (d) 5

10. If $a = \dfrac{x}{x+y}$ and $b = \dfrac{y}{x-y}$, then $\dfrac{ab}{a+b}$ is equal to

(a) $\dfrac{xy}{x^2+y^2}$ (b) $\dfrac{x^2+y^2}{xy}$

(c) $\dfrac{x}{x+y}$ (d) $\left(\dfrac{y}{x+y}\right)^2$

11. If $a^2 + b^2 = 117$ and $ab = 54$, then $\dfrac{a+b}{a-b}$ is

(a) 3   (b) 5

(c) 6   (d) 4

12. If $a^2 + \dfrac{1}{a^2} = 10$, then the value of $a^4 + \dfrac{1}{a^4}$ is

(a) 90   (b) 98

(c) 100   (d) 102

13. If $a^4 + \dfrac{1}{a^4} = 1154$, then the value of $a^3 + \dfrac{1}{a^3}$ is

(a) 198   (b) 216

(c) 200   (d) 196

14. If $x + \dfrac{1}{x} = 3$, then the value of $x^6 + \dfrac{1}{x^6}$ is

(a) 927   (b) 414

(c) 364   (d) 322

15. If $a + b + c = 0$, then the value of $a^2(b+c) + b^2(c+a) + c^2(a+b)$ is :

(a) abc (b) $3abc$

(c) $-3abc$ (d) 0

16. If $x = 3^{1/3} + 3^{-1/3}$, then $3x^3 - 10$ is equal to :

(a) $-3x$ (b) $3x$

(c) $-9x$ (d) $9x$

17. The value of $\dfrac{(a-b)^3 + (b-c)^3 + (c-a)^3}{(a-b)(b-c)(c-a)}$ is

(a) 1   (b) 3

(c) $\dfrac{1}{3}$ (d) zero

18. If $a + b + c = 11$ and $ab + bc + ca = 20$, then the value of the expression $a^3 + b^3 + c^3 - 3abc$ will be

(a) 121   (b) 341

(c) 671   (d) 781

19. $\dfrac{x^2 - (y-z)^2}{(x+z)^2 - y^2} + \dfrac{y^2 - (x-z)^2}{(x+y)^2 - z^2} + \dfrac{z^2 - (x-y)^2}{(y+z)^2 - x^2} = ?$

(a) –1   (b) 0

(c) 1   (d) 2

20. If $2a - \dfrac{1}{2a} = 3$, then $16a^4 + \dfrac{1}{16a^4}$ is equal to :

(a) 11   (b) 119

(c) 117   (d) 121

21. The value of the product $\left(7 - \dfrac{12}{x}\right)\left(49 + \dfrac{84}{x} + \dfrac{144}{x^2}\right)$ at $x = 2$ is

(a) 0   (b) 559

(c) 127   (d) 128

22. The value of $64a^3 + 48a^2b + 12a^2b + b^3$ at $a = 1$ and $b = -1$ is

(a) 25   (b) 125

(c) 27   (d) 54

23. Evaluate:

$$\dfrac{(0.43)^3 + (1.47)^3 + (1.1)^3 - 3 \times 0.43 \times 1.47 \times 1.1}{(0.43)^2 + (1.47)^2 + (1.1)^2 - 0.43 \times 1.43 - 0.43 \times 1.1 - 1.47 \times 1.1}$$

(a) 1.90   (b) 2.87

(c) 3   (d) 3.47

24. If $\dfrac{x^2 + y^2 + z^2 - 64}{xy - yz - zx} = -2$ and $x + y = 3z$, then the value of z is
(a) 2   (b) 3
(c) 4   (d) –2

25. If $(a^2 + b^2)^3 = (a^3 + b^3)^2$ and $ab \neq 0$, then $\left(\dfrac{a}{b} + \dfrac{b}{a}\right)^6$ is equal to
(a) $\dfrac{a^6 + b^6}{a^3 b^3}$ (b) $\dfrac{64}{729}$
(c) 1   (d) $\dfrac{a^6 + a^3 b^3 + b^6}{a^2 b^4 + a^4 b^2}$

Answers

1. (d)	**2.** (b)	**3.** (c)	**4.** (a)	**5.** (d)	**6.** (d)	**7.** (b)	**8.** (d)	**9.** (c)	**10.** (a)
11. (b)	**12.** (b)	**13.** (a)	**14.** (d)	**15.** (c)	**16.** (d)	**17.** (b)	**18.** (c)	**19.** (c)	**20.** (b)
21. (c)	**22.** (c)	**23.** (c)	**24.** (c)	**25.** (b)					

Hints and Solutions

1. (d) $(3x - y)^2 - (5x^2 - 2xy)$
$$= 9x^2 - 6xy + y^2 - 5x^2 + 2xy$$
$$= \mathbf{4x^2 - 4xy + y^2}$$
$$(2x - y)^2 = \mathbf{4x^2 - 4xy + y^2}$$
$$(2x + y)^2 - 2xy = 4x^2 + 4xy + y^2 - 2xy$$
$$= \mathbf{4x^2 + 2xy + y^2}$$
$$(2x + 3y)^2 - 8y(2x + y)$$
$$= 4x^2 + 12xy + 9y^2 - 16xy - 8y^2$$
$$= \mathbf{4x^2 - 4xy + y^2}$$

2. (b) $n^4 - 10n^3 + 36n^2 - 49n + 24$ at $n = 1$
$$= (1)^4 - 10 \times (1)^3 + 36 \times (1)^2 - 49 \times 1 + 24$$
$$= 1 - 10 + 36 - 49 + 24 = 61 - 59 = \mathbf{2.}$$

3. (c) $x + y + z = 0 \Rightarrow x + y = -z$
$$\therefore \quad x^2 + xy + y^2 = x^2 + 2xy + y^2 - xy$$
(Note the step)
$$= (x + y)^2 - xy = (-z)^2 - xy$$
$$= \mathbf{z^2 - xy.}$$

4. (a) $a^3 = 117 + b^3 \Rightarrow a^3 - b^3 = 117$
$$a = 3 + b \Rightarrow a - b = 3$$
Now, $(a - b)^3 = a^3 - b^3 - 3ab(a - b)$
$$\Rightarrow \quad 3^3 = 117 - 3ab \times 3$$
$$\Rightarrow \quad 27 = 117 - 9ab \Rightarrow 9ab = 90 \Rightarrow ab = 10$$
$\therefore$ Using the identity $(a + b)^2 = (a - b)^2 + 4ab$, we have

$$(a + b)^2 = 3^2 + 4 \times 10 = 9 + 40 = 49$$
$$\Rightarrow \quad a + b = \sqrt{49} = \mathbf{\pm 7.}$$

5. (d) $a + b + c = 0 \Rightarrow a + b = -c, c + a = -b$
$$b + c = -a$$
$$\therefore \quad (a + b - c)^3 + (c + a - b)^3 + (b + c - a)^3$$
$$= (-c - c)^3 + (-b - b)^3 + (-a - a)^3$$
$$= (-2c)^3 + (-2b)^3 + (-2a)^3$$
$$= -8c^3 - 8b^3 - 8a^3 = -8(a^3 + b^3 + c^3)$$
$$= -8 \times 3abc = \mathbf{-24abc}$$
$$\because \quad a + b + c = 0 \Rightarrow a^3 + b^3 + c^3 = 3abc.$$

6. (d) Since $a^{1/3} + b^{1/3} + c^{1/3} = 0$,
$$(a^{1/3})^3 + (b^{1/3})^3 + (c^{1/3})^3 = 3a^{1/3} b^{1/3} c^{1/3}$$
$$\Rightarrow a + b + c = 3(abc)^{1/3} \Rightarrow (a + b + c)^3 = \mathbf{27abc}$$

7. (b) $(1 + x)(1 + x^2)(1 + x^4)(1 + x^8)(1 - x)$
$$= (1 + x)(1 - x)(1 + x^2)(1 + x^4)(1 + x^8)$$
$$= (1 - x^2)(1 + x^2)(1 + x^4)(1 + x^8)$$
$$= (1 - x^4)(1 + x^4)(1 + x^8)$$
$$= (1 - x^8)(1 + x^8) = \mathbf{(1 - x^{16})}$$

8. (d) Dividend = Divisor × Quotient + Remainder
$$= (x + 1) \times (x^2 - x + 1) + 3$$
$$= x(x^2 - x + 1) + (x^2 - x + 1) + 3$$
Now solve.

9. (c)

$$
\begin{array}{r}
3x^3 - x^2 - x - 4 \\
x-1 \overline{)\,3x^4 - 4x^3 \qquad\quad -3x - 1} \\
\underline{3x^4 - 3x^3} \\
\quad -x^3 \\
\quad\; \underline{-x^3 + x^2} \\
\qquad\quad -x^2 - 3x \\
\qquad\quad \underline{-x^2 + x} \\
\qquad\qquad\quad -4x - 1 \\
\qquad\qquad\quad \underline{-4x + 4} \\
\qquad\qquad\qquad -5
\end{array}
$$

10. (a) $\quad ab = \left(\dfrac{x}{x+y}\right) \times \left(\dfrac{y}{x-y}\right)$

$$= \dfrac{xy}{(x+y)(x-y)} = \dfrac{xy}{x^2 - y^2}$$

$$a+b = \dfrac{x}{x+y} + \dfrac{y}{x-y}$$

$$= \dfrac{x(x-y) + y(x+y)}{(x+y)(x-y)} = \dfrac{x^2 - xy + yx + y^2}{x^2 - y^2}$$

$$= \dfrac{x^2 + y^2}{x^2 - y^2}$$

$$\therefore \; \dfrac{ab}{a+b} = \dfrac{xy}{x^2 - y^2} \div \dfrac{x^2 + y^2}{x^2 - y^2}$$

$$= \dfrac{xy}{x^2 + y^2}.$$

11. (b) $(a+b)^2 = a^2 + b^2 + 2ab = 117 + 108 = 225$

$\Rightarrow a + b = 15$

$\quad (a-b)^2 = a^2 + b^2 - 2ab = 117 - 108 = 9$

$\Rightarrow a - b = 3$

$\therefore \; \dfrac{a+b}{a-b} = \dfrac{15}{3} = \mathbf{5}.$

12. (b) $a^2 + \dfrac{1}{a^2} = 10 \Rightarrow \left(a^2 + \dfrac{1}{a^2}\right)^2 = 10^2$

$\Rightarrow a^4 + 2 + \dfrac{1}{a^4} = 100 \Rightarrow a^4 + \dfrac{1}{a^4} = 98.$

13. (a) $a^4 + \dfrac{1}{a^4} = 1154 \Rightarrow a^4 + 2 + \dfrac{1}{a^4} = 1154 + 2$

$\Rightarrow \left(a^2 + \dfrac{1}{a^2}\right)^2 = 1156 \Rightarrow a^2 + \dfrac{1}{a^2} = 34$

$\Rightarrow a^2 + 2 + \dfrac{1}{a^2} = 34 + 2 \Rightarrow \left(a + \dfrac{1}{a}\right)^2 = 36$

$\Rightarrow a + \dfrac{1}{a} = 6$

$\therefore \left(a + \dfrac{1}{a}\right)^3 = 216$

$\Rightarrow a^3 + \dfrac{1}{a^3} + 3a \times \dfrac{1}{a}\left(a + \dfrac{1}{a}\right) = 216$

$\therefore \; a^3 + \dfrac{1}{a^3} + 3 \times 6 = 216$

$\Rightarrow a^3 + \dfrac{1}{a^3} = 216 - 18 = \mathbf{198}.$

14. (d) $x + \dfrac{1}{x} = 3 \Rightarrow \left(x + \dfrac{1}{x}\right)^3 = 3^3$

$\Rightarrow x^3 + \dfrac{1}{x^3} + 3\left(x + \dfrac{1}{x}\right) = 27$

$\Rightarrow x^3 + \dfrac{1}{x^3} + 3 \times 3 = 27 \Rightarrow x^3 + \dfrac{1}{x^3} = 18$

Now squaring both the sides,

$$\left(x^3 + \dfrac{1}{x^3}\right)^2 = 18^2 \Rightarrow x^6 + 2 + \dfrac{1}{x^6} = 324$$

$\Rightarrow x^6 + \dfrac{1}{x^6} = \mathbf{322}.$

15. (c) $a + b + c = 0 \Rightarrow b + c = -a,\; c + a = -b,$
$a + b = -c$

$\quad a^2(b+c) + b^2(c+a) + c^2(a+b)$

$\qquad = a^2 \times -a + b^2 \times -b + c^2 \times -c$

$\qquad = -a^3 - b^3 - c^3 = -(a^3 + b^3 + c^3)$

If $a + b + c = 0$, then $a^3 + b^3 + c^3 = 3abc$

$\Rightarrow -(a^3 + b^3 + c^3) = -3abc$

16. (d) $x = 3^{1/3} + 3^{-1/3} \Rightarrow x - 3^{1/3} = 3^{-1/3}$

$\Rightarrow (x - 3^{1/3})^3 = (3^{-1/3})^3$

$\Rightarrow x^3 - 3.x.3^{1/3}(x - 3^{1/3}) - (3^{1/3})^3 = 3^{-1}$

$\Rightarrow x^3 - 3.x.3^{1/3}.3^{-1/3} - 3 = \dfrac{1}{3} \Rightarrow x^3 - 3x = 3 + \dfrac{1}{3}$

$\Rightarrow 3x^3 - 9x = 10 \Rightarrow 3x^3 - 10 = 9x$

17. (b) Since $(a-b) + (b-c) + (c-a) = 0$

$\therefore \; (a-b)^3 + (b-c)^3 + (c-a)^3$

$\qquad\qquad\qquad = 3(a-b)(b-c)(c-a)$

$\Rightarrow \dfrac{(a-b)^3 + (b-c)^3 + (c-a)^3}{(a-b)(b-c)(c-a)}$

$\qquad\qquad\qquad \dfrac{3(a-b)(b-c)(c-a)}{(a-b)(b-c)(c-a)} = \mathbf{3}.$

18. (c) $(a+b+c)^2 = a^2+b^2+c^2+2(ab+bc+ca)$

$\Rightarrow a^2+b^2+c^2 = 11^2 - 2\times 20 = 121 - 40 = 81$

Also, we know that

$(a+b+c)(a^2+b^2+c^2-ab-bc-ca)$
$$= a^3+b^3+c^3-3abc$$

$\Rightarrow a^3+b^3+c^3-3abc = 11\times(80-20)$

$\Rightarrow a^3+b^3+c^3-3abc = 11\times 61 = \mathbf{671.}$

19. (c) Given exp. =

$$\frac{(x-y+z)(x+y-z)}{(x+z-y)(x+z+y)} + \frac{(y-x+z)(y+x-z)}{(x+y-z)(x+y+z)}$$
$$+ \frac{(z-x+y)(z+x-y)}{(y+z-x)(y+z+x)}$$

$$= \frac{(x+y-z)+(y-x+z)+(z+x-y)}{(x+y+z)}$$

$$= \frac{x+y+z}{x+y+z} = \mathbf{1.}$$

20. (b) Square both the sides of $2a - \dfrac{1}{2a} = 3$, find the value of $4a^2 + \dfrac{1}{4a^2}$. Then, again square both the sides to find the value of $16a^4 + \dfrac{1}{16a^4}$.

21. (c) $\left(7 - \dfrac{12}{x}\right)\left(49 + \dfrac{84}{x} + \dfrac{144}{x^2}\right) = (7)^3 - \left(\dfrac{12}{x}\right)^3$

$$= 343 - \left(\frac{12}{2}\right)^3 = 343 - 216 = 127$$

22. (c) $64a^3 + 48a^2b + 12a^2b + b^3$

$= (4a+b)^3 \; [\because (a+b)^2 = a^3+b^3+3a^2b+3ab^2]$

$\therefore$ Reqd. value $= (4-1)^3 = 3^3 = \mathbf{27.}$

23. (c) Reqd. exp. is of type

$$\frac{a^3+b^3+c^3-3abc}{a^2+b^2+c^2-ab-bc-ca}$$

You know that

$(a+b+c)(a^2+b^2+c^2-ab-bc-ca)$
$$= a^3+b^3+c^3-3abc$$

$\Rightarrow a+b+c = \dfrac{a^3+b^3+c^3-3abc}{a^2+b^2+c^2-ab-bc-ca}$

$\therefore$ Reqd. exp. $= 0.43 + 1.47 + 1.1 = \mathbf{3.}$

24. (c) $\dfrac{x^2+y^2+z^2-64}{xy-yz-zx} = -2$

$\Rightarrow x^2+y^2+z^2 = -2xy+2yz+2zx+64$

Now,

$(x+y+z)^2 = x^2+y^2+z^2+2xy+2yz+2zx$

$\Rightarrow (3z+z)^2 = -2xy+2yz+2zx+64$
$$+2xy+2yz+2zx$$

$\Rightarrow 16z^2 = 4yz+4xz+64$

$\Rightarrow 16z^2 = 4z(x+y)+64 \Rightarrow 16z^2 = 4z(3z)+64$

$\Rightarrow 16z^2 - 12z^2 = 64$

$\Rightarrow 4z^2 = 64 \Rightarrow z^2 = 16 \Rightarrow z = \mathbf{4.}$

25. (b) $(a^2+b^2)^3 = (a^3+b^3)^2$

$\Rightarrow a^6+b^6+3a^2b^2(a^2+b^2) = a^6+b^6+2a^3b^3$

$\Rightarrow a^2+b^2 = \dfrac{2}{3}ab$ ... (i)

Now, $\left(\dfrac{a}{b}+\dfrac{b}{a}\right)^6 = \left(\dfrac{a^2+b^2}{ab}\right)^6 = \left(\dfrac{\frac{2}{3}ab}{ab}\right)^6$

$$= \left(\frac{2}{3}\right)^6 = \frac{\mathbf{64}}{\mathbf{729}}.$$

Self Assessment Sheet–7

1. Each of the following is a term in the polynomial which is the product of $(x+1)$, $(3x^2+6x)$ and $(2x^2+6x-1)$ except.

  (a) $6x^5$ (b) $36x^4$

  (c) $-6x$ (d) -1

2. If $4x^2 + x(m+1) + 1$ is a perfect square, then a value of m is :

  (a) -5 (b) 5

  (c) 3 (d) -3

3. $\dfrac{1}{2}(a+b)(a^2+b^2) + \dfrac{1}{2}(a-b)(a^2-b^2)$ is equal to

  (a) $a^3 - b^3$

  (b) $a^3 + 3a^2b + 3ab^2 + b^3$

  (c) $a^3 + b^3$

  (d) $a^3 - 3ab(a+b) - b^3$

4. If $x = \dfrac{1+2y}{2+y}$ and $y = \dfrac{1+2t}{2+t}$, then x equals

(a) $\dfrac{1+2t}{3+t}$

(b) $\dfrac{3+2t}{2+3t}$

(c) $\dfrac{5t+4}{4t+5}$

(d) $\dfrac{5t+6}{6t+5}$

5. Simplify : $\dfrac{3y(x-y)-2x(y-2x)}{7x(x-y)-3(x^2-y^2)}$

(a) $\dfrac{x-y}{x+y}$

(b) 1

(c) $\dfrac{x+y}{x-y}$

(d) 0

6. The difference between any number of four digits and the number formed by using the digits in the reversed order is exactly divisible by :

(a) 11   (b) 10

(c) 9   (d) 5

7. If " from the square of a half the sum of two numbers we subtract the square of a half their difference ", the result is the :

(a) sum of the two numbers

(b) quotient of the two numbers

(c) difference of the two numbers

(d) product of the two numbers

8. The value of the expression

$\dfrac{(x^2-y^2)^3+(y^2-z^2)^3+(z^2-x^2)^3}{(x-y)^3+(y-z)^3+(z-x)^3}$ is

(a) $(x^2-y^2)(y^2-z^2)(z^2-x^2)$

(b) $3(x-y)(y-z)(z-x)$

(c) $(x+y)(y+z)(z+x)$

(d) $3(x+y)(y+z)(z+x)$

9. If $x=2a-1$, $y=(2a-2)$ and $z=3-4a$, then the value of $x^3+y^3+z^3$ will be :

(a) $6(3-13a+18a^2-8a^3)$

(b) $6(3+13a-18a^2+8a^3)$

(c) $6(3+13a+18a^2-8a^3)$

(d) $6(3-13a-18a^2-8a^3)$

10. If $(5x^2+14x+2)^2-(4x^2-5x+7)^2$ is divided by x^2+x+1, then the quotient q and the remainder r are given by :

(a) $q=(x^2+19x-5)$, $r=1$

(b) $q=9(x^2+19x-5)$, $r=0$

(c) $q=(x^2+19x-5)$, $r=0$

(d) $q=9(x^2+19x-5)$, $r=1$

Answers

1. (d) 2. (c) 3. (c) 4. (c) 5. (c) 6. (c) 7. (d) 8. (c) 9. (a) 10. (c)

Chapter 8

FACTORIZATION OF ALGEBRAIC EXPRESSIONS

KEY FACTS

1. When an algebraic expression is the product of two or more expressions, then each of these expressions is called a factor of the given expression.

   For example,
   (i) $7, a, b, 7a, 7b, ab$ are all factors of the expression $7ab$.
   (ii) a and $(x^2 + y^2)$ are factors of the expression $a(x^2 + y^2)$

2. The process of writing a given algebraic expression as a product of two or more factors is called **factorisation**.

   For example, $ax + bx + x^2$ can be factorised as $x(a + b + x)$

3. The greatest common factor of two or more monomials is the product of the greatest common factors of the numerical coefficients and the common letters with the smallest powers.

   For example, greatest common factor of $15x^2y^5$ and $10x^7y^9$ is $5x^2y^5$.

4. A binomial may be factorised by taking out the greatest common factor of the terms of the binomial.

   For example, $12a^2b^6 + 21ab^7 = 3ab^6(4a + 7b)$

5. Sometimes, it is not possible to take out a common factor from all the terms of a given expression. we group the terms to take out a common factor.

   For example, $3px + 4qy + 4qx + 3py = x(3p + 4q) + y(3p + 4q)$
   $$= (3p + 4q)(x + y)$$

6. If the given expression is the difference of two squares, then we use the formula $a^2 - b^2 = (a + b)(a - b)$
   For example, $9a^2x^2 - 25 b^2y^2 = (3ax)^2 - (5by)^2$
   $$= (3ax + 5by)(3ax - 5by)$$

7. If the given expression is a perfect square, we use the following formulae :

   (i) $a^2 + 2ab + b^2 = (a + b)^2$
   (ii) $a^2 - 2ab + b^2 = (a - b)^2$

   For example, $16x^2 + 40xy + 25y^2 = (4x)^2 + 2 \times 4x \times 5y + (5y)^2$
   $$= (4x + 5y)^2$$
   $$4a^2 - 28ab + 49b^2 = (2a)^2 - 2 \times 2a \times 7b + (7b)^2$$
   $$= (2a - 7b)^2$$

8. **Factorisation by splitting the middle term :** Let the quadratic polynomial be $x^2 + px + q$. Then $x^2 + px + q = (x + a)(x + b) = x^2 + (a + b)x + ab$, where $(x + a)$ and $(x + b)$ are two linear factors of $x^2 + px + q$. So factorise the given quadratic polynomial we have to find two numbers a and b such the $p = a + b$, i.e., coefficient of x = sum of the two numbers and $q = a + b$, i.e., constant term = product of the two numbers.

For example,

(i) $x^2 + 10x + 16 = (x+2)(x+8)$ $\because 8 + 2 = 10, 8 \times 2 = 16$

(ii) $x^2 + 5x - 84 = (x+12)(x-7)$ $\because 12 + (-7) = 5, 12 \times (-7) = -84$

(iii) $x^2 - 7x + 10 = (x-5)(x-2)$ $\because (-5) + (-2) = -7, (-5) \times (-2) = 10$

(iv) $x^2 - x - 12 = (x-4)(x+3)$ $\because (-4) + 3 = -1, (-4) \times 3 = -12$

9. Factorisation of the quadratic polynomial $ax^2 + bx + c$:

To factorise the quadratic polynomial $ax^2 + bx + c$, we have to find members p and q such that $p + q = b$, $pq = ac$. Split the middle term bx as $px + qx$.

Factorise by grouping , $ax^2 + bx + c = ax^2 + px + qx + pq$

For example,

(i) $6x^2 + 19x + 3 = 6x^2 + 18x + x + 3$
$= 6x(x+3) + 1(x+3)$
$= (6x+1)(x+3)$

(ii) $2x^2 - 3x - 20 = 2x^2 - 8x + 5x - 20$
$= 2x(x-4) + 5(x-4)$
$= (x-4)(2x+5)$

10. Some important formulae

> (i) $a^2 + b^2 + c^2 + 2ab + 2bc + 2ca = (a+b+c)^2$ (ii) $a^3 + b^3 + 3a^2b + 3ab^2 = (a+b)^3$
>
> (iii) $a^3 - b^3 - 3a^2b + 3ab^2 = (a-b)^3$ (iv) $a^3 + b^3 = (a+b)(a^2 - ab + b^2)$
>
> (v) $a^3 - b^3 = (a-b)(a^2 + ab + b^2)$
>
> (vi) $a^3 + b^3 + c^3 - 3abc = (a+b+c)(a^2 + b^2 + c^2 - ab - bc - ca)$
>
> (vii) If $a + b + c = 0$, then $a^3 + b^3 + c^3 = 3abc$

Solved Examples

Ex. 1. *Factorise :* $(x^2 + y^2 - z^2)^2 - 4x^2y^2$

Sol. $(x^2 + y^2 - z^2)^2 - 4x^2y^2 = (x^2 + y^2 - z^2)^2 - (2xy)^2$

     $= (x^2 + y^2 - z^2 + 2xy)(x^2 + y^2 - z^2 - 2xy)$ $[\because a^2 - b^2 = (a-b)(a+b)]$

     $= (x^2 + 2xy + y^2 - z^2)(x^2 - 2xy + y^2 - z^2)$

     $= \left[(x+y)^2 - z^2\right]\left[(x-y)^2 - z^2\right]$

     $= (x+y+z)(x+y-z)(x-y+z)(x-y-z).$

Ex. 2. *Factorise :* $x^4 + x^2y^2 + y^2.$

Sol. $x^4 + x^2y^2 + y^4 = x^4 + 2x^2y^2 + y^4 - x^2y^2$ (**Note :** Add and subtract x^2y^2)

     $= (x^2 + y^2)^2 - x^2y^2$

     $= (x^2 + y^2)^2 - (xy)^2$

     $= (x^2 + y^2 + xy)(x^2 + y^2 - xy).$

Ex. 3. *Factorise :* $16(2x-y)^2 - 24(4x^2 - y^2) + 9(2x+y)^2$

Sol. $16(2x-y)^2 - 24(4x^2 - y^2) + 9(2x+y)^2$

     $= (4(2x-y))^2 - 2 \times 4 \times 3x(2x-y) \times (2x+y) + (3(2x+y))^2$

     $= [4(2x-y) - 3(2x+y)]^2$ $[\because a^2 - 2ab + b^2 = (a-b)^2]$

     $= [8x - 4y - 6x - 3y]^2$

     $= (2x - 7y)(2x - 7y).$

Ex. 4. *Factorise :* $4b^2 - 1 - 2a - a^2.$

Sol. $4b^2 - 1 - 2a - a^2 = 4b^2 - (1 + 2a + a^2)$

     $= (2b)^2 - (1+a)^2$

     $= (2b + 1 + a)(2b - 1 - a).$

Ex. 5. *Factorise : $x^8 - x^4 - 72$*

Sol. Let $x^4 = a$. Then, $x^8 - x^4 - 72 \quad = a^2 - a - 72$

$$= a^2 - 9a + 8a - 72 = a(a-9) + 8(a-9)$$
$$= (a-9)(a+8) = (x^4 - 9)(x^4 + 8)$$
$$= ((x)^2 - 3^2)(x^4 + 8) = \mathbf{(x^2 + 3)(x^2 - 3)(x^4 + 8)}.$$

Ex. 6. *Factorise : $(x^2 - 5x)(x^2 - 5x - 20) + 84$*

Sol. $(x^2 - 5x)(x^2 - 5x - 20) + 84 \quad = (x^2 - 5x)(x^2 - 5x) - 20(x^2 - 5x) + 84$

$$= (x^2 - 5x)^2 - 20(x^2 - 5x) + 84$$

Let $x^2 - 5x = a$. Then, the given expression

$$= a^2 - 20a + 84 = a^2 - 6a - 14a + 84$$
$$= a(a-6) - 14(a-6) = (a-6)(a-14)$$
$$= (x^2 - 5x - 6)(x^2 - 5x - 14)$$
$$= (x^2 - 6x + x - 6)(x^2 - 7x + 2x - 14)$$
$$= [x(x-6) + 1(x-6)][x(x-7) + 2(x-7)]$$
$$= \mathbf{(x-6)(x+1)(x-7)(x+2)}.$$

Ex. 7. *Factorise : $2x^2 - \dfrac{5}{6}x + \dfrac{1}{12}$*

Sol. $2x^2 - \dfrac{5}{6}x + \dfrac{1}{12} = \dfrac{24x^2 - 10x + 1}{12}$

$$= \frac{1}{12}(24x^2 - 10x + 1) = \frac{1}{12}(24x^2 - 6x - 4x + 1)$$
$$= \frac{1}{12}\{6x(4x-1) - 1(4x-1)\} = \frac{1}{12}(6x-1)(4x-1)$$
$$= \left(\frac{x}{2} - \frac{1}{12}\right)(4x-1).$$

Ex. 8. *Factorise : $a^{12} - b^{12}$*

Sol. $a^{12} - b^{12} = (a^6)^2 - (b^6)^2 = (a^6 + b^6)(a^6 - b^6)$ $\qquad [\because a^2 - b^2 = (a-b)(a+b)]$

$$= (a^6 + b^6)((a^3)^2 - (b^3)^2)$$
$$= (a^6 + b^6)(a^3 + b^3)(a^3 - b^3)$$
$$= \mathbf{(a^6 + b^6)(a+b)(a^2 - ab + b^2)(a-b)(a^2 + ab + b^2)}.$$

Ex. 9. *Factorise : $\dfrac{1}{64}a^3 - \dfrac{1}{16}a^2b + \dfrac{1}{12}ab^2 - \dfrac{1}{27}b^3$*

Sol. $\dfrac{1}{64}a^3 - \dfrac{1}{16}a^2b + \dfrac{1}{12}ab^2 - \dfrac{1}{27}b^3 = \left(\dfrac{1}{4}a\right)^3 - 3\times\left(\dfrac{1}{4}a\right)^2\times\dfrac{1}{3}b + 3\times\dfrac{1}{4}a\times\left(\dfrac{1}{3}b\right)^2 - \left(\dfrac{1}{3}b\right)^3$

$$= \left(\frac{1}{4}a - \frac{1}{3}b\right)^3.$$ $\qquad [\because (a-b)^3 = a^3 - 3a^2b + 3ab^2 - b^3]$

Ex. 10. *Factorise : $a^6 - 7a^3 - 8$*

Sol. $(a^3)^2 - 7a^3 - 8 = (a^3)^2 - 8a^3 + a^3 - 8$

$$= a^3(a^3 - 8) + 1(a^3 - 8) = (a^3 - 8)(a^3 + 1)$$
$$= \mathbf{(a-2)(a^2 + 2a + 4)(a+1)(a^2 - a + 1)}.$$

Ex. 11. *Factorise :* $2\sqrt{2}\,x^3 + 3\sqrt{3}\,y^3 + \sqrt{5}\,(5 - 3\sqrt{6}\,xy)$

Sol. $2\sqrt{2}\,x^3 + 3\sqrt{3}\,y^3 + \sqrt{5}(5 - 3\sqrt{6}\,xy) = 2\sqrt{2}\,x^3 + 3\sqrt{3}\,y^3 + 5\sqrt{5} - 3\sqrt{6} \times \sqrt{5}\,xy$

$$= (\sqrt{2}x)^3 + (\sqrt{3}y)^3 + (\sqrt{5})^3 - 3 \times \sqrt{2}x \times \sqrt{3}y \times \sqrt{5}$$

$$= (\sqrt{2}\,x + \sqrt{3}\,y + \sqrt{5})\,(2x^2 + 3y^2 + 5^2 - \sqrt{6}xy - \sqrt{15}y - \sqrt{10}x).$$

$$[\because a^3 + b^3 + c^3 - 3abc = (a + b + c)\,(a^2 + b^2 + c^2 - ab - bc - ca)]$$

Ex. 12. *Factorise :* $(5x - y)^3 + (y - 4z)^3 + (4z - 5x)^3$

Sol. Let $a = 5x - y$, $b = y - 4z$, $c = 4z - 5x$

$a + b + c = 5x - y + y - 4z + 4z - 5x = 0$

$\therefore\ a^3 + b^3 + c^3 = 3abc$

$\Rightarrow\ (5x - y)^3 + (y - 4z)^3 + (4z - 5x)^3 = 3(5x - y)(y - 4z)(4z - 5x)$

Ex. 13. *Factorise :* $x^8 + x^4 + 1$

Sol. $x^8 + x^4 + 1 = x^8 + 2x^4 + 1 - x^4$

$$= (x^4 + 1)^2 - (x^2)^2 = (x^4 + 1 + x^2)(x^4 + 1 - x^2)$$

$$= (x^4 + 2x^2 + 1 - x^2)\,(x^4 - x^2 + 1)$$

$$= \left[(x^2 + 1)^2 - x^2\right](x^4 - x^2 + 1)$$

$$= (x^2 + 1 + x)\,(x^2 + 1 - x)\,(x^4 - x^2 + 1).$$

Ex. 14. *What is the simplified value of :* $(1.25)^3 - 2.25\,(1.25)^2 + 3.75\,(0.75)^2 - (0.75)^3$

Sol. $(1.25)^3 - 2.25\,(1.25)^2 + 3.75\,(0.75)^2 - (0.75)^3$

$$= (1.25)^3 - 3 \times 0.75 \times (1.25)^2 + 3 \times 1.25 \times (0.75)^2 - (0.75)^3$$

$$= (1.25 - 0.75)^3 = (0.5)^3 = \left(\frac{1}{2}\right)^3 = \frac{1}{8}. \qquad [\because a^3 - 3a^2b + 3ab^2 - b^3 = (a - b)^3]$$

Question Bank–8

1. Factorise : $ab\,(c^2 + 1) + c\,(a^2 + b^2)$
  (a) $(ab + c)\,(a + bc)$ (b) $(ac + b)\,(ab + c)$
  (c) $(a + bc)\,(ac + b)$ (d) $(a + b)\,(ac + b)$

2. The factors of $9a^2 - 6\sqrt{5}a + 5$ are
  (a) $(3a + \sqrt{5})(3a - \sqrt{5})$ (b) $(3a - 5)\,(3a - 5)$
  (c) $(3a - \sqrt{5})\,(3a - \sqrt{5})$ (d) $(3a + \sqrt{5})\,(3a - 5)$

3. Factorise :
  $49\,(2x + 3y)^2 - 70\,(4x^2 - 9y^2) + 25\,(2x - 3y)^2$
  (a) $4\,(x - 9y)^2$ (b) $9\,(x + 4y)^2$
  (a) $16\,(x + 9y)^2$ (d) $16\,(x - 9y)^2$

4. $a^2 - b^2 - c^2 + 2bc + a + b - c$ when factorised equals
  (a) $(a - b - c)\,(a - b + c + 1)$
  (b) $(a + b - c)\,(a - b + c + 1)$
  (c) $(a - b + c)\,(a - b + c + 1)$
  (d) $(a + b + c)\,(a - b + c + 1)$

5. Factorise : $1 + 2ab - (a^2 + b^2)$
  (a) $(1 - a + b)\,(1 - a - b)$
  (b) $(1 + a + b)\,(1 - a + b)$
  (c) $(1 + a - b)\,(1 - a + b)$
  (d) $(1 + a - b)\,(1 + a + b)$

6. The factors of $x^4 + 4$ are
  (a) $(x^2 + 2)^2$
  (b) $(x^2 + 2)(x^2 - 2)^2$
  (c) $(x^2 + 2x + 2)\,(x^2 - 2x + 2)$
  (d) $(x^2 - 2)^2$

7. The factors of $a^2 - b^2 - 4c^2 + 4d^2 - 4(ad - bc)$ are :
  (a) $(a + 2d + b + 2c)(a - 2d - b + 2c)$
  (b) $(a - 2d + b - 2c)(a + 2d - b + 2c)$
  (c) $(a - 2d + b - 2c)(a - 2d - b + 2c)$
  (d) $(a - 2d - b - 2c)\,(a + 2d + b + 2c)$

8. Factors of $x^4 + 5x^2 + 9$ are :

  (a) $(x^2 + 2x + 3)(x^2 + 3x + 3)$

  (b) $(x^2 - x + 3)(x^2 - x - 3)$

  (c) $(x^2 - x - 3)(x^2 + x + 3)$

  (d) $(x^2 - x + 3)(x^2 + x + 3)$

9. The factors of $625a^{12} - 81b^{12}$ are :

  (a) $(25a^6 + 9b^6)(5a^3 - 3b^3)(5a^3 + 3b^3)$

  (b) $(5a^3 + 3b^3)^2 (5a^3 - 3b^3)^2$

  (c) $(5a^3 - 3b^3)^4$

  (d) $(25a^6 - 9b^6)^2$

10. The factors of $x^8 - x^4 - 30$ are :

  (a) $(x^4 - 6)$ and $(x^4 - 5)$

  (b) $(x^4 - 6)$ and $(x^4 + 5)$

  (c) $(x^4 + 6)$ and $(x^4 - 5)$

  (d) $(x^4 + 6)$ and $(x^4 + 5)$

11. The factors of $(x^4 - 7x^2 y^2 + y^4)$ are

  (a) $(x^2 + y^2 - 3xy)(x^2 + y^2 + 3xy)$

  (b) $(x^2 - y^2 + 3xy)(x^2 + y^2 - 3xy)$

  (c) $(x^2 - 3xy + y^2)(x^2 - 3xy - y^2)$

  (d) $(x^2 + 3xy + y^2)(x^2 - 3xy - y^2)$

12. Factors of $3\,m^5 - 48m$ are

  (a) $3m(m-1)(m-3)$

  (b) $3m(m-2)(m+2)(m^2 + 4)$

  (c) $3m(m-1)(m-2)(m+1)$

  (d) $m(m-1)(m+2)(m^2 + 4)$

13. Factorise : $x^2 - xy + y - x$

  (a) $(x-1)(x+y)$ (b) $(x+1)(x+y)$

  (c) $(x-1)(x-y)$ (d) None of these

14. Factorise : $4x^4 + 3x^2 + 1$

  (a) $(2x^2 + x + 1)(2x^2 + x - 1)$

  (b) $(2x^2 + x + 1)(2x^2 - x + 1)$

  (c) $(2x^2 + x - 1)(2x^2 - x + 1)$

  (d) $(2x^2 - x + 1)(2x^2 - x - 1)$

15. Factorise : $x^2 + 5\sqrt{3}x + 12$

  (a) $(x + 2\sqrt{3})(x + 3\sqrt{3})$

  (b) $(x + \sqrt{3})(x + 4\sqrt{3})$

  (c) $(x + 4\sqrt{3})(x - \sqrt{3})$

  (d) $(x - \sqrt{3})(x - 4\sqrt{3})$

16. Factorise : $y^{16} - 63y^8 - 64$

  (a) $(y^8 - 1)(y^4 + 8)(y^4 - 8)$

  (b) $(y^4 + 8)^2 (y^8 + 1)$

  (c) $(y^4 - 8)^2 (y^8 - 1)$

  (d) $(y^4 + 8)(y^4 - 8)(y^8 + 1)$

17. Factorise : $a^4 - 20a^2 + 64$

  (a) $(a+2)(a-2)(a+4)(a-4)$

  (b) $(a-2)^2 (a-4)^2$

  (c) $(a-2)^2 (a+4)^2$

  (d) None of these

18. Factorise : $27 + 125a^3 + 135a + 225a^2$

  (a) $(3 + 5a)(3 + 5a)(3 - 5a)$

  (b) $(3 - 5a)(3 - 5a)(3 + 5a)$

  (c) $(3 + 5a)(3 + 5a)(3 + 5a)$

  (d) $(3 - 5a)(3 - 5a)(3 - 5a)$

19. Factorise : $(5x - 3)^2 - (5x - 3) - 20$

  (a) $(5x + 8)(5x - 1)$ (b) $(5x - 8)(5x + 1)$

  (c) $(5x - 8)(5x - 1)$ (d) $(5x + 8)(5x + 1)$

20. $6\sqrt{5}x^2 - 2x - 4\sqrt{5}$ is equal to

  (a) $(\sqrt{5}x - 2)(6x + 2\sqrt{5})$

  (b) $(\sqrt{5}x + 2)(6x + 2\sqrt{5})$

  (c) $(\sqrt{5}x + 2)(6x - 2\sqrt{5})$

  (d) $(\sqrt{5}x + 2)(6x - 2\sqrt{5})$

21. $10\left(3x - \dfrac{4}{x}\right)^2 - 3\left(3x - \dfrac{4}{x}\right) - 7$ is equal to:

  (a) $\left(3x - \dfrac{4}{x} + 7\right)\left(3x - \dfrac{4}{x} - 10\right)$

  (b) $(3x - \dfrac{4}{x} - 14)(3x - \dfrac{4}{x} + 5)$

  (c) $\left(3x - \dfrac{4}{x} - 1\right)\left(30x - \dfrac{40}{x} + 7\right)$

  (d) $\left(3x - \dfrac{4}{x} - 1\right)\left(3x - \dfrac{4}{x} + 7\right)$

22. $\dfrac{0.86 \times 0.86 \times 0.86 + 0.14 \times 0.14 \times 0.14}{0.86 \times 0.86 - 0.86 + 0.14 + 0.14 \times 0.14}$ is equal to:

  (a) 1   (b) 0

  (c) 2   (d) 10

23. Factorise : $a^6 - 26a^3 - 27$

  (a) $(a-3)(a+1)(a^2 + a + 1)(a^2 + 3a + 9)$

  (b) $(a-3)(a-1)(a^2 + a + 1)(a^2 + 3a + 9)$

  (c) $(a-3)(a+1)(a^2 + 3a + 9)(a^2 - a + 1)$

  (d) $(a+3)(a-1)(a^2 + 3a + 9)(a^2 + a + 1)$

24. Find the square root of :

$$4a^2 + 9b^2 + c^2 - 12ab + 6bc - 4ac$$

(a) $(2a + 3b - c)$ (b) $(2a - 3b + c)$

(c) $(-2a + 3b + c)$ (d) $(-2a + 3b + c)$

25. Factorise : $27a^3 - b^3 - 1 - 9ab$

(a) $(3a + b - 1)(9a^2 + b^2 + 1 - 3ab + 3a - b)$

(b) $(3a - b - 1)(9a^2 - b^2 + 1 + 3ab + 3a - b)$

(c) $(3a + b + 1)(9a^2 + b^2 + 1 + 3ab + 3a - b)$

(d) $(3a - b - 1)(9a^2 + b^2 + 1 + 3ab + 3a - b)$

26. Factorise : $x^3 + \dfrac{1}{x^3} - 2$

(a) $\left(x - \dfrac{1}{x}\right)\left(x^2 + 1 + \dfrac{1}{x^2}\right)$

(b) $\left(x + \dfrac{1}{x} + 1\right)\left(x^2 + \dfrac{1}{x^2} - \dfrac{1}{x} - x\right)$

(c) $\left(x + \dfrac{1}{x}\right)\left(x^2 - 1 + \dfrac{1}{x^2}\right)$

(d) $\left(x + \dfrac{1}{x} - 1\right)\left(x^2 + \dfrac{1}{x^2} + \dfrac{1}{x} + x\right)$

27. Factors of $(2x - 3y)^3 + (3y - 5z)^3 + (5z - 2x)^3$ are :

(a) $3(2x - 3y)(3y - 5z)(5z - 2x)$

(b) $3(3x - 2y)(3y - 5z)(5z - 2x)$

(c) $3(2x - 3y)(5y - 3z)(5z - 2x)$

(d) $3(2x - 3y)(3y - 5z)(2z - 5x)$

28. If $(x^{3/2} - xy^{1/2} + x^{1/2}y - y^{3/2})$ is divided by $(x^{1/2} - y^{1/2})$, the quotient is :

(a) $x + y$ (b) $x - y$

(c) $x^{1/2} + y^{1/2}$ (d) $x^2 - y^2$

29. Factorise : $a^2 + \dfrac{1}{a^2} + 3 - 2a - \dfrac{2}{a}$

(a) $\left(a + \dfrac{1}{a} - 1\right)\left(a - \dfrac{1}{a} + 1\right)$

(b) $\left(a + \dfrac{1}{a} - 1\right)\left(a + \dfrac{1}{a} + 1\right)$

(c) $\left(a + \dfrac{1}{a} + 1\right)\left(a + \dfrac{1}{a} + 1\right)$

(d) $\left(a + \dfrac{1}{a} - 1\right)\left(a + \dfrac{1}{a} - 1\right)$

30. Factorise : $x^3 - 3x^2 + 3x + 7$

(a) $(x - 1)(x^2 - 4x + 7)$

(b) $(x + 1)(x^2 - 4x + 7)$

(c) $(x + 1)(x^2 + 4x + 7)$

(d) $(x - 1)(x^2 + 4x + 7)$

Answers

1. (c)	**2.** (c)	**3.** (c)	**4.** (b)	**5.** (c)	**6.** (c)	**7.** (c)	**8.** (d)	**9.** (a)	**10.** (b)
11. (a)	**12.** (b)	**13.** (c)	**14.** (b)	**15.** (b)	**16.** (d)	**17.** (a)	**18.** (c)	**19.** (b)	**20.** (a)
21. (c)	**22.** (a)	**23.** (c)	**24.** (c)	**25.** (d)	**26.** (b)	**27.** (a)	**28.** (a)	**29.** (d)	**30.** (b)

Hints and Solutions

1. (c) $ab(c^2 + 1) + c(a^2 + b^2)$

$= abc^2 + ab + ca^2 + cb^2$

$= abc^2 + ca^2 + ab + cb^2$

$= ac(bc + a) + b(a + bc)$

$= (a + bc)(ac + b)$

2. (c) $9a^2 - 6\sqrt{5}a + 5$

$= (3a)^2 - 2 \times 3a \times \sqrt{5} + (\sqrt{5})^2$

$= (3a - \sqrt{5})^2 = (3a - \sqrt{5})(3a - \sqrt{5})$

3. (c) $49(2x + 3y)^2 - 70(4x^2 - 9y^2) + 25(2x - 3y)^2$

$= (7(2x + 3y))^2 - 2 \times 7(2x + 3y) \times 5(2x - 3y)$

$+ (5(2x - 3y))^2$

$= [7(2x + 3y) - 5(2x - 3y)]^2$

$= [14x + 21y - 10x + 15y]^2$

$= (4x + 36y)^2 = (4(x + 9y))^2 = 16(x + 9y)^2$

4. (b) $a^2 - b^2 - c^2 + 2bc + a + b - c$

$= [a^2 - (b^2 + c^2 - 2bc)] + (a + b - c)$

$= [a^2 - (b - c)^2)] + (a + b - c)$

$= [(a + b - c)(a - b + c) + (a + b - c)]$

$= (a + b - c)[a - b + c + 1]$

5. (c) $1 + 2ab - (a^2 + b^2)$

$= 1 + 2ab - a^2 - b^2$

$= 1-(a^2-2ab+b^2)$

$= 1-(a-b)^2$

$= (1+a-b)(1-a+b)$

6. (c) x^4+4

$= x^4+4+4x^2-4x^2$ (Note the step)

$= (x^2+2)^2-(2x)^2$

$= (x^2+2x+2)(x^2-2x+2)$

7. (c) $a^2-b^2-4c^2+4d^2-4(ad-bc)$

$= a^2-b^2-4c^2+4d^2-4ad+4bc$

$= (a^2-4ad+4d^2)-(b^2-4bc+4c^2)$

$= (a-2d)^2-(b-2c)^2$

$= (a-2d+b-2c)(a-2d-b+2c)$

8. (d) x^4+5x^2+9

$= x^4+6x^2+9-x^2$ (Note the step)

$= (x^2+3)^2-x^2$

Now, factorise using $x^2-y^2=(x+y)(x-y)$

9. (a) $625a^{12}-81b^{12}$

$= (25a^6)^2-(9b^6)^2$

$= (25a^6-9b^6)(25a^6+9b^6)$

$= [(5a^3)^2-(3b^3)^2](25a^6+9b^6)$

$= (5a^3-3b^3)(5a^3+3b^3)(25a^6+9b^6)$

10. (b) x^8-x^4-30

$= x^8-6x^4+5x^4-30$

$= x^4(x^4-6)+5(x^4-6)$

$= (x^4+5)(x^4-6)$

11. (a) $x^4-7x^2y^2+y^4$

$= x^4+2x^2y^2+y^4-9x^2y^2$

$= (x^2+y^2)^2-(3xy)^2$

$= (x^2+y^2+3xy)(x^2+y^2-3xy)$

12. (b) $3m^5-48m$

$= 3m(m^4-16)$

$= 3m[(m^2)^2-4^2]$

$= 3m(m^2-4)(m^2+4)$

$= 3m(m+2)(m-2)(m^2+4)$

13. (c) $x^2-xy+y-x$

$= x(x-y)-1(x-y)$

$= (x-y)(x-1)$

14. (b) $4x^4+3x^2+1$

$= 4x^4+4x^2+1-x^2$

$= (2x^2+1)^2-x^2$

Now, factorise using $x^2-y^2=(x+y)(x-y)$

15. (b) $x^2+5\sqrt{3}x+12$

$= x^2+4\sqrt{3}x+\sqrt{3}x+12$

$= x(x+4\sqrt{3})+\sqrt{3}(x+4\sqrt{3})$

$= (x+4\sqrt{3})(x+\sqrt{3})$

16. (d) $y^{16}-63y^8-64$

$= y^{16}-64y^8+y^8-64$

$= y^8(y^8-64)+1(y^8-64)$

$= (y^8-64)(y^8+1)$

$= (y^4-8)(y^4+8)(y^8+1)$

17. (a) a^4-20a^2+64

$= a^4-16a^2-4a^2+64$

$= a^2(a^2-16)-4(a^2-16)$

$= (a^2-16)(a^2-4)$

$= (a+4)(a-4)(a+2)(a-2)$

18. (c) $27+125a^3+135a+225a^2$

$= (3)^3+(5a)^3+3\times3^2\times5a+3\times3\times(5a)^2$

$= (3+5a)^3$

19. (b) $(5x-3)^2-(5x-3)-20$

Let $(5x-3)=a$. Then, the expression becomes

$a^2-a-20 = a^2-5a+4a-20$

$= a(a-5)+4(a-5)=(a-5)(a+4)$

$= (5x-3-5)(5x-3+4)$

$= (5x-8)(5x+1)$

20. (a) $6\sqrt{5}x^2-2x-4\sqrt{5}$ Prod. $= 6\sqrt{5}\times-4\sqrt{5}=-120$

$= 6\sqrt{5}x^2-12x+10x-4\sqrt{5}$

$= 6x(\sqrt{5}x-2)+2\sqrt{5}(\sqrt{5}x-2)$

$= (\sqrt{5}x-2)(6x+2\sqrt{5})$

21. (c) $10\left(3x-\dfrac{4}{x}\right)^2-3\left(3x-\dfrac{4}{x}\right)-7$

Let $3x-\dfrac{4}{x}=a$. Then, the expression becomes

$10a^2-3a-7$

$$= 10a^2 - 10a + 7a - 7$$

$$= 10a(a-1) + 7(a-1)$$

$$= (10a+7)(a-1)$$

$$= \left[10\left(3x-\frac{4}{x}\right)+7\right]\left[\left(3x-\frac{4}{x}\right)-1\right]$$

$$= \left[30x - \frac{40}{x} + 7\right]\left[3x - \frac{4}{x} - 1\right]$$

22. (a) Given exp. $= \dfrac{(0.86)^3 + (0.14)^3}{(0.86)^2 - 0.86 \times 0.14 + (0.14)^2}$

$$= \frac{(0.86+0.14)\left[(0.86)^2 - 0.86 \times 0.14 + (0.14)^2\right]}{(0.86)^2 - 0.86 \times 0.14 + (0.14)^2}$$

$$[\because (a^3 + b^3) = (a+b)(a^2 - ab + b^2)]$$

$$= 0.86 + 0.14 = 1.$$

23. (c) $a^6 - 26a^3 - 27$

$$= a^6 - 27a^3 + a^3 - 27$$

$$= a^3(a^3 - 27) + 1(a^3 - 27)$$

$$= (a^3 - 27)(a^3 + 1)$$

$$= (a-3)(a^2 + 3a + 9)(a+1)(a^2 - a + 1)$$

24. (c) $4a^2 + 9b^2 + c^2 - 12ab + 6bc - 4ac$

$$= (-2a)^2 + (3b)^2 + c^2 + 2 \times (-2a) \times 3b + 2$$
$$\times 3b \times c + 2 \times (-2a) \times c$$

$$= (-2a + 3b + c)^2$$

$$\therefore \text{ Reqd. sq. root} = -2a + 3b + c$$

25. (d) $27a^3 - b^3 - 1 - 9ab$

$$= (3a)^3 + (-b)^3 + (-1)^3 - 3 \times 3a \times (-b) \times (-1)$$

$$= [3a + (-b) + (-1)][(3a)^2 + (-b)^2 + (-1)^2$$
$$- 3a \times (-b) - (-b) \times (-1) - 3a \times (-1)]$$

$$= (3a - b - 1)(9a^2 + b^2 + 1 + 3ab + b - 3a)$$

$$[\because a^3 + b^3 + c^3 - 3abc = (a+b+c)$$
$$(a^2 + b^2 + c^2 - ab - bc - ca)]$$

27. (a) Since $(2x - 3y) + (3y - 5z) + (5z - 2x) = 0$

$$(2x - 3y)^3 + (3y - 5z)^3 + (5z - 2x)^3$$
$$= 3(2x - 3y)(3y - 5z)(5z - 2x)$$

28. (a) $x^{3/2} - xy^{1/2} + x^{1/2}y - y^{3/2}$

$$= x(x^{1/2} - y^{1/2}) + y(x^{1/2} - y^{1/2})$$

$$= (x+y)(x^{1/2} - y^{1/2})$$

$$\therefore (x^{3/2} - xy^{1/2} + x^{1/2}y - y^{3/2}) \div (x^{1/2} - y^{1/2})$$

$$= \frac{(x+y)(x^{1/2} - y^{1/2})}{(x^{1/2} - y^{1/2})} = (x+y)$$

29. (d) $a^2 + \dfrac{1}{a^2} + 3 - 2a - \dfrac{2}{a}$

$$= a^2 + \frac{1}{a^2} + 1 + 2 - 2a - \frac{2}{a}$$

$$= (a)^2 + \left(\frac{1}{a}\right)^2 + (-1)^2 + 2 \times a \times \frac{1}{a}$$

$$+ 2 \times a \times (-1) + 2 \times \frac{1}{a} \times -1$$

$$= \left(a + \frac{1}{a} - 1\right)^2$$

30. (b) $x^3 - 3x^2 + 3x + 7$

$$= x^3 - 3x^2 + 3x - 1 + 8$$

$$= (x-1)^3 + (2)^3$$

$$= (x - 1 + 2)\left[(x-1)^2 - (x-1) \times 2 + 4\right]$$

$$= (x+1)(x^2 - 2x + 1 - 2x + 2 + 4)$$

$$= (x+1)(x^2 - 4x + 7)$$

Self Assessment Sheet–8

1. The absolute difference between two linear factors of $x^2 + 4xy + 4y^2 + x + 2y$ is :

  (a) 0   (b) 1

  (c) 2   (d) 3

2. Factors of $(2x^2 - 3x - 2)(2x^2 - 3x) - 63$ are :

  (a) $(x-3)(2x+3)(x-1)(x-7)$

  (b) $(x+3)(2x+3)(x-1)(x-7)$

  (c) $(x+3)(2x+3)(2x^2 - 3x + 7)$

  (d) $(x-3)(2x+3)(2x^2 - 3x + 7)$

3. The factors of $(2x^2 - x - 6)^2 - (2x^2 - 9x + 10)$ are:

  (a) $8(x-2)(2x-1)^2$

  (b) $16(x-2)^2(2x-1)$

  (c) $16(x-2)(2x-1)$

  (d) $16(x-2)^2(2x-1)^2$

4. The factors of
$(12x - 8y)(a - 2b) + (6b - 2a)(3x - 2y)$ are:

  (a) $(3x + 2y)(a - 2b)$

  (b) $2(3x - 2y)(a - b)$

(c) $(3x-2y)(a+2b)$

(d) $(3x+2y)(a+2b)$

5. The positive square root of $(2x^2+5x+2)(x^2-4)$ $(2x^2-3x-2)$ expressed as factors is:

(a) $(2x+3)(x+2)(x-1)$

(b) $(x+2)(x-2)(2x-1)$

(c) $(2x+1)(x+2)(x-2)$

(d) $(2x+1)(x-2)(x-1)$

6. The factors of $1-(a^2+b^2)+a^2b^2$ are:

(a) $(1+a)(1-a)(1+b^2)$

(b) $(1+a^2)(1+b^2)$

(c) $(1+a^2)(1-b)(1+b)$

(d) $(1-a)(1+a)(1-b)(1+b)$

7. Simplify, giving the answers in factors :

$(a+1)(a-18)+(a-2)(a+15)$

(a) $2(a-6)(a-4)$ (b) $2(a-6)(a+4)$

(c) $2(a-4)(a+6)$ (d) $2(a+6)(a+4)$

8. The positive square root of

$$\frac{(a+b)^2-(c+d)^2}{(a+b)^2-(c-d)^2}\times\frac{(a+b+c)^2-d^2}{(a+b-c)^2-d^2}$$ using factorisation is:

(a) $\dfrac{a+b+c+d}{a+b-c+d}$ (b) $\dfrac{a+b+c-d}{a+b+c+d}$

(c) $\dfrac{a+b-c+d}{a+b+c-d}$ (d) $\dfrac{a-b-c+d}{a+b+c+d}$

9. One factor of x^3-7x+6 is $x-1$. The other factors are:

(a) $(x-3)(x+2)$ (b) $(x+3)(x-2)$

(c) $(x-3)(x-2)$ (d) $(x+3)(x+2)$

10. Factorise : $(m+a)(n-b)-(m+b)(n-a)$

(a) $(a-b)(m-n)$

(b) $(a+b)(m-n)$

(c) $(a-b)(m+n)$

(d) $(a-b)(m-n)$

Answers

1. (b)	2. (d)	3. (b)	4. (b)	5. (c)	6. (d)	7. (b)	8. (a)	9. (b)	10. (c)

Chapter
9 HCF AND LCM OF POLYNOMIALS AND RATIONAL EXPRESSIONS

KEY FACTS

1. The HCF of polynomials is the polynomial of highest degree and greatest numerical coefficient which divides both the polynomial exactly.

2. **HCF of monomials :**

    (i) Find the HCF of the numerical coefficients of all the monomials.

    (ii) Find the highest power of each of the variables common to all the monomials. Omit the variables that are not common.

    (iii) The HCF of the given monomials is the product of (i) and (ii)

    For example, to find the HCF of $12\,a^2b^4$, $15\,ab^2c$ and $21a^3\,b$. The HCF of 12, 15, 21 = 3

    Highest power of variables common to the three variables = ab

    $\therefore$ Reqd. HCF = $3ab$

3. **HCF of polynomials :**

    Step 1. *Factorise the polynomials.*

    Step 2. *Find the HCF of numerical factors.*

    Step 3. *The product of the common factors is the required HCF.*

    For example: HCF of $2x^2 + 2x - 4$ and $x^2 - 1$ can be found as :

    $2x^2 + 2x - 4 = 2(x^2 + x - 2) = 2(x + 2)(x - 1)$

    $x^2 - 1 = (x + 1)(x - 1)$ $\therefore$ HCF $= (x - 1)$

4. The LCM of polynomials is the polynomial of the lowest degree and smallest numerical coefficient which is exactly divisible by the given polynomials.

5. **To find the LCM of polynomials that can be easily factorised.**

    Step 1. *Write each polynomial in factorised form.*

    Step 2. *Include in the LCM, the factors that are common to the given polynomials and then the remaining factors that are not common.*

    Step 3. *The LCM is the product of all the common factors and the remaining factors.*

    For example: LCM of $8x\,(x^2 - 1)$ and $6x^2\,(x - 1)^2$ can be found as: $8x(x^2 - 1) = 8 \times x \times (x - 1)\,(x + 1)$

    $6x^2\,(x - 1)^2 = 2 \times 3 \times x \times x \times (x - 1)\,(x - 1)$

    $\therefore$ LCM $= 8 \times 3 \times x^2 \times (x - 1)^2 \times (x + 1)$

    $= 24x^2\,(x - 1)^2\,(x + 1)$

6. Fractions like $\dfrac{x+2}{x-5}, \dfrac{x^2-4x+4}{6x^2-3x-5}, \dfrac{4}{4x^2-7}$ etc, having polynomials in the numerator or denominator or both are called algebraic fractions or **rational expressions**.

7. An algebraic expression is in the simplest form when the polynomials in the numerator and denominator do not have a common factor.

8. Addition, subtraction, multiplication and division of rational expressions is done in the same way as we do of rational numbers.

Thus if $\dfrac{m}{n}$ and $\dfrac{p}{q}$, $n \neq 0$, $q \neq 0$ are rational expressions. Then,

(i) $\dfrac{m}{n} + \dfrac{p}{q} = \dfrac{mq + pn}{nq}$ (ii) $\dfrac{m}{n} - \dfrac{p}{q} = \dfrac{mq - pn}{nq}$

(iii) $\dfrac{m}{n} \times \dfrac{p}{q} = \dfrac{m \times p}{n \times q} = \dfrac{mp}{nq}$ (iv) $\dfrac{m}{n} \div \dfrac{p}{q} = \dfrac{m}{n} \times \dfrac{q}{p} = \dfrac{mq}{np}$

9. An algebraic expression is simplified by removing the brackets in the order: (i) Bar (ii) Parenthesis (iii) Curly Brackets (iv) Square Brackets and following the rule of **BODMAS**.

Solved Examples

Ex. 1. *What is HCF of $8x^2y^2$, $12x^3y^2$ and $24x^4y^3z^2$?*

Sol. $8x^2y^2 = \underline{2 \times 2} \times 2 \times \underline{x \times x} \times \underline{y \times y}$

$12x^3y^2 = \underline{2 \times 2} \times 3 \times \underline{x \times x} \times x \times \underline{y \times y}$

$24x^4y^3z^2 = \underline{2 \times 2} \times 2 \times 3 \times \underline{x \times x} \times x \times x \times \underline{y \times y} \times y \times z \times z.$

HCF of $8x^2y^2$, $12x^3y^2$ and $24x^4y^3z^2 = 2 \times 2 \times x \times x \times y \times y = \mathbf{4x^2y^2}$.

Ex. 2. *Find the HCF of $x^2 - 5x + 6$ and $x^2 - 9$.*

Sol. $x^2 - 5x + 6 = x^2 - 2x - 3x + 6$

$\qquad\qquad = x(x-2) - 3(x-2) = (x-2)(x-3)$

$\quad x^2 - 9 = (x-3)(x+3)$

$\therefore$ HCF of $(x^2 - 5x + 6)$ and $(x^2 - 9) = \mathbf{x - 3}$

Ex. 3. *Find the LCM of $14a^2b^3c^4$, $20ab^4c^3$ and $35a^5b^3c$.*

Sol. LCM. of 14, 20 and 35 = $2 \times 5 \times 7 \times 2 = 140$

LCM of a^2, a and $a^5 = a^5$

LCM of b^3, b^4 and $b^2 = b^4$

LCM of c^4, c^3 and $c = c^4$

LCM of the given monomials = $\mathbf{140\ a^5\ b^4\ c^4}$

2	14, 20, 35
5	7, 10, 35
7	7, 2, 7
	1, 2, 1

Ex. 4. *Find the LCM of $3y + 12$, $y^2 - 16$ and $y^4 - 64y$.*

Sol. $3y + 12 = 3(y + 4)$

$y^2 - 16 = (y + 4)(y - 4)$

$y^4 - 64y = y(y^3 - 64) = y(y - 4)(y^2 + 4y + 16)$

$\therefore$ LCM of given polynomials = $\mathbf{3y\ (y - 4)\ (y + 4)\ (y^2 + 4y + 16)}$

Ex. 5. *The HCF of two expressions is x and their LCM is $x^3 - 9x$. If one of the expressions is $x^2 + 3x$, then find the other expression.*

Sol. HCF = x,

LCM = $x^3 - 9x = x(x^2 - 9) = x(x + 3)(x - 3)$

Given expression = $x^2 + 3x = x(x + 3)$

$$\therefore \text{Other expression} = \frac{\text{HCF} \times \text{LCM}}{\text{Given expression}}$$

$$= \frac{x \times x \times (x+3) \times (x-3)}{x \times (x+3)}$$

$$= x(x-3) = x^2 - 3x$$

Ex. 6. *If the HCF of $x^3 - 343$ and $x^2 - 9x + 14$ and $(x - 7)$, then find their LCM.*

Sol. $x^3 - 343 = x^3 - (7)^3 = (x-7)(x^2 + 7x + 49)$

$x^2 - 9x + 14 = x^2 - 2x - 7x + 14$
$\qquad = x(x-2) - 7(x-2) = (x-7)(x-2)$

HCF of the polynomials $= (x - 7)$

$\therefore$ Reqd. LCM $= \dfrac{(x^3 - 343) \times (x^2 - 9x + 14)}{\text{HCF}} = \dfrac{(x-7)(x^2 + 7x + 49) \times (x-7)(x-2)}{(x-7)}$

$\qquad = (x-7)(x-2) \times (x^2 + 7x + 49)$

Ex. 7. *Simplify the expression $\dfrac{6p^2 - 150}{p^2 - 3x - 40}$*

Sol. $\dfrac{6p^2 - 150}{p^2 - 3x - 40} = \dfrac{6(p^2 - 25)}{p^2 - 8x + 5x - 40}$

$\qquad = \dfrac{6(p-5)(p+5)}{p(p-8) + 5(p-8)} = \dfrac{6(p-5)(p+5)}{(p-8)(p+5)} = \dfrac{6(p-5)}{p-8}.$

Ex. 8. *Add : $\dfrac{a}{3xy} + \dfrac{2b}{6yz} + \dfrac{3c}{15xz}$*

Sol. The least common denominator (LCD) of the given expression is $30\,xyz$.

3	3, 6, 15
	1, 2, 5

$\therefore \dfrac{a}{3xy} + \dfrac{2b}{6yz} + \dfrac{3c}{15xz} = \dfrac{10az + 10bx + 6cy}{30\,xyz}$

$\qquad = \dfrac{5az + 5bx + 3cy}{15\,xyz}.$

Ex. 9. *Simplify : $\dfrac{1}{x^2 - 8x + 15} - \dfrac{1}{x^2 - 25}$*

Sol. $x^2 - 8x + 15 = x^2 - 5x - 3x + 15$
$\qquad = x(x-5) - 3(x-5) = (x-5)(x-3)$

$x^2 - 25 = (x-5)(x-5)$

$\therefore \dfrac{1}{x^2 - 8x + 15} - \dfrac{1}{x^2 - 25} = \dfrac{1}{(x-5)(x-3)} - \dfrac{1}{(x-5)(x-5)}$

$\qquad = \dfrac{(x-5) - (x-3)}{(x-3)(x-5)(x-5)}$ (Note the step)

$\qquad = \dfrac{x-5-x+3}{(x-3)(x-5)(x-5)} = \dfrac{-2}{(x-3)(x-5)(x-5)}$

Ex. 10. *Simplify the expression :*

$$\left[\frac{x^3 + y^3}{(x-y)^2 + 3xy} \right] \div \left[\frac{(x+y)^2 - 3xy}{x^3 - y^3} \right] \times \frac{xy}{x^2 - y^2}$$

Sol. $\left(\dfrac{x^3+y^3}{(x-y)^2+3xy}\right) \div \left(\dfrac{(x+y)^2-3xy}{x^3-y^3}\right) \times \dfrac{xy}{x^2-y^2}$

$= \left(\dfrac{(x+y)(x^2-xy+y^2)}{x^2-2xy+y^2+3xy}\right) \div \left(\dfrac{x^2+y^2+2xy-3xy}{(x-y)(x^2+xy+y^2)}\right) \times \dfrac{xy}{x^2-y^2}$

$= \dfrac{(x+y)(x^2-xy+y^2)}{(x^2+xy+y^2)} \div \dfrac{(x^2-xy+y^2)}{(x-y)(x^2+xy+y^2)} \times \dfrac{xy}{(x+y)(x-y)}$

$= \dfrac{(x+y)(x^2-xy+y^2)}{(x^2+xy+y^2)} \times \dfrac{(x-y)(x^2+xy+y^2)}{(x^2-xy+y^2)} \times \dfrac{xy}{(x+y)(x-y)}$

$= \boldsymbol{xy}$

Question Bank–9

1. HCF of the polynomials
   $20x^2y(x^2-y^2)$ and $35xy^2(x-y)$ is :
   (a) $5x^2y^2(x-y)$ (b) $5xy(x-y)$
   (c) $5x^2y^2(x+y)$ (d) $5xy(x^2-y^2)$

2. HCF of x^3-1 and x^4+x^2+1 will be
   (a) $(x-1)$ (b) x^2+x+1
   (c) x^2-x+1 (d) x^2-x-1

3. The HCF of the polynomials x^3-3x^2+x-3 and x^3-x^2-9x+9 is :
   (a) $x-3$ (b) $x-1$
   (c) x^2+1 (d) $(x-1)(x-3)$

4. The LCM of the polynomials $xy+yz+zx+y^2$ and $x^2+xy+yz+zx$ is :
   (a) $x+y$ (b) $y+z$
   (c) $(x+y)(y+z)(z+x)$ (d) x^2+y^2

5. The LCM of $x^2-10x+16$, $x^2-9x+14$ and $x^2-10x+21$ is :
   (a) $(x-2)^2(x-3)(x-7)^2(x-8)$
   (b) $(x-2)^2(x-3)(x-7)(x-8)$
   (c) $(x-2)(x-3)(x-7)^2(x-8)$
   (d) $(x-2)(x-3)(x-7)(x-8)$

6. The LCM of $6(x^2+xy), 8(xy-y^2), 12(x^2-y^2)$
   and $20(x+y)^2$ is :
   (a) $120x\,(x+y)(x-y)$
   (b) $120xy\,(x+y)(x-y)$
   (c) $120xy\,(x+y)^2\,(x-y)$
   (d) $120xy\,(x+y)(x-y)^2$

7. The HCF of x^4-y^4 and x^6-y^6 is :
   (a) x^2-y^2 (b) x^2+y^2
   (c) x^3+y^3 (d) x^3-y^3

8. The LCM of the polynomials x^3+3x^2+3x+1, x^2+2x+1 and x^2-1 is :
   (a) $(x^2-1)(x+1)^3$ (b) $(x^2+1)(x-1)^2$
   (c) $(x^2-1)(x-1)^2$ (d) $(x+1)^3$

9. The product of two expression is $x^3+x^2-44x-84$. If the HCF of these two expressions is $x+6$, then their LCM will be:
   (a) $(x+2)(x+7)$ (b) $(x+2)(x-7)$
   (c) $(x-2)(x+7)$ (d) $(x-2)(x-7)$

10. The HCF of x^4-11x^2+10, x^2-5x+4 and x^3-3x^2+3x-1 is
    (a) $x+1$ (b) $x-4$
    (c) $x+2$ (d) $x-1$

11. The HCF of two polynomials $4x^2(x^2-3x+2)$ and $12x(x-2)(x^2-4)$ is $4x(x-2)$. The LCM of the two polynomials is :
    (a) $12x(x^2-4)$
    (b) $12x^2(x^2-3x+4)\,(x^2-2)$
    (c) $12x^2(x^2-3x+2)\,(x^2-4)$
    (d) $12x(x^2-3x-2)\,(x^2-4)$

12. The rational expression $\dfrac{8x^3-125}{4x^2+10x+25}$ in its simplest form is :

(a) $2x$ (b) 5

(c) $2x+5$ (d) $2x-5$

13. $\sqrt{\dfrac{(x^2+3x+2)(x^2+5x+6)}{x^2(x^2+4x+3)}}$ is equal to :

(a) $x(x+1)$ (b) $\dfrac{x+2}{x}$

(c) $\dfrac{x}{x+2}$ (d) $x(x+2)$

14. If $A=\dfrac{2x+1}{2x-1}$ and $B=\dfrac{2x-1}{2x+1}$, then $A-B$ is equal to :

(a) $\dfrac{1}{4x^2-1}$ (b) $\dfrac{8x}{4x^2-1}$

(c) $\dfrac{-2}{2x^2-1}$ (d) $\dfrac{4x}{4x^2-1}$

15. $\dfrac{1}{x+1}-\dfrac{1}{x-1}-\dfrac{x^2}{x+1}+\dfrac{x^2}{x-1}$, when simplified is equal to :

(a) 0 (b) 1

(c) 2 (d) -2

16. The product of the rational expressions $\dfrac{x^2-y^2}{x^2+2xy+y^2}$ and $\dfrac{xy+y^2}{x^2-xy}$ is :

(a) xy (b) y/x

(c) x/y (d) 1

17. $\left(\dfrac{2x+y}{x+y}-1\right)\div\left(1-\dfrac{y}{x+y}\right)$ is equal to :

(a) x (b) y

(c) xy (d) 1

18. $\dfrac{x^3+y^3+z^3-3xyz}{a^3+b^3+c^3-3abc}\times\dfrac{a^2+b^2+c^2-ab-bc-ca}{x^2+y^2+z^2-xy-yz-zx}$ equals

(a) 1 (b) $\dfrac{x^2+y^2+z^2}{a^2+b^2+c^2}$

(c) $\dfrac{x+y+z}{a+b+c}$ (d) $\dfrac{xyz}{abc}$

19. What should be added to $\dfrac{a}{a-b}+\dfrac{b}{a+b}$ to get 1 ?

(a) $\dfrac{-2ab}{a^2+b^2}$ (b) $\dfrac{2ab}{a^2-b^2}$

(c) $\dfrac{2ab}{b^2-a^2}$ (d) $\dfrac{-2ab}{b^2-a^2}$

20. Simplify $\left[\dfrac{1}{1+a}+\dfrac{2a}{1-a^2}\right]\times\dfrac{(a^2+4a-5)}{(a^2+10a+25)}$

(a) $\dfrac{-1}{a+1}$ (b) $\dfrac{1}{1-a}$

(c) $\dfrac{1}{a+5}$ (d) $\dfrac{-1}{a+5}$

Answers

1. (b)	**2.** (b)	**3.** (a)	**4.** (c)	**5.** (d)	**6.** (c)	**7.** (a)	**8.** (a)	**9.** (b)	**10.** (d)
11. (c)	**12.** (d)	**13.** (b)	**14.** (b)	**15.** (c)	**16.** (b)	**17.** (d)	**18.** (c)	**19.** (a)	**20.** (d)

Hints and Solutions

1. (b) HCF of 20 and 35 = 5

HCF of x^2y and $xy^2=xy$

HCF of (x^2-y^2), i.e., $(x-y)(x+y)$

and $(x-y)=(x-y)$

∴ Reqd. HCF $=5xy(x-y)$

2. (b) $x^3-1=(x-1)(x^2+x+1)$

$x^4+x^2+1=x^4+2x^2+1-x^2=(x^2+1)-x^2$

$=(x^2+1-x)(x^2+1+x)$

∴ Reqd. HCF $=x^2+x+1$

3. (a) $x^3-3x^2+x-3=x^2(x-3)+1(x-3)$

$=(x-3)(x^2+1)$

$x^3-x^2-9x+9=x^2(x-1)-9(x-1)$

$=(x^2-9)(x-1)=(x+3)(x-3)(x-1)$

∴ Reqd. HCF $=(x-3)$

4. (c) $xy+yz+zx+y^2=x(y+z)+y(y+z)$

$=(y+z)(x+y)$

$x^2+xy+yz+zx=x(x+y)+z(y+x)$

$=(x+y)(x+z)$

∴ Reqd. LCM $=(x+y)(y+z)(x+z)$

5. (d) $x^2-10x+16=(x-8)(x-2)$

$x^2-9x+14=(x-7)(x-2)$

$x^2-10x+21=(x-7)(x-3)$

∴ Reqd. LCM $=(x-2)(x-3)(x-7)(x-8)$

6. (c) $6(x^2 + xy) = 6x(x+y)$

$8(xy - y^2) = 8y(x-y)$

$12(x^2 - y^2) = 12(x-y)(x+y)$

$20(x+y)^2 = 20(x+y)(x+y)$

2	6, 8, 12, 20
2	3, 4, 6, 10
3	3, 2, 3, 5
	1, 2, 1, 5

LCM of 6, 8, 12, 20 = 120

∴ Reqd. LCM $= 120\,xy(x-y)(x+y)^2$

7. (a) $x^4 - y^4 = (x^2 - y^2)(x^2 + y^2)$

$= (x-y)(x+y)(x^2+y^2)$

$x^6 - y^6 = (x^3)^2 - (y^3)^2 = (x^3 - y^3)(x^3 + y^3)$

$= (x-y)(x^2+xy+y^2)(x+y)(x^2-xy+y^2)$

∴ Reqd. HCF $= (x-y)(x+y) = x^2 - y^2$

8. (a) $x^3 + 3x^2 + 3x + 1 = (x+1)^3$

$x^2 + 2x + 1 = (x+1)^2$

$x^2 - 1 = (x+1)(x-1)$

∴ Reqd. LCM $= (x-1)(x+1)^3$

$= (x^2 - 1)(x+1)^3$

9. (b) LCM × HCF = Product of the expressions

$\Rightarrow$ LCM $= \dfrac{\text{Prod. of expressions}}{\text{HCF}}$

$= \dfrac{x^3 + x^2 - 44x - 84}{(x+6)}$

Performing long division, we have

$$
\begin{array}{r}
x^2 - 5x - 14 \\
x+6\,\overline{\smash{\big)}\,x^3 + x^2 - 44x - 84} \\
\underline{x^3 + 6^3 } \\
-5x^2 - 44x \\
\underline{-5x^2 + 30x } \\
-14x - 84 \\
\underline{-14x - 84} \\
0
\end{array}
$$

∴ Reqd. LCM $= x^2 - 5x - 14 = (x-7)(x+2)$

10. (d) $x^4 - 11x^2 + 10 = x^4 - 10x^2 - x^2 + 10$

$= x^2(x^2 - 10) - 1(x^2 - 10)$

$= (x^2 - 10)(x^2 - 1)$

$= (x^2 - 10)(x+1)(x-1)$

$x^2 - 5x + 4 = (x-4)(x-1)$

$x^3 - 3x^2 + 3x + 1 = (x-1)^3$

∴ Reqd. HCF $= (x-1)$

11. (c) LCM $= \dfrac{\text{product of the polynomials}}{\text{HCF}}$

$= \dfrac{4x^2(x^2 - 3x + 2) \times 12x(x-2)(x^2 - 4)}{4x(x-2)}$

$= \dfrac{4x^2(x-2)(x-1) \times 12x(x-2)(x-2)(x+2)}{4x(x-2)}$

$= 12x^2(x-1)(x-2)(x-2)(x+2)$

$= 12x^2(x^2 - 3x + 2)(x^2 - 4)$

12. (d) $\dfrac{8x^3 - 125}{4x^2 + 10x + 25} = \dfrac{(2x)^3 - 5^3}{4x^2 + 10x + 25}$

$= \dfrac{(2x-5)(4x^2 + 10x + 25)}{4x^2 + 10x + 25}$

$= 2x - 5$

13. (b) $\sqrt{\dfrac{(x^2 + 3x + 2)(x^2 + 5x + 6)}{x^2(x^2 + 4x + 3)}}$

$= \sqrt{\dfrac{(x+1)(x+2)(x+2)(x+3)}{x^2(x+1)(x+3)}}$

$= \sqrt{\dfrac{(x+2)^2}{x^2}} = \dfrac{x+2}{x}$

14. (b) $A - B = \dfrac{2x+1}{2x-1} - \dfrac{2x-1}{2x+1}$

$= \dfrac{(2x+1)^2 - (2x-1)^2}{(2x-1)(2x+1)}$

$= \dfrac{(4x^2 + 4x + 1) - (4x^2 - 4x + 1)}{4x^2 - 1}$

$= \dfrac{8x}{4x^2 - 1}$

15. (c) $\dfrac{1}{x+1} - \dfrac{1}{x-1} - \dfrac{x^2}{x+1} + \dfrac{x^2}{x-1}$

$= \dfrac{(x-1) - (x+1) - x^2(x-1) + x^2(x+1)}{(x-1)(x+1)}$

$= \dfrac{x - 1 - x - 1 - x^3 + x^2 + x^3 + x^2}{x^2 - 1}$

$= \dfrac{2x^2 - 2}{x^2 - 1} = \dfrac{2(x^2 - 1)}{(x^2 - 1)} = 2$

16. (b) Reqd. product = $\dfrac{x^2-y^2}{x^2+2xy+y^2}\times\dfrac{xy+y^2}{x^2-xy}$

$=\dfrac{(x+y)(x+y)}{(x+y)^2}\times\dfrac{y(x+y)}{x(x-y)}$

$=\dfrac{y}{x}$

17. (d) $\left(\dfrac{2x+y}{x+y}-1\right)\div\left(1-\dfrac{y}{x+y}\right)$

$=\left[\dfrac{2x+y-(x+y)}{x+y}\right]\div\left[\dfrac{x+y-y}{x+y}\right]$

$=\dfrac{x}{x+y}\times\dfrac{x+y}{x}=1$

18. (c) Reqd. exp.

$=\dfrac{(x+y+z)(x^2+y^2+z^2-xy-yz-zx)}{(a+b+c)(a^2+b^2+c^2-ab-bc-ca)}$

$\times\dfrac{(a^2+b^2+c^2-ab-bc-ca)}{(x^2+y^2+z^2-xy-yz-zx)}=\dfrac{x+y+z}{a+b+c}$

19. (c) Reqd. exp. $=1-\left[\dfrac{a}{a-b}+\dfrac{b}{a+b}\right]$

$=1-\left[\dfrac{a(a+b)+b(a-b)}{a^2-b^2}\right]$

$=1-\left[\dfrac{a^2+ab+ab-b^2}{a^2-b^2}\right]$

$=\dfrac{(a^2-b^2)-(a^2+2ab-b^2)}{a^2-b^2}$

$=\dfrac{-2ab}{a^2-b^2}=\dfrac{2ab}{b^2-a^2}$

20. (d) $\left(\dfrac{1}{1+a}+\dfrac{2a}{1-a^2}\right)\times\dfrac{(a^2+4a-5)}{a^2+10a+25}$

$=\left(\dfrac{1-a+2a}{(1-a^2)}\right)\times\dfrac{(a+5)(a-1)}{(a+5)^2}$

$=\dfrac{1+a}{1-a^2}\times\dfrac{(a+5)(a-1)}{(a+5)^2}$

$=\dfrac{(1+a)}{(1+a)(1-a)}\times\dfrac{-(a+5)(1-a)}{(a+5)^2}=\dfrac{-1}{a+5}$

Self Assessment Sheet–9

1. If p, m and n are prime numbers, none of which is equal to the other two, what is the greatest common factor of $24pm^2n^2$, $9pmn^2$ and $36p(mn)^3$?

   (a) $3pmn$ (b) $3p^2m^2n^2$
   (c) $3pmn^2$ (d) $3pmn^3$

2. The HCF of $x^5+2x^4+x^3$ and x^7-x^5 is

   (a) x (b) $x(x+1)$
   (c) x^3 (d) $x^3(x+1)$

3. The LCM of x^2-3x+2 and x^3-2x^2-3x is :

   (a) $x(x-2)(x+3)(x^2-1)$
   (b) $x(x-2)(x-3)(x^2+1)$
   (c) $x(x-2)(x-3)(x^2-1)$
   (d) $x(x-2)(x+3)(x^2+1)$

4. The LCM and HCF of two polynomials are respectively $(2a-5)^2(a+1)$ and $(2a-5)$. If one of the polynomials is $4a^2-20a+25$, the other one is :

   (a) $4a^2+20a+5$ (b) $4a^2-25$
   (c) $2a^2+3a-5$ (d) $2a^2-3a-5$

5. $\dfrac{a+1}{a^2+5a}\times\dfrac{a^2-25}{a^2-a-20}\div\dfrac{a^2-a-2}{a^2+2a-8}$ when simplified is equal to :

   (a) 1   (b) a
   (c) $\dfrac{1}{a}$ (d) a^2

6. The value of $\dfrac{x+y}{x-y}+\dfrac{x-y}{x+y}-\dfrac{2(x^2+y^2)}{x^2-y^2}$ is :

   (a) 1   (b) x
   (c) y (d) 0

7. Evaluate

   $\dfrac{x+2}{(x+1)(2x+3)}-\dfrac{2x+3}{(x+1)(x+2)}+\dfrac{3x+5}{(2x+3)(x+2)}$

   (a) $2x$ (b) -1
   (c) 0   (d) x

8. The rational expression

   $\dfrac{(x^2-xy-12y^2)(x^2+xy-12y^2)}{(x^2-16y^2)(x^2-9y^2)}$ when simplified equals.

   (a) 1   (b) xy
   (c) $(x+y)$ (d) $(x-y)$

9. $\left[\dfrac{x+1}{x-1} - \dfrac{x-1}{x+1} - \dfrac{4x}{x^2+1}\right] \div \dfrac{4}{x^4-1}$ when simplified is equal to :

(a) 1   (b) 0

(c) x^2-1 (d) 2

10. The positive square root of the rational expression.

$\left[y^3 - \dfrac{1}{y^3} - 3\left(y - \dfrac{1}{y}\right)\right] \div \left(y - \dfrac{1}{y}\right)$ is :

(a) $y + \dfrac{1}{y}$ (b) 1

(c) $y - \dfrac{1}{y}$ (d) 2

Answers

1. (c)	2. (d)	3. (c)	4. (d)	5. (c)	6. (d)	7. (c)	8. (a)	9. (d)	10. (c)

Chapter

10 LINEAR EQUATIONS IN ONE VARIABLE (REVISION)

KEY FACTS

1. An equation involving only one variable of degree 1 (power of the variable does not exceed 1) is called a linear equation.
2. The value of the variable that makes both the sides of the equation equal is called the **solution** or **root** of the equation.
3. The equation remains unchanged by
   (i) adding the same number to both the sides of an equation or
   (ii) subtracting the same number from both the sides of the equation or
   (iii) multiplying both the sides of the equation by the same non-zero number or
   (iv) dividing both the sides of the equation by the same non-zero number.
4. Transposing a term from LHS to RHS changes the sign of the term from (+ve) to (−ve) and (−ve) to (+ve).
5. To **Solve a linear equation** in one variable, take all the terms involving the variable on one side and the constant terms to the other. Reduce the equation to the from $cx = d$, where $x = \dfrac{d}{c}$.

Question Bank–10

1. The solution of $\dfrac{2x+3}{2x-1} = \dfrac{3x-1}{3x+1}$ is

   (a) $\dfrac{1}{8}$
   (b) $\dfrac{-1}{8}$
   (c) $\dfrac{8}{3}$
   (d) $\dfrac{-8}{3}$

2. If $\dfrac{x+3}{2} + 3x = 5(x-3) + \dfrac{x+23}{5}$, then what is the value of x ?
   (a) 7
   (b) 8
   (c) 10
   (d) 12

3. The solution for $\dfrac{2}{x+3} - \dfrac{4}{x-3} = \dfrac{-6}{x+3}$ is :
   (a) 9
   (b) −1
   (c) −3
   (d) 3

4. If $(3x + 4)^2 + (3x - 2)^2 = (6x + 5)(3x - 2) + 12$, then the value of x is :
   (a) 2
   (b) 1
   (c) −3
   (d) −2

5. Given $\dfrac{(x^2 - 5x + 6)}{(x^2 - 7x + 12)} = \dfrac{(x^2 - x - 6)}{(x^2 + 7x + 10)}$, the value of x is :
   (a) 2.4
   (b) −2.0
   (c) 2.2
   (d) 2.3

6. A number consists of two digits. The digit in the ten's place exceeds the digit in the unit's place by 4. The sum of the digits is $\dfrac{1}{7}$ of the number. The number is :
   (a) 27
   (b) 72
   (c) 48
   (d) 84

7. The sum of the numerator and the denominator of a fraction is 11. If 1 is added to the numerator and 2 is subtracted from the denominator it becomes $\dfrac{2}{3}$. The fraction is :

   (a) $\dfrac{5}{6}$ (b) $\dfrac{3}{8}$

   (c) $\dfrac{4}{7}$ (d) $\dfrac{1}{10}$

8. If Dennis is $\dfrac{1}{3}$rd the age of his father Keith now, and was $\dfrac{1}{4}$th the age of his father 5 years ago, then how old will his father Keith be 5 years from now?

   (a) 20 years (b) 45 years
   (c) 40 years (d) 50 years

9. A daily wage worker was paid Rs 1700 during a period of 30 days. During this period he was absent for 4 days and was fined Rs 15 per day for absence. He was paid the full salary only for 18 days as he came late on the other days. Those who came late were given only half the salary for that day. What was the total salary paid per month to a worker who came on time every day and was never absent ?

   (a) Rs 2400 (b) Rs 3000
   (c) Rs 2700 (d) Rs 2250

10. A person spends $\dfrac{1}{3}$ of the money with him on clothes, $\dfrac{1}{5}$ of the remaining on food and $\dfrac{1}{4}$ of the remaining on travel. Now he is left with Rs 100. How much did he have with him in the beginning ?

   (a) Rs 200 (b) Rs 250
   (c) Rs 300 (d) Rs 450

11. In a three digit number, the digit in the units' place is four times the digit in the hundreds' place. If the digit in the units' place and tens' place and interchanged, the new number so formed is 18 more than the original number. If the digit in the hundreds place is one-third of the digit in the tens place, what is 25% of the original number ?

   (a) 67 (b) 84
   (c) 73 (d) 64

12. My grandfather was 8 times older to me 16 years ago. He would be 3 time of my age 8 years from now. Eight years ago, what was the ratio of my age to that of my grandfather ?

   (a) 1 : 2 (b) 1 : 5
   (c) 13 : 18 (d) 11 : 53

13. If a scooterist drives at the rate of 25 km per hour, he reaches his destination 7 minutes late, and if he drives at the rate of 30 km per hour, he reaches his destination 5 minutes earlier. How far is his destination ?

   (a) 20 km (b) 25 km
   (c) 30 km (d) 32 km

14. The length of a rectangle is 3 times it breadth. If the length is decreased by 3 cm and the breadth increased by 5 cm, the area of the rectangle is increased by 57 cm^2. The perimeter of the rectangle is :

   (a) 18 cm (b) 48 cm
   (c) 24 cm (d) 20 cm

15. In an examination, a candidate attempts 90% of the total questions. Out of these 70% of his answers are correct. Each question carries 3 marks for the correct answer and (–1) mark for the wrong answer. If the marks secured by the candidate is 243, what is the total number of questions ?

   (a) 110 (b) 140
   (c) 150 (d) 200

Answers

1. (b)	2. (a)	3. (a)	4. (d)	5. (c)	6. (d)	7. (b)	8. (d)	9. (a)	10. (b)
11. (a)	12. (d)	13. (c)	14. (b)	15. (c)					

Hints and Solutions

1. (b) $\dfrac{2x+3}{2x-1} = \dfrac{3x-1}{3x+1}$

   $\Rightarrow (2x+3)(3x+1) = (3x-1)(2x-1)$

   $\Rightarrow 6x^2 + 11x + 3 = 6x^2 - 5x + 1$

   $\Rightarrow 11x + 5x = 1 - 3 \Rightarrow 16x = -2 \Rightarrow x = \dfrac{-2}{16} = \dfrac{-1}{8}$

2. (a) $\dfrac{x+3}{2} + 3x = 5(x-3) + \dfrac{x+23}{5}$

   Multiplying all the terms by LCM = 10, we get

   $10 \times \dfrac{(x+3)}{2} + 10 \times 3x = 10 \times 5(x-3) + 10 \times \dfrac{(x+23)}{5}$

   $\Rightarrow 5(x+3) + 30x = 50(x-3) + 2(x+23)$

   $\Rightarrow 5x + 15 + 30x = 50x - 150 + 2x + 46$

$\Rightarrow 35x - 52x = -150 + 46 - 15$

$\Rightarrow -17x = -119 \Rightarrow x = 7$

3. (a) $\dfrac{2}{x+3} - \dfrac{4}{x-3} = \dfrac{-6}{x+3}$

$\Rightarrow \dfrac{2(x-3) - 4(x+3)}{(x+3)(x-3)} = \dfrac{-6}{x+3}$

$\Rightarrow \dfrac{2x - 6 - 4x - 12}{(x+3)(x-3)} = \dfrac{-6}{(x+3)}$

$\Rightarrow \dfrac{-2x - 18}{x-3} = -6 \Rightarrow -2x - 18 = -6x + 18$

$\Rightarrow 4x = 36 \Rightarrow x = 9$

4. (d) $(3x+4)^2 + (3x-2)^2 = (6x+5)(3x-2) + 12$

$\Rightarrow 9x^2 + 24x + 16 + 9x^2 - 12x + 4$

$\qquad\qquad = 18x^2 + 3x - 10 + 12$

$\Rightarrow 18x^2 + 12x + 20 = 18x^2 + 3x + 2$

$\Rightarrow 9x + 18 \Rightarrow x = -2$

5. (c) $\dfrac{x^2 - 5x + 6}{x^2 - 7x + 12} = \dfrac{x^2 - x - 6}{x^2 + 7x + 10}$

$\Rightarrow \dfrac{(x-3)(x-2)}{(x-3)(x-4)} = \dfrac{(x-3)(x+2)}{(x+5)(x+2)}$

$\Rightarrow \dfrac{x-2}{x-4} = \dfrac{x-3}{x+5} \Rightarrow (x-2)(x+5) = (x-3)(x-4)$

$\Rightarrow x^2 + 3x - 10 = x^2 - 7x + 12$

$\Rightarrow 3x + 7x = 12 + 10 \Rightarrow 10x = 22 \Rightarrow x = 2.2$

6. (d) Let the digit in the unit's place $= x$

Let the digit in the ten's place $= x + 4$

$\therefore$ The number is

$\qquad 10(x+4) + x = 10x + 40 + x = 11x + 40$

Sum of the digits $= x + x + 4 = 2x + 4$

Given, $2x + 4 = \dfrac{1}{7}(11x + 40)$

$\Rightarrow 14x + 28 = 11x + 40 \Rightarrow 3x = 12 \Rightarrow x = 4$

$\therefore$ The number is $11 \times 4 + 40 = 44 + 40 = 84$

7. (b) Let the numerator be x. Then,

denominator $= 11 - x$

Given, $\dfrac{x+1}{11-x-2} = \dfrac{2}{3} \Rightarrow 3x + 3 = 2(9-x)$

$\Rightarrow 3x + 3 = 18 - 2x \Rightarrow 5x = 15 \Rightarrow x = 3$

$\therefore$ Numerator $= 3$, Denominator $= 11 - 3 = 8$

$\therefore$ Fraction $= \dfrac{3}{8}$

8. (d) Let Keith's age now be x years. Then,

Dennis's age now $= \dfrac{x}{3}$ years

Keith's age 5 years ago $= (x - 5)$ years

Dennis's age 5 years ago $= \left(\dfrac{x}{3} - 5\right)$ years

Given, $\left(\dfrac{x}{3} - 5\right) = \dfrac{1}{4}(x-5)$

$\Rightarrow \dfrac{x-15}{3} = \dfrac{x-5}{4} \Rightarrow 4x - 60 = 3x - 15$

$\Rightarrow x = 45$

$\therefore$ Keith's age 5 years from now $= (45 + 5)$ years
$\qquad = 50$ years.

9. (a) Let the salary of the worker per day be Rs x.

Then, $18 \times x + 8 \times \dfrac{x}{2} - 4 \times 15 = 1700$

$\Rightarrow 18x + 4x - 60 = 1700 \Rightarrow 22x = 1760$

$\Rightarrow x = \dfrac{1760}{22} = 80$

$\therefore$ Total salary of a worker who came everyday on time for 30 days $= 30 \times$ Rs $80 =$ Rs 2400

10. (b) Let the total money with the person be Rs x.

Then, money spent on clothes $=$ Rs $\dfrac{x}{3}$

Remaining money $=$ Rs $\left(x - \dfrac{x}{3}\right) =$ Rs $\dfrac{2x}{3}$

$\therefore$ Money spent on food $= \dfrac{1}{5} \times$ Rs $\dfrac{2x}{3} =$ Rs $\dfrac{2x}{15}$

Now, remaining money $= \dfrac{2x}{3} - \dfrac{2x}{15} =$ Rs $\dfrac{8x}{15}$

$\therefore$ Money spent on travel $= \dfrac{1}{4} \times$ Rs $\dfrac{8x}{15} =$ Rs $\dfrac{2x}{15}$

Given, $\dfrac{x}{3} + \dfrac{2x}{15} + \dfrac{2x}{15} + 100 = x$

$\Rightarrow \dfrac{5x + 2x + 2x + 1500}{15} = x$

$\Rightarrow 9x + 1500 = 15x \Rightarrow 6x = 1500$

$\Rightarrow x = \dfrac{1500}{6} =$ Rs 250

11. (a) Let the digit in ten's place be x.

Then, the digit in the hundred's place $= \dfrac{x}{3}$ and

digit in the unit's place $= 4 \times \dfrac{x}{3} = \dfrac{4x}{3}$

$\therefore$ The original number $= 100 \times \dfrac{x}{3} + 10 \times x + \dfrac{4x}{3}$

$= \dfrac{100x}{3} + 10x + \dfrac{4x}{3} = \dfrac{100x + 30x + 4x}{3} = \dfrac{134x}{3}$

In the new number

Digit in the ten's place $= \dfrac{4x}{3}$

Digit in the unit's place $= x$

Digit in the hundred's place $= \dfrac{x}{3}$

$\therefore$ New number $= 100 \times \dfrac{x}{3} + 10 \times \dfrac{4x}{3} + x$

$$= \dfrac{100x + 40x + 3x}{3} = \dfrac{143x}{3}$$

Given, $\dfrac{143x}{3} - \dfrac{134x}{3} = 18$

$\Rightarrow \dfrac{9x}{3} = 18 \quad \Rightarrow \quad x = 6$

$\therefore$ Original number $= \dfrac{134}{\cancel{3}_1} \times \cancel{6}^2 = 268$

$\therefore$ 25% of 268 $= \dfrac{25}{100} \times 268 = \dfrac{268}{4} = \mathbf{67}$

12. (d) Let my age 16 years ago be x years.

Then, my grandfather's age 16 years ago
$= 8x$ years

At present, my age $= (x + 16)$ years,

grandfather's age $= (8x + 16)$ years,

8 years from now, my age $= (x + 16 + 8)$ years
$= (x + 24)$ years

grandfather's age $= (8x + 16 + 8)$ years
$= (8x + 24)$ years

Given , $8x + 24 = 3(x + 24)$

$\Rightarrow 8x + 24 = 3x + 72 \quad \Rightarrow \quad 5x = 48$

$\Rightarrow x = \dfrac{48}{8} = 9.6$ years

$\therefore$ My age 16 years ago $= 9.6$ years

Grand father's age 16 years ago $= 8 \times 9.6$ years
$= 76.8$ years

Required ratio $= \dfrac{9.6 + 8}{76.8 + 8} = \dfrac{17.6}{84.8} = \dfrac{11}{53} = 11 : 53$

13. (c) Let the required distance be x km.

Difference in the times take at the two speeds
$= 12$ min $= \dfrac{1}{5}$ hrs

Given,

$\dfrac{x}{25} - \dfrac{x}{30} = \dfrac{1}{5} \Rightarrow \dfrac{6x - 5x}{150} = \dfrac{1}{5} \Rightarrow x = 30 \, \text{km}$

14. (b) Let the breadth of the rectangle be x cm.

Then, length $= 3x$ cm

New breadth $= (x + 5)$ cm

New length $= (3x - 3)$ cm

Then, $(x + 5)(3x - 3) - 3x \times x = 57$

$\Rightarrow 3x^2 + 12x - 15 - 3x^2 = 57$

$\Rightarrow 12x = 57 + 15 = 72 \Rightarrow x = 6$

$\therefore$ Breadth $= 6$ cm, Length $= 18$ cm

Perimeter $= 2(6 \, \text{cm} + 18 \, \text{cm}) = 2 \times 24$ cm
$= 48$ cm

15. (c) Let the total number of questions be x.

Number of questions attempted $= 90\%$ of $x = 0.9x$

Number of correct answers $= 70\%$ of $0.9x$
$= 0.7 \times 0.9x = 0.63x$

Number of incorrect answers $= 30\%$ of $0.9x$
$= 0.3 \times 0.9x = 0.27x$

According to the question,

$0.63x \times 3 + 0.27x \times (-1) = 243$

$\Rightarrow 1.89x - 0.27x = 243$

$\Rightarrow 1.62x = 243 \Rightarrow x = \dfrac{243}{1.62} = \dfrac{24300}{162} = 150$

Self Assessment Sheet–10

1. The solution of $\dfrac{x + a}{x - 2a} = \dfrac{x + 3a}{x - 4a}$ is

  (a) $x = a$ (b) $x = -a/2$

  (c) $x = a/2$ (d) $x = -a$

2. Solve : $\dfrac{t + 2}{3} + \dfrac{1}{t + 1} = \dfrac{t + 3}{2} - \dfrac{t - 1}{6}$

  (a) $t = 1$ (b) $t = -2$

  (c) $t = 2$ (d) $t = 0$

3. The digits of a three-figure number are consecutive odd numbers. The number is 51 less than thirty times the sum of its digits. What is the number ?

  (a) 975   (b) 579

  (c) 759   (d) 597

4. A wire is bent so as to form four sides of a square. A length of 4 cm is cut from it and the remainder is again bent to form the four sides of a square. If the difference in areas of the squares is 25 cm², how long was the wire before being cut ?

  (a) 52 cm   (b) 48 cm

  (c) 12 cm   (d) 13 cm

5. The distance between two towns is 20 km less by road than by rail. A train takes three hours for the journey, a car four hours. If the average speed of the car is 15 km /hr less than that of the train, what

is the average speeds of the car ?

(a) 40 km/hr   (b) 35 km/hr

(c) 25 km/hr   (d) 50 km/hr

6. A man's age is now four times that of his son and it is also three times that of his daughter. In six years time it will be three times that of his son. How old was he when his daughter was born ?

(a) 48 years   (b) 26 years

(c) 32 years   (d) 40 years

7. A man saves the same amount each year and now he has Rs 1200 saved. In four years' time he will have three times as much saved as he had four years ago. How much does he same each year ?

(a) Rs 120   (b) Rs 100

(c) Rs 150   (d) Rs 200

8. A square carpet when placed in a room leaves a strip 1 m wide uncarpeted along one pair of opposite walls and a strip 2 m wide along the other two walls.

If the area uncarpeted is 62 sq m, what is the length of the carpet ?

(a) 10 m   (b) 20 m

(c) 8 m   (d) 7 m

9. ABCD is a rectangle, sides 36 cm and 90 cm. P is a point on BC which is one of the longer sides such that PA = 2PD. The length of PB is

(a) 80 cm   (b) 18 cm

(c) 72 cm   (d) 64 cm

10. A line is divided into three parts. The first part is two-thirds the length of the second part; the third part is $\frac{1}{2}$ cm shorter than the first part, and 2 cm shorter than the second part. Find the length of the line ?

(a) 12 cm   (b) 10 cm

(c) 9 cm   (d) 8 cm

Answers

1. (c)   2. (d)   3. (b)   4. (a)   5. (c)   6. (c)   7. (b)   8. (b)   9. (c)   10. (b)

Chapter
11
SIMULTANEOUS LINEAR EQUATIONS

KEY FACTS

1. A linear equation in two variables is represented algebraically as $ax + by + c = 0$, where $a \neq 0$, $b \neq 0$.
   Graphically it represents a straight line.

2. A pair of linear equations in two variables x and y (simultaneous linear equations) can be represented algebraically as follows:

   $$a_1x + b_1y + c_1 = 0$$
   $$a_2x + b_2y + c_2 = 0$$

   where $a_1, a_2, b_1, b_2, c_1, c_2$ are real numbers such that $a_1^2 + b_1^2 \neq 0$ and $a_2^2 + b_2^2 \neq 0$.

3. Simultaneous linear equations in two variables can be solved by :
   (i) Algebraic method   (ii) Graphical method

4. To solve the simultaneous linear equations algebraically. We use the following methods :
   (i) Substitution method   (ii) Elimination method
   Both the methods have been explained with the help of solved examples.

5. If $a_1x + b_1y + c_1 = 0$
   $a_2x + b_2y + c_2 = 0$
   is a pair of linear equations in two variables x and y such that :

   (i) $\dfrac{a_1}{a_2} \neq \dfrac{b_1}{b_2}$, then the pair of linear equations is consistent with a unique solution, *i.e.*, they intersect at a point.

   (ii) $\dfrac{a_1}{a_2} = \dfrac{b_1}{b_2} \neq \dfrac{c_1}{c_2}$, then the pair of linear equations inconsistent with no solution, *i.e.*, they represent a pair of parallel lines.

   (iii) $\dfrac{a_1}{a_2} = \dfrac{b_1}{b_2} = \dfrac{c_1}{c_2}$, then the pair of linear equations are consistent with infinitely many solutions, *i.e.*, they represent coincident lines.

Solved Examples

Ex. 1. *If $3x + 7y = 75$ and $5x - 5y = 25$, then what is the value of $x + y$?*

Sol. $3x + 7y = 75$ ...(1)

         $5x - 5y = 25$ ...(2)

In this example, we have to transform each equation to obtain the same numerical coefficient of one of the variables. To do this we multiply the first equation by 5 and the second equation by 7. Then, the variable y has the same coefficient.

$$15x + 35y = 375 \qquad \text{Multiply (1) by 5}$$
$$35x - 35y = 175 \qquad \text{Multiply (2) by 7}$$
$$\overline{50x = 550} \qquad \text{Add}$$

$$x = \frac{550}{50} = 11$$

Putting $x = 11$ in (1), we get

$$3 \times 11 + 7y = 75 \quad \Rightarrow \quad 33 + 7y = 75 \quad \Rightarrow \quad 7y = 75 - 33 = 42 \quad \Rightarrow \quad y = 6$$

$\therefore \ x + y = 11 + 6 = \mathbf{17}.$

Ex. 2. *Find the solution set of the system of equations:* $\dfrac{4}{x} + 5y = 7$ *and* $\dfrac{3}{x} + 4y = 5.$

Sol. Let $\dfrac{1}{x} = a.$ Then the system of equations become

$$4a + 5y = 7 \qquad \qquad \qquad \qquad \qquad \qquad \qquad \dots (1)$$
$$3a + 4y = 5 \qquad \qquad \qquad \qquad \qquad \qquad \qquad \dots (2)$$

Multiplying equation (1) by 3 and equation (2) by 4 and subtracting equation (2) from equation (1), we get

$$12a + 15y = 21 \qquad \text{Multiply (1) by 3}$$
$$12a + 16y = 20 \qquad \text{Multiply (2) by 4}$$
$$\overline{\ -\ \ -\ \qquad \qquad -}$$
$$-y = +1 \qquad \text{Subtract}$$

$\Rightarrow \qquad \qquad \qquad \quad y = -1$

Putting $y = -1$ in (1), we get

$$4a - 5 = 7 \Rightarrow 4a = 12 \ \Rightarrow \ a = 3 \Rightarrow x = \frac{1}{3}$$

$\therefore$ The solution set is $\left(\dfrac{1}{3}, -1 \right).$

Ex. 3. *Which one of the following is the solution of the system of linear equations* $\left(\dfrac{5}{x} \right) - \left(\dfrac{4}{y} \right) = 3$ *and* $\left(\dfrac{9}{x} \right) - \left(\dfrac{8}{y} \right) = 7$ *?*

    (a) x is +ve and y is –ve   *(b) y is +ve and x is –ve*   *(c) Both x and y are +ve*   *(d) Both x and y are –ve*

Sol. Let $\dfrac{1}{x} = a$ and $\dfrac{1}{y} = b.$ Then, the system of linear equations is

$$5a - 4b = 3 \qquad \qquad \qquad \qquad \qquad \qquad \qquad \dots (1)$$
$$9a - 8b = 7 \qquad \qquad \qquad \qquad \qquad \qquad \qquad \dots (2)$$

$$45a - 36b = 27 \qquad \text{Multiply (1) by 9}$$
$$45a - 40b = 35 \qquad \text{Multiply (2) by 5}$$
$$\overline{-\ \ +\ \qquad \qquad -}$$
$$4b = -8 \qquad \text{Subtract}$$
$$b = -2$$

$\Rightarrow \qquad \qquad \qquad \quad y = -\dfrac{1}{2}$

Putting $b = -2$ in (1), we get

$$5a - 4 \times -2 = 3 \ \Rightarrow \ 5a + 8 = 3 \Rightarrow 5a = -5 \Rightarrow \ a = -1 \Rightarrow x = -1$$

$\therefore \ (d)$ is the correct option. Both x and y are negative.

Ex. 4. *If $2^a + 3^b = 17$ and $2^{a+2} - 3^{b+1} = 5$, then find the value of a and b.*

Sol. $2^a + 3^b = 17$ ...(1)

$$2^{a+2} - 3^{b+1} = 5 \Rightarrow 2^2 \cdot 2^a - 3 \cdot 3^b = 5$$
$$\Rightarrow \qquad 4 \cdot 2^a - 3 \cdot 3^b = 5 \qquad\qquad ...(2)$$

Let $2^a = x$ and $3^b = y$. Then the system of equations is
$$x + y = 17 \qquad\qquad ...(3)$$
$$4x - 3y = 5 \qquad\qquad ...(4)$$

Multiplying equations (3) by 3 and adding to equation (4), we get
$$3x + 3y = 51$$

$$\dfrac{4x - 3y = 5}{7x \qquad = 56} \qquad \text{Add}$$

$$x = \dfrac{56}{7} = 8$$

Putting $x = 8$ in (3), we get
$$8 + y = 17 \Rightarrow y = 17 - 8 = 9$$
$$\Rightarrow 2^a = 8 \text{ and } 3^b = 9 \Rightarrow 2^a = 2^3 \text{ and } 3^b = 3^2$$
$$\Rightarrow a = \mathbf{3} \text{ and } b = \mathbf{2}.$$

Ex. 5. *For what value of k does the system of equations $2x + ky = 11$ and $5x - 7y = 5$ has no solution ?*

Sol. The given equations are : $2x + ky - 11 = 0$ and $5x - 7y - 5 = 0$

Here, $a_1 = 2, b_1 = k, c_1 = -11$ and $a_2 = 5, b_2 = -7, c_2 = -5$

The two equations will have no solution if,

$$\dfrac{a_1}{a_2} = \dfrac{b_1}{b_2} \neq \dfrac{c_1}{c_2}, \text{ i.e., } \dfrac{2}{5} = \dfrac{k}{-7} \neq \dfrac{-11}{-5}$$

$$\Rightarrow \dfrac{2}{5} = \dfrac{k}{-7} \Rightarrow k = \dfrac{-14}{5}.$$

Ex. 6. *If the equations $4x + 7y = 10$ and $10x + ky = 25$ represent coincident lines, then find the value of k ?*

Sol. The given equations are : $4x + 7y - 10 = 0$ and $10x + ky - 25 = 0$

Here, $a_1 = 4, b_1 = 7, c_1 = -10$
$$a_2 = 10, b_2 = k, c_2 = -25$$

For the given equation to represent coincident lines,

$$\dfrac{a_1}{a_2} = \dfrac{b_1}{b_2} = \dfrac{c_1}{c_2}, \text{ i.e., } \dfrac{4}{10} = \dfrac{7}{k} = \dfrac{-10}{-25}$$

$$\Rightarrow \dfrac{4}{10} = \dfrac{7}{k} \Rightarrow k = \dfrac{70}{4} = \dfrac{\mathbf{35}}{\mathbf{2}}.$$

Ex. 7. *Find the condition that the system of equations $ax + by = c$ and $lx + my = n$ has a unique solution ?*

Sol. The given equations are : $ax + by - c = 0$ and $lx + my - n = 0$

Here, $a_1 = a, b_1 = b, c_1 = -c,$ and $a_2 = l, b_2 = m, c_2 = -n$

For unique solution,

$$\dfrac{a_1}{a_2} \neq \dfrac{b_1}{b_2}, \text{ i.e., } \dfrac{a}{l} \neq \dfrac{b}{m} \Rightarrow am \neq bl.$$

Ex. 8. *Find the values of p and q for which the following system of linear equations has infinite number of solutions : $2x + 3y = 1$, $(p + q)x + (2p - q)y = 21$*

Sol. The given equations are :
$$2x + 3y - 1 = 0 \text{ and } (p + q)x + (2p - q)y - 21 = 0$$

Here, $a_1 = 2$, $b_1 = 3$, $c_1 = -1$ and $a_2 = p + q$, $b_2 = 2p - q$, $c_2 = -21$

For infinite solutions,

$$\frac{a_1}{a_2} = \frac{b_1}{b_2} = \frac{c_1}{c_2} \Rightarrow \frac{2}{p+q} = \frac{3}{2p-q} = \frac{-1}{-21} \Rightarrow \frac{2}{p+q} = \frac{1}{21} \text{ and } \frac{3}{2p-q} = \frac{1}{21}$$

$\Rightarrow p + q = 42$ \qquad ...(1) \quad and \quad $2p - q = 63$ \qquad ...(2)

Adding (1) and (2), we get

$3p = 105 \Rightarrow p = 35$

Putting $p = 35$ in (1), we get

$35 + q = 42 \Rightarrow q = 42 - 35 = 7$

$\therefore p = 35$, $q = 7$.

Ex. 9. *3 chairs and 2 tables cost Rs 700 while 5 chairs and 3 tables cost Rs 1100. What is the cost of 2 chairs and 2 tables ?*

Sol. Let the cost of one chair be Rs x and that of a table be Rs y.

Then, $3x + 2y = 700$ \hfill ...(1)

$5x + 3y = 1100$ \hfill ...(2)

$9x + 6y = 2100$ \qquad Multiply (1) by 3

$10x + 6y = 2200$ \qquad Multiply (2) by 2

$- \quad - \quad - \quad -$

$-x = -100$ \qquad Subtract

$x = 100$

$\therefore$ Putting $x = 100$ in (1), we get

$300 + 2y = 700 \Rightarrow 2y = 400 \Rightarrow y = 200$

$\therefore$ Cost of one chair = Rs 100, Cost of one table = Rs 200.

$\Rightarrow$ Cost of 2 chairs and 2 tables = $2 \times$ Rs $100 + 2 \times$ Rs 200

$= $ Rs $200 +$ Rs $400 = $ Rs **600**.

Ex. 10. *The average of two numbers is 6 and four times the difference between them is 16. Find the numbers.*

Sol. Let the numbers be x and y. Given,

$$\frac{x+y}{2} = 6 \Rightarrow x + y = 12$$ \hfill ...(1)

$4(x - y) = 16 \Rightarrow 4x - 4y = 16$ \hfill ...(2)

Multiplying equation (1) by 4 and adding to equation (2), we get

$4x + 4y = 48$ \quad Multiplying (1) by 4

$\underline{4x - 4y = 16}$

$8x \quad = 64$ \quad Add

$\Rightarrow \qquad x = \dfrac{64}{8} = 8$

Putting $x = 8$ in (1), we get

$8 + y = 12 \Rightarrow y = 12 - 8 = 4$

$\therefore$ The numbers are 8 and 4.

Ex. 11. *A part of monthly expenses of a family is constant and the remaining part varies with the price of wheat. When the rate of wheat is Rs 250 a quintal, the total monthly expenses of the family are Rs 1000 and when it is Rs 240 a quintal, the total monthly expenses are Rs 980. Find the total monthly expenses of the family, when the cost of wheat is Rs 350 a quintal.*

Sol. Let the constant part be Rs c and quantity of wheat consumed per month be q quintals. Then,

$$c + 250q = 1000 \qquad \qquad ....(1)$$
$$\frac{c + 240q = 980}{10q = 20} \qquad \text{Subtract} \qquad ....(2)$$

$$q = 2$$

Putting $q = 2$ in equation (1), we get

$$c + 250 \times 2 = 1000 \Rightarrow c + 500 = 1000 \Rightarrow c = 500$$

$\therefore$ Total monthly by expenses of the family when the rate of wheat is Rs 350 a quintal

$$= 500 + 350 \times 2 = 500 + 700 = \textbf{Rs 1200}.$$

Ex. 12. *If the numerator of a certain fraction is increased by 2 and the denominator increased by 1, then the resulting fraction equals* $\dfrac{1}{2}$. *If however the numerator is increased by 1 and the denominator decreased by 2, then the resulting fraction equal* $\dfrac{3}{5}$. *Find the fraction.*

Sol. Let the given fraction be $\dfrac{x}{y}$. Then,

$$\frac{x+2}{y+1} = \frac{1}{2} \quad \Rightarrow \quad 2x + 4 = y + 1 \quad \Rightarrow \quad 2x - y = -3 \qquad ...(1)$$

$$\frac{x+1}{y-2} = \frac{3}{5} \quad \Rightarrow \quad 5x + 5 = 3y - 6 \quad \Rightarrow \quad 5x - 3y = -11 \qquad ...(2)$$

Now,
$$6x - 3y = -9 \qquad \text{Multiply (1) by (3)}$$
$$\frac{5x - 3y = -11}{}$$
$$\begin{array}{ccc} - & + & + \end{array}$$
$$\overline{x = 2} \qquad \text{Subtract}$$

Putting $x = 2$ in equation (1), we get

$$4 - y = -3 \Rightarrow -y = -3 - 4 \Rightarrow y = 7$$

$\therefore$ The given fraction is $\dfrac{2}{7}$.

Ex. 13. *Ram buys 4 horses and 9 cows for Rs 13400. If he sells the horses at 10% profit and cows at 20% profit, then he earns a total profit of Rs 1880. What is the cost of a horse ?*

Sol. Let H be the cost of a horse and C the cost of a cow.

Then, $4H + 9C = 13400$ \qquad\qquad ...(1)

Also, $\dfrac{10}{100} \times 4H + \dfrac{20}{100} \times 9C = 1880$

$\Rightarrow 4H + 18C = 18800$ \qquad\qquad ...(2)

Subtracting equation (1) from equation (2), we get

$$9C = 5400 \Rightarrow C = \frac{5400}{9} = 600$$

Putting $C = 600$ in (1), we get

$$4H + 9 \times 600 = 13400 \Rightarrow 4H = 8000 \Rightarrow H = \frac{8000}{4} = 2000.$$

$\therefore$ The cost of a horse is **Rs 2000**.

Ex. 14. *Students of a class are made to stand in rows. If 4 students are extra in each row, then there would be 2 rows less. If four students are less in each row, then there would be 4 more rows. What is the number of students in the class ?*

Sol. Let the number of students in each row be x and the number of rows be y. Then total number of students = xy

∴ By the given conditions,

$(x + 4)(y - 2) = xy \Rightarrow xy + 4y - 2x - 8 = xy$

$\Rightarrow -2x + 4y = 8$ ...(1)

and $(x - 4)(y + 4) = xy$

$\Rightarrow xy - 4y + 4x - 16 = xy$

$\Rightarrow 4x - 4y = 16$ ...(2)

Adding equations (1) and (2), we get

$2x = 24 \Rightarrow x = 12$

Putting $x = 12$ in equation (1), we get

$-2 \times 12 + 4y = 8 \Rightarrow 4y = 8 + 24 = 32 \Rightarrow y = 8$

∴ Total number of students in the class = $x \times y = 12 \times 8 =$ **96**.

Question Bank–11

1. The solution of the two simultaneous equations $2x + y = 8$ and $3y = 4 + 4x$ is :
   (a) $x = 4, y = 1$ (b) $x = 1, y = 4$
   (c) $x = 2, y = 4$ (d) $x = 3, y = -4$

2. The solution of the simultaneous equation $\dfrac{x}{2} + \dfrac{y}{3} = 4$ and $x + y = 10$ is given by
   (a) $(6, 4)$ (b) $(4, 6)$
   (c) $(-6, 4)$ (d) $(6, -4)$

3. The course of an enemy submarine as plotted on a set of rectangular axes is $2x + 3y = 5$. On the same axes the course of the destroyer is indicated by $x - y = 10$. The point (x, y) at which the submarine can be destroyed is :
   (a) $(-7, 3)$ (b) $(-3, 7)$
   (c) $(3, -7)$ (d) $(7, -3)$

4. The values of x and y satisfying $(x + 3)(y - 5) = xy + 39$ and $(x - 2)(y + 3) = xy - 40$ respectively are :
   (a) 6 and 8 (b) 6 and –8
   (c) –6 and 8 (d) –6 and –8

5. If the sum and the difference of two expressions is $5x^2 - x - 4$ and $x^2 + 9x - 10$ respectively, then the expressions are :
   (a) $(4x^2 + 8x - 6)$ and $(4x^2 - 10x + 2)$
   (b) $(2x^2 + 4x - 3)$ and $(3x^2 - 10x - 6)$
   (c) $(3x^2 + 4x - 7)$ and $(2x^2 - 5x + 3)$
   (d) $(3x^2 + 4x + 7)$ and $(2x^2 - 5x - 3)$

6. If $\dfrac{5x + 6}{(2 + x)(1 - x)} = \dfrac{a}{2 + x} + \dfrac{b}{1 - x}$, then the values of a and b respectively are :
   (a) $-\dfrac{5}{3}, \dfrac{6}{5}$ (b) $\dfrac{5}{3}, -\dfrac{6}{5}$
   (c) $-\dfrac{4}{3}, \dfrac{11}{3}$ (d) $\dfrac{4}{3}, -\dfrac{11}{3}$

7. If $(4)^{x+y} = 1$ and $(4)^{x-y} = 4$, then the value of x and y will be respectively :
   (a) $\dfrac{1}{2}$ and $-\dfrac{1}{2}$ (b) $\dfrac{1}{2}$ and $\dfrac{1}{2}$
   (c) $-\dfrac{1}{2}$ and $-\dfrac{1}{2}$ (d) $-\dfrac{1}{2}$ and $\dfrac{1}{2}$

8. If $\dfrac{2}{x} + \dfrac{3}{y} = \dfrac{9}{xy}$ and $\dfrac{4}{x} + \dfrac{9}{y} = \dfrac{21}{xy}$, where $x \neq 0$, $y \neq 0$, then the value of x and y are respectively :
   (a) 0 and 1 (b) 1 and 2
   (c) 2 and 3 (d) 1 and 3

9. If $\dfrac{2}{x} + \dfrac{3}{y} = 2$ and $\dfrac{6}{x} + \dfrac{18}{y} = 9$, then the values of x and y respectively are :
   (a) 3 and 2 (b) 2 and 3
   (c) 4 and 3 (d) 3 and 4

10. The solution of the equations $\dfrac{p}{x} + \dfrac{q}{y} = m, \dfrac{q}{x} + \dfrac{p}{y} = n$ is
    (a) $x = \dfrac{q^2 - p^2}{mp - nq}, y = \dfrac{p^2 - q^2}{np - mq}$
    (b) $x = \dfrac{p^2 - q^2}{mp - nq}, y = \dfrac{q^2 - p^2}{np - mq}$

(c) $x = \dfrac{p^2 - q^2}{mp - nq}, \; y = \dfrac{p^2 - q^2}{np - mq}$

(d) $x = \dfrac{q^2 - p^2}{mp - nq}, \; y = \dfrac{q^2 - p^2}{np - mq}$

11. For what value of k, the following system of equations has a unique solution : $2x + 3y - 5 = 0$, $kx - 6y - 8 = 0$?
    (a) $k = -4$ (b) $k \neq -4$
    (c) $k \neq 4$ (d) $k = 4$

12. The value of k for which the equations $9x + 4y = 9$ and $7x + ky = 5$, have no solution is :
    (a) $\dfrac{9}{5}$ (b) $\dfrac{9}{7}$
    (c) $\dfrac{9}{28}$ (d) $\dfrac{28}{9}$

13. For what value of k will the given equations in two variables represent coincident lines : $2x + 32y + 3 = 0$ and $3x + 48y + k = 0$?
    (a) $\dfrac{2}{3}$ (b) $\dfrac{3}{2}$
    (c) $\dfrac{9}{2}$ (d) 1

14. The simultaneous equatoins $2x + 3y = 5$, $4x + 6y = 10$ have :
    (a) no solution   (b) only one solution
    (c) only two solutions   (d) several solutions

15. If $2a = b$, the pair of equations $ax + by = 2a^2 - 3b^2$, $x + 2y = 2a - 6b$ possess :
    (a) no solution
    (b) only one solution
    (c) only two solutions
    (d) an infinite number of solutions

16. A bill for Rs 40 is paid by means of Rs 5 notes and Rs 10 notes. Seven notes are used in all. If x is the number of Rs 5 notes and y is the number of Rs 10 notes, then
    (a) $x + y = 7$ and $x + 2y = 40$
    (b) $x + y = 7$ and $x + 2y = 8$
    (c) $x + y = 7$ and $2x + y = 8$
    (d) $x + y = 7$ and $2x + y = 40$

17. The total cost of 8 apples and 5 oranges is Rs 92 and the total cost of 5 apples and 8 oranges is Rs 77, Find the cost of 2 oranges and 3 apples.
    (a) Rs 30   (b) Rs 35
    (c) Rs 38   (d) Rs 70

18. In a group of buffaloes and ducks, the number of legs are 24 more than twice the number of heads. What is the number of buffaloes in the group ?

(a) 6   (b) 12
(c) 8   (d) 10

19. X has pens and pencils which together are 40 in number. If he had 5 more pencils and 5 less pens, the number of pencils would have become 4 times the number of pens. Find the original number of pens.
    (a) 10   (b) 11
    (c) 12   (d) 13

20. If 1 is added to the age of the elder sister, then the ratio of the ages of the two sisters becomes 0.5 : 1, but if 2 is subtracted from the age of the younger one, the ratio becomes 1 : 3. The age of the younger sister will be
    (a) 9 years   (b) 5 years
    (c) 18 years   (d) 15 years

21. Ram and Mohan are friends. Each has some money. If Ram gives Rs 30 to Mohan, then Mohan will have twice the money left with Ram. But if Mohan gives Rs 10 to Ram, then Ram will have thrice as much as is left with Mohan. How much money does each have ?
    (a) Rs 62, Rs 34   (b) Rs 6, Rs 2
    (c) Rs 170, Rs 124   (d) Rs 43, Rs 26

22. A fraction becomes 2 when 1 is added to both the numerator and the denominator, and it beocmes 3 when 1 is subtracted from both the numerator and the denominator. The numerator of the given fraction is
    (a) 7   (b) 4
    (c) 3   (d) 2

23. On selling a pen at 5% loss and a book at 15% gain, Karim gains Rs 7. If he sell the pen at 5% gain and the book at 10% gain, then he gains Rs 13. The actual price of the book is
    (a) Rs 100   (b) Rs 80
    (c) Rs 10   (d) Rs 400

24. If three times the larger of the two numbers is divided by the smaller one, we get 4 as quotient and 3 as the remainder. Also if seven times the smaller number is divided by the larger one, we get 5 as quotient and 1 as remainder. Find the sum of the numbers.
    (a) 34   (b) 43
    (c) 47   (d) 74

25. If the two digits of the ages of Mr. Manoj are reversed, then the new age so obtained is the age of his wife. $\dfrac{1}{11}$ of the sum of their ages is equal to the difference between their ages. If Mr. Manoj is elder than his wife, then find the difference between their ages.
    (a) 10 years   (b) 8 years
    (c) 7 years   (d) 9 years

26. I had Rs 14.40 in one-rupee coins and 20 paise coins when I went out shopping. When I returned, I had as many one rupee coins as I originally had 20 paise coins and as many 20 paise coins as I originally had one rupee coins. Briefly, I came back with about one-third of what I had started out with. How many one-rupee coins did I have initially ?

    (a) 10 (b) 12
    (c) 14 (d) 16

27. Five times A's money added to B's money is more than Rs 51.00. Three times A's money minus B's money is Rs 21.00. If a represents A's money in Rs and b represents B's money in Rs, then :

    (a) $a > 9, b > 6$
    (b) $a > 9, b < 6$
    (c) $a > 9, b = 6$
    (d) $a > 9$, but we can put no bounds on b

28. In a triangle ABC, $\angle A = x°$, $\angle B = y°$ and $\angle C = (y + 20)°$. If $4x - y = 10$, then the triangle is

    (a) Right angled (b) Obtuse angled
    (c) Equilateral (d) None of these

29. A two digit number is obtained by either multiplying sum of the digits by 8 and adding 1 or by multiplying the difference of the digits by 13 and adding 2. The number is

    (a) 14 (b) 42
    (c) 24 (d) 41

30. The population of a town is 53,000. If in a year the number of males was to increase by 6% and that of the females by 4%, the population will grow to 55,630. Find the difference between the number of males and females in the town at present.

    (a) 3000 (b) 4000
    (c) 2000 (d) 5000

31. Places A and B are 100 km apart from each other on a highway. A car starts from A another from B at the same time. If they move in the same direction, they meet in 10 hours and if they move in opposite

direction, they meet in 1 hour 40 minutes. Find the speed of the cost from places A and B respectively are :

    (a) 45 km/hr, 25 km/hr
    (b) 65 km/hr, 75 km/hr
    (c) 35 km/hr, 25 km/hr
    (d) 60 km/hr, 45 km/hr

32. A person invested some amount at the rate of 10% simple interest and some amount at the rate of 12% simple interest. He received an yearly interest of Rs 130. But if he had interchanged the amounts invested, he would have received Rs 4 more as interest. How much amounts did he invest at 12% and 10% respectively ?

    (a) Rs 600, Rs 550 (b) Rs 800, Rs 450
    (c) Rs 700, Rs 500 (d) Rs 500, Rs 700

33. Solve the following system of equations for x and y, $x, y \neq 0$.
    $$\frac{a}{x} - \frac{b}{y} = 0, \frac{ab^2}{x} + \frac{a^2b}{y} = a^2 + b^2$$
    (a) $x = b, y = a$ (b) $x = a, y = b$
    (c) $x = -a, y = -b$ (d) $x = -b, y = -a$

34. A man had a certain number of oranges. He divides them into two lots A and B. He sells the first lot at the rate of Rs 2 for 3 oranges and the second lot at the rate Re 1 per orange and gets a total of Rs 400. If he had sold the first lot at the rate of Re 1 per orange and the second lot at the rate of Rs 4 for 5 oranges, his total collection would have been Rs 460. Find the total number of oranges he had ?

    (a) 50 (b) 100
    (c) 500 (d) 400

35. Two men and 7 children complete a certain piece of work in 4 days while 4 men and 4 children complete the same work in only 3 days. The number of days required by 1 man to complete the work is

    (a) 60 days (b) 15 days
    (c) 6 days (d) 51 days

Answers

1. (c)	**2.** (b)	**3.** (d)	**4.** (c)	**5.** (c)	**6.** (c)	**7.** (a)	**8.** (d)	**9.** (b)	**10.** (c)
11. (b)	**12.** (d)	**13.** (c)	**14.** (d)	**15.** (d)	**16.** (b)	**17.** (b)	**18.** (b)	**19.** (d)	**20.** (b)
21. (a)	**22.** (a)	**23.** (b)	**24.** (b)	**25.** (d)	**26.** (c)	**27.** (a)	**28.** (a)	**29.** (d)	**30.** (c)
31. (c)	**32.** (d)	**33.** (a)	**34.** (c)	**35.** (b)					

Hints and Solutions

1. (c) $2x + y = 8$ ...(i)

$-4x + 3y = 4$ ...(ii)

Multiplying eqn (i) by 2 and adding to eqn (ii), we get

$4x + 2y - 4x + 3y = 16 + 4$

$\Rightarrow 5y = 20 \Rightarrow y = 4$

Putting $y = 4$ in (i), we get $2x + 4 = 8$

$\Rightarrow 2x = 4 \Rightarrow x = \mathbf{2}.$

2. (b) $\dfrac{x}{2} + \dfrac{y}{3} = 4 \Rightarrow 6 \times \dfrac{x}{2} + 6 \times \dfrac{y}{3} = 6 \times 4$

$\Rightarrow 3x + 2y = 24$ ...(i)

Given, $x + y = 10 \Rightarrow y = 10 - x$ ...(ii)

Substituting the value of y in (i), we get

$3x + 2(10 - x) = 24$

$\Rightarrow 3x + 20 - 2x = 24 \Rightarrow x = 24 - 20 = 4$

$\therefore$ From eqn (ii), $y = 10 - 4 = \mathbf{6}.$

3. (d) $x - y = 10 \Rightarrow x = y + 10$

Substitute this value of x in eqn $2x + 3y = 5$.

Then solve yourself.

4. (c) $(x + 3)(y - 5) = xy + 39$

$\Rightarrow xy + 3y - 5x - 15 = xy + 39$

$\Rightarrow -5x + 3y = 54 \Rightarrow 5x - 3y = -54$ ...(i)

$(x - 2)(y + 3) = xy - 40$

$\Rightarrow xy - 2y + 3x - 6 = xy - 40$

$\Rightarrow -2y + 3x = -34 \Rightarrow 3x - 2y = -34$ ...(ii)

Multiply (i) by 3 and (ii) by 5 and then subtract eqn (ii) from eqn (i).

Solve yourself now.

5. (c) Let the two expressions be A and B.

Then, $A + B = 5x^2 - x - 4$ ...(i)

$A - B = x^2 + 9x - 10$ ...(ii)

Adding eqn (i) and (ii), we get

$2A = (5x^2 + x^2) + (-x + 9x) + (-4 - 10)$

$= 6x^2 + 8x - 14$

$\Rightarrow A = 3x^2 + 4x - 7$

$\therefore$ From (i), $B = (5x^2 - x - 4) - A$

$= (5x^2 - x - 4) - (3x^2 + 4x - 7)$

$= 2x^2 - 5x + 3$

6. (c) $\dfrac{5x + 6}{(2 + x)(1 - x)} = \dfrac{a}{(2 + x)} + \dfrac{b}{(1 - x)}$

$\Rightarrow \dfrac{a(1 - x) + b(2 + x)}{(2 + x)(1 - x)} = \dfrac{5x + 6}{(2 + x)(1 - x)}$

$\Rightarrow \dfrac{a - ax + 2b + bx}{(2 + x)(1 - x)} = \dfrac{5x + 6}{(2 + x)(1 - x)}$

$\Rightarrow \dfrac{(a + 2b) - x(a - b)}{(2 + x)(1 - x)} = \dfrac{5x + 6}{(2 + x)(1 - x)}$

Equating the coefficient of x and constant term on both the sides of the equation, we get

$a + 2b = 6$ ...(i)

$a - b = -5$ ...(ii)

Subtracting eqn (ii) from eqn (i), we get

$3b = 11 \Rightarrow b = \dfrac{11}{3}$

Putting the value of b in (ii), we get

$a - \dfrac{11}{3} = -5$

$\Rightarrow a = -5 + \dfrac{11}{3} = \dfrac{-15 + 11}{3} = \dfrac{-4}{3}.$

7. (a) $(4)^{x+y} = 1 \Rightarrow (4)^{x+y} = 4^0 \Rightarrow x + y = 0$ ...(i)

$(4)^{(x-y)} = 4 \Rightarrow (4)^{x-y} = 4^1 \Rightarrow x - y = 1$ ...(ii)

Adding eqn (i) and (ii), $2x = 1 \Rightarrow x = \dfrac{1}{2}$

$\therefore$ From eqn (i), $\dfrac{1}{2} + y = 0 \Rightarrow y = -\dfrac{1}{2}.$

8. (d) $\dfrac{2}{x} + \dfrac{3}{y} = \dfrac{9}{xy} \Rightarrow \dfrac{2y + 3x}{xy} = \dfrac{9}{xy}$

$\Rightarrow 3x + 2y = 9$ ...(i)

$\dfrac{4}{x} + \dfrac{9}{y} = \dfrac{21}{xy} \Rightarrow \dfrac{4y + 9x}{xy} = \dfrac{21}{xy}$

$\Rightarrow 9x + 4y = 21$ ...(ii)

Multiplying eqn (i) by 3 and subtracting from eqn (ii), we get

$(9x - 9x) + (4y - 6y) = 21 - 27$

$\Rightarrow -2y = -6 \Rightarrow y = 3$

Now substitute the value of y in eqn (i) and find x.

9. (b) Let $\dfrac{1}{x} = a, \dfrac{1}{y} = b,$ Then, the given equations reduce to: $2a + 3b = 2$ ...(i)

$6a + 18b = 9$ ...(ii)

Now solve eqn (i) and (ii) to find the values of a and b, and then x and y.

10. (c) Let $\dfrac{1}{x} = a$ and $\dfrac{1}{y} = b$. Then, the given equations

reduce to

$$pa + qb = m \qquad\qquad …(i)$$
$$qa + pb = n \qquad\qquad …(ii)$$

Multiplying eqn (i) by q and eqn (ii) by p, we get

$$pqa + q^2 b = mq \qquad\qquad …(iii)$$
$$qpa + p^2 b = np \qquad\qquad …(iv)$$

Now, subtracting eqn (iii) from eqn (iv),

$$(p^2 - q^2)b = np - mq$$

$$\Rightarrow b = \frac{np - mq}{p^2 - q^2} \quad\Rightarrow\quad y = \frac{1}{b} = \frac{p^2 - q^2}{np - mq}$$

Substituting this value of b in (i), we have

$$pa + \frac{q(np - mq)}{p^2 - q^2} = m$$

$$\Rightarrow pa = m - \frac{(pqn - mq^2)}{p^2 - q^2}$$

$$\Rightarrow pa = \frac{mp^2 - mq^2 - pqn + mq^2}{p^2 - q^2}$$

$$\Rightarrow pa = \frac{p(mp - qn)}{p^2 - q^2}$$

$$\Rightarrow a = \frac{mp - qn}{p^2 - q^2} \Rightarrow x = \frac{1}{a} = \frac{p^2 - q^2}{mp - qn}.$$

11. (b) In the given system of equations,

$$a_1 = 2, b_1 = 3, c_1 = -5$$
$$a_2 = k, b_2 = -6, c_2 = -8$$

For a unique solution.

$$\frac{a_1}{a_2} \neq \frac{b_1}{b_2} \Rightarrow \frac{2}{k} \neq \frac{3}{-6} \Rightarrow 3k \neq -12 \Rightarrow k \neq -4.$$

12. (d) Here, $a_1 = 9, b_1 = 4, c_1 = -9$
$$a_2 = 7, b_2 = k, c_2 = -5$$

For no solution $\dfrac{a_1}{a_2} = \dfrac{b_1}{b_2} \neq \dfrac{c_1}{c_2}$

$$\Rightarrow \frac{9}{7} = \frac{4}{k} \Rightarrow 9k = 28 \Rightarrow k = \frac{28}{9}.$$

13. (c) Here, $a_1 = 2, b_1 = 32, c_1 = 3$
$$a_2 = 3, b_2 = 48, c_2 = k$$

For coincident lines,

$$\frac{a_1}{a_2} = \frac{b_1}{b_2} = \frac{c_1}{c_2}$$

$$\Rightarrow \frac{2}{3} = \frac{32}{48} = \frac{3}{k} \quad\Rightarrow \frac{3}{k} = \frac{2}{3} \quad\Rightarrow\quad 2k = 9$$

$$\Rightarrow k = \frac{9}{2}.$$

14. (d) Check for $\dfrac{a_1}{a_2} \neq \dfrac{b_1}{b_2}, \dfrac{a_1}{a_2} = \dfrac{b_1}{b_2} \neq \dfrac{c_1}{c_2}$

and $\dfrac{a_1}{a_2} = \dfrac{b_1}{b_2} = \dfrac{c_1}{c_2}$

Here $a_1 = 2, b_1 = 3, c_1 = -5$
$$a_2 = 4, b_2 = 6, c_2 = -10.$$

15. (d) Here $a_1 = a, b_1 = b, c_1 = -(2a^2 - 3b^2) = 3b^2 - 2a^2$
$$a_2 = 1, b_2 = 2, c_2 = -(2a - 6b) = 6b - 2a$$

$$\frac{a_1}{a_2} = \frac{a}{1}, \frac{b_1}{b_2} = \frac{b}{2}, \frac{c_1}{c_2} = \frac{3b^2 - 2a^2}{6b - 2a}$$

$$2a = b \quad\Rightarrow\quad a = \frac{b}{2}$$

$$\therefore \frac{a_1}{a_2} = \frac{b}{2}, \qquad \frac{b_1}{b_2} = \frac{b}{2} \text{ and}$$

$$\frac{c_1}{c_2} = \frac{3b^2 - 2 \times \dfrac{b^2}{4}}{6b - 2 \times \dfrac{b}{2}} = \frac{3b^2 - \dfrac{b^2}{2}}{6b - b} = \frac{\dfrac{5b^2}{2}}{5b} = \frac{b}{2}$$

$$\therefore \frac{a_1}{a_2} = \frac{b_1}{b_2} = \frac{c_1}{c_2}$$

16. (b) Total number of notes $= 7 \Rightarrow x + y = 7$

Total value of notes $= $ Rs 40

$$\Rightarrow 5x + 10y = 40$$
$$\Rightarrow x + 2y = 8$$

17. (b) Let the cost of one apple and one orange be Rs x and Rs y respectively.

Then, according to the given question,

$$8x + 5y = 92 \qquad\qquad …(i)$$
$$5x + 8y = 77 \qquad\qquad …(ii)$$

Multiplying eqn (i) by 5 and eqn (ii) by 8, we get

$$40x + 25y = 460 \qquad\qquad …(iii)$$
$$40x + 64y = 616 \qquad\qquad …(iv)$$

Subtracting eqn (iii) from eqn (iv),

$$39y = 156 \quad\Rightarrow\quad y = \frac{156}{39} = 4$$

From (i) putting the value of y, we get

$$8x + 20 = 92 \Rightarrow 8x = 72 \Rightarrow x = 9$$

$\therefore$ Cost of 2 oranges and 3 apples

$$= 2 \times \text{Rs } 4 + 3 \times \text{Rs } 9 = \text{Rs } 8 + \text{Rs } 27 = \textbf{Rs 35}.$$

18. (b) Let the number of buffaloes be b and the number of ducks be d. Then,

Number of legs of buffaloes = $4 \times b = 4b$

Number of legs of ducks = $2 \times d = 2d$

Total number of heads = $2(b + d)$

According to the question,

$4b + 2d = 2(b + d) + 24$

$\Rightarrow 4b + 2d = 2b + 2d + 24 \Rightarrow 2b = 24 \Rightarrow b = \mathbf{12}.$

19. (d) Let the number of pencils and pen with X be x and y respectively.

Given, $x + y = 40$ ...(i)

Also $(x + 5) = 4(y - 5)$

$\Rightarrow x + 5 = 4y - 20 \Rightarrow x - 4y = -25$ ...(ii)

Subtracting eqn (ii) from eqn (i), we get

$5y = 65 \Rightarrow y = \mathbf{13}.$

20. (b) Let the ages of the younger sister and elder sister be x yrs and y yrs respectively. Then,

$\dfrac{x}{y+1} = \dfrac{0.5}{1} \quad \Rightarrow \quad \dfrac{x}{y+1} = \dfrac{1}{2}$

$\Rightarrow 2x = y + 1 \quad \Rightarrow \quad 2x - y = 1$ ...(i)

$\dfrac{x-2}{y} = \dfrac{1}{3} \quad \Rightarrow \quad 3x - 6 = y$

$\Rightarrow 3x - y = 6$ ...(ii)

Now solve for x and y.

21. (a) Let Ram have Rs x and Mohan have Rs y.

If Ram gives Rs 30 to Mohan, then

Ram has = Rs $(x - 30)$

Mohan has = Rs $(y + 30)$

According to the question, $y + 30 = 2(x - 30)$

$\Rightarrow y + 30 = 2x - 60 \Rightarrow 2x - y = 90$ ...(i)

If Mohan gives Rs 10 to Ram, then

Ram has Rs $(x + 10)$

Mohan has Rs $(y - 10)$

According to the question, $(x + 10) = 3(y - 10)$

$\Rightarrow x + 10 = 3y - 30 \Rightarrow x - 3y = -40$ ...(ii)

Now solve yourself for x and y.

22. (a) Let the fraction be $\dfrac{x}{y}$.

Given, $\dfrac{x+1}{y+1} = 2$

$\Rightarrow x + 1 = 2y + 2 \quad \Rightarrow \quad x - 2y = 1$ ...(i)

and $\dfrac{x-1}{y-1} = 3$

$\Rightarrow x - 1 = 3y - 3 \Rightarrow x - 3y = -2$ ...(ii)

Now solve eqn (i) and (ii) for x and y.

23. (b) Let the cost price of the pen and the book be Rs x and Rs y respectively.

Case I: When the pen is sold at 5% loss and book at 15% gain,

Loss on pen = Rs $\dfrac{5x}{100}$ = Rs $\dfrac{x}{20}$

Gain on book = Rs $\dfrac{15y}{100}$ = Rs $\dfrac{3y}{20}$

$\therefore$ Net gain = Rs $\dfrac{3y}{20}$ – Rs $\dfrac{x}{20}$

Given, $\dfrac{3y}{20} - \dfrac{x}{20} = 7 \Rightarrow 3y - x = 140$ ...(i)

Case II: When the pen is sold at 5% gain and book at 10% gain

Gain on pen = Rs $\dfrac{5x}{100}$ = Rs $\dfrac{x}{20}$

Gain on book = Rs $\dfrac{10y}{100}$ = Rs $\dfrac{y}{10}$

$\therefore$ Net gain = $\dfrac{x}{20} + \dfrac{y}{10}$

Given, $\dfrac{x}{20} + \dfrac{y}{10} = 13 \Rightarrow x + 2y = 260$ ...(ii)

Now, solve equation (i) and (ii) for the value of x and y.

24. (b) Let the larger number be x and the smaller one be y. We know that,

Dividend = (Divisor × Quotient) + Remainder

By the first condition,

$3x = 4y + 3 \quad \Rightarrow \quad 3x - 4y = 3$ ...(i)

By the second condition,

$7y = 5x + 1 \quad \Rightarrow \quad 5x - 7y = -1$ ...(ii)

Now solve for x and y.

25. (d) Mr. Manoj's age be $(10x + y)$ yrs.

Then, his wife's age = $(10y + x)$ years

Given,

$\dfrac{1}{11}(10x + y + 10y + x) = (10x + y) - (10y + x)$

$\Rightarrow \dfrac{1}{11}(11x + 11y) = 9x - 9y$

$\Rightarrow x + y = 9x - 9y \Rightarrow 8x = 10y \Rightarrow \dfrac{x}{y} = \dfrac{5}{4}$

$\therefore x = 5, y = 4$ (because any other multiple of 5 will make x two digits)

$\therefore$ Difference = $9x - 9y = 9(x - y) = 9(5 - 4) = 9$ yrs.

26. (c) Suppose I have x, one-rupee coins and y, 20 – paise coins.

$x \times 1 + y \times 0.2 = 14.40 \Rightarrow x + 0.2y = 14.4$ ...(i)

After shopping, I had y one-rupee coins and x 20-paise coins.

Also, $x \times 0.2 + y \times 1 = \dfrac{1}{3} \times 14.4$

$\Rightarrow 0.2x + y = 4.8$ ...(ii)

[**Note:** To solve the equations of the form $ax + by = c$ and $bx + ay = d$, where $a \neq b$, we can use the following method also.]

Adding eqn (i) and (ii), we get

$1.2\,x + 1.2\,y = 19.2 \Rightarrow x + y = \dfrac{19.2}{1.2} = 16$ ...(iii)

and subtracting eqn (ii) from eqn (i), we get

$-0.8x + 0.8y = -9.6 \Rightarrow x - y = 12$ ...(iv)

Now adding (iii) and (iv), we get

$2x = 28 \Rightarrow x = \mathbf{14}.$

27. (a) $5a + b > 51$ ...(i)

$3a - b = 21$ ...(ii) $\Rightarrow b = 3a - 21$

Putting the value of b in (i), we get

$5a + 3a - 21 > 51 \Rightarrow 8a > 72 \Rightarrow a > 9$

Now, $b = 3a - 21$

If $\left.\begin{array}{l} a = 9, b = 6; \\ a = 10, b = 9; \\ a = 11, b = 12; \end{array}\right\} \Rightarrow$ If $a > 9, b > 6$

28. (a) By the angle sum property of a triangle,

$\angle A + \angle B + \angle C = 180°$

$\Rightarrow x + y + y + 20° = 180° \Rightarrow x + 2y = 160$ ...(i)

Given, $4x - y = 10$ ...(ii)

Multiplying (ii) by 2, we get

$8x - 2y = 20$ ...(iii)

Adding eqn (i) and (iii),

$9x = 180 \Rightarrow x = 20$

$\therefore$ From (i), $2y = 160 - 20 = 140 \Rightarrow y = 70$

$\Rightarrow \angle A = 20°, \angle B = 70°, \angle C = 90°.$

$\therefore$ The triangle is right angled.

29. (d) Let the two digit number be $10x + y$.

According to the question,

$10x + y = 8(x + y) + 1 \Rightarrow 2x - 7y = 1$ ...(i)

and $10x + y = 13(x - y) + 2 \Rightarrow 3x - 14y = -2$ ...(ii)

Now solve yourself for x and y.

30. (c) Let the number of females be x and males be y.

Then, $x + y = 53000$

$1.04x + 1.06y = 55630$

$\left[\begin{array}{l} \text{No. of females after increase} = x + \dfrac{4}{100}x = \dfrac{104}{100}x = 1.04x \\ \text{Similarly, no. of males} = 1.06\,y \end{array}\right.$

Now solve for x and y and find the difference.

31. (c) Let X and Y be two cars starting from points A and B respectively.

Let the speed of the car X be x km/hr and that of car Y be y km/hr.

Case I : *When the two cars move in the same direction.*

Suppose the two cars meet at point P after 10 hours.

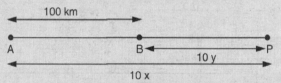

Then, Dist. travelled by car $X = AP = 10x$ km

and Dist. travelled by car $Y = BP = 10y$ km

Given, $AP - BP = AB$

$\Rightarrow 10x - 10y = 100 \Rightarrow x - y = 10$ ...(i)

Case II: *When the two cars move in the opposite direction :*

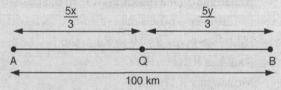

Suppose the two cars meet at point Q of 1 hrs

$40 \text{ min} = 1\dfrac{40}{60} \text{hrs} = 1\dfrac{2}{3} \text{hrs} = \dfrac{5}{3} \text{hrs.}$

Then, dist. travelled by car $X = AQ = \dfrac{5x}{3}$ km

and dist. travelled by car $Y = BQ = \dfrac{5y}{3}$ km.

Given, $AQ + BQ = 100$

$\Rightarrow \dfrac{5x}{3} + \dfrac{5y}{3} = 100 \Rightarrow (x + y) = \dfrac{300}{5} = 60$ ...(ii)

Adding (i) and (ii)

$2x = 70 \Rightarrow x = \mathbf{35}$

$\therefore$ From (i), $35 - y = 10 \Rightarrow y = \mathbf{25 \text{ km/hr}}.$

32. (d) Suppose the person invested Rs a at 12% simple interest and Rs b at 10% simple interest. Then,

$$\frac{12a}{100}+\frac{10b}{100}=130 \Rightarrow 6a+5b=6500 \quad …(i)$$

If the amounts invested are interchanged, then the yearly interest = Rs 134

$$\therefore \frac{10a}{100}+\frac{12b}{100}=134 \Rightarrow 5a+6b=6700 \quad …(ii)$$

Now solve the equations yourself for a and b.

33. (a) Let $\frac{1}{x}=p,\frac{1}{y}=q.$ Then,

$$ap-bq=0 \quad …(i)$$
$$ab^2p+a^2bq=a^2+b^2 \Rightarrow ab(bp+aq)=a^2+b^2$$

$$\Rightarrow bp+aq=\frac{a^2+b^2}{ab} \quad …(ii)$$

Multiplying eqn (i) by b and eqn (ii) by a, we get

$$abp-b^2q=0 \quad …(iii)$$

$$abp+a^2q=\frac{a^2+b^2}{b} \quad …(iv)$$

Now subtracting eqn (iii) from eqn (iv), we get

$$(a^2+b^2)q=\frac{a^2+b^2}{b} \Rightarrow q=\frac{1}{b} \Rightarrow \frac{1}{x}=\frac{1}{b} \Rightarrow x=b$$

Putting in (i), $ap-1=0 \Rightarrow ab=1$

$$\Rightarrow p=\frac{1}{a} \Rightarrow \frac{1}{y}=\frac{1}{a} \Rightarrow y=a$$

$$\therefore x=\boldsymbol{b},\ y=\boldsymbol{a}.$$

34. (c) Let the number of oranges in the lots A and B be a and b respectively.

According to the question,

$$\frac{2}{3}a+b=400 \quad …(i)$$

$$a+\frac{4}{5}b=460 \quad …(ii)$$

Now solve for a and b.

35. (b) (2 Men + 7 Children)'s 1 days' work = $\frac{1}{4}$

(4 Men + 4 children)'s 1 days' work = $\frac{1}{3}$

$$\Rightarrow 2M+7C=\frac{1}{4} \quad …(i)$$

$$4M+4C=\frac{1}{3} \quad …(ii)$$

Now solve for the value of man.

Self Assessment Sheet–11

1. If the equations $4x+7y=10$ and $10x+ky=25$ represent coincident lines, then the value of k is :

(a) 5   (b) $\frac{17}{2}$

(c) $\frac{27}{2}$ (d) $\frac{35}{2}$

2. The solution of the equations $\frac{m}{3}+\frac{n}{4}=12$ and $\frac{m}{2}-\frac{n}{3}=1$ is

(a) $m=8, n=6$ (b) $m=18, n=24$
(c) $m=24, n=18$ (d) $m=6, n=8$

3. Given that : $5=\frac{5W+2\omega}{5+2}$ and $5.1=\frac{7W+3\omega}{7+3}$, find W and ω.

(a) $W=3, \omega=10$ (b) $W=10, \omega=3$
(c) $W=-10, \omega=3$ (d) W $=3, \omega=-10$

4. Solve for a and b : $2(a+b)-(a-b)=6$, $4(a-b)=2(a+b)-9$

(a) $a=\frac{3}{4}, b=1\frac{3}{4}$ (b) $a=\frac{1}{2}, b=1\frac{1}{2}$

(c) $a=1\frac{3}{4}, b=\frac{3}{4}$ (d) $a=\frac{3}{2}, b=\frac{3}{4}$

5. Solve : $\frac{x+2}{y+2}+2=0, \frac{x-4}{y-2}=\frac{x-1}{y+7}$

(a) $x=2, y=4$ (b) $x=-2, y=4$
(c) $x=2, y=-4$ (d) $x=-2, y=-4$

6. The solution of the pair of equations: $0.25x+0.6y=0.7$ and $0.3x-3.5y=2.95$ is

(a) $x=4, y=0.5$ (b) $x=-4, y=-0.5$
(c) $x=4, y=0.5$ (d) $x=4, y=-0.5$

7. Three times Diana's age is 17 years more than twice Jim's age. The sum of their ages is 13 years less than their father's age which is three times Jim's age. What are the children's ages ?

(a) Diana 21 years, Jim 16 years
(b) Diana 15 years, Jim 14 years

(c) Diana 15 years, Jim 16 years

(d) Diana 20 years, Jim 14 years

8. A number of two digits is equal to six times the sum of its digits. If the digits are reversed the number so formed is equal to :

   (a) six times the sum of its digits.

   (b) five times the sum of its digits.

   (c) ten times the sum of its digits.

   (d) nine times the sum of digits.

9. A sports club has 130 members. An increase of 10% in the number of men and 20% in the number of ladies brought up the membership to 148. How many men and ladies were there originally ?

   (a) 90 men, 40 women   (b) 80 men, 50 women

   (c) 60 men, 70 women   (d) 50 men, 80 women

10. The difference between two angles of a triangle whose magnitude is in the ratio 10 : 7 is 20° less than the third angle. The third angle is:

   (a) 80°   (b) 56°

   (c) 44°   (d) 70°

Answers

| 1. (d) | 2. (b) | 3. (a) | 4. (a) | 5. (c) | 6. (d) | 7. (b) | 8. (b) | 9. (b) | 10. (c) |

Chapter 12

QUADRATIC EQUATIONS

KEY FACTS

1. A polynomial of the form $ax^2 + bx + c$ where $a \neq 0$ and a, b, c are real constants and x is a real variable is called a **quadratic polynomial.**

2. A **quadratic equation** is a polynomial equation of second degree, *i.e.*, of the form $ax^2 + bx + c = 0$, where $a \neq 0$; $a, b, c \in R$.

   Examples of quadratic equations are $x^2 - 25 = 0$, $x^2 + 10x + 16 = 0$, $9y^2 - 3y - 2 = 0$ etc.

3. Those values of the x, for which the quadratic polynomial $ax^2 + bx + c = 0$, are called the **roots** or **zeros** of the polynomial.

   For example, $x^2 - 25 = 0$ for $x = 5$ or -5

   Hence 5 and –5 are the roots of the quadratic equations $x^2 - 25 = 0$.

4. To solve a quadratic equation by factorisation,

   Step 1 : *Express the given equation in the standard form* $ax^2 + bx + c = 0$

   Step 2 : *Factorise the quadratic polynomial given on the left, i.e.,* $ax^2 + bx + c = (Ax + B)(Cx + D)$

   Step 3 : *Let each factor = 0* $(\therefore ab = 0 \Rightarrow a = 0 \text{ or } b = 0)$

   $\therefore \quad ax^2 + bx + c = 0$

   $\Rightarrow (Ax + B)(Cx + D) = 0 \Rightarrow Ax + B = 0 \text{ or } Cx + D = 0 \Rightarrow x = -\dfrac{B}{A} \text{ or } x = -\dfrac{D}{C}$

   $\therefore$ The two roots are $-\dfrac{B}{A}$ and $-\dfrac{D}{C}$.

   Step 4 : *Check each root by putting the value of x obtained in the original equation.*

   For example : Solve $x^2 - 3x - 10 = 0$

   $x^2 - 3x - 10 = 0$

   $\Rightarrow x^2 - 5x + 2x - 10 = 0$ *(Factorising the quadratic polynomial)*

   $\Rightarrow x(x - 5) + 2(x - 5) = 0 \Rightarrow (x - 5)(x + 2) = 0$

   $\Rightarrow x - 5 = 0 \text{ or } x + 2 = 0$ *(Let each factor = 0)*

   $\Rightarrow x = 5 \text{ or } x = -2$ *(Solve resulting equation)*

   $\therefore$ The roots of the quadratic equation $x^2 - 3x - 10 = 0$ are **5** and **–2**.

Solved Examples

Ex. 1. *Solve :* $\dfrac{x^2 - 4}{3} = 20.$

Sol . $\dfrac{x^2 - 4}{3} = 20$

$\Rightarrow \quad x^2 - 4 = 60 \;\Rightarrow\; x^2 - 64 = 0 \;\Rightarrow\; (x+8)(x-8) = 0$

$\Rightarrow \quad (x+8) = 0 \text{ or } (x-8) = 0 \;\Rightarrow\; x = -8 \text{ or } 8$

$\therefore \;\; x = \textbf{–8, 8}$

Ex. 2. *Solve : x (2x + 5) = 3.*

Sol. $x\,(2x+5) = 3 \;\Rightarrow\; 2x^2 + 5x = 3$

$\Rightarrow \; 2x^2 + 5x - 3 = 0 \;\Rightarrow\; 2x^2 + 6x - x - 3 = 0 \;\Rightarrow\; 2x(x+3) - 1\,(x+3) = 0$

$\Rightarrow \; (2x - 1)(x + 3) = 0 \;\Rightarrow\; (2x - 1) = 0 \text{ or } (x + 3) = 0$

$\Rightarrow \; x = \dfrac{1}{2} \;\text{ or }\; x = -3$

$\therefore \; x = \dfrac{\textbf{1}}{\textbf{2}}, \textbf{–3}.$

Ex. 3. Solve : $\dfrac{x+1}{x-1} - \dfrac{x-1}{x+1} = \dfrac{5}{6}$, $x \neq 1, -1.$

Sol . $\dfrac{x+1}{x-1} - \dfrac{x-1}{x+1} = \dfrac{5}{6} \Rightarrow \dfrac{(x+1)^2 - (x-1)^2}{(x-1)(x+1)} = \dfrac{5}{6}$

$\Rightarrow \dfrac{(x^2 + 2x + 1) - (x^2 - 2x + 1)}{x^2 - 1} = \dfrac{5}{6} \;\Rightarrow\; \dfrac{4x}{x^2 - 1} = \dfrac{5}{6} \Rightarrow 24x = 5x^2 - 5$

$\Rightarrow 5x^2 - 24x - 5 = 0 \Rightarrow 5x^2 - 25x + x - 5 = 0$

$\Rightarrow 5x(x - 5) + 1\,(x - 5) = 0 \;\Rightarrow\; (x - 5)\,(5x + 1) = 0$

$\Rightarrow (x - 5) = 0 \text{ or } (5x + 1) = 0$

$\Rightarrow x = \textbf{5}, \dfrac{\textbf{–1}}{\textbf{5}}.$

Ex. 4. *Solve :* $3^{4x+1} - 2 \times 3^{2x+2} - 81 = 0.$

Sol. $3^{4x+1} - 2 \times 3^{2x+2} - 81 = 0$

$\Rightarrow 3 \times 3^{4x} - 2 \times 3^2 \times 3^{2x} - 81 = 0 \;\Rightarrow\; 3 \times 3^{4x} - 18 \times 3^{2x} - 81 = 0$

$\Rightarrow 3a^2 - 18a - 81 = 0 \text{ (Put } 3^{2x} = a) \Rightarrow a^2 - 6a - 27 = 0$

$\Rightarrow a^2 - 9a + 3a - 27 = 0 \;\Rightarrow\; a(a - 9) + 3(a - 9) = 0$

$\Rightarrow (a - 9)(a + 3) = 0 \;\Rightarrow\; a - 9 = 0 \text{ or } a + 3 = 0$

$\Rightarrow a = 9 \text{ or } a = -3 \;\Rightarrow\; 3^{2x} = 9 \text{ or } 3^{2x} = -3$

$\Rightarrow 3^{2x} = 3^2 \; (\because 3^{2x} \neq -3) \Rightarrow 2x = 2 \Rightarrow x = \textbf{1}.$

Ex. 5. *Solve :* $\sqrt{2x+7} = x + 2.$

Sol . $\sqrt{2x+7} = x + 2$

Squaring both the sides, we have

$\left(\sqrt{2x+7}\right)^2 = (x + 2)^2$

$\Rightarrow \quad 2x+7 = x^2 + 4x + 4 \Rightarrow x^2 + 2x - 3 = 0 \Rightarrow x^2 + 3x - x - 3 = 0$

$\Rightarrow \quad x(x+3) - 1(x+3) = 0 \Rightarrow (x+3)(x-1) = 0 \Rightarrow (x+3) = 0 \text{ or } (x-1) = 0$

$\Rightarrow \quad x = -3, 1.$

Ex. 6. *The sum of two numbers is 18 and their product is 56. Find the numbers.*

Sol. Let one number be x. Given, sum = 18

$\therefore$ Other number = $(18 - x)$

$\therefore \quad x(18-x) = 56 \quad \Rightarrow \quad 18x - x^2 = 56 \quad \Rightarrow \quad x^2 - 18x + 56 = 0$

$\qquad x - 14x - 4x + 56 = 0$

$\Rightarrow \quad x(x-14) - 4(x-14) = 0$

$\Rightarrow \quad (x-4)(x-14) = 0 \Rightarrow (x-4) = 0 \text{ or } (x-14) = 0 \Rightarrow x = 4, 14$

$\therefore$ The numbers are **4** and **14**.

Ex. 7. *The numerator of a fraction is one more than its denominator. If its reciprocal is subtracted from it, the difference is $\dfrac{11}{30}$. Find the fraction.*

Sol. Let the denominator of the fraction be x. Then, the numerator = $x + 1$

$\therefore$ Given fraction = $\dfrac{x+1}{x}$; Its reciprocal = $\dfrac{x}{x+1}$

According to the question, $\dfrac{x+1}{x} - \dfrac{x}{x+1} = \dfrac{11}{30}$

$\Rightarrow \quad \dfrac{(x+1)^2 - x^2}{x(x+1)} = \dfrac{11}{30} \Rightarrow \dfrac{x^2 + 2x + 1 - x^2}{x^2 + x} = \dfrac{11}{30} \Rightarrow 30(2x+1) = 11(x^2 + x)$

$\Rightarrow \quad 60x + 30 = 11x^2 + 11x \Rightarrow 11x^2 - 49x - 30 = 0 \Rightarrow 11x^2 - 55x + 6x - 30 = 0$

$\Rightarrow \quad 11x(x-5) + 6(x-5) = 0 \Rightarrow (11x+6)(x-5) = 0 \Rightarrow x = \dfrac{-6}{11} \text{ or } 5$

Then, if $x = \dfrac{-6}{11}, x+1 = \dfrac{-6}{11} + 1 = \dfrac{5}{11}$

Hence the given fraction = $\dfrac{5}{11} \Big/ \dfrac{-6}{11} = -\dfrac{5}{6}$

If $x = 5, x+1 = 6$

Hence the given fraction = $\dfrac{6}{5}$

$\therefore$ The fraction is $\dfrac{6}{5}$ or $-\dfrac{5}{6}$.

Ex. 8. *In a group of children each child gives a gift to every other child. If the number of gifts are 132, find the number of children.*

Sol. Let the number of children be x. Then, by the given condition, each child exchanges gift with $(x - 1)$ children, so $x(x-1) = 132$

$\Rightarrow \quad x^2 - x - 132 = 0 \Rightarrow x^2 - 12x + 11x - 132 = 0 \Rightarrow x(x-12) + 11(x-12) = 0$

$\Rightarrow \quad (x-12)(x+11) = 0 \Rightarrow (x-12) = 0 \text{ or } (x+11) = 0$

$\Rightarrow \quad x = \mathbf{12} \text{ or } \mathbf{-11}.$

Ex. 9. *The age of father is equal to the square of the age of his son. The sum of the age of the father and five times the age of the son is 66 years. Find their ages.*

Sol. Let the age of the son be x years. Then, age of father = x^2 years

Given , $x^2 + 5x = 66$

$\Rightarrow$ $x^2 + 5x - 66 = 0 \Rightarrow x^2 + 11x - 6x - 66 = 0 \Rightarrow x(x+11) - 6(x+11) = 0 \Rightarrow (x-6)(x+11) = 0$

$\Rightarrow$ $(x-6) = 0$ or $(x+11) = 0 \Rightarrow x = 6$ or $x = -11$

Neglecting –ve value, Son's age = **6 years**, Father's age = **36 years**.

Ex. 10. *In a flight of 2800 km, an aircraft was slowed down due to bad weather. Its average speed for the trip was reduced by 100 km/hour and time increased by 30 minutes. Find the original duration of flight.*

Sol . Let the original speed = x km/hr

$\therefore$ Normal time of flight = $\dfrac{2800}{x}$ hours

New speed = $(x-100)$ km/hr

$\therefore$ Time of flight, when speed is reduced = $\dfrac{2800}{(x-100)}$ hours

According to the given condition, $\dfrac{2800}{(x-100)} - \dfrac{2800}{x} = \dfrac{30}{60} = \dfrac{1}{2}$

$\Rightarrow$ $\dfrac{2800x - 2800(x-100)}{x(x-100)} = \dfrac{1}{2} \Rightarrow \dfrac{2800x - 2800x + 280000}{x^2 - 100x} = \dfrac{1}{2}$

$\Rightarrow$ $x^2 - 100x = 560000 \Rightarrow x^2 - 100x - 560000 = 0$

$\Rightarrow$ $x^2 - 800x + 700x - 560000 = 0 \Rightarrow x(x-800) + 700(x-800) = 0$

$\Rightarrow$ $(x-800)(x+700) = 0 \Rightarrow x - 800 = 0$ or $x + 700 = 0$

$\Rightarrow$ $x = 800, -700$

Neglecting the –ve value.

Original speed of the air craft = 800 km/hr

Original duration of flight = $\dfrac{2800}{800} = 3\dfrac{1}{2}$ **hours.**

Ex. 11. *Two trains leave a railway station at the same time. The first train travels due west and the second train due north. The first train travels 5 km/hr faster than the second train. If after two hours, they are 50 km apart, find the average speed of each train.*

Sol . Let the speed of the second train be x km/hr. Then, speed of the first train is $(x + 5)$ km/hr. Distance covered by the first train after two hours = $2(x + 5)$ km

Distance covered by the second train after two hours = $2x$ km

$\therefore$ $(2x)^2 + 4(x+5)^2 = (50)^2$ (By Pythagoras' Theorem)

$\Rightarrow$ $4x^2 + 4(x^2 + 10x + 25) = 2500 \Rightarrow 8x^2 + 40x + 100 = 2500$

$\Rightarrow$ $8x^2 + 40x - 2400 = 0 \Rightarrow x^2 + 5x - 300 = 0$

$\Rightarrow$ $x^2 + 20x - 15x - 300 = 0 \Rightarrow x(x+20) - 15(x+20) = 0$

$\Rightarrow$ $(x-15)(x+20) = 0$

$\Rightarrow$ $x - 15 = 0$ or $x + 20 = 0$

$\Rightarrow$ $x = 15, -20$

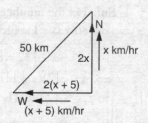

Neglecting the –ve value, *i.e.,* –20.

Speed of second train = 15 km/ hr

Speed of first train = **20 km /hr.**

Ex. 12. *The angry Arjun carried some arrows for fighting with Bheeshm. With half the arrows, he cut down the arrows thrown by Bheeshm on him and with six other arrows he killed the rath driver of Bheeshm. With one arrow each he knocked down respectively the rath, flag and the bow of Bheeshm. Finally, with one more than four times the square root of arrows he laid Bheeshm unconscious on an arrow bed. Find the total number of arrows Arjun had.*

Sol. Let x be the total number of arrows Arjun had. Then,

$$\frac{x}{2} + 6 + 3 + (4\sqrt{x} + 1) = x$$

$$\Rightarrow -\frac{x}{2} + 10 + 4\sqrt{x} = 0 \Rightarrow -x + 20 + 8\sqrt{x} = 0$$

$$\Rightarrow x - 8\sqrt{x} - 20 = 0 \Rightarrow y^2 - 8y - 20 = 0 \quad (\text{Let } \sqrt{x} = y)$$

$$\Rightarrow y^2 - 10y + 2y - 20 = 0 \Rightarrow y(y - 10) + 2(y - 10) = 0 \Rightarrow (y - 10)(y + 2) = 0$$

$$\Rightarrow y - 10 = 0 \text{ or } y + 2 = 0 \Rightarrow y = 10, -2$$

Neglecting –ve value

$$y = \sqrt{x} = 10 \Rightarrow x = (10)^2 = \mathbf{100}.$$

Question Bank–12

1. Which of the following is a quadratic equation ?

   (a) $x^{\frac{1}{2}} + 2x + 3 = 0$

   (b) $(x - 1)(x + 4) = x^2 + 1$

   (c) $x^4 - 3x + 5 = 0$

   (d) $(2x + 1)(3x - 4) = 2x^2 + 3$

2. If $a^2 - ab = 0$, which of the following is the correct conclusion ?

   (a) $a = 0$

   (b) $a = b$

   (c) $a^2 = b$

   (d) either $a = 0$ or $a = b$

3. The roots of the equation $x^2 - 8x + 15 = 0$ are :

   (a) 2, 3

   (b) 3, 5

   (c) 8, 15

   (d) 6, 5

4. The roots of the equation $2x^2 - 11x + 15 = 0$ are :

   (a) $3, \frac{5}{2}$

   (b) $5, \frac{3}{2}$

   (c) $-3, -\frac{5}{2}$

   (d) None of these

5. Find a if $a - 3 = \frac{10}{a}$

   (a) $\sqrt{7}, 7$

   (b) $5, -2$

   (c) $-5, 2$

   (d) $-\sqrt{7}, 7$

6. Of the following quadratic equations, which is the one whose roots are 2 and – 15 ?

   (a) $x^2 - 2x + 15 = 0$

   (b) $x^2 + 15x - 2 = 0$

   (c) $x^2 + 13x - 30 = 0$

   (d) $x^2 - 30 = 0$

7. The common root of the equations $x^2 - 7x + 10 = 0$ and $x^2 - 10x + 16 = 0$ is :

   (a) –2

   (b) 3

   (c) 5

   (d) 2

8. An equation equivalent to the quadratic equation $x^2 - 6x + 5 = 0$ is :

   (a) $6x^2 - 5x + 1 = 0$

   (b) $x^2 - 5x + 6 = 0$

   (c) $5x^2 - 6x + 1 = 0$

   (d) $|x - 3| = 2$

9. The roots of the equation $\frac{x}{x-1} + \frac{x-1}{x} = 2\frac{1}{2}$ are :

   (a) 1, 2

   (b) 2, 1

   (c) –2, 1

   (d) 2, –1

10. Find the roots of the equation $\frac{1}{a+b+x} - \frac{1}{x} = \frac{1}{a} + \frac{1}{b}$?

    (a) $a, -b$

    (b) $-a, b$

    (c) a, b

    (d) $-a, -b$

11. The values of x satisfying the equation

$5^{2x} - 5^{x+3} + 125 = 5^x$ are :

(a) 0 and 2   (b) –1 and 3

(c) 0 and –3   (d) 0 and 3

12. The non-zero root of the equation $3^{2x} + 9 = 10.3^x$ is :

(a) a positive fraction   (b) a negative fraction

(c) a positive integer   (d) a negative integer

13. If one root of the quadratic equation

$3x^2 - 10x + p = 0$ is $\dfrac{1}{3}$, then the value of p and

the other root respectively is :

(a) $3, \dfrac{1}{3}$ (b) 3, 3

(c) $-\dfrac{1}{3}, -\dfrac{1}{3}$ (d) –3, –3

14. Thrice the square of a natural number decreased by 4 times the number is equal to 50 more than the number. The number is :

(a) 4   (b) 5

(c) 6   (d) 10

15. The product of two successive natural numbers is 1980. Which is the smaller number ?

(a) 34   (b) 35

(c) 44   (d) 45

16. A two-digit number is such that the product of its digits is 8. When 18 is added to the number, the digits are reversed. The number is :

(a) 18   (b) 24

(c) 42   (d) 81

17. The difference of mother's age and her daughter's age is 21 years and the twelfth part of the product of their ages is less than the mother's age by 18 years. The mother's age is :

(a) 22 years   (b) 32 years

(c) 24 years   (d) 42 years

18. The sum of a number and its positive square root is 6/25. The number is :

(a) 5   (b) $\dfrac{1}{5}$

(c) 25   (d) $\dfrac{1}{25}$

19. In a school hall, 460 students were sitting in rows and columns in such a way that the number of students sitting in each column was three more than the number of students sitting in each row. The number of students in each column was :

(a) 20   (b) 23

(c) 24   (d) None of these

20. The length of the hypotenuse of a right triangle exceeds the length of the base by 2 cm and exceeds twice the length of the attitude by 1 cm. The perimeter of the triangle is :

(a) 18 cm   (b) 17 cm

(c) 25 cm   (d) 40 cm

21. If the price of an article is increased by Rs 2 per dozen than at present, the number of things available for Rs 56 is 8 less than before. The price per dozen at present is :

(a) Rs 14   (b) Rs 12

(c) Rs 10   (d) Rs 28

22. If the perimeter of a rectangular plot is 34 metres and its area is 60 square metres, what is the length of each of the shorter side ?

(a) 10 m   (b) 15 m

(b) 17 m   (d) 5 m

23. Two little bands of monkeys were at play. An eighths of them squared were jabbering wildly in the thicket when twelve shouted loudly with glee. How many monkeys were there in the thicket ?

(a) 16   (b) 32

(c) 64   (d) 24

24. One-fourth of a herd of cows is in the forest. Twice the square root of the herd has gone to mountains and the remaining 15 are on the banks of a river. The total number of cows is :

(a) 6   (b) 100

(c) 63   (d) 36

25. A positive number when decreased by 4 becomes 21 times its reciprocal. The number is

(a) 8   (b) 7

(c) 6   (d) 5

26. A man in 1900s realised that in 1980 his age was the square root of the year of his birth. What is his birth year ?

(a) 1929   (b) 1949

(c) 1936   (d) 1946

27. A swimming pool is to be built in the shape of the letter L. The shape is formed from two squares with side dimensions x and $\sqrt{x}$ as shown. If the area of the pool is 30 m², what is the value of x.

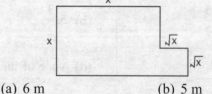

(a) 6 m   (b) 5 m

(c) 16 m   (d) 9 m

28. The speed of a boat in still water is 15 km/hr. It can go 30 km upstream and return downstream to the original point in 4 hrs 30 min. The speed of the stream is :

(a) 5 km /hr   (b) 8 km/ hr

(c) 10 km/hr   (d) 15 km/hr

29. The diagonal of a rectangular field is 15 metres and the difference between its length and width is 3 metres. The area of the rectangular field is :

(a) 9 m^2 (b) 12 m^2

(c) 21 m^2 (d) 108 m^2

30. In the following question, two equations numbered I and II are given. You have to solve both the equations and give answer if

(a) $X > Y$ (b) $X \geq Y$

(c) $X < Y$ (d) $X \leq Y$

(e) $X = Y$

    I. $x^2 + 11x + 28 = 0$ II. $y^2 + 15y + 56 = 0$

Answers

1. (d)	2. (d)	3. (b)	4. (a)	5. (b)	6. (c)	7. (d)	8. (d)	9. (d)	10. (d)
11. (d)	12. (c)	13. (b)	14. (b)	15. (c)	16. (b)	17. (c)	18. (d)	19. (a)	20. (d)
21. (b)	22. (d)	23. (a)	24. (d)	25. (b)	26. (c)	27. (b)	28. (a)	29. (d)	30. (b)

Hints and Solutions

1. (d) We reduce each of the given parts to standard form:

(a) $x^{1/2} + 2x + 3 = 0 \Rightarrow$ degree = 1
$\Rightarrow$ Not a quad. eqn.

(b) $(x - 1)(x + 4) = x^2 + 1$

$\Rightarrow x^2 + 3x - 4 = x^2 + 1 \Rightarrow 3x - 5 = 0$
$\Rightarrow$ Degree = 0
$\Rightarrow$ Not a quad. eqn.

(c) $x^4 - 3x + 5 = 0 \Rightarrow$ Degree = 4
$\Rightarrow$ Not a quad. eqn.

(d) $(2x + 1)(3x - 4) = 2x^2 + 3$

$\Rightarrow 6x^2 - 5x - 4 = 2x^2 + 3$

$\Rightarrow 4x^2 - 5x - 4 = 0 \Rightarrow$ Degree = 2
$\Rightarrow$ Hence a quad. eqn.

2. (d) $a^2 - ab = 0 \Rightarrow a(a - b) = 0$

$\Rightarrow a = 0$ or $(a - b) = 0 \Rightarrow a = 0$ or $a = b$.

3. (b) $x^2 - 8x + 15 = 0$

$\Rightarrow x^2 - 5x - 3x + 15 = 0$

$\Rightarrow x(x - 5) - 3(x - 5) = 0$

$\Rightarrow (x - 5)(x - 3) = 0$

$\Rightarrow x - 5 = 0$ or $x - 3 = 0$

$\Rightarrow x = 5, 3$.

4. (a) $2x^2 - 11x + 15 = 0$

$\Rightarrow 2x^2 - 6x - 5x + 15 = 0$

$\Rightarrow 2x(x - 3) - 5(x - 3) = 0$

$\Rightarrow (x - 3)(2x - 5) = 0$

$\Rightarrow x - 3 = 0$ or $2x - 5 = 0$

$\Rightarrow x = 3, \dfrac{5}{2}.$

5. (b) $a - 3 = \dfrac{10}{a} \Rightarrow a^2 - 3a = 10$

$\Rightarrow a^2 - 3a - 10 = 0 \Rightarrow a^2 - 5a + 2a - 10 = 0$

$\Rightarrow a(a - 5) + 2(a - 5) = 0$

$\Rightarrow (a - 5)(a + 2) = 0$

$\Rightarrow a - 5 = 0$ or $a + 2 = 0$

$\Rightarrow a = 5, -2.$

6. (c) • $x^2 - 2x + 15 = 0$
$\Rightarrow (x - 5)(x + 2) = 0 \Rightarrow x = 5, -2$

• $x^2 + 15x - 2 = 0$. Solving $x \neq 2, -15$

• $x^2 + 13x - 30 = 0$

$\Rightarrow (x + 15)(x - 2) = 0 \Rightarrow x = -15, 2$

• $x^2 - 30 = 0 \Rightarrow x^2 = 30 \Rightarrow x = \pm\sqrt{30}.$

7. (d) $x^2 - 7x + 10 = 0$
$\Rightarrow (x - 5)(x - 2) = 0 \Rightarrow x = 5, 2$

$x^2 - 10x + 16 = 0 \Rightarrow (x - 8)(x - 2) = 0$

$\Rightarrow x = 8, 2$

$\therefore$ Common root = 2.

8. (d) Two equations are equivalent if they have equal roots.

$$x^2 - 6x + 5 = 0 \Rightarrow (x-1)(x-5) = 0 \Rightarrow x = 1, 5$$

Now, • $6x^2 - 5x + 1 = 0$

$\Rightarrow 6x^2 - 3x - 2x + 1 = 0$

$\Rightarrow 3x(2x-1) - 1(2x-1) = 0$

$\Rightarrow (3x-1)(2x-1) = 0 \Rightarrow x = \dfrac{1}{3}, \dfrac{1}{2}$

• $x^2 - 5x + 6 = 0$

$\Rightarrow (x-3)(x-2) = 0 \Rightarrow x = 3, 2$

• $5x^2 - 6x + 1 = 0 \Rightarrow 5x^2 - 5x - x + 1 = 0$

$\Rightarrow 5x(x-1) - 1(x-1) = 0$

$\Rightarrow (5x-1)(x-1) = 0 \Rightarrow x = \dfrac{1}{5}, 1$

• $|x-3| = 2 \Rightarrow (x-3) = 2$ or $-(x-3) = 2$

$\Rightarrow x = 5 \quad$ or $\quad x = 1$

$\therefore$ Part (d) is equivalent.

9. (d) $\dfrac{x}{x-1} + \dfrac{x-1}{x} = \dfrac{5}{2} \Rightarrow \dfrac{x^2 + (x-1)^2}{x(x-1)} = \dfrac{5}{2}$

$\Rightarrow \dfrac{x^2 + x^2 - 2x + 1}{x^2 - x} = \dfrac{5}{2}$

$\Rightarrow 2(2x^2 - 2x + 1) = 5(x^2 - x)$

$\Rightarrow 4x^2 - 4x + 2 = 5x^2 - 5x$

$\Rightarrow x^2 - x - 2 = 0 \Rightarrow x^2 - 2x + x - 2 = 0$

$\Rightarrow x(x-2) + 1(x-2) = 0$

$\Rightarrow (x-2)(x+1) = 0 \Rightarrow x = 2, -1.$

10. (d) $\dfrac{1}{a+b+x} - \dfrac{1}{x} = \dfrac{1}{a} + \dfrac{1}{b}$

$\Rightarrow \dfrac{x - (a+b+x)}{x(a+b+x)} = \dfrac{a+b}{ab}$

$\Rightarrow \dfrac{-(a+b)}{xa + xb + x^2} = \dfrac{a+b}{ab}$

$\Rightarrow -ab = xa + xb + x^2$

$\Rightarrow x^2 + (a+b)x + ab = 0$

$\Rightarrow (x+a)(x+b) = 0 \Rightarrow x = -a, -b.$

11. (d) $5^{2x} - 5^{x+3} + 125 = 5^x$

$\Rightarrow 5^{2x} - 5^x \times 5^3 + 125 = 5^x$

$\Rightarrow 5^{2x} - 5^x(125+1) + 125 = 0$

$\Rightarrow 5^{2x} - 126 \times 5^x + 125 = 0$

Let $5^x = y$

Then, the given equation reduces to

$$y^2 - 126y + 125 = 0$$

$\Rightarrow y^2 - 125y - y + 125 = 0$

$\Rightarrow y(y-125) - 1(y-125) = 0$

$\Rightarrow (y-125)(y-1) = 0$

$\Rightarrow y - 125 = 0$ or $y - 1 = 0$

$\Rightarrow y = 125$ or $1 \Rightarrow 5^x = 125$ or $5^x = 1$

$\Rightarrow 5^x = 5^3$ or $5^x = 5^0 \Rightarrow x = 3, 0.$

12. (c) $3^{2x} + 9 = 10.3^x$

$\Rightarrow 3^{2x} - 10 \times 3^x + 9 = 0$

$\Rightarrow y^2 - 10y + 9 = 0 \quad$ Let $3^x = y$

$\Rightarrow y^2 - 9y - y + 9 = 0 \Rightarrow y(y-9) - 1(y-9) = 0$

$\Rightarrow (y-9)(y-1) = 0 \Rightarrow y = 9, 1$

$\Rightarrow 3^x = 9$ or $3^x = 1$

$\Rightarrow 3^x = 3^2$ or $3^x = 3^0 \Rightarrow x = 2, 0$

$\therefore$ The non zero root is a positive integer.

13. (b) $\dfrac{1}{3}$ is a root of the equation $3x^2 - 10x + p = 0$

$\Rightarrow 3\left(\dfrac{1}{3}\right)^2 - 10 \times \dfrac{1}{3} + p = 0$

$\Rightarrow \dfrac{1}{3} - \dfrac{10}{3} + p = 0 \Rightarrow -\dfrac{9}{3} + p = 0 \Rightarrow p = 3$

$\therefore$ The equation becomes $3x^2 - 10x + 3 = 0$

$\Rightarrow 3x^2 - 9x - x + 3 = 0$

$\Rightarrow 3x(x-3) - 1(x-3) = 0 \Rightarrow (3x-1)(x-3) = 0$

$\Rightarrow x = \dfrac{1}{3}, 3$

$\therefore p = 3$ and the other root $= 3$.

14. (b) Let the natural number be x. Then,

$$3x^2 - 4x = 50 + x$$

$\Rightarrow 3x^2 - 5x - 50 = 0$

$\Rightarrow 3x^2 - 15x + 10x - 50 = 0$

$\Rightarrow 3x(x-5) + 10(x-5) = 0$

$\Rightarrow (3x+10)(x-5) = 0$

$\Rightarrow x = -\dfrac{10}{3}, 5$

Rejecting $x = -\dfrac{10}{3}$ (not a natural number), $x = 5.$

15. (c) $x(x+1) = 1980$

$\Rightarrow x^2 + x - 1980 = 0$

$\Rightarrow x^2 + 45x - 44x - 1980 = 0$

$\Rightarrow x(x+45) - 44(x+45) = 0$

$\Rightarrow (x+45)(x-44) = 0$

$\Rightarrow x = -45, 44$

Rejecting -45, $x = 44$.

16. (b) Let the tens' digit be x. Then, ones' digit $= \dfrac{8}{x}$

Original number $= 10x + \dfrac{8}{x} = \dfrac{10x^2 + 8}{x}$

Number after reversing

$= 10 \times \dfrac{8}{x} + x = \dfrac{80}{x} + x = \dfrac{80 + x^2}{x}$

Given, $\dfrac{10x^2 + 8}{x} + 18 = \dfrac{80 + x^2}{x}$

$\Rightarrow \dfrac{10x^2 + 8 + 18x}{x} = \dfrac{80 + x^2}{x}$

$\Rightarrow 9x^2 + 18x - 72 = 0$

$\Rightarrow 9x^2 + 36x - 18x - 72 = 0$

$\Rightarrow 9x(x+4) - 18(x+4) = 0$

$\Rightarrow (9x-18)(x+4) = 0$

$\Rightarrow 9x - 18 = 0$ or $x + 4 = 0$

$\Rightarrow x = 2, -4$

Rejecting the negative value, $x = 2$

$\therefore$ Original number $= 10 \times 2 + \dfrac{8}{2} = 20 + 4 = 24$.

17. (c) Let the daughter's age be x years.

Then, the mother's age $= (x + 21)$ years

According to the question,

$(x+21) - \dfrac{1}{12} \times x \times (x+21) = 18$

$\Rightarrow 12(x+21) - (x^2 + 21x) = 216$

$\Rightarrow 12x + 252 - x^2 - 21x = 216$

$\Rightarrow x^2 + 9x - 36 = 0$

$\Rightarrow (x+12)(x-3) = 0$

$\Rightarrow x = -12, 3$

Rejecting negative value, daughter's age $= 3$ years

$\therefore$ Mother's age $= 24$ years

18. (d) Let the number be x.

Then $x + \sqrt{x} = \dfrac{6}{25}$

$\Rightarrow 25x + 25\sqrt{x} = 6$

Let $\sqrt{x} = y$. Then, the equation becomes

$25y^2 + 25y - 6 = 0$

$\Rightarrow 25y^2 + 30y - 5y - 6 = 0$

$\Rightarrow 5y(5y+6) - 1(5y+6) = 0$

$\Rightarrow (5y+6)(5y-1) = 0 \Rightarrow y = -\dfrac{6}{5}, \dfrac{1}{5}$

Rejecting the negative value, $\sqrt{x} = \dfrac{1}{5} \Rightarrow x = \dfrac{1}{25}$.

19. (a) Suppose the number of students in each row $= x$

Then, number of students in each column $= x + 3$

$\Rightarrow x(x+3) = 460 \Rightarrow x^2 + 3x - 460 = 0$

$\Rightarrow (x+23)(x-20) = 0 \Rightarrow x = -23, 20$

Rejecting negative value, number of students in each row $= 20$

$\therefore$ No. of students in each column $= 23$.

20. (d) Let the length of the hypotenuse be x cm.

Then, length of base $= (x - 2)$ cm

$x - 2 \times$ length of altitude $= 1$

$\Rightarrow$ Length of altitude $= \dfrac{1}{2}(x-1)$ cm

Applying Pythagoras' Theorem,

$(\text{Hyp.})^2 = (\text{Base})^2 + (\text{Perp.})^2$

$\Rightarrow x^2 = (x-2)^2 + \left(\dfrac{1}{2}(x-1)\right)^2$

$\Rightarrow x^2 = x^2 - 4x + 4 + \dfrac{1}{4}(x^2 - 2x + 1)$

$\Rightarrow 4x^2 = 4(x^2 - 4x + 4) + (x^2 - 2x + 1)$

$\Rightarrow 4x^2 = 4x^2 - 16x + 16 + x^2 - 2x + 1$

$x^2 - 18x + 17 = 0$

$\Rightarrow (x-17)(x-1) = 0 \Rightarrow x = 17, 1$

$x = 1$ is not possible.

$\therefore$ Length of hypotenuse $= 17$ cm

Length of base $= 15$ cm

Length of altitude $= \dfrac{1}{2} \times 16$ cm $= 8$ cm

$\therefore$ Perimeter of the triangle $= 17$ cm $+ 15$ cm $+ 8$ cm

$= 40$ cm.

21. (b) Let the present price be Rs k per dozen.

$\therefore$ Price per unit $=$ Rs $\dfrac{k}{12}$

$\therefore$ Increased price $=$ Rs $(k+2)$ per dozen

$\therefore$ Increase price per unit = Rs $\dfrac{(k+2)}{12}$

Given, $\dfrac{56}{\frac{k}{12}} - \dfrac{56}{\frac{(k+2)}{12}} = 8$

$\Rightarrow \dfrac{12 \times 56}{k} - \dfrac{12 \times 56}{k+2} = 8$

$\Rightarrow 672 \times (k+2) - 672k = 8k(k+2)$

$\Rightarrow 8k^2 + 16k - 1344 = 0$

$\Rightarrow k^2 + 2k - 168 = 0 \Rightarrow (k+14)(k-12) = 0$

$\Rightarrow k = 12$

22. (d) Let the length of one side be x m.

$\therefore$ Area of the plot = 60 m, length of the other side $= \dfrac{60}{x}$

Given, $2\left(x + \dfrac{60}{x}\right) = 34$

$\Rightarrow x^2 + 60 = 17x$

$\Rightarrow x^2 - 17x + 60 = 0$

$\Rightarrow x^2 - 12x - 5x + 60 = 0$

$\Rightarrow (x-12)(x-5) = 0$

$\Rightarrow x = 12, 5$

$\therefore$ Length of each shorter side = 5 m.

23. (a) Let the two little bands of monkeys have x monkeys. Then,

$\left(\dfrac{x}{8}\right)^2 + 12 = x \quad \Rightarrow \quad \dfrac{x^2}{64} + 12 = x$

$\Rightarrow x^2 - 64x + 768 = 0$

$\Rightarrow (x-16)(x-48) = 0$

$\Rightarrow x = 16, 48$

Seeing the options, $x = 16$.

24. (d) Suppose the total number of cows $= x$

$\dfrac{1}{4}x$ of the cows are in the forest.

$2\sqrt{x}$ have gone to mountains.

15 are on the banks of a river.

$\therefore \ 2\sqrt{x} + \dfrac{1}{4}x + 15 = x$

$\Rightarrow 2\sqrt{x} - \dfrac{3}{4}x + 15 = 0$

$\Rightarrow 8\sqrt{x} - 3x + 60 = 0$

$\Rightarrow 3y^2 - 8y - 60 = 0$ (Let $\sqrt{x} = y$)

$\Rightarrow 3y^2 - 18y + 10y - 60 = 0$

$\Rightarrow 3y(y-6) + 10(y-6) = 0$

$\Rightarrow (3y+10)(y-6) = 0 \Rightarrow y = -\dfrac{10}{3}, 6$

$\Rightarrow \sqrt{x} = y = 6$ (Rejecting –ve value)

$\Rightarrow x = 36.$

25. (b) $x - 4 = 21 \times \dfrac{1}{x}$

$\Rightarrow x^2 - 4x = 21$

$\Rightarrow x^2 - 4x - 21 = 0 \Rightarrow (x-7)(x+3) = 0$

$\Rightarrow x = 7, -3 \Rightarrow x = 7.$

26. (c) Let the year of his birth be x.

$\therefore$ His age (in years) $= 1980 - x$

Given, $1980 - x = \sqrt{x}$

$\Rightarrow x + \sqrt{x} - 1980 = 0$

$\Rightarrow y^2 + y - 1980 = 0$ (Let $\sqrt{x} = y$)

$\Rightarrow (y+45)(y-44) = 0 \Rightarrow y = -45, 44$

Rejecting –ve value, $y = 44$

$\Rightarrow \sqrt{x} = 44 \Rightarrow x = (44)^2 = 1936.$

27. (b) The total area of the pool = Sum of areas of two squares $= (x)^2 + (\sqrt{x})^2 = x^2 + x$

Given, $x^2 + x = 30$

$\Rightarrow x^2 + x - 30 = 0$

$\Rightarrow (x+6)(x-5) = 0$

$\Rightarrow x = -6, 5 \quad \Rightarrow \quad x = 5$

($\therefore$ Length cannot be –ve)

28. (a) Let the speed of the stream be x km/hr. Speed downstream $= (15 + x)$ km/hr

Speed upstream $= (15 - x)$ km/hr

$\therefore \ \dfrac{30}{(15+x)} + \dfrac{30}{(15-x)} = 4\dfrac{30}{60}$

$\Rightarrow \dfrac{30(15-x) + 30(15+x)}{(15+x)(15-x)} = \dfrac{9}{2}$

$\Rightarrow 900 \times 2 = 9(225 - x^2)$

$\Rightarrow x^2 - 225 = 200 \Rightarrow x^2 = 25 \Rightarrow x = 5.$

29. (d) Let the length of the field be x m

Then, width of the field $= (x + 3)$ m

Then, $x^2 + (x+3)^2 = 15^2$ (Pyth. Th.)

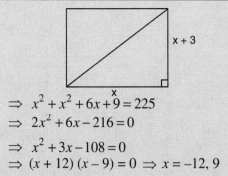

$\Rightarrow x^2 + x^2 + 6x + 9 = 225$

$\Rightarrow 2x^2 + 6x - 216 = 0$

$\Rightarrow x^2 + 3x - 108 = 0$

$\Rightarrow (x + 12)(x - 9) = 0 \Rightarrow x = -12, 9$

$\therefore$ Rejecting the negative value, length $= 9$ m

$\therefore$ Width $= 12$ m

$\Rightarrow$ Area of the field $= 12$ m $\times 9$ m $= 108$ m^2.

30. (b) $X^2 + 11X + 28 = 0 \Rightarrow (X + 7)(X + 4) = 0$

$\Rightarrow X = -7, -4$

$Y^2 + 15Y + 56 = 0 \Rightarrow (Y + 7)(Y + 4) = 0$

$\Rightarrow Y = -7, -8$

$\therefore$ $X \geq Y$.

Self Assessment Sheet–12

1. The roots of the equation $(x + 3)(x - 3) = 160$ are:

   (a) ± 13 (b) 13, 13

   (c) ± 12 (d) 12, 12

2. The roots of the equation $(2 - x)^2 = 16$ are:

   (a) ± 4 (b) $-6, +2$

   (c) $6, -2$ (d) $6, 2$

3. Solve for p: $2p(p - 3) + 5(p - 2) = 0$

   (a) $2, 2\frac{1}{2}$ (b) $2\frac{1}{2}, -2$

   (c) $-2\frac{1}{2}, 2$ (d) $-2\frac{1}{2}, -2$

4. Simplify the expression $\dfrac{2}{t} - 3 - 2t$ and find the values of t for which the expression is 0.

   (a) $-2, -1$ (b) $-2, \dfrac{1}{2}$

   (c) $-2, -\dfrac{1}{2}$ (d) $2, 1$

5. The sum of twice a whole number and three times the square of the next largest whole number is 83. What is the number ?

   (a) 6   (b) 8

   (c) 4   (d) -4

6. Two numbers differ by 8 and their product is 153. The numbers are :

   (a) $9, -17$ (b) $-17, 9$

   (c) $17, 9$ (d) $11, 19$

7. Form an equation in x for the following figure and hence find the length of the sides containing the right angle.

8. Two lawns have the same area, 121 sq m. One is a square, and the other is rectangular, and four times the perimeter of the rectangle is equal to five times that of the square. The length and breadth of the rectangle are :

   (a) 11m , 11m   (b) 22 m , $5\dfrac{1}{2}$ m

   (c) 55m, $2\dfrac{1}{5}$ m   (d) 33 m, $3\dfrac{2}{3}$ m

9. The difference between the quotients when 192 is divided by two numbers one of which is a square of the other is 21. The numbers are :

   (a) 4, 16   (b) 16, 256

   (c) 9, 81   (d) 8, 64

10. The units digit of a two-digit number is 2 more than the tens digit. If the number is subtracted from the sum of the squares of its digits the result is two-thirds of the product of the digits. What is the number ?

    (a) 62   (b) 58

    (c) 68   (d) 86

Answers

1. (a)   **2.** (c)   **3.** (b)   **4.** (b)   **5.** (c)   **6.** (c)   **7.** (b)   **8.** (b)   **9.** (d)   **10.** (c)

Chapter

LINEAR INEQUALITIES

KEY FACTS

1. Mathematical statements, such as $6 < 7$, $10 > 5$, $x \leq 2$, $5y \geq 12$, which state that the two quantities either are not equal or may not be equal are called **inequalities**. They are formed by placing an inequality symbol between two expressions.

2. An inequality uses one of the following symbols :

Symbol	Meaning
$<$	is less than
$>$	is greater than
$\leq$	is less than or equal to
$\geq$	is greater than or equal to

3. An **algebraic inequality** is an inequality containing a variable. The value of the variable that makes the inequality true is called the **solution** of the inequality.

   An inequality can have more than one solution. All the solutions together form the **solution set**.

4. The solutions of an inequality can be graphed on a number line using the following conventions.

   (i) If the variable is "less than" or "greater than" a number, then that number is indicated with an **open circle**.

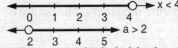

   In both these case, the numbers 4 and 2 are not included in the solution.

   (ii) If the variable is "less than or equal to" or "greater than or equal to" a number, then that number is indicated with a closed circle.

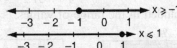

   Here the numbers –1 and 1 are included in the given solutions.

5. A **compound inequality** is the result of combining two inequalities. The words *and* and *or* are used to describe how two parts are related

<table>
<tr><td>

$x > 2$ or $x < -3$

Means

x is either greater than 2 or less than -3

</td><td>

On the number line, it is represented as

</td><td>

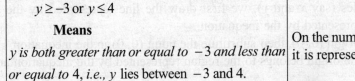

</td></tr>
</table>

<table>
<tr><td>

$y \geq -3$ or $y \leq 4$

Means

y is both greater than or equal to -3 *and less than or equal to 4, i.e., y lies between* -3 *and 4.*

</td><td>

On the number line, it is represented as

</td><td>

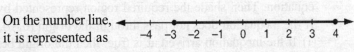

</td></tr>
</table>

Ch 13–1

6. An inequation can be solved using the same rules as for linear equations. Inequalities that have the same solutions are **equivalent** inequalities.

(i) Adding or subtracting the same number on each side of an inequality produces an equivalent inequality.

If $a > b$, then $a + c > b + c$

If $a < b$, then $a - c < b - c$

(ii) Multiplying or dividing each side of an inequality by the same positive number produces an equivalent inequality.

If $a > b$, then $a \times c > b \times c$, where $c > 0$

If $a < b$, then $\dfrac{a}{c} < \dfrac{b}{c}$, where $c > 0$

(iii) Multiplying or dividing each side of the inequality by the same negative number reverses the direction of the inequality.

If $a > b$, then $a \times c < b \times c$, where $c < 0$

If $a < b$, then $\dfrac{a}{c} > \dfrac{b}{c}$, where $c < 0$

7. Absolute value inequality.

(i) If $|x| \le a$, then either $x \le a$ or $-x \le a$, i.e. $x \ge -a$

$$\Rightarrow \quad \boxed{-a \le x \le a}$$

(ii) If $|x| \ge a$, then either $x \ge a$ or $-x \ge a$, i.e. $x \le -a$

$$\Rightarrow \quad \boxed{x \le -a \text{ or } x \ge a}$$

8. Representing inequalities on a cartesian plane.

A linear inequality can also be shown on the cartesian plane. The result is a **region** that lies on one side of the straight line or the other depending on the symbol of inequality used.

Method:

First draw the boundary line that defines the inequality. The nature of the boundary line depends upon the symbol of inequality.

If symbol '<' or '>' are involved, then a dashed or dotted line is drawn to indicate that the particular value is not included in the range of values.

But when the symbols '$\le$' or '$\ge$' are used to state the inequality, the boundary line should be drawn as a solid line to show that the particular value is included in the range of values.

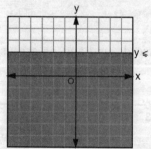

$y \le 2$

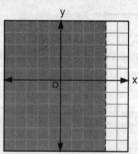

$x < 4$

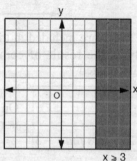

$x \ge 3$

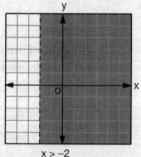

$x > -2$

9. To graph a linear inequality in two variables (say x and y), we first draw the line represented by the linear equation. Then shade the required region represented by the inequation.

If the inequation does not pass through the origin $(0, 0)$, then substitute the point $(0, 0)$ in the inequation.

(i) If the inequation arrived at is true, then the origin belongs to the region represented by the inequation and we shade origin side of the plane.

(ii) If the inequation arrived at is false, then the origin does not belong to the region represented by the inequation, and we shade the other half of the plane that does not contain the origin.

Solved Examples

Ex. 1. Solve 5 (2x + 1) ≤ 35, where x is a natural number.

Sol. $5(2x + 1) \leq 35 \Rightarrow 2x + 1 \leq \dfrac{35}{5} \Rightarrow 2x + 1 \leq 7 \Rightarrow 2x \leq 6 \Rightarrow x \leq 3$

Since $x \in N$, the solution set = {1, 2, 3}.

Ex. 2. Solve the inequality 4 (3 – x) ≤ 16.

Sol. $4(3 - x) \leq 16 \Rightarrow 12 - 4x \leq 16 \Rightarrow -4x \leq 4 \Rightarrow -x \leq 1 \Rightarrow x \geq -1$

[**Note:** Changing the sign means changing the direction of inequality.]

Ex. 3. Solve and show the given inequality on the number line :

   – 5 ≤ 3x – 2 < 7, where x ∈ R

Sol. The given inequality can be split into two inequalities :
$-5 \leq 3x - 2$ and $3x - 2 < 7$

$\Rightarrow -5 + 2 \leq 3x$ and $3x < 7 + 2 \Rightarrow -3 \leq 3x$ and $3x < 9 \Rightarrow -1 \leq x$ and $x < 3 \Rightarrow -1 \leq x < 3$
This can be shown on a number line as :

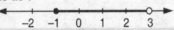

Ex. 4. Solve | 2x + 6 | ≤ 16.

Sol. $|2x + 6| \leq 16$

$\Rightarrow -16 \leq (2x + 6) \leq 16$

$\Rightarrow -16 - 6 \leq 2x \leq 16 - 6$ (Subtracting 6 throughout the continued inequation)

$\Rightarrow -22 \leq 2x \leq 10$

$\Rightarrow -11 \leq x \leq 5$ (Dividing by 2 throughout the continued inequation)

Ex. 5. Solve | 4x – 3 | ≥ 5.

Sol. $|4x - 3| \geq 5$

$\Rightarrow 4x - 3 \geq 5$ or $-(4x - 3) \geq 5$, i.e., $4x - 3 \leq -5$

$\Rightarrow 4x \geq 8$ or $4x \leq -2 \Rightarrow x \geq 2$ or $x \leq -\dfrac{1}{2}$

$\Rightarrow$ Solution set = $\left\{ x \mid x \geq 2 \text{ or } x \leq -\dfrac{1}{2} \right\} = (-\infty, -1/2) \cup (2, \infty)$

Ex. 6. Show the region defined by –1 < y ≤ 4 on the cartesian.

Sol.

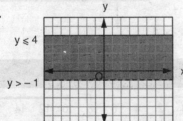

Ex. 7. Draw the graph of the inequality y ≤ x.

Sol. We first draw the line $y = x$.

x	–1	0	1
y	–1	0	1

Plot the points (–1, –1) (0, 0), (1, 1) to get the required line.

∴ The sign ≤ is used the line has to be solid.

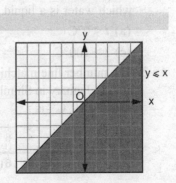

Ex. 8. *Draw the graph of x + 2y ≥ –4.*

Sol. We first draw $x + 2y = -4$

x	–6	–4	0
y	1	0	–2

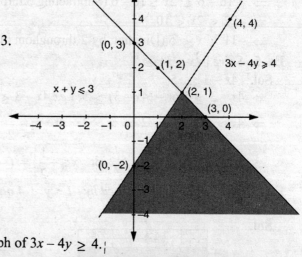

Now to check as to which side of the line to shade

We substitute (0, 0) in the inequation $x + 2y \geq -4$

$\Rightarrow 0 + 2 \times 0 \geq -4$

$\Rightarrow 0 \geq -4$, which is true.

∴ (0, 0) belongs to graph represented by $x + 2y \geq -4$, so shade the part of plane on that side of the line that contains the origin.

Ex. 9. *Shade the region represented by the inequations*

  x + y ≤ 3 and 3x – 4y ≥ 4.

Sol. Draw the graphs of the two equations $x + y = 3$ and $3x - 4y = 4$ on the same co-ordinate plane.

To draw the graph of x + y = 3

$x + y = 3 \Rightarrow y = 3 - x$

x	0	1	3
y	3	2	0

Plot the points (0, 3), (1, 2) and (3, 0) and join.

Substitute (0, 0) in $x + y \leq 3$

$\Rightarrow 0 + 0 \leq 3 \Rightarrow 0 \leq 3$, which is true.

∴ The side on which the origin lies is the graph of $x + y \leq 3$.

To draw the graph of 3x – 4y = 4

$3x - 4y = 4 \Rightarrow y = \dfrac{3}{2}x - 2$

x	0	2	4
y	–2	1	4

Plot the points (0, –2), (2, 1) and (4, 4) and join.

Now substitute (0, 0) in $3x - 4y \geq 4$

$\Rightarrow 0 - 0 \leq 4 \Rightarrow 0 \geq 4$, which is false.

∴ The half plane that does not contain the origin is the graph of $3x - 4y \geq 4$.

Question Bank–13

1. Water freezes at 0°C and boils at 100°C. Write an inequality to show the range of temperature (*t*) for which water is a liquid.

   (a) $t < 0°C$ (b) $0°C \leq t \leq 100°C$

   (c) $t > 100°C$ (d) $0°C < t < 100°C$

2. The number line matching the statement "It was so cold in January in Shimla that the temperature never reached 10°C" is :

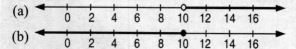

3. The solution set of $x + 2 < 9$ over a set of positive even integers is

   (a) {8, 10, 12,...} (b) {2, 4, 6}

   (c) {1, 2, 3, 4, 5, 6} (d) {2, 4, 6, 8}

4. The solution set of $3x - 4 < 8$ over the set of square numbers is

   (a) {1, 2, 3} (b) {1, 4}

   (c) {1} (d) {16}

5. The solution to the inequality $-5x > 4$ is

(a) $x < \dfrac{4}{5}$ (b) $x > -\dfrac{4}{5}$

(c) $x < -\dfrac{4}{5}$ (d) $x > \dfrac{4}{5}$

6. Which inequality has the following number line solution :

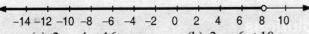

(a) $2x - 4 < 16$ (b) $2x - 6 < 10$

(c) $2x - 6 > 12$ (d) $2x - 4 > 16$

7. Which graph shows the solution to the inequality $-3.5x - 12 \le 58$?

(a) $x \le -20$ (b) $x \ge -70$

(c) $x \le -70$ (d) $x \ge -20$

8. The range of x giving the solution set of $-1 < 5x + 4 \le 19$ is

(a) $-1 \le x < 3$ (b) $-1 < x \le 3$

(c) $-1 \le x \le 3$ (d) $-1 \le x \le 3$

9. The inequality $-1 \le 2x + 4 < 5$, where x is an integer can be represented on the number line as :

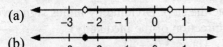

10. The solution to the inequality $|10 - 2x| > 6$ is

(a) $2 < x < 8$ (b) $x < -2$ and $x > 8$

(c) $x > 2$ and $x < -8$ (d) $x < 2$ or $x > 8$

11. How many integers are there in the solution set of $|2x + 6| < \dfrac{19}{2}$?

(a) None (b) Two

(c) Fourteen (d) Nine

12. The solution to the inequality $-2x + (3^3 - 5^2) \ge 4$ is

(a) $x \ge -1$ (b) $x \le -1$

(c) $x > -2$ (d) $x < 2$

13. The inequality $|3 - p| - 4 < 1$ can be represented on the number line as :

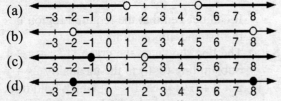

14. The solution set of the inequality $2(4x - 1) \le 3(x + 4)$ is

15. (a) $x > \dfrac{14}{5}$ (b) $x < 7$

(c) $x \le \dfrac{14}{5}$ (d) $x \ge 7.5$

15. Which of the following is the solution set of $\left|\dfrac{2}{3}x - 5\right| > 8$?

(a) $\left\{x : -\dfrac{39}{2} < x < \dfrac{9}{2}\right\}$ (b) $\left\{x : -\dfrac{9}{2} < x < \dfrac{39}{2}\right\}$

(c) $\left\{x : x > \dfrac{39}{2} \text{ or } x < -\dfrac{9}{2}\right\}$

(d) $\left\{x : x > \dfrac{9}{2} \text{ or } x < \dfrac{-39}{2}\right\}$

16. If $(2x - y < 7)$ and $(x + 4y < 11)$, then which one of the following is corect ?

(a) $x + y < 5$ (b) $x + y < 6$

(c) $x + y \le 5$ (d) $x + y \ge 6$

17. The region for which $x \ge 4$ is a part of the:

(a) first and second quadrants

(b) second and third quadrants

(c) third and fourth quadrants

(d) fourth and first quadrants

18. Which of the following inequations represents the shaded region ?

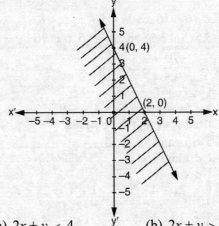

(a) $2x + y \le 4$ (b) $2x + y \ge 4$

(c) $x + 2y \le 4$ (d) $x + 2y \ge 4$

19. The shaded region is represented by the inequation :

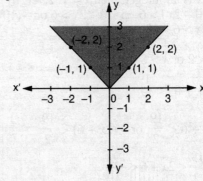

(a) $y \geq x$
(b) $y \geq -x$
(c) $y \geq |x|$
(d) $y \leq |x|$

20. The area of the plane region $|x| \leq 5$; $|y| \leq 3$ is
(a) 15 sq unit
(b) 34 sq unit
(c) 60 sq unit
(d) 120 sq unit

Answers

1. (d)	**2.** (d)	**3.** (b)	**4.** (c)	**5.** (c)	**6.** (b)	**7.** (d)	**8.** (b)	**9.** (c)	**10.** (d)
11. (d)	**12.** (b)	**13.** (b)	**14.** (c)	**15.** (c)	**16.** (b)	**17.** (d)	**18.** (a)	**19.** (c)	**20.** (c)

Hints and Solutions

1. (d) Water will be in the liquid state between the temperatures 0° C and 100° C, but not including them. At 0° C, water will freeze at 100° C, it will convert to vapour.

Therefore, required inequality = 0° C < t < 100° C

2. (d) Temperature was less than 10° C, it never touched 10° C.

3. (b) $x + 2 < 9 \Rightarrow x < 9 - 2 \Rightarrow x < 7$

∴ The positive even integers less than 7 are 2, 4, 6.

4. (c) $3x - 4 < 8 \Rightarrow 3x < 12 \Rightarrow x < \dfrac{12}{3} \Rightarrow x < 4$

⇒ The solution set = {1} as 1 is the only square number less than 4.

5. (c) $-5x > 4 \Rightarrow 5x < -4 \Rightarrow x < -\dfrac{4}{5}$

6. (b) $2x - 6 < 10 \Rightarrow 2x < 16 \Rightarrow x < 8$

7. (d) $-3.5x - 12 \leq 58 \Rightarrow -3.5x \leq 70$

$\Rightarrow x \geq -\dfrac{70}{3.5} \Rightarrow x \geq -20$

8. (b) $-1 < 5x + 4 \leq 19$
$\Rightarrow -1 < 5x + 4$ and $5x + 4 \leq 19$
$\Rightarrow -1 - 4 < 5x$ and $5x \leq 19 - 4$
$\Rightarrow -5 < 5x$ and $5x \leq 15$
$\Rightarrow -1 < x$ and $x \leq 3$
$\Rightarrow -1 < x \leq 3$

9. (c) $-1 \leq 2x + 4 < 5 \Rightarrow -1 - 4 \leq 2x < 5 - 4$

$\Rightarrow -5 \leq 2x < 1 \Rightarrow -\dfrac{5}{2} \leq x < \dfrac{1}{2}$

$\Rightarrow x$ can take the values between $-2\dfrac{1}{2}$ and $\dfrac{1}{2}$

$\Rightarrow$ If x is an integer, then $x = \{-2, -1, 0\}$

10. (d) $|10 - 2x| > 6 \Rightarrow 10 - 2x > 6$ or $-(10 - 2x) > 6$
$\Rightarrow -2x > 6 - 10$ or $2x > 6 + 10$
$\Rightarrow -2x > -4$ or $2x > 16$
$\Rightarrow x < 2$ or $x > 8$

11. (d) $|2x + 6| < \dfrac{19}{2}$

$\Rightarrow 2x + 6 < \dfrac{19}{2}$ or $-(2x + 6) < \dfrac{19}{2}$

$\Rightarrow -\dfrac{19}{2} < 2x + 6 < \dfrac{19}{2}$

$\Rightarrow -\dfrac{19}{2} - 6 < 2x < \dfrac{19}{2} - 6 \Rightarrow -\dfrac{31}{2} < 2x < \dfrac{7}{2}$

$\Rightarrow -\dfrac{31}{4} < x < \dfrac{7}{4} \Rightarrow -7.75 < x < 1.75$

∴ The integers between –7.75 and 1.75 are –7, –6, –5, –4, –3, –2, –1, 0, 1, *i.e.,* nine in number.

12. (b) $-2x + (3^3 - 5^2) \geq 4$
$\Rightarrow -2x + (27 - 25) \geq 4$
$\Rightarrow -2x + 2 \geq 4 \Rightarrow -2x \geq 2 \Rightarrow x \leq -1$

13. (b) $|3 - p| - 4 < 1 \Rightarrow |3 - p| < 5$
$\Rightarrow -5 < 3 - p < 5$
$\Rightarrow -8 < -p < 2$
$\Rightarrow -2 < p < 8$
(Note the step: Dividing by minus sign reverses the direction of the inequality)

14. (c) $2(4x - 1) \leq 3(x + 4)$

$\Rightarrow 8x - 2 \leq 3x + 12 \Rightarrow 5x \leq 14 \Rightarrow x \leq \dfrac{14}{5}$

15. (c) Similar to Q. 10.

16. (b) $2x - y < 7$ ...(i)
$x + 4y < 11$ ...(ii)
Adding (i) and (ii), we get
$3x + 3y < 18 \Rightarrow x + y < 6$

17. (d) The graphical representation of $x \geq 4$ is

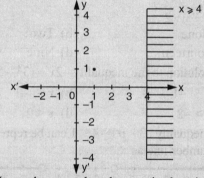

It can be seen that the required region lies in the 1st and 4th quadrant.

18. (a) The line shown on the graph is $2x + y = 4$
Substituting (0, 0) in the inequality $2x + y \leq 4$.
$2 \times 0 + 0 \leq 4 \Rightarrow 0 \leq 4$, which hold true
$\Rightarrow$ (0, 0) lies in the required region.
∴ (a) is the correct option.

19. (c) $y \geq |x| \Rightarrow y \geq x$ or $y \geq -x$

$$\left[\because |x| = \begin{cases} -x, x < 0 \\ x, x > 0 \end{cases}\right]$$

$\therefore x - y \leq 0$ or $y + x \geq 0$

Plotting the points for $x - y = 0$, we get the points $(0, 0)$ $(1, 1)$ $(2, 2)$ etc.

Plotting the points for $x + y = 0$, we get the points $(0, 0)$ $(1, -1)$ $(2, -2)$ etc.

$\therefore$ (c) is the required option.

20. (c) $|x| \leq 5 \Rightarrow -5 \leq x \leq 5$

$|y| \leq 3 \Rightarrow -3 \leq y \leq 3$

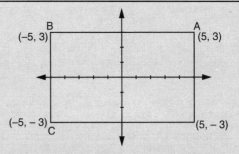

$\therefore$ Area bounded by the rectangle so formed $= AB \times AD = 10 \times 6 = 60$ sq units.

Self Assessment Sheet–13

1. The greatest value of x that satisfies the inequality $2x + 3 < 25$, where x is a prime number is
   (a) 11
   (b) 7
   (c) 10
   (d) 2

2. Which graph represents the solution of the inequality "x subtracted from 7 is less than 2".

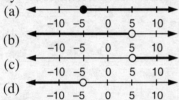

3. You are buying a carpet for a rectangular room. The carpet can be at most 12 m long and 6 m wide. Which inequality represents the area of the carpet is square metres ?
   (a) $A \leq 36$
   (b) $A \geq 36$
   (c) $A \leq 72$
   (d) $A \geq 72$

4. The absolute value of a number is its distance from 0 on a number line. The number line representing the inequality $|x| < 4$ is

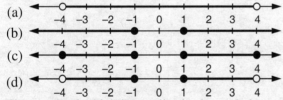

5. The graph of which inequality is shown below :

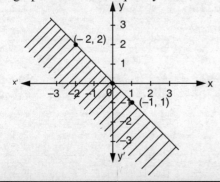

(a) $y - x \leq 0$
(b) $x - y \leq 0$
(c) $y + x \leq 0$
(d) None of the above

6. The solution set of $x \geq 5, y \geq 0$ and $x \leq 5, y \leq 0$ is
   (a) $x \geq -5, y = 0$
   (b) $x = 5, y = 0$
   (c) $x \geq -5, y \leq 0$
   (d) $x \leq -5, y \geq 0$

7. Given $a > 0, b > 0, a > b$ and $c \neq 0$. Which inequality is not always correct ?
   (a) $a + c > b + c$
   (b) $a - c > b - c$
   (c) $ac > bc$
   (d) $\dfrac{a}{c^2} > \dfrac{b}{c^2}$

8. The shaded region is represented by the inequality:
   (a) $y - 2x \leq -1$
   (b) $x - 2y \leq -1$
   (c) $y - 2x \geq -1$
   (d) $x - 2y \geq -1$

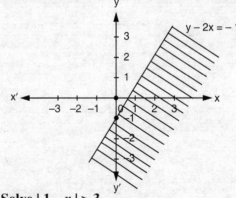

9. Solve $|1 - x| > 3$.
   (a) $x > 4$ or $x < -1$
   (b) $x > 2$ or $x < -2$
   (c) $x > 5$ or $x < -2$
   (d) $x > 4$ or $x < -2$

10. If x is an integer greater than -10, but less than 10 and $|x - 2| < 3$, then the values of x are
   (a) $-10, -9, -8, -7, -6, -5, -4, -3, -2, -1, 0, 1, 2, 3, 4$
   (b) $0, 1, 2, 3, 4, 5, 6, 7, 8, 9, 10$
   (c) $0, 1, 2, 3, 4$
   (d) $-1, 0, 1$

Answers

1. (b)	2. (c)	3. (c)	4. (a)	5. (c)	6. (b)	7. (c)	8. (a)	9. (d)	10. (c)

Chapter 14

MATRICES

1. A matrix (plural matrices) is an array of real numbers (or other suitable entities) arranged in rows and columns.

   Examples of matrices are $\begin{bmatrix} 1 & 2 & \frac{1}{2} \\ 3 & 4 & 5 \end{bmatrix}$, $\begin{bmatrix} \frac{1}{4} \end{bmatrix}$, $[4 -5\ 6]$, $\begin{bmatrix} -2 & 0 \\ -3 & 4 \end{bmatrix}$ etc.

2. Each number or entity in a matrix is called an **element** or **entry** of the matrix. The horizontal lines of numbers are called **rows** and the vertical lines of numbers are called **columns**. We denote matrices by capital letters A, B, C, ... and elements by small letters a, b, c, ...

3. **Order of the matrix :** A matrix with m rows and n columns is an **$m \times n$ matrix** (read "m by n")
   The first number always denotes the number of rows and the second denotes the number of columns.
   A 3×4 matrix means a matrix having 3 rows and 4 columns.
   Special Types of Matrices

4. **Row Matrix :** *A matrix having only one row is called a **row matrix**.*
   For example, $A = [\ 4\ \ 1]$ is a row matrix of order 1×2 (1 row, 2 columns)
   $B = [\ 5\ -1\ \frac{1}{2}\ ]$ is a row matrix of order 1×3.

5. **Column Matrix :** *A matrix having only one column is called a **column matrix**.*
   For example, $A = \begin{bmatrix} 1 \\ 2 \end{bmatrix}$ is a column matrix of order 2×1 (2 rows, 1 column)
   $B = \begin{bmatrix} -1 \\ 4 \\ 5 \\ 3 \end{bmatrix}$ is a column matrix of order 4×1.

6. **Square Matrix :** *A matrix having the same number of columns as it has rows is called a **square matrix**.*
   For example, $A = \begin{bmatrix} 4 & 7 \\ -2 & 9 \end{bmatrix}$ having 2 rows and 2 columns is 2×2 square matrix or a square matrix of order 2. It
   may also be denoted by A_2.
   $B_3 = \begin{bmatrix} 1 & 4 & 9 \\ 2 & -4 & 5 \\ -3 & 4 & 8 \end{bmatrix}$ is a square matrix of order 3.

7. Zero Matrix : *A matrix each of whose elements is zero is called a **zero matrix** or **null matrix**.*
For example,

$$O_2 = \begin{bmatrix} 0 & 0 \\ 0 & 0 \end{bmatrix}, \quad O_{2\times4} = \begin{bmatrix} 0 & 0 & 0 & 0 \\ 0 & 0 & 0 & 0 \end{bmatrix}$$

8. Diagonal Matrix : *A square matrix having all the elements zero, except the principal diagonal elements is called a **diagonal matrix**.*
For example,

$$A = \begin{bmatrix} 4 & 0 & 0 \\ 0 & 7 & 0 \\ 0 & 0 & 1 \end{bmatrix}, \quad B = \begin{bmatrix} 2 & 0 \\ 0 & 1 \end{bmatrix}$$

9. Identity Matrix or Unit Matrix : *A square matrix in which each diagonal element is unity, all other elements being zero, is called a **Unit matrix** or **Identity matrix**.*

$$I_2 = \begin{bmatrix} 1 & 0 \\ 0 & 1 \end{bmatrix}, \quad I_3 = \begin{bmatrix} 1 & 0 & 0 \\ 0 & 1 & 0 \\ 0 & 0 & 1 \end{bmatrix}$$

10. General form of a Matrix:

A matrix of the order $m \times n$ in the general format is written as :

$$A = \begin{bmatrix} a_{11} & a_{12} & a_{13} & \cdots & a_{1n} \\ a_{21} & a_{22} & a_{23} & \cdots & a_{2n} \\ \cdots & \cdots & \cdots & \cdots & \cdots \\ a_{m_1} & a_{m_2} & a_{m_3} & \cdots & a_{mn} \end{bmatrix} = [a_{ij}]_{m\times n} \text{ or } [a_{ij}]$$

where m represents the number of rows and n the number of columns. Hence, the element a_{ij} belongs to the ith row and jth column and is called the (i, j)th element of the matrix $A = [a_{ij}]$

Thus, in the matrix

$$\begin{bmatrix} 1 & 2 & 4 \\ -7 & 3 & 2 \end{bmatrix} \qquad \begin{aligned} a_{11} &= 1, & a_{12} &= 2, & a_{13} &= 4 \\ a_{21} &= -7, & a_{22} &= 3, & a_{23} &= 2 \end{aligned}$$

11. Equality of Matrices : Two matrices are equal if and only if :
(i) both the matrices are of the some order.
(ii) their corresponding elements are equal.

Thus, $\begin{bmatrix} 2 & 4 \\ 8 & -3 \end{bmatrix} = \begin{bmatrix} 3-1 & \dfrac{10-2}{2} \\ \dfrac{16}{2} & -4+1 \end{bmatrix}$ and $\begin{bmatrix} 3 \\ 2 \\ 4 \end{bmatrix} \neq \begin{bmatrix} 3 & 2 & 4 \end{bmatrix}$

12. Addition of Matrices : Two matrices can only be added if they are of the same order. The sum of two matrices A and B, i.e., $A + B$ is found by **adding** the **corresponding elements** and therefore the sum is also a matrix of the same order.
For example,

$$\begin{bmatrix} a & b \\ c & d \end{bmatrix} + \begin{bmatrix} e & f \\ g & h \end{bmatrix} = \begin{bmatrix} a+e & b+f \\ c+g & d+h \end{bmatrix}$$

13. Negative of a Matrix: The negative of a matrix A denoted by $-A$ is the matrix formed by replacing each entry in matrix A with the additive inverse.

Thus, if $A = \begin{bmatrix} 6 & 1 \\ 7 & -2 \\ -9 & 3 \end{bmatrix}$, then $-A = \begin{bmatrix} -6 & -1 \\ -7 & 2 \\ 9 & -3 \end{bmatrix}$

The sum of a matrix and its negative is always a zero matrix.

$$A + (-A) = \begin{bmatrix} 6 & 1 \\ 7 & -2 \\ -9 & 3 \end{bmatrix} + \begin{bmatrix} -6 & -1 \\ -7 & 2 \\ 9 & -3 \end{bmatrix} = \begin{bmatrix} 6+(-6) & 1+(-1) \\ 7+(-7) & -2+2 \\ -9+9 & 3+(-3) \end{bmatrix} = \begin{bmatrix} 0 & 0 \\ 0 & 0 \\ 0 & 0 \end{bmatrix}$$

14. Subtraction of Matrices : The subtraction of matrices is defined in the same manner as the subtraction of real numbers. Thus, for matrices A and B of the same order,
$A - B = A + (-B)$

If $A = \begin{bmatrix} a & b \\ c & d \end{bmatrix}$ and $B = \begin{bmatrix} e & f \\ g & h \end{bmatrix}$, then

$$A - B = A + (-B) = \begin{bmatrix} a & b \\ c & d \end{bmatrix} + \begin{bmatrix} -e & -f \\ -g & -h \end{bmatrix} = \begin{bmatrix} a-e & b-f \\ c-g & d-h \end{bmatrix}$$

15. Properties of Addition of Matrices:
  (i) **Addition of matrices is commutative,** *i.e., if A and B are any two matrices of the same order, then* $A + B = B + A$.
 (ii) **Addition of matrices is associative,** *i.e., if A, B and C are three matrices of the same order, then* $(A + B) + C = A + (B + C)$.
(iii) **Existence of additive - identity :** *If O is the null matrix of the same order as the matrix A, then* $A + O = O + A = A$.
  The null matrix O is the *additive identity.*
(iv) **Existence of additive - inverse :** *For every matrix A, the matrix $-A$ has the property $A + (-A) = 0$ and $(-A) + A = 0$.*
  The matrix $-A$ is called the *additive inverse* of matrix A.

16. Multiplication of a Matrix by a Scalar.

If K is a real number (or scalar). We define the product of K and the matrix $\begin{bmatrix} a & b \\ c & d \end{bmatrix}$ as $K\begin{bmatrix} a & b \\ c & d \end{bmatrix} = \begin{bmatrix} Ka & Kb \\ Kc & Kd \end{bmatrix}$

Solved Examples

Ex. 1. *Find p, q, r and s if* $\begin{bmatrix} p+4 & 2q-7 \\ s-3 & r+2s \end{bmatrix} = \begin{bmatrix} 6 & -3 \\ 2 & 14 \end{bmatrix}$

Sol. Two matrices are equal, if the corresponding elements are equal, *i.e.,* $p + 4 = 6 \Rightarrow p = 6 - 4 = 2$

$2q - 7 = -3 \Rightarrow 2q = -3 + 7 = 4 \Rightarrow q = 2$

$s - 3 = 2 \Rightarrow s = 5$

$r + 2s = 14 \Rightarrow r + 10 = 5 \Rightarrow r = -5$

Ex. 2. *If* $P = \begin{bmatrix} -3 & 1 \\ 2 & 5 \end{bmatrix}$, $Q = \begin{bmatrix} 1 & 6 \\ -4 & 0 \end{bmatrix}$ *and* $R\begin{bmatrix} 4 & -1 \\ 2 & 3 \end{bmatrix}$ *find the value of 4P – 2Q + 3R.*

Sol. $4P - 2Q + 3R$

$$= 4\begin{bmatrix} -3 & 1 \\ 2 & 5 \end{bmatrix} - 2\begin{bmatrix} 1 & 6 \\ -4 & 0 \end{bmatrix} + 3\begin{bmatrix} 4 & -1 \\ 2 & 3 \end{bmatrix} = \begin{bmatrix} -12 & 4 \\ 8 & 20 \end{bmatrix} - \begin{bmatrix} 2 & 12 \\ -8 & 0 \end{bmatrix} + \begin{bmatrix} 12 & -3 \\ 6 & 9 \end{bmatrix}$$

$$= \begin{bmatrix} -12-2+12 & 4-12-3 \\ 8-(-8)+6 & 20-0+9 \end{bmatrix} = \begin{bmatrix} -2 & -11 \\ 22 & 29 \end{bmatrix}$$

Ex. 3. *Find the matrix X such that* $-A + 3B + X = 0$, *where* $A = \begin{bmatrix} -2 & 6 \\ 5 & 8 \end{bmatrix}$ *and* $B = \begin{bmatrix} 1 & 2 \\ -2 & 3 \end{bmatrix}$

Sol. Given, $-A + 3B + X = 0$

$\Rightarrow X = A - 3B$

$\Rightarrow X = \begin{bmatrix} -2 & 6 \\ 5 & 8 \end{bmatrix} - 3\begin{bmatrix} 1 & 2 \\ -2 & 3 \end{bmatrix}$

$= \begin{bmatrix} -2 & 6 \\ 5 & 8 \end{bmatrix} - \begin{bmatrix} 3 & 6 \\ -6 & 9 \end{bmatrix} = \begin{bmatrix} -2-3 & 6-6 \\ 5-(-6) & 8-9 \end{bmatrix} = \begin{bmatrix} -5 & 0 \\ 11 & -1 \end{bmatrix}$

Question Bank–14

1. The order of the matrix $\begin{bmatrix} -1 \\ 3 \\ 4 \end{bmatrix}$ is :

   (a) 1×3 (b) 3×1
   (c) 1×1 (d) 3×3

2. Which of the following pair of matrices are equal ?

   (a) $\begin{bmatrix} 1 & 0 \\ 0 & 1 \end{bmatrix}, \begin{bmatrix} 0 & 1 \\ 1 & 0 \end{bmatrix}$

   (b) $\begin{bmatrix} 4 & 5 & 6 \end{bmatrix}, \begin{bmatrix} 4 \\ 5 \\ 6 \end{bmatrix}$

   (c) $\begin{bmatrix} 4 & 7 \\ 3 & 2 \end{bmatrix}, \begin{bmatrix} 3+1 & \sqrt{49} \\ \dfrac{7-1}{2} & \sqrt[3]{8} \end{bmatrix}$

   (d) $\begin{bmatrix} 4 & 0 \\ 0 & 4 \end{bmatrix}, \begin{bmatrix} 4 & 0 \\ 4 & 4 \end{bmatrix}$

3. A square matrix A has 9 elements. What is the possible order of A ?

   (a) 1×9 (b) 9×9
   (c) 3×3 (d) 2×7

4. The matrix $\begin{bmatrix} -12 \\ 10 \\ 13 \\ 4 \end{bmatrix}$ is a :

   (a) square matrix   (b) row matrix
   (c) column matrix   (d) null matrix

5. $\begin{bmatrix} 1 & 0 & 0 \\ 0 & 4 & 0 \\ 0 & 0 & 5 \end{bmatrix}$ is an:

   (a) unit matrix   (b) null matrix
   (c) diagonal matrix   (d) row matrix

6. Which of the following statements is true?

   (a) Every zero matrix is a square matrix
   (b) A unit matrix is a diagonal matrix
   (c) $\begin{bmatrix} 1 & 0 \\ 0 & 1 \end{bmatrix}$ is the identity matrix for addition of 2×2 matrix
   (d) $\begin{bmatrix} 4 & x \\ 1 & 3 \end{bmatrix} = \begin{bmatrix} 4 & 5 \\ 1 & 0 \end{bmatrix}$, if $x = 5$

7. If $\begin{bmatrix} 2x & 3 \\ 0 & y-1 \end{bmatrix} = \begin{bmatrix} x-3 & 3 \\ 0 & 2 \end{bmatrix}$, then the values of x and y respectively are :

   (a) $3, -3$ (b) $-3, 3$
   (c) $-3, -3$ (d) $3, 3$

8. If $A = [a_{ij}]_{m \times n}$, $B = [b_{ij}]_{m \times n}$, then the element C_{23} of the matrix $C = A + B$ is :

   (a) $a_{13} + b_{13}$ (b) $a_{23} + b_{32}$
   (c) $a_{23} + b_{23}$ (d) $a_{32} + b_{23}$

9. Given, matrix $A = \begin{bmatrix} 3 \\ 2 \end{bmatrix}$ and $B = \begin{bmatrix} -2 \\ -1 \end{bmatrix}$, find the matrix X such that $X - A = B$.

   (a) $\begin{bmatrix} 0 \\ 0 \end{bmatrix}$ (b) $\begin{bmatrix} 1 \\ 1 \end{bmatrix}$

   (c) $\begin{bmatrix} 4 \\ 0 \end{bmatrix}$ (d) $\begin{bmatrix} 1 \\ -1 \end{bmatrix}$

10. If $A = \begin{bmatrix} x & y \\ z & w \end{bmatrix}$, $B = \begin{bmatrix} x & -y \\ -z & w \end{bmatrix}$

    and $C = \begin{bmatrix} -2x & 0 \\ 0 & -2w \end{bmatrix}$, then $A + B + C$ is a

    (a) identity matrix (b) null matrix
    (c) row matrix (d) column matrix

11. If $\begin{bmatrix} 2 & -1 \\ 2 & 0 \end{bmatrix} + 2A = \begin{bmatrix} -3 & 5 \\ 4 & 3 \end{bmatrix}$, the matrix A equals

    (a) $\begin{bmatrix} -5 & 6 \\ 2 & 3 \end{bmatrix}$ (b) $\begin{bmatrix} -\frac{5}{2} & 3 \\ 1 & \frac{3}{2} \end{bmatrix}$

    (c) $\begin{bmatrix} -\frac{5}{2} & 6 \\ 2 & 3 \end{bmatrix}$ (d) $\begin{bmatrix} -5 & 8 \\ 1 & 3 \end{bmatrix}$

12. Given $A = \begin{bmatrix} 2 & -1 \\ 2 & 0 \end{bmatrix}$, $B = \begin{bmatrix} -3 & 2 \\ 4 & 0 \end{bmatrix}$ and $C = \begin{bmatrix} 1 & 0 \\ 0 & 2 \end{bmatrix}$, find the matrix X such that $A + X = 2B + C$.

    (a) $\begin{bmatrix} 5 & -7 \\ 2 & 6 \end{bmatrix}$ (b) $\begin{bmatrix} -7 & 2 \\ 6 & 5 \end{bmatrix}$

    (c) $\begin{bmatrix} -7 & 5 \\ 6 & 2 \end{bmatrix}$ (d) $\begin{bmatrix} -7 & 6 \\ 5 & 2 \end{bmatrix}$

13. A 2×2 matrix whose elements a_{ij} are given by $a_{ij} = i - j$ is :

    (a) $\begin{bmatrix} 0 & 1 \\ 1 & 0 \end{bmatrix}$ (b) $\begin{bmatrix} 0 & -1 \\ 1 & 0 \end{bmatrix}$

    (c) $\begin{bmatrix} -1 & 0 \\ 0 & 1 \end{bmatrix}$ (d) $\begin{bmatrix} 0 & 1 \\ -1 & 0 \end{bmatrix}$

14. The values of x and y respectively if $\begin{bmatrix} 2x + y \\ 3x - 2y \end{bmatrix} = \begin{bmatrix} 5 \\ 4 \end{bmatrix}$ are :

    (a) $x = -2, y = 1$ (b) $x = -2, y = -1$
    (c) $x = 2, y = 1$ (d) $x = 2, y = -1$

15. Solve the matrix equation $\begin{bmatrix} 2 & 1 \\ 5 & 0 \end{bmatrix} - 3X = \begin{bmatrix} -7 & 4 \\ 2 & 6 \end{bmatrix}$

    (a) $X = \begin{bmatrix} 3 & 1 \\ -1 & 2 \end{bmatrix}$ (b) $X = \begin{bmatrix} 3 & -1 \\ 1 & -2 \end{bmatrix}$

    (c) $X = \begin{bmatrix} 3 & -1 \\ -1 & -2 \end{bmatrix}$ (d) $X = \begin{bmatrix} -3 & -1 \\ 1 & -2 \end{bmatrix}$

Answers

1. (b)	2. (c)	3. (c)	4. (c)	5. (c)	6. (b)	7. (b)	8. (c)	9. (b)	10. (b)
11. (b)	12. (c)	13. (b)	14. (c)	15. (b)					

Hints and Solutions

1. (b) Since there are 3 rows and 1 column, the order is 3×1.

2. (c) For equality of matrices, the order should be same and corresponding elements should be equal. option (c) follows both.

3. (c) The factors of 9 are 1, 3 and 9 so, the possible orders of a matrix containing 9 elements is 1×9, 9×1, 3×3. In a square matrix the number of rows is equal to the number of columns so the required order is 3×3.

6. (b) • Every zero matrix is not necessarily a square matrix.

   • A unit matrix is a diagonal matrix whose diagonal elements are all 1.

   • The null matrix is the identity matrix for addition.

   If $x = 5$, then the matrix on

   $$\text{LHS} = \begin{bmatrix} 4 & 5 \\ 1 & 3 \end{bmatrix} \neq \begin{bmatrix} 4 & 5 \\ 1 & 0 \end{bmatrix}$$

7. (b) Equating corresponding elements, we get

$2x = x - 3 \Rightarrow x = -3$

and $y - 1 = 2 \Rightarrow y = 3$

8. (c) $A = [a_{ij}]_{m \times n}, B = [b_{ij}]_{m \times n}$

$\therefore C = A + B \Rightarrow [c_{ij}]_{m \times n} = [a_{ij} + b_{ij}]_{m \times n}$

Hence, $c_{23} = a_{23} + b_{23}$

9. (b) $X - A = B \Rightarrow X = A + B$

$\Rightarrow X = \begin{bmatrix} 3 \\ 2 \end{bmatrix} + \begin{bmatrix} -2 \\ -1 \end{bmatrix} = \begin{bmatrix} 3-2 \\ 2-1 \end{bmatrix} = \begin{bmatrix} 1 \\ 1 \end{bmatrix}$

10. (b) $A + B + C = \begin{bmatrix} x & y \\ z & w \end{bmatrix} + \begin{bmatrix} x & -y \\ -z & w \end{bmatrix} + \begin{bmatrix} -2x & 0 \\ 0 & -2w \end{bmatrix}$

$= \begin{bmatrix} x + x + (-2x) & y + (-y) + 0 \\ z + (-z) + 0 & w + w + (-2w) \end{bmatrix}$

$= \begin{bmatrix} 0 & 0 \\ 0 & 0 \end{bmatrix}$

11. (b) $\begin{bmatrix} 2 & -1 \\ 2 & 0 \end{bmatrix} + 2A = \begin{bmatrix} -3 & 5 \\ 4 & 3 \end{bmatrix}$

$\Rightarrow 2A = \begin{bmatrix} -3 & 5 \\ 4 & 3 \end{bmatrix} - \begin{bmatrix} 2 & -1 \\ 2 & 0 \end{bmatrix} = \begin{bmatrix} -3-2 & 5-(-1) \\ 4-2 & 3-0 \end{bmatrix}$

$= \begin{bmatrix} -5 & 6 \\ 2 & 3 \end{bmatrix}$

$\Rightarrow A = \frac{1}{2} \begin{bmatrix} -5 & 6 \\ 2 & 3 \end{bmatrix} = \begin{bmatrix} -\frac{5}{2} & 3 \\ 1 & \frac{3}{2} \end{bmatrix}$

12. (c) $A + X = 2B + C \Rightarrow X = 2B + C - A$

$\Rightarrow X = 2 \begin{bmatrix} -3 & 2 \\ 4 & 0 \end{bmatrix} + \begin{bmatrix} 1 & 0 \\ 0 & 2 \end{bmatrix} - \begin{bmatrix} 2 & -1 \\ 2 & 0 \end{bmatrix}$

$= \begin{bmatrix} -6 & 4 \\ 8 & 0 \end{bmatrix} + \begin{bmatrix} 1 & 0 \\ 0 & 2 \end{bmatrix} - \begin{bmatrix} 2 & -1 \\ 2 & 0 \end{bmatrix}$

$= \begin{bmatrix} -6+1-2 & 4+0-(-1) \\ 8+0-2 & 0+2-0 \end{bmatrix} = \begin{bmatrix} -7 & 5 \\ 6 & 2 \end{bmatrix}$

13. (b) Given, $a_{ij} = i - j$

$\therefore a_{11} = 1-1 = 0, a_{12} = 1-2 = -1$

$a_{21} = 2-1 = 1, a_{22} = 2-2 = 0$

$\therefore$ Required matrix $= \begin{bmatrix} 0 & -1 \\ 1 & 0 \end{bmatrix}$

14. (c) On equating the corresponding elements, we get a pair of simultaneous equations.

$2x + y = 5$

$3x - 2y = 4$

Now solve for x and y.

15. (b) $\begin{bmatrix} 2 & 1 \\ 5 & 0 \end{bmatrix} - 3X = \begin{bmatrix} -7 & 4 \\ 2 & 6 \end{bmatrix}$

$\Rightarrow -3X = \begin{bmatrix} -7 & 4 \\ 2 & 6 \end{bmatrix} - \begin{bmatrix} 2 & 1 \\ 5 & 0 \end{bmatrix}$

$= \begin{bmatrix} -7-2 & 4-1 \\ 2-5 & 6-0 \end{bmatrix} = \begin{bmatrix} -9 & 3 \\ -3 & 6 \end{bmatrix}$

$\Rightarrow X = -\frac{1}{3} \begin{bmatrix} -9 & 3 \\ -3 & 6 \end{bmatrix} = \begin{bmatrix} \frac{-9}{-3} & \frac{3}{-3} \\ \frac{-3}{-3} & \frac{6}{-3} \end{bmatrix} = \begin{bmatrix} 3 & -1 \\ 1 & -2 \end{bmatrix}$

Self Assessment Sheet–14

1. If the matrix is a square matrix and it contains 36 elements, then the order of the matrix is :

(a) 4×4 (b) 8×8

(c) 6×6 (d) 3×3

2. The element in the second row and third column of

the matrix $\begin{bmatrix} 4 & 5 & -6 \\ 3 & -4 & 3 \\ 2 & 1 & 0 \end{bmatrix}$ is :

(a) 3   (b) 1

(c) 2   (d) – 4

3. Given that $M = \begin{bmatrix} 3 & -2 \\ -4 & 0 \end{bmatrix}$ and $N = \begin{bmatrix} -2 & 2 \\ 5 & 0 \end{bmatrix}$, then

$M + N$ is a

(a) null matrix   (b) unit matrix

(c) $\begin{bmatrix} 0 & 1 \\ 1 & 0 \end{bmatrix}$ (d) $\begin{bmatrix} -1 & 0 \\ 0 & -1 \end{bmatrix}$

4. If $2 \begin{bmatrix} 3 & 4 \\ 5 & x \end{bmatrix} + \begin{bmatrix} 1 & y \\ 0 & 1 \end{bmatrix} = \begin{bmatrix} 7 & 0 \\ 10 & 5 \end{bmatrix}$, then the values of

x and y are :

(a) $x = 0, y = -2$ (b) $x = 2, y = -8$

(c) $x = -2, y = -8$ (d) $x = 2, y = 8$

5. If $\begin{bmatrix} x+y & a+b \\ a-b & x-y \end{bmatrix} = \begin{bmatrix} 5 & -1 \\ 3 & -5 \end{bmatrix}$, then the values of x,

y, a, b are respectively.

(a) 0, –5, 1, 2   (b) 0, 5, 1, –2

(c) 0, 5, –1, 2   (d) 0, – 5, 1, –2

Unit Test–2

1. The simplified form of the rational expression $\left(\dfrac{x^2}{x^2-y^2}-1\right)\left(\dfrac{x-y}{y}+2\right)$ is :

(a) $\dfrac{x}{x+y}$

(b) $\dfrac{y}{x+y}$

(c) $\dfrac{y}{x-y}$

(d) $\dfrac{x}{x-y}$

2. If $x+y=a$ and $xy=b$, then the value of $\dfrac{1}{x^3}+\dfrac{1}{y^3}$ is :

(a) a^3-3ab

(b) $\dfrac{a^3+3ab}{b^3}$

(c) $\dfrac{a^3-3ab}{b^3}$

(d) a^3+3ab

3. If $a^2=(b+c)$, $b^2=(c+a)$, $c^2=(a+b)$, then the value of $\dfrac{1}{a+1}+\dfrac{1}{b+1}+\dfrac{1}{c+1}$ is equal to :

(a) 1

(b) –1

(c) 0

(d) $\left(\dfrac{1}{a}+\dfrac{1}{b}+\dfrac{1}{c}\right)$

4. If $\dfrac{a}{x}+\dfrac{y}{b}=1$ and $\dfrac{b}{y}+\dfrac{z}{c}=1$, then $\dfrac{x}{a}+\dfrac{c}{z}$ will be equal to :

(a) $\dfrac{y}{b}$

(b) $\dfrac{b}{y}$

(c) 0

(d) 1

5. The factors of x^4+x^2+25 are :

(a) $(x^2+5)(x^2-x+5)$

(b) $(x^2+5+x)(x^2+5-x)$

(c) $(x^2+5+3x)(x^2+5-3x)$

(d) $(x^2+5)^2$

6. The factors of $(a^2+36b^2)^2-169a^2b^2$ are :

(a) $(a+13b)(a-13b)$

(b) $(a+4b)(a+9b)(a-4b)(a-9b)$

(c) $(a+6b)(a-13b)(a+13b)(a+6b)$

(d) $(a-13b)(a+6b)$

7. One of the factors of $a^6+b^6-a^2b^4-a^4b^2$ is :

(a) $a+1$

(b) a^2+b^2

(c) a^3+b^3

(d) a^4+b^4

8. In a number of three digits the units digit is double the tens' digit. The sum of the number and the number formed by reversing the digits is 1191 and the average of three digits is 5. What is the number ?

(a) 663

(b) 348

(c) 924

(d) 843

9. A batsmen's average before the last two innings of a season was 66. He failed to score in those innings and his average dropped to 55. How many innings did he play that season.

(a) 10

(b) 11

(c) 12

(d) 8

10. The solution of the equations :

$3(x+3y-1)=2(y-3x+1)$ and

$3(5y-7x-3)=2(5x+7y+12)$ are :

(a) $x=1, y=2$

(b) $x=-1, y=2$

(c) $x=1, y=-2$

(d) $x=-1, y=-2$

11. Some bananas are to be shared among a number of children. To give each child 9 bananas would required 15 more bananas. But if the share of each is 8 there are 10 bananas left over. How many bananas are there ?

(a) 200

(b) 210

(c) 120

(d) 140

12. The value of K for which the system of equations $x+2y+7=0$ and $2x+ky+14=0$ have infinitely many solutions is :

(a) 2

(b) 4

(c) 6

(d) 8

13. Roots of the equation $x^2+x(2-p^2)-2p^2=0$

(a) $-p^2$ and -2

(b) p^2 and -2

(c) $-p^2$ and 2

(d) p^2 and 2

14. The roots of the equation $\dfrac{4}{x^2} = 1 + \dfrac{3}{x}$ are :

    (a) $-1, -4$ (b) $1, 4$

    (c) $1, -4$ (d) $-1, 4$

15. Find the value of x if the shaded area is a half of the whole area in the given diagram.

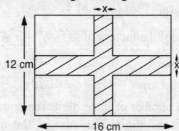

    (a) 4 cm   (b) 6 cm

    (c) 3 cm   (d) 2 cm

16. Find two numbers which are such that the sum of the first and twice the second and also the sum of their squares are equal to 5.

    (a) 1, 3   (b) 1, 2

    (c) 2, 3   (d) 3, 4

17. If x and y are positive integers that satisfy $1 \le x \le 4$ and $3x + 2y \le 13$, then the possible values of y are:

    (a) 1, 2, 3, 4, 5   (b) 1, 2, 3, 4, 5, 6

    (c) $\dfrac{1}{2}, 3, 5$ (d) 2, 3, 4, 5

18. If $6 \le x \le 8$, then :

    (a) $(x-6)(x-8) \ge 0$ (b) $(x-6)(x-8) > 0$

    (c) $(x-6)(x-8) \le 0$ (d) $(x-6)(x-8) < 0$

19. If $x^3 - \dfrac{1}{x^3} = 14$, than the value of $x - \dfrac{1}{x}$ will be

    (a) 2   (b) 3

    (c) 4   (d) 5

20. The LCM of the polynomials $(x^8 - y^8)$ and $(x^4 - y^4)(x + y)$ is :

    (a) $(x^8 - y^8)$

    (b) $(x^8 - y^8)(x^4 - y^4)(x + y)$

    (c) $(x^8 - y^8)(x + y)$

    (d) $(x^4 - y^4)(x^8 - y^8)$

Answers

1. (c)	2. (c)	3. (a)	4. (d)	5. (c)	6. (b)	7. (b)	8. (b)	9. (c)	10. (b)
11. (b)	12. (b)	13. (b)	14. (c)	15. (a)	16. (b)	17. (c)	18. (c)	19. (a)	20. (c)

UNIT-3

COMMERCIAL ARITHMETIC

- *Ratio and Proportion*
- *Percentage and Its Applications*
- *Average*
- *Simple Interest*
- *Compound Interest*
- *Time and Work*
- *Distance, Time and Speed*

Chapter 15

RATIO AND PROPORTION

1. **Ratio :** It is a tool to compare two or more quantities (or numbers) having same units by division. First term is called **antecedent** and second term is called **consequent.**

2. **To divide a given number in a given ratio :** If we have to divide A in the ratio $a : b : c$, then

   First part $= \dfrac{a}{a+b+c} \times A$, Second part $= \dfrac{b}{a+b+c} \times A$, Third part $= \dfrac{c}{a+b+c} \times A$

3. Four quantities are said to be in proportion if $a : b = c : d$ or $ad = bc$.

4. If $a : b = b : c$, then b is called the mean proportional and $b = \sqrt{ac}$.

Solved Examples

Ex. 1. *If $p : q = r : s = t : u = 2 : 3$, then what is $(mp + nr + ot) : (mq + ns + ou)$ equal to ?*

Sol. Given, $p : q = r : s = t : u = 2 : 3$

Let $p = r = t = 2x$ and $q = s = u = 3x$

Then, $\dfrac{mp + nr + ot}{mq + ns + ou} = \dfrac{m \times 2x + n \times 2x + o \times 2x}{m \times 3x + n \times 3x + o \times 3x} = \dfrac{2x(m+n+o)}{3x(m+n+o)} = \dfrac{2}{3} = 2 : 3.$

Ex. 2. *If $x : y = 3 : 1$, then find the ratio $X^3 - Y^3 : X^3 + Y^3$.*

Sol. Let $x = 3k$ and $y = k$

Then, $\dfrac{x^3 - y^3}{x^3 + y^3} = \dfrac{(3k)^3 - k^3}{(3k)^3 + k^3} = \dfrac{27k^3 - k^3}{27k^3 + k^3} = \dfrac{26k^3}{28k^3} = \dfrac{13}{14} = 13 : 14.$

Ex. 3. *If $(x + y) : (x - y) = 4 : 1$, then find the ratio $(X^2 + Y^2) : (X^2 - Y^2)$.*

Sol. Given, $\dfrac{x+y}{x-y} = \dfrac{4}{1} \Rightarrow x + y = 4x - 4y \Rightarrow 3x = 5y \Rightarrow \dfrac{x}{y} = \dfrac{5}{3}$

$\therefore \dfrac{x^2 + y^2}{x^2 - y^2} = \dfrac{\dfrac{x^2}{y^2} + 1}{\dfrac{x^2}{y^2} - 1} = \dfrac{\dfrac{25}{9} + 1}{\dfrac{25}{9} - 1} = \dfrac{\dfrac{34}{9}}{\dfrac{16}{9}} = \dfrac{34}{16} = \dfrac{17}{8}$

$\Rightarrow (x^2 + y^2) : (x^2 - y^2) = 17 : 8.$

Ex. 4. *Find the third proportional to $(x^2 - y^2)$ and $(x + y)$.*

Sol. Let the third proportional to $(x^2 - y^2)$ and $(x + y)$ be A. Then

$(x^2 - y^2) : (x + y) :: (x + y) : A$

$\Rightarrow (x^2 - y^2) \times A = (x + y)^2 \Rightarrow A = \dfrac{(x+y)^2}{x^2 - y^2} \Rightarrow A = \dfrac{(x+y)^2}{(x+y)(x-y)} = \dfrac{x+y}{x-y}.$

Ex. 5. *Three numbers A, B and C are in the ratio 1 : 2 : 3. Their average is 600. If A is increased by 10% and B is decreased by 20%, then by how much will C be increased to get the average increased by 5% ?*

Sol. Let $A = x$, $B = 2x$ and $C = 3x$

Given, $\dfrac{x + 2x + 3x}{3} = 600 \Rightarrow 6x = 1800 \Rightarrow x = 300$

$\therefore$ The numbers are $A = 300$, $B = 600$, $C = 900$

New average = 105% of 600 = $\dfrac{105}{100} \times 600 = 630$

New value of $A = \dfrac{110}{100} \times 300 = 330$

New value of $B = \dfrac{80}{100} \times 600 = 480$

Let new value of $C = y$

Then, $\dfrac{330 + 480 + y}{3} = 630 \Rightarrow 810 + y = 1890$

$\Rightarrow y = 1080 \quad \therefore$ Increase in $C = 1080 - 900 = 180$.

Ex. 6. *A began a business with Rs 45000 and B joined after wards with Rs 30000. At the end of a year, the profit is divided in the ratio 2 : 1. When did B join ?*

Sol. Let B join after x months. Then profit is divided in the ratio $45000 \times 12 : 30000 \times x$

Given, $\dfrac{45000 \times 12}{30000 \times x} = \dfrac{2}{1}$

$\Rightarrow 45000 \times 12 = 60000\,x \Rightarrow x = \dfrac{45000 \times 12}{60000} = 9$ months

$\therefore$ B joined after 9 months.

Ex. 7. *An employer reduces the number of his employees in the ratio 7 : 5 and increases their wages in the ratio 15 : 28. State whether his bill of total wages increase or decrease and in what ratio.*

Sol. Let the number of employees be $7x$ and $5x$ and

let their wages be $15y$ and $28y$ respectively.

Then, Original wage bill = $7x \times 15y$ = Rs $105xy$

New wage bill = $5x \times 28y$ = Rs $140xy$

$\therefore$ The wage bill increases in the ratio $140 : 105 = 4 : 3$.

Ex. 8. *In three vessels, the ratio of water and milk is 6 : 7, 5 : 9 and 8 : 7 respectively. If the mixtures of the three vessels are mixed together, then what will be the ratio of water and milk ?*

Sol. Vessel	Water	Milk
1	$\dfrac{6}{13}$	$\dfrac{7}{13}$
2	$\dfrac{5}{14}$	$\dfrac{9}{14}$
3	$\dfrac{8}{15}$	$\dfrac{7}{15}$

$$\therefore \text{ Proportion of water} = \frac{6}{13} + \frac{5}{14} + \frac{8}{15} = \frac{1260 + 975 + 1456}{2730} = \frac{3691}{2730}$$

$$\text{Proportion of milk} = \frac{7}{13} + \frac{9}{14} + \frac{7}{15} = \frac{1470 + 1755 + 1274}{2730} = \frac{4499}{2730}$$

$$\therefore \text{ Required ratio} = \frac{3691}{2730} : \frac{4499}{2730} = 3691 : 4499.$$

Ex. 9. *In an express train, the number of passengers travelling in AC sleeper class, first class and sleeper class are in the ratio 1 : 2 : 3 and the fares of each classes are in the ratio 5 : 4 : 2. If the total income from this train is Rs 54000, then what is the income from AC sleeper class ?*

Sol. Let the number of passengers in AC sleeper class, first class and sleeper class be x, $2x$ and $3x$ respectively.

Let the fares of AC sleeper class, first class and sleeper class be $5y$, $4y$ and $2y$ respectively.

Then, the total incomes from the three classes $= 5xy + 8xy + 6xy = 19xy$

Given, $19xy = $ Rs 54000 $\Rightarrow xy = $ Rs $\dfrac{54000}{19}$

$\therefore$ Income from AC sleeper $=$ Rs $\dfrac{54000}{19} \times 5 = $ Rs 14210 (approx).

Ex. 10. *A drum contains 20 litres of a paint. From this, 2 litres of paint is taken out and replaced by 2 litres of oil. Again 2 litres of this mixture is taken out and replaced by 2 litres of oil. If this operation is performed once again, then what would be the final ratio of paint and oil in the drum ?*

Sol. After the first operation,
Paint $= 18$ litres, Oil $= 2$ litres
After the second operation,
Paint $= 18 - \dfrac{18}{20} \times 2 = 16.2$ litres,

Oil $= 2 - \dfrac{2}{20} \times 2 + 2 = 3.8$ litres
After the third operation,
Paint $= 16.2 - \dfrac{16.2}{20} \times 2 = 14.58$ litres

Oil $= 3.8 - \dfrac{3.8}{20} \times 2 + 2 = 5.42$ litres

$\therefore$ Required ratio $= \dfrac{14.58}{5.42} = \dfrac{729}{271}$

Question Bank–15

1. If $a : b : c = 7 : 3 : 5$, then $(a + b + c) : (2a + b - c)$ is equal to
   (a) $1 : 2$ (b) $2 : 3$
   (c) $3 : 4$ (d) $5 : 4$

2. If $a : b = c : d = e : f = 1 : 2$, then $(3a + 5c + 7e) : (3b + 5d + 7f)$ is equal to
   (a) $8 : 7$ (b) $2 : 1$
   (c) $1 : 4$ (d) $1 : 2$

3. If $a : b = b : c$, then $a^4 : b^4$ is equal to
   (a) $ac : b^2$ (b) $a^2 : c^2$
   (c) $c^2 : a^2$ (d) $b^2 : ac$

4. If $x : y = 3 : 2$, then the ratio $2x^2 + 3y^2 : 3x^2 - 2y^2$ is equal to
   (a) $12 : 5$ (b) $6 : 5$
   (c) $30 : 19$ (d) $5 : 3$

5. If $a : b = c : d = e : f$, then $(pa + qc + re) : (pb + qd + rf)$ is equal to
   (a) $p : (q + r)$ (b) $(p + q) : r$
   (c) $2 : 3$ (d) $1 : 2$

6. If $a : (b + c) = 1 : 3$ and $c : (a + b) = 5 : 7$, then $b : (a + c)$ is equal to
   (a) $1 : 2$ (b) $2 : 3$
   (c) $1 : 3$ (d) $2 : 1$

7. If $p:q:r=1:2:4$, then $\sqrt{5p^2+q^2+r^2}$ is equal to

    (a) 5

    (b) $2q$

    (c) $5p$

    (d) $4r$

8. The mean proportional between $(3+\sqrt{2})$ and $(12-\sqrt{32})$ is

    (a) $\sqrt{7}$

    (b) $2\sqrt{7}$

    (c) 6

    (d) $\dfrac{15-3\sqrt{2}}{2}$

9. If b is the mean proportional of a and c, then $(a-b)^3:(b-c)^3$ equals

    (a) $a^{3/2}:c^{3/2}$

    (b) $b^2:c^2$

    (c) $a^2:c^2$

    (d) $a^3:b^3$

10. If $x:a=y:b=z:c$ then

    $$\dfrac{ax-by}{(a+b)(x-y)}+\dfrac{by-cz}{(b+c)(y-z)}+\dfrac{cz-ax}{(c+a)(z-x)}$$

    is equal to

    (a) 1

    (b) 2

    (c) 3

    (d) 0

11. A sum of money is divided among A, B, C and D in the ratio $3:5:7:11$ respectively. If the share of C is Rs 1668 more than the share of A, then what is the total amount of money of B and D together ?

    (a) Rs 6762

    (b) Rs 6672

    (c) Rs 7506

    (d) Rs 6255

12. If $a+b:b+c:c+a=6:7:8$ and $a+b+c=14$, then the value of c is

    (a) 6

    (b) 7

    (c) 8

    (d) 14

13. A profit of Rs 6000 is to be distributed among A, B and C in the ratio $3:4:5$ respectively. How much more will C get than B ?

    (a) Rs 500

    (b) Rs 1200

    (c) Rs 2000

    (d) Rs 2500

14. Rs 2010 are to be divided among A, B and C in such a way that if A gets Rs 5, than B must get Rs 12 and if B gets Rs 4, then C must get Rs 5.50. The share of C will exceed that of B by

    (a) Rs 620

    (b) Rs 430

    (c) Rs 360

    (d) Rs 270

15. Rs 180 is to be divided among 66 persons (men and women). The ratio of the total amuont of money received by men and women is $5:4$. But the ratio of the money received by each man and woman is $3:2$. The number of men is

    (a) 20

    (b) 24

    (c) 30

    (d) 36

16. Rs 53 are divided among A, B and C such that A gets Rs 7 more than B and B gets Rs 8 more than C. What is the ratio of their shares ?

    (a) $16:9:18$

    (b) $25:18:10$

    (c) $18:25:10$

    (d) $15:8:30$

17. 15 men, 18 women and 12 boys working together earned Rs 2070. If the daily wages of a man, a woman and a boy are in the ratio $4:3:2$, the daily wages (in Rs) of 1 man, 2 women and 3 boys are

    (a) 135

    (b) 180

    (c) 240

    (d) 205

18. Krishnamurthy started business investing Rs 6000. Three months later Madhavan joined him investing Rs 4000. If they make a profit of Rs 5100 at the end of the year, how much should be Madhavan's share ?

    (a) Rs 1700

    (b) Rs 1400

    (c) Rs 1300

    (d) Rs 1732.75

19. The incomes of A, B and C are in the ratio $7:9:12$ and their spendings are in the ratio $8:9:15$. If A saves $\dfrac{1}{4}$th of his income, then the savings of A, B and C are in the ratio

    (a) $56:99:69$

    (b) $69:56:99$

    (c) $99:56:69$

    (d) $99:69:56$

20. The employer reduces the number of employees in the ratio $9:8$ and increases their wages in the ratio $14:15$. If the previous wage bill was Rs 189000, what is the amount by which the new wage bill will increase or decrease ?

    (a) Rs 21000

    (b) Rs 10000

    (c) Rs 9000

    (d) Rs 25000

21. Ratio of the incomes of A, B and C last year was $3:4:5$. The ratio of their individual incomes of the last year and this year are $4:5$, $2:3$ and $3:4$ respectively. If the sum of their present incomes is Rs 78800, then find the present individual income of A, B and C.

    (a) Rs 10000, Rs 40000, Rs 28800

    (b) Rs 18000, Rs 28800, Rs 32000

    (c) Rs 45000, Rs 12800, Rs 21000

    (d) Rs 20000, Rs 14600, Rs 44200

22. The ratio of the age of a man and his wife is $4:3$. After 4 years, this ratio will be $9:7$. If at the time of marriage the ratio was $5:3$, then how many years ago were they married ?

(a) 12 years   (b) 8 years

(c) 10 years   (d) 15 years

23. A barrel contains a mixture of wine and water in the ratio 3 : 1. How much fraction of the mixture must be drawn off and substituted by water so that the ratio of wine and water in the resultant mixture in the barrel becomes 1 : 1 ?

(a) $\dfrac{1}{4}$ (b) $\dfrac{1}{3}$

(c) $\dfrac{3}{4}$ (d) $\dfrac{2}{3}$

24. There are 3 containers of equal capacity. The ratio of sulphuric acid to water in the first container is 3 : 2, that in the second container is 7 : 3 and in the third container it is 11 : 4. If all the liquids are mixed together then the ratio of sulphuric acid to water in the mixture will be

(a) 61 : 29   (b) 61 : 28

(c) 60 : 29   (d) 59 : 29

25. The students in three classes are in the ratio 2 : 3 : 5. If 40 students are increased in each class, the ratio changes to 4 : 5 : 7. Originally the total number of students was

(a) 100   (b) 180

(c) 200   (d) 400

26. Vessels A and B contain mixtures of milk and water in the ratio 4 : 5 and 5 : 1 respectively. In what ratio should quantities of mixtures be taken from A to B to form a mixture in which milk to water is in the ratio 5 : 4 ?

(a) 2 : 5   (b) 2 : 3

(c) 4 : 3   (d) 5 : 2

27. A bottle is full of dettol. One-third of it is taken out and then an equal amount of water is poured into the bottle to fill it. This operation is done four times. Find the final ratio of dettol and water in the bottle

(a) 13 : 55   (b) 20 : 74

(c) 16 : 65   (d) 10 : 48

28. In two alloys, copper and zinc are related in the ratio of 4 : 1 and 1 : 3. 10 kg of 1st alloy, 16 kg of 2nd alloy and some of pure copper are melted together. An alloy was obtained in which the ratio of copper to zinc was 3 : 2. Find the weight of the new alloy ?

(a) 35 kg   (b) 45 kg

(c) 40 kg   (d) 50 kg

29. Railway fares of 1st, 2nd and 3rd classes between two stations were in the ratio 8 : 6 : 3. The fares of 1st and 2nd class were subsequently reduced by $\dfrac{1}{6}$ and $\dfrac{1}{12}$ respectively. If during a year, the ratio between the passengers of 1st, 2nd and 3rd classes was 9 : 12 : 26 and the total amount collected by the sale of tickets was Rs 1088, the collection from the passengers of 1st class was

(a) Rs 260   (b) Rs 280

(c) Rs 300   (d) Rs 320

30. Three containers A, B and C are having mixtures of milk and water in the ratio 1 : 5, 3 : 5 and 5 : 7 respectively. If the capacities of the containers are in the ratio 5 : 4 : 5, then find the ratio of the milk to the water if the mixtures of all the three containers are mixed together.

(a) 51 : 115   (b) 52 : 115

(c) 53 : 115   (d) 54 : 115

Answers

1. (d)	2. (d)	3. (b)	4. (c)	5. (d)	6. (a)	7. (c)	8. (b)	9. (a)	10. (c)
11. (b)	12. (a)	13. (a)	14. (d)	15. (c)	16. (b)	17. (c)	18. (a)	19. (a)	20. (c)
21. (b)	22. (a)	23. (b)	24. (a)	25. (c)	26. (d)	27. (c)	28. (a)	29. (d)	30. (c)

Hints and Solutions

1. (d) $a : b : c = 7 : 3 : 5$

$\Rightarrow a = 7k, b = 3k, c = 5k$

$\therefore (a + b + c) : (2a + b - c)$

$= (7k + 3k + 5k) : (14k + 3k - 5k)$

$= 15k : 12k = 5 : 4$

2. (d) Given , $\dfrac{a}{b} = \dfrac{c}{d} = \dfrac{e}{f} = \dfrac{1}{2}$

$\Rightarrow a = \dfrac{1}{2}b, c = \dfrac{1}{2}d, e = \dfrac{1}{2}f$

$\therefore \dfrac{3a + 5c + 7e}{3b + 5d + 7f} = \dfrac{3 \times \frac{1}{2}b + 5 \times \frac{1}{2}d + 7 \times \frac{1}{2}f}{3b + 5d + 7f}$

$= \dfrac{1}{2}\left[\dfrac{3b + 5d + 7f}{3b + 5d + 7f}\right] = \dfrac{1}{2}$

$\Rightarrow (3a + 5c + 7e) : (3b + 5d + 7f) = 1 : 2$

3. (b) $\dfrac{a}{b} = \dfrac{b}{c} \Rightarrow b^2 = ac$

$$\therefore \frac{a^4}{b^4} = \frac{a^4}{(b^2)^2} = \frac{a^4}{(ac)^2} = \frac{a^4}{a^2 c^2} = \frac{a^2}{c^2}$$

4. (c) $\dfrac{x}{y} = \dfrac{3}{2} \Rightarrow \dfrac{x^2}{y^2} = \dfrac{9}{4}$

$$\therefore \frac{2x^2 + 3y^2}{3x^2 - 2y^2} = \frac{2\dfrac{x^2}{y^2} + 3}{3\dfrac{x^2}{y^2} - 2} = \frac{2 \times \dfrac{9}{4} + 3}{3 \times \dfrac{9}{4} - 2} = \frac{\dfrac{18 + 12}{4}}{\dfrac{27 - 8}{4}}$$

$$= \frac{30}{19} = 30 : 19$$

5. (d) Given, $\dfrac{a}{b} = \dfrac{c}{d} = \dfrac{e}{f} = \dfrac{1}{2}$

So, let $a = c = e = x$ and $b = d = f = 2x$

$$\therefore \frac{(pa + qc + re)}{(pb + qd + rf)} = \frac{p \times x + q \times x + r \times x}{p \times 2x + q \times 2x + r \times 2x}$$

$$= \frac{x(p + q + r)}{2x(p + q + r)} = \frac{1}{2}$$

6. (a) $\dfrac{a}{b + c} = \dfrac{1}{3} \Rightarrow a = \dfrac{b + c}{3}$ ... (i)

$$\frac{c}{a + b} = \frac{5}{7} \Rightarrow \frac{c}{\dfrac{b + c}{3} + b} = \frac{5}{7}$$

$$\Rightarrow \frac{3c}{b + c + 3b} = \frac{5}{7} \Rightarrow \frac{3c}{4b + c} = \frac{5}{7}$$

$$\Rightarrow 21c = 20b + 5c \Rightarrow 16c = 20b \Rightarrow b = \frac{4c}{5}$$

Putting $b = \dfrac{4c}{5}$ in (i) we get

$$\frac{a}{\dfrac{4c}{5} + c} = \frac{1}{3} \Rightarrow \frac{5a}{9c} = \frac{1}{3} \Rightarrow a = \frac{3c}{5}$$

$$\frac{b}{a + c} = \frac{\dfrac{4c}{5}}{\dfrac{3c}{5} + c} = \frac{4c}{8c} = \frac{1}{2} = 1 : 2$$

7. (c) Let $p = x$, $q = 2x$, $r = 4x$

Then, $\sqrt{5p^2 + q^2 + r^2} = \sqrt{5x^2 + 4x^2 + 16x^2}$

$$= \sqrt{25x^2} = 5x$$

$\therefore p = x \quad \therefore 5p = 5x$

Hence, $\sqrt{5p^2 + q^2 + r^2} = 5p$

8. (b) Required mean proportional

$$= \sqrt{(3 + \sqrt{2})(12 - \sqrt{32})}$$

$$= \sqrt{36 + 12\sqrt{2} - 3\sqrt{32} - \sqrt{64}}$$

$$= \sqrt{36 + 12\sqrt{2} - 3\sqrt{16 \times 2} - 8}$$

$$= \sqrt{28 + 12\sqrt{2} - 12\sqrt{2}} = \sqrt{28} = \sqrt{4 \times 7}$$

$$= 2\sqrt{7}$$

9. (a) Since b is the mean proportional of a and c,

$b^2 = ac$, i.e., $b = \sqrt{ac}$.

$$\therefore \frac{(a - b)^3}{(b - c)^3} = \frac{(a - \sqrt{ac})^3}{(\sqrt{ac} - c)^3} = \frac{[\sqrt{a}(\sqrt{a} - \sqrt{c})]^3}{[\sqrt{c}(\sqrt{a} - \sqrt{c})]^3}$$

$$= \frac{(\sqrt{a})^3 (\sqrt{a} - \sqrt{c})^3}{(\sqrt{c})^3 (\sqrt{a} - \sqrt{c})^3}$$

$$= \frac{(a^{1/2})^3}{(c^{1/2})^3} = \frac{a^{3/2}}{c^{3/2}}$$

10. (c) $\dfrac{x}{a} = \dfrac{y}{b} = \dfrac{z}{c} = k$ (say)

$\Rightarrow x = ak, y = bk, z = ck$

$$\therefore \frac{ax - by}{(a + b)(x - y)} + \frac{by - cz}{(b + c)(y - z)} + \frac{cz - ax}{(c + a)(z - x)}$$

$$= \frac{a \times ak - b \times bk}{(a + b)(ak - bk)} + \frac{b \times bk - c \times ck}{(b + c)(bk - ck)}$$

$$+ \frac{c \times ck - a \times ak}{(c + a)(ck - ak)}$$

$$= \frac{k(a^2 - b^2)}{k(a^2 - b^2)} + \frac{k(b^2 - c^2)}{k(b^2 - c^2)} + \frac{k(c^2 - a^2)}{k(c^2 - a^2)}$$

$$= 1 + 1 + 1 = 3$$

11. (b) Let the shares of A, B, C, D be Rs $3x$, Rs $5x$, Rs $7x$ and Rs $11x$ respectively.

Then, $7x - 3x = 1668 \Rightarrow 4x = 1668 \Rightarrow x = 417$

$\therefore$ B's share + D's share = $5x + 11x =$ Rs $16x$

$$= 16 \times 417 = \text{Rs } 6672$$

12. (a) Let $a + b = 6k$, $b + c = 7k$ and $c + a = 8k$.

Then $2(a + b + c) = 6k + 7k + 8k = 21k$

$$\Rightarrow 2 \times 14 = 21k \Rightarrow k = \frac{2 \times 14}{21} = \frac{4}{3}$$

Also $a + b + c = \dfrac{21k}{2}$ and $a + b = 6k$

$$\Rightarrow c = \frac{21k}{2} - 6k = \frac{9}{2}k = \frac{9}{2} \times \frac{4}{3} = 6$$

13. (a) C's share $= \frac{5}{12} \times$ Rs 6000 = Rs 2500

B's share $= \frac{4}{12} \times$ Rs 6000 = Rs 2000

∴ C gets Rs 500 more than B.

14. (d) $A : B = 5 : 12$

$B : C = 4 : 5.5 = 12 : 16.5$

$\Rightarrow A : B : C = 5 : 12 : 16.5$

∴ The share of C will exceed that of B by

$$\frac{(16.5 - 12)}{5 + 12 + 16.5} \times \text{Rs } 2010 = \frac{4.5 \times 2010}{33.5} = \text{Rs } 270$$

15. (c) Let the number of men be x.

Then, number of women $= 66 - x$

Money received by x men $= \frac{5}{(5+4)} \times \text{Rs } 180$

$= \text{Rs } 100$

∴ Money received by $(66 - x)$ women

$= \text{Rs } (180 - 100) = \text{Rs } 80$

Given, $\frac{100}{x} : \frac{80}{66-x} = 3 : 2$

$$\Rightarrow \frac{100}{x} \times \frac{66-x}{80} = \frac{3}{2}$$

$$\Rightarrow \frac{200}{x} = \frac{240}{66-x} \Rightarrow 240x = 13200 - 200x$$

$$\Rightarrow 440x = 13200 \Rightarrow x = \frac{13200}{440} = 30$$

16. (b) Given, $A + B + C = 53$ ... (i)

Also, $A = B + 7$ and $B = C + 8$

∴ From (i), we get $(B + 7) + B + (B - 8) = 53$

$\Rightarrow 3B = 54 \Rightarrow B = 18$

$\Rightarrow A = 25$ and $C = 10$

∴ $A : B : C = 25 : 18 : 10$

17. (c) Suppose the daily wages of a man, a woman and a boy are $4k$, $3k$ and $2k$ respectively.

∴ $15 \times 4k + 18 \times 3k + 12 \times 2k = 2070$

$\Rightarrow 60k + 54k + 24k = 2070$

$\Rightarrow 138k = 2070 \Rightarrow k = 15$.

∴ Daily wages of a man, a woman and a boy are Rs 60, Rs 45 and Rs 30 respectively.

∴ Daily wages of 1 man + 2 women + 3 boys

$= \text{Rs } 60 + 2 \times \text{Rs } 45 + 3 \times \text{Rs } 30$

$= \text{Rs } 60 + \text{Rs } 90 + \text{Rs } 90 = \text{Rs } 240$

18. (a) The profit of Rs 5100 is to be distributed among Krishnamurthy and Madhavan in the ratio $12 \times 6000 : 9 \times 4000$, i.e., $72000 : 36000 = 2 : 1$

∴ Madhavan's share $= \frac{5100 \times 1}{3} = \text{Rs } 1700$.

19. (a) Let the incomes of A, B and C be $7x$, $9x$ and $12x$ respectively.

Let the expenditures of A, B and C be $8y$, $9y$ and $15y$ respectively.

∴ Savings of A, B and C are $7x - 8y$, $9x - 9y$ and $12x - 15y$ respectively.

Given, $7x - 8y = \frac{1}{4} \times 7x \Rightarrow 7x - \frac{7x}{4} = 8y$

$$\Rightarrow \frac{21x}{4} = 8y \Rightarrow y = \frac{21}{32}x$$

∴ Savings of A, B and C are

$$\frac{7x}{4} : 9x - \frac{9 \times 21x}{32} : 12x - 15 \times \frac{21x}{32}$$

i.e., $\frac{7x}{4} : \frac{(288x - 189x)}{32} : \frac{(384x - 315x)}{32}$

$$\Rightarrow \frac{56}{32} : \frac{99x}{32} : \frac{69x}{32} \Rightarrow 56 : 99 : 69$$

20. (c) Original wage bill of $9x$ employees = Rs 189000

Given, decrease in number of employees in the ratio $9 : 8$.

∴ Wage bill of $8x$ employees $= \text{Rs } \frac{189000 \times 8x}{9x}$

$= \text{Rs } 168000$

Given, increase in the wages in the ratio $14 : 15$

∴ Present wage bill $= \text{Rs } \frac{168000 \times 15}{14}$

$= \text{Rs } 180000$.

∴ The wage bill will be decreased by Rs 9000.

21. (b) Let the incomes of A, B and C last year be Rs $3x$, Rs $4x$ and Rs $5x$ respectively.

A's income last year : this year = 4 : 5

∴ A's income this year $= \frac{5}{4} \times \text{Rs } 3x = \frac{\text{Rs } 15x}{4}$

B's income last year : this year = 2 : 3

∴ B's income this year $= \frac{3}{2} \times \text{Rs } 4x = \text{Rs } 6x$

C's income last year : this year = 3 : 4

∴ C's income this year $= \frac{4}{3} \times \text{Rs } 5x = \text{Rs } \frac{20x}{3}$

∴ Ratio of present incomes of A, B, C

$$= \frac{15x}{4} : 6x : \frac{20x}{3}$$

$$= \frac{15x \times 12}{4} : 6x \times 12 : \frac{20x \times 12}{3}$$

$$= 45 : 72 : 80$$

Given, $45x + 72x + 80x = 78800$

$\Rightarrow 197x = 78800 \Rightarrow x = \dfrac{78800}{197} = 400$

∴ A's present income $= 45 \times 400 = $ Rs 18000

B's present income $= 72 \times 400 = $ Rs 28800

C's present income $= 80 \times 400 = $ Rs 32000

22. (a) Let the man's age $= 4x$ years. Then,

Wife's age $= 3x$ years

Given, $\dfrac{4x+4}{3x+4} = \dfrac{9}{7}$

$\Rightarrow 28x + 28 = 27x + 36 \Rightarrow x = 8$

∴ Man's age $= 32$ years, wife's age $= 24$ years

Suppose they were married x years ago.

Then, $\dfrac{32-x}{24-x} = \dfrac{5}{3}$

$\Rightarrow 96 - 3x = 120 - 5x \Rightarrow 2x = 24 \Rightarrow x = 12$

23. (b) Let the barrel contains wine $3x$ litres and water x litres.

Then, total mixture $= 4x$ litres

Let the part of mixture drawn out be p litres.

∴ $(4x-p) \times \dfrac{3}{4} : (4x-p) \times \dfrac{1}{4} + p = 1 : 1$

$\Rightarrow 3x - \dfrac{3p}{4} : x - \dfrac{p}{4} + p = 1 : 1$

$\Rightarrow 3x - \dfrac{3p}{4} = x + \dfrac{3p}{4} \Rightarrow 2x = \dfrac{6p}{4}$

$\Rightarrow p = \dfrac{2x \times 4}{6} = \dfrac{1}{3}(4x)$

∴ $\dfrac{1}{3}$ part of mixture is drawn out.

24. (a) Let the quantity of mixture in each container be x. Then,

Sulphuric acid in 1st container $= \dfrac{3x}{5}$

Water in 1st container $= \dfrac{2x}{5}$

Sulphuric acid in 2nd container $= \dfrac{7x}{10}$

Water is 2nd container $= \dfrac{3x}{10}$

Sulphuric acid in 3rd container $= \dfrac{11x}{15}$

Water in 3rd container $= \dfrac{4x}{15}$

∴ Required ratio

$$= \left(\frac{3x}{5} + \frac{7x}{10} + \frac{11x}{15}\right) : \left(\frac{2x}{5} + \frac{3x}{10} + \frac{4x}{15}\right)$$

$$= \left(\frac{36+42+44}{60}\right) : \left(\frac{24+18+16}{60}\right)$$

$$= 122 : 58 = 61 : 29.$$

25. (c) In the beginning, the three classes had $2x, 3x$ and $5x$ students where x is a constant of proportionality. 40 students were added in each section.

$\Rightarrow$ There are $2k + 40, 3k + 40$ and $5k + 40$ number of students in each section.

Given, $2k + 40 : 3k + 40 : 5k + 40 = 4 : 5 : 7$

$\Rightarrow \dfrac{2k+40}{3k+40} = \dfrac{4}{5} \Rightarrow 10k + 200 = 12k + 160$

$\Rightarrow 2k = 40 \Rightarrow k = 20$

∴ Originally the total number of students was $2 \times 20 + 3 \times 20 + 5 \times 20 = 200$.

26. (d) Let the required ratio be $x : y$.

Milk in x litres of 1st mixture $= \left(x \times \dfrac{4}{9}\right)$ litres

$$= \dfrac{4x}{9} \text{ litres}$$

Water in x litres of 1st mixture $= \left(x - \dfrac{4x}{9}\right)$ litres

$$= \dfrac{5x}{9} \text{ litres}$$

Milk in y litres of 2nd mixture $= \left(y \times \dfrac{5}{6}\right)$ litres

$$= \dfrac{5y}{6} \text{ litres}$$

Water in y litres of 2nd mixture $= \left(y - \dfrac{5y}{6}\right)$ litres

$$= \dfrac{y}{6} \text{ litres}$$

∴ Milk : Water $= \left(\dfrac{4x}{9} + \dfrac{5y}{6}\right) : \left(\dfrac{5x}{9} + \dfrac{y}{6}\right)$

$$= \left(\frac{8x+15y}{18}\right):\left(\frac{10x+3y}{18}\right)$$
$$= (8x+15y):(10x+3y)$$

Given, $\dfrac{8x+15y}{10x+3y}=\dfrac{5}{4} \Rightarrow 32x+60y=50x+15y$

$$\Rightarrow 18x=45y \Rightarrow \frac{x}{y}=\frac{45}{18}=\frac{5}{2}$$

∴ Required ratio = 5 : 2

27. (c) Let the original quantity of dettol in the bottle be x litres.

Then, quantity of water in the bottle = 0 litres

After the 1st operation :
Quantity of dettol in the bottle

$$= \left(x-\frac{x}{3}\right) \text{ litres} = \frac{2x}{3} \text{ litres}$$

Quantity of water in the bottle $=\dfrac{x}{3}$ litres

After the 2nd operation :
Quantity of dettol in the bottle

$$= \left(\frac{2x}{3}-\frac{1}{3}\times\frac{2x}{3}\right) \text{ litres} = \frac{4x}{9} \text{ litres}$$

∴ Quantity of water in the bottle $=\left(\dfrac{x}{3}+\dfrac{2x}{9}\right)$ litres

After the third operation :
Quantity of dettol in the bottle

$$= \left(\frac{4x}{9}-\frac{1}{3}\times\frac{4x}{9}\right) \text{ litres} = \frac{8x}{27} \text{ litres}$$

∴ Quantity of water in the bottle

$$= \left(\frac{x}{3}+\frac{2x}{9}+\frac{4x}{27}\right) \text{ litres}$$

After the fourth operation :
Quantity of dettol in the bottle

$$= \left(\frac{8x}{27}-\frac{1}{3}\times\frac{8x}{27}\right) \text{ litres} = \frac{16x}{81} \text{ litres}$$

Quantity of water in the bottle

$$= \left(\frac{x}{3}+\frac{2x}{9}+\frac{4x}{27}+\frac{8x}{81}\right) \text{ litres}$$
$$= \left(\frac{27x+18x+12x+8x}{81}\right) \text{ litres}$$
$$= \frac{65x}{81} \text{ litres}$$

∴ Required ratio $= \dfrac{16x}{81}:\dfrac{65x}{81}=16:65$

28. (a) In 10 kg of 1st alloy the ratio of copper and zinc = 4 : 1

∴ Copper $=\dfrac{4}{5}\times10\text{ kg}=8\text{ kg}$,

Zinc $=\dfrac{1}{5}\times10\text{ kg}=2\text{ kg}$

In 16 kg of 2nd alloy the ratio of copper and zinc = 1 : 3

∴ Copper $=\dfrac{1}{4}\times16\text{ kg}=4\text{ kg}$,

Zinc $=\dfrac{3}{4}\times16\text{ kg}=12\text{ kg}$

Let x kg be the amount of pure copper added.
Given, Copper : Zinc = 3 : 2 in the new alloy,
i.e., $(8+4+x):(2+12)=3:2$

$$\Rightarrow \frac{12+x}{14}=\frac{3}{2}$$
$$\Rightarrow 24+2x=42 \Rightarrow 2x=18 \Rightarrow x=9\text{ kg}.$$

∴ Weight of the new alloy
$$= (8+4+9+2+12)=35\text{ kg}$$

29. (d) New ratio of fares = 1st class : 2nd class : 3rd class $=8\times\dfrac{5}{6}:6\times\dfrac{11}{12}:3\times1$

$$= \frac{20}{3}:\frac{11}{2}:3=\frac{20}{3}\times6:\frac{11}{2}\times6:3\times6$$
$$= 40:33:18$$

Ratio of passengers = 9 : 12 : 26

∴ Ratio of amounts collected
$$= 40\times9:33\times12:18\times26$$
$$= 360:390:468$$
$$= 90:99:117$$

∴ Amount collected from 1st class fares
$$= \frac{90}{306}\times\text{Rs }1088=\text{Rs }320.$$

30. (c) Ratio of milk and water
$$= \left(\frac{1}{6}\times5+\frac{3}{8}\times4+\frac{5}{12}\times5\right):\left(\frac{5}{6}\times5+\frac{5}{8}\times4+\frac{7}{12}\times5\right)$$
$$= \left(\frac{5}{6}+\frac{3}{2}+\frac{25}{12}\right):\left(\frac{25}{6}+\frac{5}{2}+\frac{35}{12}\right)$$
$$= \left(\frac{10+18+25}{12}\right):\left(\frac{50+30+35}{12}\right)$$
$$= \frac{53}{12}:\frac{115}{12}=53:115$$

Self Assessment Sheet–15

1. A sum of Rs 4830 was divided among three persons P, Q and R such that if their shares were diminished by Rs 5, Rs 10 and Rs 15 respectively, the remainders would be in the ratio 3 : 4 : 5. What was P's share?

   (a) Rs 1200   (b) Rs 1210

   (c) Rs 1205   (d) Rs 1215

2. A man divided two sums of money among 4 sons R, S, T, U the first in the ratio 4 : 3 : 2 : 1 and the second in the ratio 5 : 6 : 7 : 8. If the second sum of money is twice the first sum, which son receives the largest part?

   (a) R   (b) S

   (c) T   (d) U

3. The contents of two vessels containing water and milk in the ratio 1 : 2 and 2 : 5 are mixed in the ratio 1 : 4. The resulting mixture will have water and milk in the ratio:

   (a) 30 : 75   (b) 31 : 74

   (c) 31 : 75   (d) 30 : 74

4. Ratio of incomes of A, B and C last year was 3 : 4 : 5. The ratios of their individual incomes of last year and this year are 4 : 5, 2 : 3 and 3 : 4 respectively. If the sum of their present incomes is Rs 78800. Find the present individual income of B.

   (a) Rs 18000

   (b) Rs 32000

   (c) Rs 28800

   (d) Cannot be determined.

5. The expenses on rice, fish and oil of a family are in the ratio 12 : 17 : 13. The prices of these articles are increased by 20%, 30% and 50% respectively. The total expenses of the family are increased by:

   (a) $14\dfrac{1}{8}\%$ (b) $7\dfrac{1}{8}\%$

   (c) $56\dfrac{1}{8}\%$ (d) $28\dfrac{1}{8}\%$

6. The ratio of the age of a man and his wife is 4 : 3. After 4 years this ratio will be 9 : 7. If at the time of marriage, the ratio was 5 : 3, then how many years ago were they married?

   (a) 12 years   (b) 8 years

   (c) 10 years   (d) 15 years

7. Vessels A and B contain mixtures of milk and water in the ratio 4 : 5 and 5 : 1 respectively. In what ratio should quantities of mixtures be taken from A to B to form a mixture in which milk to water is in the ratio 5 : 4?

   (a) 2 : 5   (b) 2 : 3

   (c) 4 : 3   (d) 5 : 2

8. Concentrations of three solutions A, B and C are 20%, 30% and 40% respectively. They are mixed in the ratio 3 : 5 : x resulting in a solution of 30% concentration. Find x.

   (a) 5   (b) 2

   (c) 3   (d) 4

9. A precious stone worth Rs 6800 is accidently dropped and breaks into three pieces: The weight of three pieces are in the ratio 5 : 7 : 8. The value of the stone is proportional to the square of its weight. Find the loss.

   (a) Rs 4260   (b) Rs 4273

   (c) Rs 4454   (d) Rs 3250

10. A, B and C spend 70%, 75% and 80% respectively of their salaries. Their actual savings are in the ratio 4 : 5 : 6. What is the ratio of their salaries?

    (a) 4 : 5 : 6   (b) 4 : 6 : 9

    (c) 6 : 5 : 4   (d) 9 : 6 : 4

Answers

1. (c)   2. (a)   3. (b) [**Hint:** In vessel I , water $=\dfrac{1}{3}$, milk $=\dfrac{2}{3}$; In vessel II, water $=\dfrac{2}{7}$, milk $=\dfrac{5}{7}$. From vessel I,

   1/5 is taken and from vessle II, 4/5 is taken. Now find reqd. ratio]

4. (c) [**Hint:** Let last years incomes be l_1, l_2 and l_3 and present incomes be P_1, P_2, P_3. Then $l_1 : l_2 : l_3 = 3 : 4 : 5$ .....(i)

   $\dfrac{l_1}{P_1}=\dfrac{4}{5}, \dfrac{l_2}{P_2}=\dfrac{2}{3}, \dfrac{l_3}{P_3}=\dfrac{3}{4}$ ...(ii) and $P_1 + P_2 + P_3 = 78,800$ ...(iii)

   Now using (i) and (ii) find $P_1 + P_2 + P_3$ and the substitute in (iii)]

5. (d)   6. (a)   7. (d)   8. (c)   9. (c)   10. (b)

Chapter

16 PERCENTAGE AND ITS APPLICATIONS

Section-A
PERCENTAGE

Solved Examples

Ex. 1. *Shatabdi Express has a capacity of 500 seats of which 10% are in Executive class and the rest being Chair cars. During one journey, the train was booked 85% of its capacity. If Executive class was booked to 96% of its capacity, then how many chair car seats were empty during that journey ?*

Sol. Seats in Executive class = 10% of total seats = $\dfrac{10}{100} \times 500 = 50$

∴ Seats in Chair car = 450

Total seats booked = 85% of 450 = $\dfrac{85}{100} \times 450 = 425$

Total seats booked in Executive class = 96% of 50 = $\dfrac{96}{100} \times 50 = 48$

∴ Seats booked in Chair car = (425 – 48) = 377
Empty seats in Chair car = 450 – 377 = 73.

Ex. 2. *In an examination, a student who gets 20% of maximum marks fail by 5 marks. Another student who scores 30% of the maximum marks gets 20 marks more than the pass marks. What is the necessary percentage required for passing ?*

Sol. Let the maximum marks be x. Then,

Pass mark for the 1st student = $\dfrac{20}{100} \times x + 5$

Pass mark for the 2nd student = $\dfrac{30}{100} \times x - 20$

Since pass marks are same for both the students, $\dfrac{20x}{100} + 5 = \dfrac{30x}{100} - 20$

⇒ $\dfrac{10x}{100} = 25$ ⇒ $x = 250$

∴ Pass mark = $\dfrac{20}{100} \times 250 + 5 = 55$

Pass percentage = $\dfrac{55}{250} \times 100 = 22\%$.

Ex. 3. *What per cent of numbers from 1 to 70 have squares that end in the digit 1 ?*

Sol. The numbers from 1 to 70 that have their squares ending in digit 1 are :

1, 9, 11, 19, 21, 29, 31, 39, 41, 49, 51, 59, 61, 69 *i.e.*, 14 in numbers.

∴ Required percentage $= \dfrac{14}{70} \times 100 = 20\%$.

Ex. 4. *Entry fee in an exhibition was Re 1. Later this was reduced by 25% which increased the sale by 20%. Find the percentage increase in the number of visitors.*

Sol. Let the number of visitors be x. Then, as entry fee = Re 1, Sale = $x \times$ Re 1 = Re x

Reduced entry fee = Re 0.75, Increased sale = x + 20% of x = Rs 1.2 x

Then, increased number of visitors $= \dfrac{\text{Rs } 1.2x}{\text{Re } 0.75} = 1.6x$

∴ % increase $= \left(\dfrac{(1.6x - x)}{x} \times 100 \right)\% = (0.6 \times 100)\% = 60\%$.

Ex. 5. *The cost of manufacturing a TV set is made up of material costs, labour costs and overhead costs. These costs are in the ratio 4 : 3 : 2. If materials costs and labour costs rise by 10% and 8% respectively, while the overhead costs reduce by 5%, what is the percentage increase in the total cost of the TV set ?*

Sol. Let the cost of the T.V. be Rs x. Then,

Material cost $= \dfrac{4x}{9}$, Labour cost $= \dfrac{3x}{9}$, Overhead costs $= \dfrac{2x}{9}$

New material cost $= \dfrac{110}{100} \times \dfrac{4x}{9} = \dfrac{44x}{90}$

New labour cost $= \dfrac{108}{100} \times \dfrac{3x}{9} = \dfrac{9x}{25}$

New overhead costs $= \dfrac{95}{100} \times \dfrac{2x}{9} = \dfrac{19x}{90}$

Increase in the cost of T.V. $= \left(\dfrac{44x}{90} - \dfrac{4x}{9} \right) + \left(\dfrac{9x}{25} - \dfrac{3x}{9} \right) + \left(\dfrac{19x}{90} - \dfrac{2x}{9} \right)$

$= \dfrac{4x}{90} + \dfrac{2x}{75} - \dfrac{x}{90} = \dfrac{20x + 12x - 5x}{450} = \dfrac{27x}{450}$

∴ % increase in cost $= \dfrac{\frac{27x}{450}}{x} \times 100\% = \dfrac{2700}{450}\% = 6\%$.

Ex. 6. *The price of rice is reduced by 2%. How many kilograms of rice can now be bought for the money which was sufficient to buy 49 kg of rice earlier ?*

Sol. Let the price of rice be Rs x per kg. Then,

Cost of 49 kg rice = Rs 49 x

New price of rice = 98% of Rs x = Rs $\dfrac{98x}{100}$

∴ New quantity of rice bought $= \dfrac{49x \times 100}{98x} = 50$ kg.

Ex. 7. *What is the ratio in which two sugar solutions of concentrations 15% and 40% are to be mixed to get a solution of concentration 30% ?*

Sol. Let the required ratio be $x : y$. Then, $\dfrac{15x}{100} + \dfrac{40y}{100} = \dfrac{30(x+y)}{100}$

$\Rightarrow 15x + 40y = 30x + 30y \Rightarrow 10y = 15x \Rightarrow \dfrac{x}{y} = \dfrac{10}{15} = \dfrac{2}{3} \Rightarrow x : y = 2 : 3.$

Ex. 8. *A man spends 75% of his income. This income is increased by 20% and he increases his expenditure by 10%. By what per cent are his savings increased ?*

Sol. Let the income of the person be Rs x.

Then, his savings = Rs $\dfrac{x}{4}$

Expenditure $= x - \dfrac{x}{4} = $ Rs $\dfrac{3x}{4}$

Increased income $= \dfrac{120}{100} \times x = $ Rs $\dfrac{6x}{5}$

Increased expenditure $= \dfrac{110}{100} \times \dfrac{3x}{4} = $ Rs $\dfrac{33x}{40}$

$\therefore$ Increased savings $= \dfrac{6x}{5} - \dfrac{33x}{40} = \dfrac{48x - 33x}{40} = \dfrac{3x}{8}$

$\therefore$ % increase in savings $= \dfrac{\left(\dfrac{3x}{8} - \dfrac{x}{4}\right)}{\dfrac{x}{4}} \times 100 = \left(\dfrac{x}{8} \times \dfrac{4}{x} \times 100\right)\% = 50\ \%.$

Ex. 9. *A number is increased by 20% and then again by 20%. By what per cent should the increased number be reduced so as to get back the original number ?*

Sol. Let the original number be 100.

An increase of 20% followed by another increase

of 20% $= \dfrac{120}{100} \times \dfrac{120}{100} \times 100 = 144$

$\therefore$ Required decrease = 44

Required % decrease $= \dfrac{44}{144} \times 100\% = \dfrac{275}{9}\% = 30\dfrac{5}{9}\%.$

Ex. 10. *A reduction of 20% in the price of sugar enables a purchaser to obtain $2\dfrac{1}{2}$ kg more for Rs 160. Find the original price per kg of sugar.*

Sol. Let the original price per kg of sugar be Rs x.

Then, the reduced price per kg = Rs $\dfrac{4x}{5}$

Quantity of sugar that can be bought for Rs 160 originally = Rs $\dfrac{160}{x}$

Quantity of sugar that can be bought for Rs 160 with reduced price $=$ Rs $\dfrac{160}{4x/5} = 160 \times \dfrac{5}{4x}$

Given, $160 \times \dfrac{5}{4x} = \dfrac{160}{x} + \dfrac{5}{2}$

$\Rightarrow \dfrac{200}{x} - \dfrac{160}{x} = \dfrac{5}{2} \Rightarrow \dfrac{40}{x} = \dfrac{5}{2} \Rightarrow x = 16$

$\therefore$ Original price per kg = Rs 16.

1. The sum of two numbers is 4000. 10% of one number is $6\frac{2}{3}$% of the other. The difference of the number is
   (a) 600   (b) 800
   (c) 1025   (d) 1175

2. The difference of the squares is of two numbers is 80% of the sum of their squares. The ratio of the larger number to the smaller number is
   (a) 5 : 2   (b) 2 : 5
   (c) 3 : 1   (d) 1 : 3

3. 50 g of an alloy of gold and silver contains 80% gold (by weight). The quantity of gold, that is to be mixed up with this alloy, so that it may contain 95% gold is
   (a) 200 g   (b) 150 g
   (c) 50 g   (d) 10 g

4. A candidate who gets 20% marks in an examination fails by 30 marks but another candidate who gets 32%, gets 42 marks more than the pass marks. The percentage of pass marks is
   (a) 52%   (b) 50%
   (c) 33%   (d) 25%

5. In expressing a length 81.472 km as nearly as possible with three significant digits, the per cent error is
   (a) 0.34%   (b) 0.034%
   (c) 0.0034%   (d) 0.0038%

6. A reduction of 25% in the price of an article enables a man to buy 50 kilograms more for Rs 500. What is the reduced price per kilogram ?
   (a) Rs 3   (b) Rs 2.50
   (c) Rs 2.05   (d) Rs 2.40

7. A reduction of 20% in the price of oranges enables a man to buy 5 oranges more for Rs 10. The price per orange before reduction was :
   (a) 20 paise   (b) 40 paise
   (c) 50 paise   (d) 60 paise

8. 37.85% and 92% alcoholic solutions are mixed to get 35 litres of an 89% alcoholic solution. How many litres of each solution are there in the new mixture ?
   (a) 10 L of the 1st, 25 L of the 2nd
   (b) 20 L of the 1st, 15 L of the 2nd
   (c) 15 L of the 1st, 20 L of the 2nd
   (d) None

9. From 5 litres of 20% solution of alcohol in water, 2 litres of solution is taken out and 2 litres water is added to it. The strength of alcohol in the new solution is
   (a) 12%   (b) 15%
   (c) 16%   (d) 18%

10. The number of employees working in a farm is increased by 25% and the wages per head are decreased by 25%. If it results in x% decrease in total wages, then the value of x is
    (a) 0   (b) 25
    (c) 20   (d) $\frac{25}{4}$

11. If the price of a book is first decreased by 25% and then increased by 20%, the net change in the price of the book is
    (a) 10% decrease   (b) 5% decrease
    (c) No change   (d) 5% increase

12. The price of an article was increased by r%. Later the new price was decreased by r%. If the latest price was Re 1, then the original price was
    (a) Re 1   (b) Rs $\left(\frac{1-r^2}{100}\right)$
    (c) Rs $\frac{\sqrt{1-r^2}}{100}$ (d) Rs $\frac{10000}{10000-r^2}$

13. The price of a commodity has been increased by 60%. By what per cent must a consumer reduce the consumption of the commodity so as not to increase his expenditure on the commodity
    (a) 30%   (b) 35%
    (c) 27.5%   (d) 37.5%

14. The price of an article is reduced by 25% but the daily sale of the article is increased by 30%. The net effect on the daily sale receipts is
    (a) $2\frac{1}{2}$% decrease   (b) $2\frac{1}{2}$% increase
    (c) 2% decrease   (d) 2% increase

15. If the altitude of a triangle is increased by 10% while its area remains the same, its corresponding base will have to be decreased by
    (a) 10%   (b) 9%
    (c) $9\frac{1}{11}$%   (d) $11\frac{1}{9}$%

16. In the expression xy^2, the values of both variables x and y are decreased by 20%. By this the value of the expression will be decreased by
    (a) 40%   (b) 80%
    (c) 48.8%   (d) 51.2%

17. In a hotel, 60% had vegetarian lunch while 30% had non-vegetarian lunch and 15% had both types of lunch. If 96 people were present, how many did

not eat either type of lunch ?
(a) 20   (b) 24
(c) 26   (d) 28

18. A student took five papers in an examination, where the full marks were the same for each papers, this marks in these papers were in the proportion 6 : 7 : 8 : 9 : 10. In all the papers together, the candidate obtained 60% of the total marks. Then, the number of papers in which he got more than 50% marks is
(a) 1   (b) 3
(c) 4   (d) 5

19. In an examination, 70% of the candidates passed in English, 80% passed in Mathematics and 10% failed in both these subjects. If 144 candidates passed in both, then the total number of candidates was
(a) 125   (b) 200
(c) 240   (d) 375

20. In a medical certificate, by mistake a candidate gave his height as 25% more than the actual. In the interview panel, he clarified his height was 5 feet 5 inches. Find the percentage correction made by the candidate from his stated height to his actual height.
(a) 28.56   (b) 20
(c) 25   (d) 24

21. A 20% hike in bus fare resulted in a 10% fall in passenger traffic, still the daily collection at the bus depot increased by Rs 150. The daily collection at the depot after the fare hike is
(a) Rs 1600   (b) Rs 1750
(c) Rs 2025   (d) Rs 1875

22. If A's salary is 25% higher than B's salary, then how much per cent is B's salary lower than that of A's ?
(a) 16%   (b) 20%
(c) 25%   (d) $33\frac{1}{3}\%$

23. In an examination in which full marks were 800, A gets 20% more than B, B gets 20% more than C, and C gets 15% less than D. If A got 576, what percentage of full marks did D get (approximately) ?
(a) 45.7   (b) 51.2
(c) 58.8   (d) 61.7

24. A father gives 1% of his monthly salary to his two sons as pocket money. The elder son gets 80% of the total amount given to the two sons and he spends

80% of his share. If he saves Rs 20 every month, then the monthly salary of the father is
(a) Rs 10000   (b) Rs 11500
(c) Rs 12000   (d) Rs 12500

25. In an examination Mohit obtained 20% more marks than Sushant but are 10% less than Rajesh. If the marks obtained by Sushant are 1080, find the percentage of marks obtained by Rajesh, if the full marks are 2000.
(a) 72%   (b) 86.66%
(c) 78.33%   (d) 75%

26. Mira's expenditure and savings are in the ratio 3 : 2. Her income increases by 10%. Her expenditure also increases by 12%. By how much % do her savings increase ?
(a) 7%   (b) 9%
(c) 10%   (d) 13%

27. A tax payer is exempted of income tax for the first Rs 100000 of his annual income but for the rest of the income, he has to pay a tax at the rate of 20%. If he paid Rs 3160 as income tax for a year, his monthly income is
(a) Rs 11580   (b) Rs 103160
(c) Rs 13610   (d) Rs 9650

28. Mrs. Sharma invests 15% of her monthly salary, i.e., Rs 4428 in Mutual funds. Later she invests 18% of her monthly salary on Pension plans; also she invests another 9% of her salary on Insurance policies. What is the total monthly amount invested by Mrs. Sharma ?
(a) Rs 1,13,356.80   (b) Rs 12,398.40
(c) Rs 56, 678.40   (d) None of these

29. A house-owner was having his house painted. He was advised that he would require 25 kg of paint. Allowing for 15% wastage and assuming that the paint is available in 2 kg cans, what would be the cost of paint purchased, if one can costs Rs 16 ?
(a) Rs 240   (b) Rs 180
(c) Rs 120   (d) Rs 360

30. Vishal goes to a shop to buy a radio costing Rs 2568. The rate of sales tax is 7%. He tells the shop keeper to reduce the price of radio to such an extent that he has to pay Rs 2568 inclusive of sales tax. Find the reduction needed in the price of the radio.
(a) Rs 179   (b) Rs 170
(c) Rs 168   (d) Rs 169

Answers

1. (b)	2. (c)	3. (b)	4. (d)	5. (b)	6. (b)	7. (c)	8. (d)	9. (a)	10. (d)
11. (a)	12. (d)	13. (d)	14. (a)	15. (c)	16. (c)	17. (b)	18. (c)	19. (c)	20. (b)
21. (c)	22. (b)	23. (c)	24. (d)	25. (a)	26. (a)	27. (d)	28. (b)	29. (a)	30. (c)

Hints and Solutions

1. (b) Let one number = x

Then, the other number = $4000 - x$

Given, 10% of $x = 6\frac{2}{3}$% of $(4000 - x)$

$\Rightarrow \frac{10}{100} \times x = \frac{20}{3} \times \frac{1}{100} \times (4000 - x)$

$\Rightarrow 10x = \frac{20}{3} \times 4000 - \frac{20x}{3}$

$\Rightarrow 10x + \frac{20x}{3} = \frac{20}{3} \times 4000 \Rightarrow \frac{50x}{3} = \frac{20}{3} \times 4000$

$\Rightarrow x = \frac{20 \times 4000}{50} = 1600$

The two numbers are 1600 and 2400.

$\therefore$ Their difference = 2400 – 1600 = 800.

2. (c) Let the two numbers be x and y.

Then, $x^2 - y^2 = 80\%$ of $(x^2 + y^2)$

$\Rightarrow x^2 - y^2 = \frac{4}{5}(x^2 + y^2) \Rightarrow x^2 - \frac{4}{5}x^2 = \frac{4}{5}y^2 + y^2$

$\Rightarrow \frac{1}{5}x^2 = \frac{9}{5}y^2 \Rightarrow \frac{x^2}{y^2} = \frac{9}{1} \Rightarrow \frac{x}{y} = \frac{3}{1} \Rightarrow x : y = 3 : 1.$

3. (b) In 50 gm of an alloy of gold and silver, the quantity of gold = 40 gm and the quantity of silver = 10 gm.

Let x gm of gold be mixed in 50 gm of an alloy of gold and silver such that quantity of gold becomes 95%.

Then, $40 + x = 95\%$ of $(x + 50)$

$40 + x = \frac{19}{20}(x + 50)$

$\Rightarrow 800 + 20x = 19x + 950 \Rightarrow x = 150.$

4. (d) Let the maximum marks be M. Then,

Pass marks = 20% of M + 30 = 32% of M – 42

$\Rightarrow$ 12% of M = 72 $\Rightarrow M = \frac{72 \times 100}{12} = 600$

$\therefore$ Pass marks = 20% of 600 + 30 = 150

$\Rightarrow$ Percentage of pass marks $= \frac{150}{600} \times 100\%$

$= 25\%$.

5. (b) 81.472 km = 81472 meters

= 81500 meters with three significant digits

$\therefore$ Error% $= \frac{81500 - 81472}{81472} \times 100 = 0.034\%$.

6. (b) Let the original price per kg of the article be Rs x.

Then, the reduced price per kg $= \frac{75}{100} \times$ Rs x

$=$ Rs $\frac{3x}{4}$

Amount of article that can be bought for Rs 500 originally $= \frac{500}{x}$ kg

Amount of article that can be bought for Rs 500 by reduced price $= \frac{500}{3x/4}$ kg $= \frac{500 \times 4}{3x}$ kg

Given, $\frac{500 \times 4}{3x} = \frac{500}{x} + 50$

$\Rightarrow \frac{2000 - 1500}{3x} = 50 \Rightarrow \frac{500}{3x} = 50$

$\Rightarrow x = \frac{500}{50 \times 3} =$ Rs $\frac{10}{3}$

$\therefore$ Reduced price $= \frac{3}{4} \times$ Rs $\frac{10}{3}$ = Rs 2.50

7. (c) Similar to Q. No. 6.

8. (d) Let x litres of the 37.85% alcoholic solution and $(35 - x)$ litres of 92% alcoholic solution be required to get 35 L of 89% solution. Then,

37.85% of x + 92% of $(35 - x)$ = 89% of 35

$\Rightarrow \frac{37.85x}{100} + \frac{92 \times 35}{100} - \frac{92 \times x}{100} = \frac{89}{100} \times 35$

$\Rightarrow 54.15x = 105 \Rightarrow x = \frac{105}{54.15} = 1.94$ L (approx.)

9. (a) Quantity of alcohol in 3 litres of solution

$= \frac{20}{100} \times 3 = \frac{3}{5}$ L

Strength of alcohol in new solution $= \frac{315}{5} \times 100$

$= 12\%$

10. (d) Let the number of workers = x, wage per employee = Rs y

Then, total wages = Rs xy

Number of workers after increase = $1.25x$

Reduced wage per employee = Rs $0.75y$

$\therefore$ Total wages = $1.25x \times$ Rs $0.75y = 0.9375xy$

$\therefore$ Required % decrease $= \dfrac{xy - 0.9375xy}{xy} \times 100$

$\qquad = 0.0625 \times 100 = 6.25$

$\qquad = \dfrac{25}{4}\%$

11. (a) Let the original price of the book be Rs 100.
Decreased price of the book = Rs 75
Increased price of the book after 20% increase

$\qquad = \dfrac{120}{100} \times Rs\ 75 = Rs\ 90$

$\therefore$ Net change in price = Rs 10 decrease

$\therefore$ % change $= \dfrac{10}{100} \times 100 = 10\%$ decrease.

12. (d) Let the original price of the article be Rs x. Then,

Increased price of the article $= \left(\dfrac{100 + r}{100}\right) \times x$

Decreased price of the article after r % decrease

$\qquad = \left(\dfrac{100 - r}{100}\right)\left(\dfrac{100 - r}{100}\right) \times x$

$\qquad = \left(\dfrac{10000 - r^2}{10000}\right) \times x$

Given $\dfrac{10000 - r^2}{10000} \times x = 1$

$\Rightarrow x = \dfrac{10000}{10000 - r^2}$

13. (d) Let the price of the commodity be Rs x and its consumption be y. Then
Expenditure = Rs xy

Increased price = Rs $\dfrac{160}{100} x = Rs\ 1.6x$

Expenditure remaining the same = Rs xy

$\therefore$ Reduced consumption $= \dfrac{xy}{1.6x} = \dfrac{10y}{16} = \dfrac{5}{8}y$

% reduction in consumption $= \dfrac{y - 5/8\,y}{y} \times 100$

$\qquad = \dfrac{3}{8} \times 100\% = 37.5\%$

14. (a) Let the price of the article be Rs x and daily sale be y units.
Then, daily sale receipts = Rs xy

Reduced price of the article $= Rs\ \dfrac{3}{4}x$

Increased daily sale $= Rs\ \dfrac{13}{10}y$

$\therefore$ Daily sale receipts $= Rs\ \dfrac{3}{4}x \times \dfrac{13}{10}y = Rs\ \dfrac{39}{40}xy$

$\therefore$ % reduction $= \dfrac{xy - \dfrac{39}{40}xy}{xy} \times 100\%$

$\qquad = \dfrac{100}{40}\% = 2.5\%$.

15. (c) Let the altitude of the triangle be h and corresponding base = b. Then, its area $= \dfrac{1}{2}bh$,

Increased altitude = $1.1h$,

Area remaining same $= \dfrac{1}{2}bh$

$\therefore$ Reduced base $= \dfrac{\dfrac{1}{2}bh}{1.1h} \times 2 = \dfrac{b}{2.2} \times 2 = \dfrac{10b}{11}$

$\therefore$ % reduction $= \dfrac{b - \dfrac{10}{11}b}{b} \times 100 = \dfrac{100}{11}\%$

$\qquad = 9\dfrac{1}{11}\%$

16. (c) New value of $x = \dfrac{80}{100}x = 0.8x$

New value of $y^2 = \dfrac{80}{100}y \times \dfrac{80}{100}y = 0.64y^2$

$\therefore$ New value of expression $= 0.8x \times 0.64y^2$

$\qquad = 0.512xy^2$

$\therefore$ % reduction in the value

$\qquad = \dfrac{xy - 0.512xy^2}{xy^2} \times 100\%$

$\qquad = (0.488 \times 100)\% = 48.8\%$

17. (b) Number of people having either or both type of lunches

$= \dfrac{60}{100} \times 96 + \dfrac{30}{100} \times 96 - \dfrac{15}{100} \times 96 = \dfrac{75 \times 96}{100} = 72$

Number of people who did not eat either type of lunch = 96 – 72 = 24

18. (c) Let the marks obtained in 5 subjects be $6x$, $7x$, $8x$, $9x$ and $10x$.
Average score = 60%

$\therefore \dfrac{6x + 7x + 8x + 9x + 10x}{5} = \dfrac{60}{100} \Rightarrow \dfrac{40x}{5} = \dfrac{60}{100}$

$$\Rightarrow x = \frac{60 \times 5}{100 \times 40} = \frac{3}{40} = 0.075$$

∴ The marks are 0.45, 0.525, 0.6, 0.675 and 0.75, *i.e.*, 45%, 52.5%, 60%, 67.5% and 75%.

∴ Number of papers in which the marks exceed 50% = 4.

19. (c) Let the total number of candidates be x.
Given, 70% of x passed in English
80% of x passed in Maths
144 passed in English and Maths both
10% of x failed in English and Maths both
∴ 90% of x passed in English and Maths both.
∴ 90% of x = 70% x + 80% of x – 144

$$\Rightarrow 60\% \text{ of } x = 144 \quad \Rightarrow \quad x = \frac{144 \times 100}{60} = 240.$$

20. (b) Actual height = 5 feet 5 inches
= 5 × 12 inches + 5 inches
= 65 inches.

$$\text{Height given by mistake} = \frac{125}{100} \times 65 \text{ inches}$$
$$= 81.25 \text{ inches}$$

∴ Required percentage error

$$= \frac{(81.25 - 65)}{81.25} \times 100\%$$

$$= \left(\frac{16.25}{81.25} \times 100\right)\% = 20\%.$$

21. (c) Let the original bus fare per person be Rs x, daily passenger traffic be y.
Daily collection at the depot = Rs xy
Increased bus fare = Rs $1.2x$,
Reduced passenger traffic = $0.9y$
Daily collection at the depot = Rs $1.2x \times 0.9y$
$$= 1.08xy$$

Given, $1.08xy = xy + 150 \Rightarrow xy = \frac{150}{0.08} = 1875$

∴ Increased daily collection = Rs 1875 + Rs 150
$$= \text{Rs } 2025.$$

22. (b) Let B's salary be Rs 100. Then,
A's salary = Rs 1 25

$$\therefore \text{ Required } \% = \frac{25}{125} \times 100\% = 20\%$$

23. (c) $A = \frac{120}{100} B, B = \frac{120}{100} C, C = \frac{85}{100} D$

$$\Rightarrow B = \frac{5}{6} A, C = \frac{5}{6} B, D = \frac{20}{17} D$$

$$\therefore B = \frac{5}{6} \times 576 = 480; C = \frac{5}{6} \times 480 = 400;$$

$$D = \frac{20}{17} \times 400 = \frac{8000}{17}$$

So, required percentage $= \left(\frac{8000}{17} \times \frac{1}{800} \times 100\right)\%$
$$= 58.82\% \cdot$$

24. (d) Let the father's salary be Rs x. Then,
Part of salary given to the two sons as pocket

money $= \frac{x}{100}$

Share of the elder son $= \frac{80}{100} \times \frac{x}{100} = \frac{4x}{500}$

Expenditure of the elder son $= \frac{80}{100} \times \frac{4x}{500}$
$$= \frac{16x}{2500}$$

∴ Savings of the elder son $= \frac{4x}{500} - \frac{16x}{2500} = \frac{4x}{2500}$

Given, $\frac{4x}{2500} = $ Rs 20

$$\Rightarrow x = \text{Rs } \frac{20 \times 2500}{4} = \text{Rs } 12500.$$

25. (a) Sushant's marks = 1080

Mohit's marks $= \frac{120}{100} \times 1080 = 1296$

Let Rajesh's marks be x. Then,

$$\frac{90x}{100} = 1296 \Rightarrow x = 1440$$

∴ Percentage of Rajesh's marks $= \frac{1440}{2000} \times 100\%$
$$= 72\%.$$

26. (a) Let Mira's expenditure and savings be $3x$ and $2x$ respectively.
Then, her income = $5x$

Increased income $= \frac{110}{100} \times 5x = 5.5x$

Increased expenditure $= \frac{112}{100} \times 3x = 3.36x$

∴ Increased savings $= 5.5x - 3.36x = 2.14x$

∴ % increase in savings $= \left(\frac{2.14x - 2x}{2x} \times 100\right)\%$

$$= \left(\frac{0.14x}{2x} \times 100\right)\% = 7\%.$$

27. (d) Let the annual income of the person be Rs x. Then, 20% of $(x - 1,00,000) = 3160$

$$\Rightarrow x - 1,00,000 = \frac{3160 \times 100}{20} = 15800$$

$$\Rightarrow x = 115800$$

∴ Monthly income = Rs $\frac{115800}{12}$ = Rs 9650

28. (b) Let Mrs. Sharma's monthly salary be Rs x. Then,

15% of x = Rs 4428

$\Rightarrow \quad x = \text{Rs } \dfrac{4428 \times 100}{15} = \text{Rs } 29520$

$\therefore$ Total monthly amount invested by Mrs. Sharma

$= (15\% + 18\% + 9\%)$ of Rs 29520

$= 42\%$ of Rs 29250 $= \dfrac{42}{100} \times 29250$

$= \text{Rs } 12398.40$.

29. (a) Paint required = 25 kg + 15% of 25 kg

$= 25 \text{ kg} + 0.15 \times 25 \text{ kg} = 28.75 \text{ kg}$

$\therefore$ Number of 2 kg cans of paint required = 15

Cost of paint purchased = 15 × Rs 16 = Rs 240

30. (c) Let the reduced price of the radio be Rs x. Then,

$x + 7\%$ of $x = 2568$

$\Rightarrow \dfrac{107}{100} \times x = 2568 \Rightarrow x = \dfrac{2568 \times 100}{107} = 2400$

$\therefore$ Reduction needed in the price of radio

$= \text{Rs } 2568 - \text{Rs } 2400 = \text{Rs } 168$.

Self Assessment Sheet–16(a)

1. A solution of salt and water contains 15% salt by weight 30 kg of water evaporates and the solution now contains 20% of the salt. The original quantity of the solution is
   - (a) 100 kg
   - (b) 110 kg
   - (c) 115 kg
   - (d) 120 kg

2. Ram ordered for 6 black toys and some additional brown toys. The price of a black toy is $2\frac{1}{2}$ times that of a brown toy. While preparing the bill, the clerk interchanged the number of black toys with the number of brown toys which increased the bill by 45%. The number of brown toys is:
   - (a) 8
   - (b) 6
   - (c) 15
   - (d) 12

3. In an election, 10% of the people in the voter's list did not participate. 60 votes were declared invalid. There are only two candidates A and B. A defeated B by 308 votes. It has found that 47% of the people listed in the voters' list voted for A. Find the total number of votes polled.
   - (a) 6200
   - (b) 5580
   - (c) 6000
   - (d) 7200

4. Prices register an increase of 10% on food grains and 15% on other items of expenditure. If the ratio of an employee's expenditure on food grains and other items be 2 : 5, by how much should his salary be increased in order that he may maintain the same level of consumption as before, his present salary being Rs 2590.
   - (a) Rs 323.75
   - (b) Rs 350
   - (c) Rs 360.50
   - (d) Rs 351.50

5. In a recent survey 25% houses contained two or more people. Of those houses containing only one person 20% were having only a male. What is percentage of all houses which contain exactly one female and no males?
   - (a) 55%
   - (b) 65%
   - (c) 60%
   - (d) 50%

6. One kg of tea and one kg of sugar together cost Rs 95. If the price of tea falls by 10% and that of sugar rises by 20%, then the price of one kg of each combined comes to Rs 90. The original price of tea in Rs per kg is :
   - (a) Rs 72
   - (b) Rs 55
   - (c) Rs 60
   - (d) Rs 80

7. On decreasing the price of a colour TV by 30%, its sale is increased by 20%. The effect on the revenue is :
   - (a) 16% decrease
   - (b) 16% increase
   - (c) 20% increase
   - (d) 20% decrease

8. There are some coins and rings of either gold or silver in a box. 60% of the objects are coins. 40% of the rings are of gold and 30% of the coins are of silver. What is the percentage of gold articles?
   - (a) 16
   - (b) 27
   - (c) 58
   - (d) 70

9. The income of A is 20% higher than that of 'B'. The income of 'B' is 25% less than that of 'C'. What per cent less is 'A' s income from 'C' s income?
   - (a) 7%
   - (b) 8%
   - (c) 10%
   - (d) 12.5%

10. A reduction of 25% in the price of rice enables a person to buy 100 kg more rice for Rs 600. The reduced price per kg of rice is :
    - (a) Rs 30
    - (b) Rs 25
    - (c) Rs 35
    - (d) Rs 15

Answers

| 1. (d) | 2. (c) | 3. (b) | 4. (d) | 5. (c) | 6. (d) | 7. (a) | 8. (d) | 9. (c) | 10. (d) |

Section-B
PROFIT AND LOSS

KEY FACTS

1. Profit or Gain = S.P. – C.P.

2. $\text{Gain}\% = \dfrac{\text{Gain}}{\text{C.P.}} \times 100$

3. Loss = C.P. – S.P.

4. $\text{Loss}\% = \dfrac{\text{Loss}}{\text{C.P.}} \times 100$

5. $\text{S.P.} = \dfrac{(100 + \text{Gain}\%)}{100} \times \text{C.P.}$

6. $\text{S.P.} = \dfrac{(100 - \text{Loss}\%)}{100} \times \text{C.P.}$

7. $\text{C.P.} = \dfrac{100}{(100 + \text{Gain}\%)} \times \text{S.P.}$

8. $\text{C.P.} = \dfrac{100}{(100 - \text{Loss}\%)} \times \text{S.P.}$

9. Loss or gain is always calculated on C.P.

10. If a trader professes to sell his goods at cost price, but uses false weights, then

$$\text{Gain}\% = \left[\dfrac{\text{Error}}{\text{True value} - \text{Error}} \times 100 \right]\%$$

Solved Examples

Ex. 1. *A vendor purchased 40 dozen bananas for Rs 250. Out of these, 30 bananas were rotten and could not be sold. At what rate per dozen should he sell the remaining bananas to make a profit of 20% ?*

Sol. C.P. of bananas = Rs 250, Gain required = 20%

$\therefore$ S.P. of bananas = $\dfrac{250 \times 120}{100} = \text{Rs } 300$

Number of good bananas = $(400 \times 12 - 30) = 450$

$\therefore$ S.P. of 450 bananas = Rs 300

$\Rightarrow$ S.P. of 12 bananas = Rs $\dfrac{300}{450} \times 12 = \text{Rs } 8$

$\therefore$ The remaining bananas should be sold at Rs 8 per dozen to make a profit of 20%.

Ex. 2. *If 6 articles are sold for Re 1 then there is a loss of 20%. In order to gain 20%, what must be the number of articles sold for Re 1 ?*

Sol. S.P. of 6 articles = Re 1, Loss = 20 %

$\therefore$ C.P. of 6 articles = $\dfrac{1 \times 100}{(100 - 20)} = \dfrac{100}{80} = \dfrac{5}{4} = 1.25$

If the gain is 20%, then S.P. of 6 articles = $\dfrac{1.25 \times 120}{100} = \text{Rs } 1.50$

$\therefore$ For Re 1, 4 articles are sold.

Ex. 3. *The price of a jewel passing through three hands rises on the whole by 65%. If the first and the second sellers earned 20% and 25% profit respectively, then what is the profit earned by the third seller ?*

Sol. Let the original price of the jewel be Rs P and the profit earned by the third seller be $x\%$.

$$\text{S.P. of 1st seller} = \text{Rs } \frac{120}{100} \times P = \text{C.P. of 2nd seller}$$

$$\text{S.P. of 2nd seller} = \text{Rs } \frac{125}{100} \times \frac{120P}{100} = \text{C.P. of 3rd seller}$$

$$\therefore \text{ S.P. of 3rd seller} = \text{Rs } \frac{(100+x)}{100} \times \frac{125}{100} \times \frac{120P}{100}$$

Given, $\dfrac{(100+x)}{100} \times \dfrac{\overset{5}{\cancel{125}}}{\cancel{100}} \times \dfrac{\overset{30}{\cancel{120}}P}{\underset{20}{\cancel{100}}} = \dfrac{165P}{\cancel{100}}$

$$\Rightarrow (100+x) = \frac{2}{3} \times 165 = 110 \Rightarrow x = 110 - 100 = 10\%.$$

Ex. 4. *A sold a watch to B at 20% gain and B sold it to C at a loss of 10%. If C bought the watch for Rs 216, at what price did A purchase it ?*

Sol. Let the C.P. of A = Rs x

Gain = 20%

$$\therefore \text{ S.P. of } A = \frac{120}{100} \times \text{Rs } x = \text{Rs } \frac{120x}{100}$$

$$\text{C.P. of } B = \text{Rs } \frac{120x}{100}$$

Loss = 10%

$$\therefore \text{ S.P. of } B = \frac{90}{100} \times \frac{120x}{100} = \text{Rs } \frac{27x}{25} \Rightarrow \text{C.P. of } C = \frac{27x}{25}$$

Given, $\dfrac{27x}{25} = 216 \quad \Rightarrow \quad x = \text{Rs } 200.$

Ex. 5. *A man sold two steel chairs for Rs 500 each. On one, he gains 20% and on the other he loss 12%. How much does he gain or loses in the whole transaction ?*

Sol. Total S.P. = 2 × Rs 500 = Rs 1000

$$\text{Total C.P.} = \left(\frac{500 \times 100}{20} + \frac{500 \times 100}{80} \right) = \text{Rs } \left(\frac{1250}{3} + \frac{6250}{11} \right) = \text{Rs } \frac{32500}{33}$$

$$\therefore \text{ Gain} = 1000 - \frac{32500}{33} = \frac{33000 - 32500}{33} = \frac{500}{33}$$

$$\therefore \text{ Gain}\% = \frac{\dfrac{500}{33}}{\dfrac{32500}{33}} \times 100 = \frac{500}{32500} \times 100 = 1.53\% \text{ (approx.)}$$

Ex. 6. *A dishonest shopkeeper claims to sell his goods at cost price but uses a weight of 800 gm in place of the standard 1 kg weight. What is his gain per cent ?*

Sol. Let the C.P. = Rs 100 per kg

= Rs 100 per 1000 gm = Rs 1 per gm.

The shopkeeper uses 800 gm weight as a 1 kg weight, so

His gain = Cost of 200 gm = Rs 20

$$\therefore \text{ His gain}\% = \frac{20}{80} \times 100 = 25\%.$$

Ex. 7. *A fair price shopkeeper takes 10% profit on his goods. He lost 20% goods during theft. What is his loss per cent ?*

Sol. Suppose the number of items = 100 and C.P. of each item = Re 1
Total C.P. = Rs 100
Then, number of items remaining after theft = 80

S.P. of 1 item = $\dfrac{1 \times 110}{100}$ = Rs 1.10

∴ S.P. of 80 items = 80 × Rs 1.1 = Rs 88
⇒ Loss = Rs 100 – Rs 88 = Rs 12

Loss% = $\dfrac{12}{100} \times 100 = 12\%$.

Ex. 8. *Three items are purchased at Rs 450 each. One of them is sold at a loss of 10%. At what price should the other two be sold so as to gain 20% on the whole transaction. What is the gain% on these two items ?*

Sol. C.P. of one item = Rs 450
⇒ C.P. of three items = 3 × Rs 450 = 1350
Gain on the whole transaction = 20%

∴ S.P. of the three items = $\dfrac{1350 \times 120}{100}$ = Rs 1620

C.P. of 1 item = Rs 450, Loss = 10%

∴ S.P. of that item = $\dfrac{450 \times 90}{100}$ = Rs 405

∴ S.P. of the remaining two items = Rs 1620 – Rs 405 = Rs 1215
C.P. of these two remaining items = 2 × Rs 450 = Rs 900
∴ Gain on these two items = Rs 315

Gain % = $\dfrac{315}{900} \times 100 = 35\%$.

Ex. 9. *By selling 90 ball pens for Rs 160 a person loses 20%. How many ball pens should be sold for Rs 96 so as to have a profit of 20% ?*

Sol. S.P. of 90 ball pens = Rs 160, Loss = 20%

∴ C.P. of 90 ball pens = Rs $\dfrac{160 \times 100}{(100 - 20)}$ = Rs $\dfrac{160 \times 100}{80}$ = Rs 200

∴ C.P. of 1 ball pen = Rs $\dfrac{20}{9}$

Suppose x ball pens are sold to earn a profit of 20%.

Then, C.P. of x ball pens = Rs $\dfrac{20}{9}x$

S.P. of x ball pens = Rs 96

∴ Profit = Rs $\left(96 - \dfrac{20}{9}x\right)$

Given, profit % = 20%

∴ $\dfrac{\left(96 - \dfrac{20}{9}x\right)}{\dfrac{20x}{9}} \times 100 = 20$ ⇒ $\left(96 - \dfrac{20x}{9}\right) \times 5 = \dfrac{20x}{9}$

$$\Rightarrow 96 - \frac{20x}{9} = \frac{4x}{9} \Rightarrow \frac{24x}{9} = 96 \Rightarrow x = \frac{96 \times 9}{24} = 36.$$

∴ 36 ball pens should be sold for Rs 96 to earn a profit of 20%.

Ex. 10. *A person sells an article for Rs 75 and gains as much per cent as the cost price of the article. What is the cost price of the article ?*

Sol. Let the C.P. of the article be Rs x. Then,

Gain = x%

$$\therefore \text{ S.P.} = \frac{(100 + x) \times x}{100} = \frac{100x + x^2}{100}$$

Given, $\frac{x^2 + 100x}{100} = 75 \Rightarrow x^2 + 100x - 7500 = 0$

$\Rightarrow x^2 + 150x - 50x - 7500 = 0 \Rightarrow x(x + 150) - 50(x + 150) = 0$

$\Rightarrow (x - 50)(x + 150) = 0 \Rightarrow x = 50$ or -150

Neglecting negative value $x = 50$.

Question Bank–16(b)

1. If an article is sold for Rs 178 at a loss of 11%, what should be its selling price in order to earn a profit of 11% ?
   (a) Rs 222.50 (b) Rs 267
   (c) Rs 222 (d) Rs 220

2. If I would have purchased 11 articles for Rs 10 and sold all the articles at the rate of Rs 11 for 10, the profit per cent would have been
   (a) 10% (b) 11%
   (c) 21% (d) 100%

3. By selling 100 pencils, a shopkeeper gains the S.P. of 20 pencils. His gain per cent is
   (a) 25% (b) 20%
   (c) 15% (d) 12%

4. An article is sold at a loss of 10%. Had it been sold for Rs 9 more, there would have been a gain of $12\frac{1}{2}$% on it. The cost price of the article is
   (a) Rs 40 (b) Rs 45
   (c) Rs 50 (d) Rs 35

5. A shopkeeper sells a pair of sunglasses at a profit of 25%. If he had bought it at 25% less and sold it for Re 10 less, then be would have gained 40%. The cost price of the pair of sunglasses is
   (a) Rs 25 (b) Rs 50
   (c) Rs 60 (d) Rs 70

6. If a man reduces the selling price of a fan from Rs 400 to Rs 380, his loss increases by 4%. What is the cost price of the fan in rupees.

   (a) 600 (b) 480
   (c) 500 (d) 450

7. Mukesh purchased 40 kg of wheat at Rs 12.50 per kg and 25 kg of wheat at Rs 15.10 per kg. He mixed the two qualities of wheat for selling. At what rate should it be sold to gain 10% ?
   (a) Rs 13.25 (b) Rs 13.50
   (c) Rs 14.75 (d) Rs 14.85

8. A sells a box to B at a profit of 15%, B sells the same to C for Rs 1012 and makes a profit of 10%. A's cost price is
   (a) Rs 720 (b) Rs 680
   (c) Rs 880 (d) Rs 800

9. A wholesale dealer sold his goods to a retail dealer at a profit of $12\frac{1}{2}$%. The retail dealer gained 20% by selling the goods for Rs 3240. The cost price of the whole sale dealer was
   (a) Rs 2625 (b) Rs 2575
   (c) Rs 2500 (d) Rs 2400

10. A bought a radio set and spent Rs 110 on its repairs. He then sold it to B at 20% profit, B sold it to C at a loss of 10% and C sold it for Rs 1188 at a profit of 10%. What is the amount for which A bought the radio set ?
    (a) Rs 850 (b) Rs 890
    (c) Rs 1000 (d) Rs 950

11. A dishonest dealer uses a scale of 90 cm instead of a metre scale and claims to sell at cost price. His profit is
(a) 9%
(b) 10%
(c) $10\frac{9}{11}\%$
(d) $11\frac{1}{9}\%$

12. A dishonest shopkeeper pretends to sell his goods at cost price but uses false weights and gains $11\frac{1}{9}\%$. For a weight of 1 kg he uses
(a) a weight of 875 gm
(b) a weight of 900 gm
(c) a weight of 950 gm
(d) a weight of 850 gm

13. A horse and a carriage together cost Rs 8000. If by selling the horse at a profit of 10% and the carriage at a loss of 10%, a total profit of 2.5% is made, then what is the cost price of the horse ?
(a) Rs 3000
(b) Rs 6000
(c) Rs 2000
(d) Rs 5000

14. A television and a washing machine were sold for Rs 12500 each. If the television was sold at a gain of 30% and the washing machine at a loss of 30%, then the entire transaction resulted in
(a) 9% gain
(b) 9% loss
(c) 11% gain
(d) Neither gain nor loss

15. Shridhar bought two buffaloes for Rs 30,000. By selling one at a loss of 15% and the other at a gain of 19%, he found that selling price of both buffaloes is the same. Find the cost price of each (in Rs).
(a) 10000; 20000
(b) 15000; 15000
(c) 17500; 12500
(d) 16000, 14000

16. On selling each of the two radios for Rs 5000, a person neither gained nor lost. If he had sold one radio at 25% gain, then at what per cent loss did she sell the radio ?
(a) $16\frac{2}{3}\%$
(b) $18\frac{2}{9}\%$
(c) 25%
(d) $26\frac{2}{3}\%$

17. A person sold his watch for Rs 144. If the percentage of his profit was equal to the cost price, then the watch would have cost him
(a) Rs 100
(b) Rs 90
(c) Rs 85
(d) Rs 80

18. A shopkeeper sold one-fourth of his goods at a loss of 10%. He sold the remaining at a higher per cent of profit to get $12\frac{1}{2}\%$ profit on the whole transaction. The higher profit per cent is
(a) $17\frac{1}{2}\%$
(b) $33\frac{1}{3}\%$

(c) $22\frac{1}{2}\%$
(d) 20%

19. A man buys a field of agricultural land for Rs 3,60,000. He sells one-third at a loss of 20% and $\frac{2}{5}$th at a gain of 25%. At what price must he sell the remaining field so as to make an overall profit of 10% ?
(a) Rs 1,00,000
(b) Rs 1,15,000
(c) Rs 1,20,000
(d) Rs 1,25,000

20. A person purchases 90 clocks and sells 40 clocks at a gain of 10% and 50 clocks at a gain of 20%. If he sold all of them at a uniform profit of 15%, then he would have got Rs 40 less. The cost price of each clock is :
(a) Rs 50
(b) Rs 60
(c) Rs 80
(d) Rs 90

21. Mani bought two horses at Rs 20,000 each. He sold one horse at 15% gain. But he had to sell the second horse at a loss. If he had suffered a loss of Rs 1800 on the whole transaction, find the selling price of the second horse.
(a) 20%
(b) 10%
(c) 24%
(d) 25%

22. Vineet calculates his profit percentage on the selling price while Roshan calculates his profit on the cost price. They find that the difference of their profits is Rs 275. If the selling price of both of them are the same, and Vineet gets 25% profit and Roshan gets 15% profit, then find their selling price.
(a) Rs 2100
(b) Rs 2300
(c) Rs 2350
(d) Rs 2250

23. If 7% of the sale price of an article is equivalent to 8% of its cost price and 9% of its sale price exceeds 10% of its cost price by Re 1, then what is the cost of the article?
(a) Rs 400
(b) Rs 350
(c) Rs 300
(d) Rs 280

24. A shopkeeper sells tea at 10% profit and uses a weight which is 20% less than the actual measure. His gain per cent is
(a) 30%
(b) 35%
(c) $37\frac{1}{2}\%$
(d) 32%

25. Some lollipops are bought at 11 for a rupee and the same number at 9 a rupee. If the whole lot is sold at 10 a rupee, find the gain or loss per cent.
(a) 2% gain
(b) 2% loss
(c) 1% gain
(d) 1% loss

26. Rajeshwar bought 16 dozen ball point pens and sold them by and by. Due to a calculation mistake in fixing selling price, he lost an amount equal to S.P. of 4 dozen pens. Find the loss per cent. Find the S.P. of one dozen pens, if he purchased these 16 dozen pens for Rs 240.

(a) Rs 18 (b) Rs 10

(c) Rs 12 (d) Rs 14

27. On selling a pen at 5% loss and a book at 15% gain, Karim gains Rs 7. If he sells the pen at 5% gain and the book at 10% gain, then he gains Rs 13. The actual price of the book is

(a) Rs 100 (b) Rs 80

(c) Rs 10 (d) Rs 400

28. A man purchased a scooter for Rs 6250 and sold it at 8% profit. He purchased another scooter for Rs 3750. After selling it, he found that he has gained 2% on the whole. Then in the sale of the second scooter, he has

(a) 8% loss (b) 8% gain

(c) 10% gain (d) 6% loss

29. Albert buys 4 horses and 9 cows for Rs 13400. If he sells the horses at 10% profit and the cows at 20% profit, then he earns a total profit of Rs 1880. The cost of a horse is

(a) Rs 1000 (b) Rs 2000

(c) Rs 2500 (d) Rs 3000

30. A cloth merchant sold half of his cloth at 20% profit, half of the remaining at 20% loss and the rest was sold at the cost price. In the whole transaction, his gain or loss will be

(a) Neither gain nor loss (b) 5% loss

(c) 5% gain (d) 10% gain

Answers

1. (c)	**2.** (c)	**3.** (a)	**4.** (a)	**5.** (b)	**6.** (c)	**7.** (d)	**8.** (d)	**9.** (d)	**10.** (b)
11. (d)	**12.** (b)	**13.** (d)	**14.** (b)	**15.** (c)	**16.** (a)	**17.** (d)	**18.** (d)	**19.** (c)	**20.** (c)
21. (c)	**22.** (b)	**23.** (b)	**24.** (c)	**25.** (d)	**26.** (c)	**27.** (a)	**28.** (a)	**29.** (b)	**30.** (c)

Hints and Solutions

1. (c) S.P. = Rs 178, Loss = 11%

$$\therefore C.P. = \frac{178 \times 100}{(100-11)} = \frac{178 \times 100}{89} = 200$$

Now C.P. = Rs 200, Profit = 11%

$$\therefore S.P. = \frac{200 \times 111}{100} = Rs\ 222.$$

2. (c) C.P. of 11 articles = Rs 10

C.P. of 1 article = Rs $\frac{10}{11}$

S.P. of 10 articles = Rs 11

$\Rightarrow$ S.P. of 1 article = Rs $\frac{11}{10}$

$$\therefore Profit\ \% = \frac{\frac{11}{10}-\frac{10}{11}}{\frac{10}{11}} \times 100 = \frac{\frac{121-100}{110}}{\frac{10}{11}} \times 100$$

$$= \frac{21 \times 11}{110 \times 10} \times 100 = 21\%.$$

3. (a) S.P. of 100 pencils – C.P. of 100 pencils = S.P. of 20 pencils

$\Rightarrow$ S.P. of 80 pencils = C.P. of 100 pencils

Let C.P. of 1 pencil = Re 1. Then,

S.P. of 80 pencils = Rs 100

C.P. of 80 pencils = Rs 80

$$\therefore Profit\ \% = \frac{100-80}{80} \times 100 = \frac{20}{80} \times 100 = 25\%$$

4. (a) Let the C.P. of the article be Rs x. Then,

$$S.P.\ at\ a\ loss\ of\ 10\% = \frac{x \times 90}{100} = Rs\ \frac{90x}{100}$$

$$S.P.\ at\ a\ gain\ of\ 12\frac{1}{2}\% = \frac{x \times 112.5}{100} = Rs\ \frac{112.5x}{100}$$

Given, $\frac{112.5x}{100} - \frac{90x}{100} = 9$

$$\Rightarrow \frac{22.5x}{100} = 9 \Rightarrow x = \frac{900}{22.5} = Rs\ 40.$$

5. (b) Let the C.P. of the pair of sunglasses be Rs x.

Then S.P. $= \frac{x \times 125}{100} = Rs\ \frac{5x}{4}$

New C.P. = Rs $\left(\dfrac{75}{100} \times x\right)$ = Rs $\dfrac{3x}{4}$

New S.P. = Rs $\left(\dfrac{5x}{4} - 10\right)$

Given, $\dfrac{5x}{4} - 10 = \dfrac{140}{100} \times \dfrac{3x}{4} \Rightarrow \dfrac{5x}{4} - \dfrac{21x}{20} = 10$

$\Rightarrow \dfrac{25x - 21x}{20} = 10 \Rightarrow 4x = 200 \Rightarrow x = $ Rs 50.

6. (c) Let the C.P. = Rs x. Then,

First loss % = $\dfrac{(x - 400)}{x} \times 100 = \dfrac{100x - 40000}{x}$%

Second loss% = $\dfrac{(x - 380)}{x} \times 100 = \dfrac{100x - 38000}{x}$%

Given, $\dfrac{(100x - 38000)}{x} - \dfrac{(100x - 40000)}{x} = 4$

$\Rightarrow 100x - 38000 - 100x + 40000 = 4x$

$\Rightarrow 2000 = 4x \Rightarrow x = 500.$

7. (d) C.P. of the wheat = $40 \times$ Rs $12.50 + 25 \times$ Rs 15.10
$\qquad\qquad\qquad = $ Rs 877.5

Gain = 10%

$\therefore$ S.P. of the wheat = Rs $\dfrac{877.5 \times 110}{100}$ = Rs 965.25

$\therefore$ S.P. per kg of wheat = $\dfrac{965.25}{65}$ = Rs 14.85

8. (d) Let A's C.P. = Rs x, Profit = 15%

$\therefore$ A's S.P. = $\dfrac{x \times 115}{100}$ = Rs $\dfrac{23x}{25}$

B's C.P. = Rs $\dfrac{23x}{20}$, Profit = 10%

$\therefore$ B's S.P. = $\dfrac{23x}{20} \times \dfrac{110}{100} = \dfrac{253x}{250}$

Given, $\dfrac{253x}{200} = 1012 \Rightarrow x = \dfrac{1012 \times 200}{253} = $ Rs 800.

9. (d) Let the C.P. of the wholesale dealer = Rs x. Then,

S.P. of the wholesale dealer = $\dfrac{x \times 112.5}{100}$

$\qquad\qquad\qquad\qquad = $ Rs $\dfrac{112.5x}{100}$

C.P. of the retail dealer = Rs $\dfrac{112.5x}{100}$, Gain = 20%

$\therefore$ S.P. of the retail dealer = $\dfrac{112.5x}{100} \times \dfrac{120}{100}$

Given, $\dfrac{112.5x}{100} \times \dfrac{12}{10} = 3240$

$\Rightarrow x = $ Rs $\dfrac{3240 \times 1000}{112.5 \times 12} = $ Rs 2400.

10. (b) Let A's C.P. = Re x. Then,

A's S.P. = B's C.P. = Rs $\left(\dfrac{120}{100} \times x\right)$

B's S.P. = C's C.P. = Rs $\left(\dfrac{90}{100} \times \dfrac{120}{100} \times x\right)$

C's S.P. = Rs $\left(\dfrac{110}{100} \times \dfrac{90}{100} \times \dfrac{120}{100} \times x\right)$

Given, $\dfrac{110}{100} \times \dfrac{90}{100} \times \dfrac{120}{100} \times x = 1188$

$\Rightarrow x = \dfrac{1188 \times 1000}{11 \times 9 \times 12} = 1000.$

Amount spent by A on repairs = Rs 110

$\therefore$ Amount for which A bought the radio set
$\qquad\qquad = $ Rs 1000 – Rs 110
$\qquad\qquad = $ Rs 890.

11. (d) Let the cost price of the cloth be Rs 100 per metre, *i.e.*, Re 1 per cm.

Since the shopkeeper uses 90 cm scale for 100 cm scale,

His gain = Cost of 10 cm = Rs 10

$\therefore$ Gain % = $\dfrac{10}{90} \times 100 = \dfrac{100}{9}$% = $11\dfrac{1}{9}$%.

12. (b) Let the error (difference between 1 kg and false weight) be x gm. Then,

$\dfrac{x}{(1000 - x)} \times 100 = \dfrac{100}{9} \Rightarrow \dfrac{x}{(1000 - x)} = \dfrac{1}{9}$

$\Rightarrow 9x = 1000 - x \Rightarrow 10x = 1000 \Rightarrow x = 100$

$\therefore$ Weight used = (1000 – 100) gm = 900 gm.

13. (d) Let C.P. of the horse = Rs x. Then,

C.P. of the carriage = Rs $(8000 - x)$

S.P. of the horse = $\dfrac{x \times 110}{100} = $ Rs $\dfrac{11x}{10}$

S.P. of the carriage = $\dfrac{(8000 - x) \times 90}{100} = 7200 - \dfrac{9x}{10}$

Total C.P. = Rs 8000, Total profit = 2.5%

$\therefore$ Total S.P. = $8000 \times \dfrac{102.5}{100}$ = Rs 8200

Given, $\dfrac{11x}{10} + 7200 - \dfrac{9x}{10} = 8200$

$\Rightarrow \dfrac{2x}{10} = 1000 \Rightarrow x$ = Rs 5000.

14. (b) S.P. of the T.V. and washing machine
$\qquad = 2 \times$ Rs 12500 = Rs 25000

C.P. of the T.V. = Rs $12500 \times \dfrac{100}{130}$

C.P. of the washing machine = Rs $12500 \times \dfrac{100}{70}$

$\therefore$ Total C.P. = Rs $12500 \left(\dfrac{100}{130} + \dfrac{100}{70} \right)$

$\qquad = $ Rs $12500 \times \dfrac{200}{91}$

$\therefore$ Reqd. loss % = $\dfrac{\dfrac{2500000}{91} - 25000}{\dfrac{2500000}{91}} \times 100$

$\qquad = \dfrac{2500000 - 2275000}{2500000} \times 100$

$\qquad = \dfrac{225000}{2500000} \times 100 = 9\%.$

15. (c) Let the cost price of one buffalo be Rs x. Then,
Cost price of the other buffalo = Rs $(30000 - x)$

S.P. of first buffalo = $\dfrac{(100-15)}{100} \times x$ = Rs $\dfrac{85x}{100}$

S.P. of second buffalo = $\dfrac{(100+19)}{100} \times (30000 - x)$

$\qquad = \dfrac{119}{100} \times (30000 - x)$

$\because$ Selling price of both buffalos is the same,

$\dfrac{85x}{100} = \dfrac{119(30000 - x)}{100}$

$\Rightarrow 5x = 7(30000 - x) \Rightarrow 12x = 210000$

$\therefore x = \dfrac{210000}{12}$ = Rs 17500

$\therefore$ Cost price of the other buffalo = Rs 12500.

16. (a) **1st radio.**
S.P. = Rs 5000, Gain = 25%

$\therefore$ C.P. = Rs $\dfrac{5000 \times 100}{125}$ = Rs 4000

2nd radio.
S.P. = Rs 5000, Loss = $x\%$ (say)

$\therefore$ C.P. = $\dfrac{5000 \times 100}{(100 - x)}$

Since the person neither gained nor lost,
Total C.P. = Total S.P.

$\Rightarrow 4000 + \dfrac{500000}{(100 - x)} = 10000$

$\Rightarrow \dfrac{500000}{(100 - x)} = 6000$

$\Rightarrow 500000 = 600000 - 6000x$

$\Rightarrow 6000x = 100000$

$\Rightarrow x = \dfrac{100}{6}\% = \dfrac{50}{3}\% = 16\dfrac{2}{3}\%$

17. (d) Let C.P. = Rs x, S.P. = Rs 144, Profit = $x\%$

$\therefore x + x\%$ of $x = 144$

$\Rightarrow x + \dfrac{x}{100} \times x = 144$

$\Rightarrow x^2 + 100x - 14400 = 0$

$\Rightarrow x^2 + 180x - 80x - 14400 = 0$

$\Rightarrow x(x + 180) - 80(x + 180) = 0$

$\Rightarrow (x + 180)(x - 80) = 0$

$\Rightarrow x = -180$ or 80

Neglecting negative values, $x = 80$.

18. (d) Let C.P. = Rs c

$\therefore$ C.P. of $\dfrac{1}{4}$th of the goods = Rs $\dfrac{c}{4}$
Loss = 10%

$\therefore$ S.P. of $\dfrac{1}{4}$th of the goods = $\dfrac{\dfrac{c}{4} \times 90}{100}$ = Rs $\dfrac{9c}{40}$

C.P of $\dfrac{3}{4}$th of the goods = Rs $\dfrac{3c}{4}$

Let profit on this remaining part = $P\%$. Then,

S.P. of $\dfrac{3}{4}$th of the goods = $\dfrac{\dfrac{3c}{4} \times (100 + P)}{100}$

$\qquad = \dfrac{3c}{400}(100 + P)$

Profit on the whole transaction = 12.5%

$\therefore$ S.P. of the whole = Rs $\dfrac{c \times 112.5}{100}$

$$\therefore \quad \frac{9c}{40} + \frac{3c}{4} + \frac{3c \times P}{400} = \frac{112.5c}{100}$$

$$\Rightarrow \quad \frac{90 + 300 + 3p}{400} = \frac{112.5}{100}$$

$$\Rightarrow \quad \frac{390 + 3P}{4} = 112.5 \quad \Rightarrow \quad 390 + 3P = 450$$

$$\Rightarrow 3P = 60 \quad \Rightarrow \quad P = 20.$$

19. (c) S.P. of the whole field $= \dfrac{360000 \times 110}{100} = $ Rs 396000

S.P. of the $\dfrac{1}{3}$rd part of the field $= \dfrac{360000}{3} \times \dfrac{80}{100}$
$= $ Rs 96000

S.P. of the $\dfrac{2}{5}$th part of the field

$= 360000 \times \dfrac{2}{5} \times \dfrac{125}{100} = $ Rs 180000

S.P. of the remaining field
$= 396000 - (96000 + 180000)$
$= $ Rs 120000.

20. (c) Let the C.P. of each clock be Rs x. Then,
C.P. of 90 clocks $= $ Rs $90x$

$$\therefore \quad \left(\frac{110}{100} \times 40x\right) + \left(\frac{120}{100} \times 50x\right) - \left(\frac{115}{100} \times 90x\right) = 40$$

$$\Rightarrow 44x + 60x - 103.5x = 40$$

$$\Rightarrow 0.5x = 40 \quad \Rightarrow \quad x = 80$$

21. (c) **1st horse.**
C.P. $= $ Rs 20000, Gain $= 15\%$

$$\therefore \quad \text{S.P.} = 20000 \times \frac{115}{100} = \text{Rs } 23000$$

2nd horse.
C.P. $= $ Rs 20000, Loss $= x\%$

$$\therefore \quad \text{S.P.} = 20000 \times \frac{(100 - x)}{100} = \text{Rs } (20000 - 200x)$$

Total loss $= $ Total C.P. – Total S.P.
$= 40000 - (23000 + 20000 - 200x)$
$1800 = -3000 + 200x$

$$\Rightarrow \quad 200x = 4800 \quad \Rightarrow \quad x = 24\%$$

22. (b) Let the selling price of Vineet and Roshan be Rs x. Then,

Cost price of Vineet $= \dfrac{125x}{100}$

Cost price of Roshan $= \dfrac{100x}{115}$

According to the question,

$$\left(\frac{125x}{100} - x\right) - \left(x - \frac{100x}{115}\right) = 275$$

$$\Rightarrow \quad \frac{25x}{100} - \frac{15x}{115} = 275 = 275$$

$$\Rightarrow \quad \frac{575x - 300x}{2300} = 275$$

$$\Rightarrow \quad \frac{275x}{2300} = 275 \quad \Rightarrow \quad x = 2300.$$

23. (b) Suppose C.P. $= $ Rs x and S.P. $= $ Rs y

$\Rightarrow$ 7% of $y = 8\%$ of x, *i.e.*, $7y = 8x$ …(i)

and 9% of $y = 10\%$ of $x + 1$, *i.e.*,
$$9y = 10x + 1 \qquad \text{… (ii)}$$

From (i) we have $y = \dfrac{8x}{7}$. Putting the value of y in (2), we get

$$9 \times \frac{8x}{7} = 10x + 100 \quad \Rightarrow \quad 72x = 70x + 700$$

$$\Rightarrow 2x = 700 \quad \Rightarrow \quad x = 350$$

24. (c) Let the marked weight be 1 kg

But the real weight he uses $= 80\%$ of 1 kg
$$= 800 \text{ gm}$$

Let the C.P. of 1 gm be Rs 1. Then,
C.P. of 800 gm $= $ Rs 800 and
C.P. of 1000 gm $= $ Rs 1000

$$\therefore \quad \text{S.P.} = 1000 \times \frac{110}{100} = \text{Rs } 1100$$

Gain $= $ Rs 1100 – Rs 800 $= $ Rs 300

$$\text{Gain\%} = \frac{300}{800} \times 100 = \frac{300}{8}\% = 37\frac{1}{2}\%.$$

25. (d) LCM of 9, 10 and 11 $= 990$

So, let us assume that 990 lollipops of each kind are bought.

Now, C.P. of 990 lollipops of first kind

$$= \text{Rs}\left(\frac{1}{11} \times 990\right) = \text{Rs } 90$$

C.P. of 990 lollipops of second kind

$$= Rs\left(\frac{1}{9} \times 990\right) = Rs\ 110$$

S.P. of 1980 lollipops $= Rs\left(\frac{1}{10} \times 1980\right) = Rs\ 198$

Total C.P. = Rs 90 + Rs 110 = Rs 200

$\because$ C.P. > S.P. , Loss = Rs 200 – Rs 198 = Rs 2

$\therefore$ Loss% $= \dfrac{\text{Loss}}{\text{C.P.}} \times 100 = \dfrac{2}{200} \times 100 = 1\%$.

26. (c) Let the S.P. of one dozen pens be Re x. Then,

S.P of 16 dozen pens = Rs $16x$

Loss = S.P. of 4 dozen pens = Rs $4x$

$\therefore$ C.P. = S.P. + Loss = Rs $16x$ + Rs $4x$ = Rs $20x$

$\therefore$ Loss per cent $= \left(\dfrac{4x}{20x} \times 100\right) = 20\%$

C.P. of 16 dozen pens = Rs 240, Loss = 20%

$\therefore$ S.P. of 16 dozen pens $= Rs\ \dfrac{240 \times 80}{100} = Rs\ 192$

$\therefore$ S.P. of 1 dozen pens $= Rs\ \dfrac{192}{16} = Rs\ 12$.

27. (a) Let the C.P. of the pen be Rs x and C.P. of the book be Rs y.

Then, S.P. of pen at 5% loss $= x \times \dfrac{95}{100} = Rs\ \dfrac{95x}{100}$

S.P. of book at 15% gain $= y \times \dfrac{115}{100} = Rs\ \dfrac{115y}{100}$

Given, $\left(\dfrac{95x}{100} - x\right) + \left(\dfrac{115y}{100} - y\right) = 7$

$\Rightarrow -5x + 15y = 700$

$\Rightarrow -x + 3y = 140$ or $x - 3y = -140$ ... (i)

Again, S.P. of pen at 5% gain $= x \times \dfrac{105}{100} = Rs\ \dfrac{105x}{100}$

S.P. of book at 10% gain $= y \times \dfrac{110}{100} = Rs\ \dfrac{110y}{100}$

Given, $\left(\dfrac{105x}{100} - x\right) + \left(\dfrac{110y}{100} - y\right) = 13$

$\Rightarrow 5x + 10y = 1300 \Rightarrow x + 2y = 260$ ... (ii)

Subtracting eqn. (i) from eqn. (ii) we get

$(x + 2y) - (x - 3y) = 260 + 140$

$\Rightarrow 5y = 500 \Rightarrow y = 100$.

28. (a) Profit on the 1st scooter $= \dfrac{8}{100} \times Rs\ 6250$

$= Rs\ 500$

Profit on both the scooters $= \dfrac{2}{100} \times Rs\ 10000$

$= Rs\ 200$

$\therefore$ The man has a loss of Rs 300 on the second scooter.

$\therefore$ Loss % $= \dfrac{300}{3750} \times 100 = 8\%$

29. (b) Let the C.P. of 1 horse = Rs x and C.P. of 1 cow = Rs y. Then,

$4x + 9y = 13400$ ... (i)

Also, 10% of $4x$ + 20% of $9y$ = 1880

$\dfrac{2}{5}x + \dfrac{9}{5}y = 1880 \Rightarrow 2x + 9y = 9400$ ...(ii)

Subtracting eqn. (ii) from eqn. (i) we get

$(4x + 9y) - (2x + 9y) = 13400 - 9400$

$\Rightarrow 2x = 4000 \Rightarrow x = 2000$

30. (c) Let the C.P. of the whole stock = Rs x. Then,

C.P. of $\dfrac{1}{2}$ stock = Rs $\dfrac{x}{2}$, C.P. of $\dfrac{1}{2}$ of remaining stock = Rs $\dfrac{x}{4}$

$\therefore$ Total S.P. $= Rs\left[\left(\dfrac{120 \times x/2}{100}\right) + \left(\dfrac{80 \times x/4}{100}\right) + \dfrac{x}{4}\right]$

$= Rs\left[\dfrac{3x}{5} + \dfrac{x}{5} + \dfrac{x}{4}\right]$

$= Rs\left(\dfrac{12x + 4x + 5x}{20}\right) = Rs\ \dfrac{21x}{20}$

Gain $= Rs\left(\dfrac{21x}{20} - x\right) = Rs\ \dfrac{x}{20}$

$\therefore$ Gain% $= \left(\dfrac{x}{20} \times \dfrac{1}{x} \times 100\right)\% = 5\%$.

Self Assessment Sheet–16(b)

1. An article is sold for Rs 500 and hence a merchant loses some amount. Had the article been sold for Rs 700, the merchant would have gained three times the former loss. The cost price of the article is:
   (a) Rs 525   (b) Rs 550
   (c) Rs 600   (d) Rs 650

2. A man purchased a table and a chair for Rs 2000. He sold the table at a profit of 20% and the chair at a profit of 30%. His total profit was 23%. Find the cost price of the table.
   (a) Rs 1400   (b) Rs 600
   (c) Rs 1100   (d) Rs 1200

3. A retailer bought some apples rate of 7 apples for Rs 4 and sold them rate of 8 apples for Rs 5. If he gains Rs 30 on that day, the quantity of apples sold by him on that day is:
   (a) 555   (b) 560
   (c) 565   (d) 570

4. A merchant buys 1260 kg of corn, $\frac{1}{4}$ of which he sells at a gain of 5%, $\frac{1}{3}$ at a gain of 8% and the remainder at a gain of 12%. If he had sold the whole at a gain of 10%, he would have gained Rs 27.30 more. Find the cost price per kg.
   (a) Rs 5   (b) Rs 2
   (c) Rs 3   (d) Rs 2.50

5. An article is sold at a profit of 20%. If both the cost price and selling price were to be Rs 20 less, the profit would be 10% more. Find the cost price of the article.
   (a) Rs 120   (b) Rs 80
   (c) Rs 60   (d) Rs 75

6. A shopkeeper sells 100 kg of sugar partly at 10% profit and the remaining at 20% profit. If he gains 12% on the whole transaction, how much sugar did he sell at 20% profit?
   (a) 25 kg   (b) 40 kg
   (c) 20 kg   (d) 30 kg

7. A manufacturer sells his goods to a wholesaler at 10% gain; the whole saler to the retailer at 20% gain and the retailer to the customer at 30% gain. Find what per cent the customer has to pay more on the manufactured price?
   (a) 60%   (b) $66\frac{2}{3}\%$
   (c) $48\frac{1}{5}\%$ (d) $71\frac{3}{5}\%$

8. A woman buys toffees at Rs 2.50 a dozen and an equal number at Rs 3 a score. She sells them at Rs 3.60 a score and thus makes a profit of Rs 10. How many toffees did she buy?
   (a) 10000   (b) 12000
   (c) 5000   (d) 6000

9. Three tables are purchased for Rs 2500 each. First is sold at a profit of 8%, the second is sold at a loss of 3%. If their average selling price is Rs 2575, find the profit per cent on the third.
   (a) 6%   (b) 5%
   (c) 4%   (d) 8%

10. A man sells a TV at a profit of 25% of the cost. Had he sold it at a profit of 25% of the selling price, his profit would have increased by 5% of the cost price plus Rs 100. Find the cost of the TV?
    (a) Rs 6000   (b) Rs 8000
    (c) Rs 10,000   (d) Rs 7500

Answers

1. (b)	2. (a)	3. (b)	4. (b)	5. (c)	6. (c)	7. (d)	8. (b)	9. (c)	10. (b)

Section-C

DISCOUNT

KEY FACTS

1. **Discount** is the per cent of rebate offered on the **marked price** (printed or list price) of goods.

$$\text{Discount} = \frac{\text{Discount rate}}{100} \times \text{M.P.}$$

2. The customer or the buyer pays the difference between the marked price and the discount. Thus,
S.P. = M.P. – Discount

3. M.P. $= \dfrac{\text{S.P.} \times 100}{(100 - \text{Discount\%})}$

4. **Successive Discounts:** When two or more discounts are allowed one after the other, then such discounts are known as successive discounts. In successive discounts, one discount is subtracted from the marked price to get net price after the 1st discount. This net price becomes the marked price and the second discount is calculated on it and subtracted from it to get the net price after second discount and so on.

Solved Examples

Ex. 1. *A shopkeeper offers 10% discount on the marked price of his articles and still makes a profit of 20%. What is the actual cost of the article marked Rs 500 for him ?*

Sol. M.P. = Rs 500, Discount = 10%

∴ S.P. = Rs 500 – 10% of Rs 500 = 90% of Rs 500 $= \dfrac{90}{100} \times$ Rs 500 = Rs 450

Profit = 20%

C.P. = Rs $\left(\dfrac{450 \times 100}{120}\right)$ = Rs 375.

Ex. 2. *The marked price of a watch is Rs 1600. After two successive discounts it is sold for Rs 1224. If the rate of first discount is 10%. What is the rate of second discount ?*

Sol. M.P. = Rs 1600, First discount = 10%

∴ Net price = Rs 1600 – 10% of Rs 1600 = Rs 1600 – Rs 160 = Rs 1440

S.P. = Rs 1224

∴ Second discount = Rs 1440 – Rs 1224 = Rs 216

∴ Rate of second discount $= \dfrac{216}{1440} \times 100 = 15\%$

Ex. 3. *On an article with marked price Rs 20000, a customer has a choice between the successive discounts of 20%, 20% and 10% and three successive discounts of 40%, 5% and 5%. How much can he save by choosing the better offer ?*

Sol. M.P. = Rs 20000

S.P. after choosing 1st set of successive discounts = 80% of 80% of 90% of Rs 20000

$$= \dfrac{80}{100} \times \dfrac{80}{100} \times \dfrac{90}{100} \times \text{Rs } 20000 = \text{Rs } 11520$$

S.P. after choosing 2nd set of successive discounts = 60 % of 95% of 95% of Rs 20000

$$= \dfrac{60}{100} \times \dfrac{95}{100} \times \dfrac{95}{100} \times \text{Rs } 20000 = \text{Rs } 10830$$

∴ The second offer is better and the customer can save (Rs 11520 – Rs 10830) = Rs 690.

Ex. 4. *A discount series of p% and q% on an invoice is the same as a single discount of*

(a) $\left(p + q + \dfrac{pq}{100}\right)\%$ (b) $\left(p - q + \dfrac{pq}{100}\right)\%$ (c) $\left\{100 - \left(p + q + \dfrac{pq}{100}\right)\right\}\%$ (d) $\dfrac{pq}{100}\%$

Sol. Let the M.P. of the invoice = Rs 100. Then,

S.P. $= (100 - p)\%$ of $(100 - q)\%$ of Rs 100

$$= Rs \left\{ \frac{(100-p)}{100} \times \frac{(100-q)}{100} \times 100 \right\} = Rs \frac{(100-p)(100-q)}{100}$$

$$\therefore \text{ Single discount } = \left\{ 100 - \frac{(100-p)(100-q)}{100} \right\}\% = \left[\frac{10000 - \{10000 - 100q - 100p + pq\}}{100} \right]\%$$

$$= \frac{100q + 100p - pq}{100} = \left\{ (p+q) - \frac{pq}{100} \right\}\%.$$

Ex. 5. *A merchant fixes the sale price of his goods at 15% above the cost price. He sells his goods at 12% less than the fixed price. What is his percentage of profit ?*

Sol. Let the C.P. be Rs 100

$\therefore$ Marked Price = Rs 115

S.P. = Rs 115 – 12% of Rs 115 = 88% of Rs 115 = Rs 101.20

$\therefore$ Profit = Rs 1.20

Profit % = 1.2%.

Ex. 6. *How much per cent more than the cost price should a shopkeeper marks his goods so that after allowing a discount of 20% on the marked price, he gains 10% ?*

Sol. Let the C.P. = Rs 100, Gain = 10%

$\therefore$ S.P. = Rs 110   ... (i)

Let the M.P. = Rs x, Discount = 20%

$\therefore$ S.P. = 80% of $x = \dfrac{80x}{100}$ ... (ii)

From (i) and (ii) $\dfrac{80x}{100} = 110 \Rightarrow x = Rs \dfrac{110 \times 100}{80} = Rs \ 137.5$

$\therefore$ Marked price = $37\dfrac{1}{2}\%$ above C.P.

Ex. 7. *A shopkeeper sold a TV set for Rs 17940 with a discount of 8% and earned a profit of 19.6%. What would have been the percentage of profit earned if no discount was offered ?*

Sol. Let the marked price of the TV set be Rs x. Discount = 8%

$\therefore$ S.P. of the TV = 92 % of Rs x = Rs $\dfrac{92x}{100}$

Given, $\dfrac{92x}{100} = 17940 \Rightarrow x = Rs \dfrac{17940 \times 100}{92} = Rs \ 19500$

S.P. = Rs 17940, Profit = 19.6%

$\therefore$ C.P. = Rs $\left(\dfrac{17940 \times 100}{119.6} \right) = Rs \ 15000$

Had no discount been offered S.P. would have been Rs 19500.

$\therefore$ Profit = Rs 19500 – Rs 15000 = Rs 4500

Profit % = $\dfrac{4500}{15000} \times 100 = 30\%.$

Ex. 8. *A shopkeeper fixes the marked price of an item 35% above its cost price. What is the percentage of discount allowed to gain 8% ?*

Sol. Let the C.P. = Rs 100. Then,

M.P. = Rs 135 and S.P. = $\dfrac{108 \times 100}{100} = Rs \ 108$

$\therefore$ Discount = Rs 135 – Rs 108 = Rs 27

Discount% = $\dfrac{27}{135} \times 100 = 20\%$.

Ex. 9. *A tradesman gives 4% discount on his marked price and gives 1 article free for buying every 15 articles and thus gains 35%. By what per cent is the marked price increased above the cost price ?*

Sol. Let the C.P. of each article be Rs x.

Then, C.P. of 16 articles = Rs $16x$

S.P. of 15 articles = 135% of Rs $16x$ = Rs $\left(\dfrac{135 \times 6x}{100}\right)$ = Rs $\dfrac{108x}{5}$

S.P. of 1 article = Rs $\left(\dfrac{108x}{5} \times \dfrac{1}{15}\right)$ = Rs $\dfrac{36x}{25}$

Let M.P. = Rs 100,

Then S.P. = Rs 96 after a discount of 4%

$\therefore$ If S.P. = Rs 96, then M.P. = Rs 100

If S.P. = Rs $\dfrac{36x}{25}$, then M.P. = Rs $\left(\dfrac{100}{96} \times \dfrac{36x}{25}\right)$ = Rs $\dfrac{3x}{2}$

$\therefore$ % increase in M.P. over C.P. = $\dfrac{\dfrac{3x}{2} - x}{x} \times 100 = \left(\dfrac{x}{2} \times \dfrac{1}{x} \times 100\right)\%$ = 50%.

Ex. 10. *A manufacturer marks his goods at 40% above the cost price. He allows a discount of 10% for cash customers and 5% to credit customers. $\dfrac{3}{5}$ of the goods are sold for cash and the rest on credit. What is the percentage of profit when all the goods are sold and amount realised ?*

Sol. Let the C.P. of the goods be Rs 100.

$\therefore$ Marked price = Rs 140

Value of goods sold for cash = $\dfrac{3}{5} \times$ Rs 140 = Rs 84

Value of goods sold on credit = $\dfrac{2}{5} \times$ Rs 140 = Rs 56

S.P. of the goods sold on cash = 90% of Rs 84 = Rs 75.60

S.P. of the goods sold on credit = 95% of Rs 56 = Rs 53.20

$\therefore$ Total S.P. = Rs 128.80

$\Rightarrow$ Profit = 28.80 , *i.e.*, 28.8%.

Question Bank–16(c)

1. A man buys an article for Rs 80 and marks it at Rs 120. He then allows a discount of 40%. What is the loss or gain% ?
   - (a) 12% gain
   - (b) 12% loss
   - (c) 10% gain
   - (d) 10% loss

2. Ramesh bought a calculator with 20% discount on the tag-price. He obtained 10% profit by selling it for Rs 440. What was the tag-price ?
   - (a) Rs 500
   - (b) Rs 400
   - (c) Rs 480
   - (d) Rs 360

3. A dealer allows 25% discount on the marked price of articles and earns a profit of 20% on them. What is the marked price of the article on which he gains Rs 800?
   - (a) Rs 6000
   - (b) Rs 6400
   - (c) Rs 7200
   - (d) Rs 7000

4. Shekhar has purchased a cordless phone for Rs 3520 after getting 12% discount on the printed price. If he sold it to get 8% profit on the printed price, at what price did he sell the cordless phone ?
   - (a) Rs 3801.60
   - (b) Rs 4224
   - (c) Rs 4320
   - (d) Rs 3942.40

5. An article listed at Rs 800 is sold at successive discounts of 25% and 15%. The buyer desires to sell it off at a profit of 20% after allowing a 10% discount. What would be his list price ?

   (a) Rs 620   (b) Rs 600
   (c) Rs 640   (d) Rs 680

6. By selling an umbrella for Rs 300, a shopkeeper gains 20%. During a clearance sale, the shopkeeper allows a discount of 10% on the marked price. Find his gain per cent during the sale season.

   (a) 10%   (b) 8%
   (c) 12%   (d) 9%

7. What is more favourable for a buyer — A discount series of 20%, 15% and 10% or a discount series of 25%, 12% and 8% ?

   (a) First   (b) Second
   (c) Both first and second   (d) None

8. A dealer marks his goods 25% above the cost price and allows 10% discount to his customers. What is his gain per cent ?

   (a) 12.5   (b) 35
   (c) 15   (d) 17.5

9. By selling an article at 80% of the marked price, there is a loss of 10%. If the article is sold at the marked price, the profit per cent will be

   (a) 18.4   (b) 20
   (c) 12.5   (d) 15

10. The marked price of an electric iron is Rs 300. The shopkeeper allows a discount of 12% and still gains 10%. If no discount is allowed his gain percentage would have been

    (a) 20   (b) 25
    (c) 27   (d) 30

11. A sells a scooter priced Rs 36000. He gives a discount of 8% on the first Rs 20000 and 5% on the next Rs 10000. How much discount can he afford on the remaining Rs 6000, if he is to get as much as when 7% discount is allowed on the total ?

    (a) 5%   (b) 6%
    (c) 7%   (d) 8%

12. If 10% discount is allowed on the marked price of an article, the profit of a dealer is 20%. If he allows a discount of 20% his profit will be

    (a) $4\frac{1}{3}\%$ (b) 5%
    (c) $6\frac{2}{3}\%$ (d) 8%

13. A fan is listed at Rs 1500 and a discount of 20% is offered on the list price. What additional discount must be offered to the customer to bring the net price to Rs 1104.

    (a) 8%   (b) 10%
    (c) 12%   (d) 15%

14. At what per cent above the cost price must a shopkeeper mark his goods so that he gains 20% even after giving a discount of 10% on the marked price.

    (a) 25%   (b) 30%
    (c) $33\frac{1}{3}\%$ (d) $37\frac{1}{2}\%$

15. A shopkeeper sells a badminton racket whose marked price is Rs 30 at a discount of 15% and gives a shuttle cock costing Rs 1.50 free with each racket. Even then he makes a profit of 20%. His cost price per racket is

    (a) Rs 21   (b) Rs 21.25
    (c) Rs 20   (d) Rs 19.75

16. The price of an article is raised by 30% and then two successive discounts of 10% each are allowed. Ultimately the price of the article is

    (a) Increased by 10%   (b) Increased by 5.3%
    (c) Decreased by 3%   (d) Decreased by 5.3%

17. A dealer buys an article listed at Rs 100 and gets successive discounts of 10% and 20%. He spends 10% of the cost price on transport. At what price should he sell the article to earn a profit of 15% ?

    (a) Rs 90   (b) Rs 90.02
    (c) Rs 91.08   (d) Rs 91.28

18. A shopkeeper claims to sell his articles at a discount of 10%, but marks his articles by increasing the cost of each by 20%. His gain per cent is

    (a) 6%   (b) 8%
    (c) 10%   (d) 12%

19. The marked price of a shirt and trousers are in the ratio 1:2. The shopkeeper gives 40% discount on the shirt. If the total discount on both is 30%, the discount offered on trousers is

    (a) 15%   (b) 20%
    (c) 25%   (d) 30%

20. On reducing the marked price of an article by 32 a shopkeeper gains 15%. If the cost price of the article be Rs 320 and it is sold at the marked price.What will be the gain per cent ?

    (a) 10%   (b) 20%
    (c) 25%   (d) 12%

21. A shopkeeper earns a profit of 15% after selling a book at 20% discount on the printed price. The ratio of the cost price and the printed price of the book is
(a) 16 : 23   (b) 20 : 23
(c) 23 : 16   (d) 23 : 20

22. A shopkeeper marks his goods at 20% above the cost price. He sells three - fourths of his goods at the marked price. He sells the remaining goods at 50% of the marked price. Determine his profit per cent on the whole transaction.
(a) 10%   (b) 8%
(c) 5%   (d) 7%

23. Two shopkeepers announce the same price of Rs 700 for a shirt. The first offers successive discounts of 30% and 6% while the second offers successive discounts of 20% and 16%. The shopkeeper who offers better discounts charges — less than the other shopkeeper.
(a) 22.40   (b) 16.80
(c) 9.80   (d) 36.40

24. A trader marked the selling price of an article at 10% above the cost price. At the time of selling, he allows certain discount and suffers a loss of 1%. He allowed a discount of
(a) 9%   (b) 10%
(c) 10.5%   (d) 11%

25. Peter bought an item at 20% discount on its original price. He sold it with 40% increase on the price he bought it. The new sale price is by what per cent more than the original price ?
(a) 7.5   (b) 8
(c) 10   (d) 12

26. Kunal bought a suitcase with 15% discount on the labelled price. He sold the suitcase for Rs 2880 with 20% profit on the labelled price. At what price did he buy the suitcase ?
(a) Rs 2040   (b) Rs 2400
(c) Rs 2604   (d) Rs 2640

27. A shopkeeper sold an article offering a discount of 5% and earned a profit of 23.5%. What would have been the percentage of profit earned if no discount was offered ?
(a) 24.5   (b) 25
(c) 28.5   (d) 30

28. A shopkeeper sold an article for Rs 6750 after giving a discount of 10% on the labelled price. He would have earned a profit of 50%, had there been no discount. What was the actual percentage of profit earned ?
(a) 36   (b) 40
(c) 44   (d) 35

29. A retailer gets a discount of 40% on the printed price of an article. The retailer sells it at the printed price. What is the gain or loss per cent ?
(a) 12% gain   (b) 12% loss
(c) 10% gain   (d) 10% loss

30. A trader marks his goods at 20% above the cost price. He sold half the stock at the marked price, one quarter at a discount of 20% on the marked price and the rest at a discount of 40% on the marked price. His total gain is
(a) 2%   (b) 4.5%
(c) 13.5%   (d) 15%

Answers

1. (d)	2. (a)	3. (b)	4. (c)	5. (d)	6. (b)	7. (b)	8. (a)	9. (c)	10. (b)
11. (c)	12. (c)	13. (a)	14. (c)	15. (c)	16. (d)	17. (c)	18. (b)	19. (c)	20. (c)
21. (a)	22. (c)	23. (c)	24. (b)	25. (d)	26. (a)	27. (d)	28. (d)	29. (b)	30. (a)

Hints and Solutions

1. (d) C.P. = Rs 80, M.P. = Rs 120, Discount = 40%

∴ S.P. = 60% of Rs 120 = $\frac{60}{100}$ × Rs 120 = Rs 72

∴ Loss = Rs 80 – Rs 72 = Rs 8

Loss % = $\frac{8}{80}$ × 100 = 10%.

2. (a) Let the tag price of the calculator Rs x. Then,
C.P. of Ramesh after 20% discount = 80% of

Rs x = Rs $\frac{80x}{100}$ = Rs $\frac{4x}{5}$ ... (i)

Also, given S.P. = Rs 440 and Profit = 10%

∴ C.P. = Rs $\left(\frac{440 \times 100}{110}\right)$ = Rs 400   ... (ii)

From (i) and (ii)

∴ $\frac{4x}{5}$ = 400 ⇒ x = Rs 500

3. (b) Let the M.P. = Rs 100, Discount = 25%

$\therefore$ S.P. = Rs 75, Profit = 20%

$\Rightarrow$ C.P. = Rs $\dfrac{75 \times 100}{120}$ = Rs 62.50

$\therefore$ Profit = Rs 75 – Rs 62.50 = Rs 12.50

If the gain is Rs 12.50, M.P. = Rs 100

If the gain is Rs 800, M.P. = Rs $\dfrac{100}{12.50} \times 800$

= Rs 6400

4. (c) Let the printed price of the cordless phone be Rs x. Then,

$x - 12\%$ of $x = 3520$ $\Rightarrow$ 88% of $x = 3520$

$\Rightarrow$ $x = \dfrac{3520 \times 100}{88}$ = Rs 4000, Profit = 8%

S.P. = Rs 4000 + 8% of Rs 4000

= Rs 4000 + Rs 320 = Rs 4320.

5. (d) M.P = Rs 800

$\therefore$ C.P. of the buyer = 75% of 85% of Rs 800

= $\dfrac{75}{100} \times \dfrac{85}{100} \times$ Rs 800 = Rs 510

Profit = 20%

$\therefore$ S.P. of the buyer = Rs $\left(\dfrac{510 \times 120}{100}\right)$ = Rs 612

Discount = 10%

$\therefore$ List price of the buyer = Rs $\left(\dfrac{612 \times 100}{90}\right)$

= Rs 680.

6. (b) C.P. of the umbrella = Rs $\left(\dfrac{300 \times 100}{120}\right)$ = Rs 250

M.P. of the umbrella = Rs 300, Discount = 10%

$\therefore$ S.P. of the umbrella during sale = 90% of Rs 300

= Rs 270

$\therefore$ Gain % during sale season

= $\dfrac{\text{Rs } 270 - \text{Rs } 250}{\text{Rs } 250} \times 100$

= $\dfrac{20}{250} \times 100 = 8.$

7. (b) Let the marked price = Rs 100

S.P. for the 1st discount series

= $\dfrac{80}{100} \times \dfrac{85}{100} \times \dfrac{90}{100} \times 100$ = Rs 61.20

S.P. for the 2nd discount series

= $\dfrac{75}{100} \times \dfrac{88}{100} \times \dfrac{92}{100} \times 100$ = Rs 60.72

$\therefore$ The second discount series is more favourable.

8. (a) Let the C.P. of the goods be Rs 100. Then,

M.P. of the goods = Rs 125, Discount = 10%

$\therefore$ S.P. of the goods = 90% of Rs 125

= $\dfrac{90}{100} \times$ Rs 125 = Rs 112.5

$\therefore$ Gain% = $\dfrac{(112.5 - 100)}{100} \times 100 = 12.5\%.$

9. (c) Let M.P. = Rs 100, S.P. = 80% of M.P. = Rs 80

Loss = 10% $\Rightarrow$ C.P. = Rs $\dfrac{(80 \times 100)}{90}$ = Rs $\dfrac{800}{9}$

Had S.P. been equal to the M.P., i.e., S.P. = Rs 100, then

Profit% = $\dfrac{\left(100 - \dfrac{800}{9}\right)}{\dfrac{800}{9}} \times 100 = \dfrac{\dfrac{100}{9}}{\dfrac{800}{9}} \times 100$

= $\dfrac{10000}{800} = 12.5\%$

10. (b) M.P. = Rs 300, Discount = 12%

$\therefore$ S.P. = Rs 300 – 12% of Rs 300 = Rs 300 – Rs 36

= Rs 264

Gain = 10%

$\therefore$ C.P. = Rs $\left(\dfrac{264 \times 100}{110}\right)$ = Rs 240

Had there been no discount, S.P. would have been Rs 300

$\therefore$ Profit% = $\dfrac{(300 - 240)}{240} \times 100 = \dfrac{60}{240} \times 100 = 25\%$

11. (c) Discount on Rs 36000 at 7% = $\dfrac{7}{100} \times$ Rs 36000

= Rs 2520

Discount on Rs 20000 at 8% = $\dfrac{8}{100} \times$ Rs 20000

= Rs 1600

Discount on Rs 10000 at 5% = $\dfrac{5}{100} \times$ Rs 10000

= Rs 500

$\therefore$ Discount on remaining Rs 6000

= Rs 2520 – Rs (1600 + 500)

= Rs 2520 – Rs 2100 = Rs 420

$\therefore$ Discount % = $\dfrac{420}{6000} \times 100 = 7\%.$

12. (c) Let the M.P. of the article = Rs 100
Discount = 10%
∴ S.P. = 90% of Rs 100 = Rs 90, Profit = 20%

∴ C.P. = Rs $\dfrac{90 \times 100}{120}$ = Rs 75

If the discount is 20%, then S.P. = 80% of Rs 100
= Rs 80

∴ Required profit % = $\dfrac{(80-75)}{75} \times 100$

$= \dfrac{5}{75} \times 100 = 6\dfrac{2}{3}\%$

13. (a) M.P. = Rs 1500, Discount = 20%
∴ S.P. = 80% of Rs 1500 = Rs 1200
Final S.P. = Rs 1104
∴ Additional discount = Rs 1200 – Rs 1104 = Rs 96

∴ Additional discount rate = $\dfrac{96}{1200} \times 100$ = 8%

14. (c) Let the M.P. be Rs x. Discount = 10%

∴ S.P. = 90% of Rs x = Rs $\dfrac{9x}{10}$, Profit = 20%

C.P. = $\dfrac{\dfrac{9x}{10} \times 100}{120} = \dfrac{3}{4}x$

∴ Reqd. per cent = $\dfrac{\left(x - \dfrac{3}{4}x\right)}{\dfrac{3}{4}x} \times 100$

$= \dfrac{100}{3}\% = 33\dfrac{1}{3}\%$

15. (c) M.P. of the racket = Rs 30, Discount = 15%

∴ S.P. of the racket = $30 \times \dfrac{85}{100}$ = Rs 25.50

S.P. when a shuttle cock costing Rs 1.50 is given free = Rs 25.50 – Rs 1.50 = Rs 24
Profit = 20%

∴ C.P. of the racket = Rs $\dfrac{24 \times 100}{120}$ = Rs 20.

16. (d) Let the original cost of the article be Rs x.

Raising it by 30%, M.P. = $x \times \dfrac{130}{100}$ = Rs $\dfrac{13x}{10}$

After allowing two discounts each of 10%, the

price of the article = $\dfrac{13x}{10} \times \dfrac{90}{100} \times \dfrac{90}{100}$

= Rs $\dfrac{1053x}{1000}$

Per cent increase in the cost of the article

$= \dfrac{\left(\dfrac{1053x}{1000} - x\right)}{x} \times 100$

$= \dfrac{53x}{1000x} \times 100 = 5.3\%$

17. (c) Cost price of the article after discount
= 90% of 80% of Rs 100

$= \dfrac{90}{100} \times \dfrac{80}{100} \times$ Rs 100 = Rs 72

Amount spent on transport = 10% of Rs 72
= Rs 7.20
∴ Net C.P. = Rs 72 + Rs 7.20 = Rs 79.20
Profit = 15%

∴ S.P. = Rs $\left(\dfrac{79.2 \times 115}{100}\right)$ = Rs 91.08

18. (b) Let C.P. = Rs 100. Then M.P. = Rs 120
Discount = 10%

∴ S.P. = 90% of Rs 120 = $\dfrac{90}{100} \times$ Rs 120 = Rs 108

∴ Gain % = 8%

19. (c) Let the M.P. of a shirt be Rs x and that of trousers be Rs $2x$.
Let the discount on the trousers be $y\%$ Then,

$\dfrac{60}{100} \times x + \dfrac{(100-y)}{100} \times 2x = \dfrac{70}{100} \times (x + 2x)$

$\Rightarrow \dfrac{3}{5} + \dfrac{(100-y)}{100} = \dfrac{21}{10}$

$\Rightarrow \dfrac{100-y}{100} = \dfrac{21}{10} - \dfrac{3}{5} = \dfrac{21-6}{10} = \dfrac{15}{20} = \dfrac{3}{2}$

$\Rightarrow (100 - y) = \dfrac{3}{2} \times 50 = 75 \Rightarrow y = 25\%$

20. (c) C.P. = Rs 320, Gain = 15%

∴ S.P. = Rs $\left(320 \times \dfrac{115}{100}\right)$ = Rs 368

M.P. = Rs (368 + 32) = Rs 400
⇒ New S.P. = Rs 400

∴ Gain% = $\left(\dfrac{80}{320} \times 100\right)\%$ = 25%.

21. (a) Let the printed price be Rs 100. Then S.P. = Rs 80
Now S.P. = Rs 80, Gain% = 15%

∴ C.P. = Rs $\left(\dfrac{100}{115} \times 80\right)$ = Rs $\dfrac{1600}{23}$

∴ C.P. : Printed price = $\dfrac{1600}{23}$: 100 = 1600 : 2300
= 16 : 23

22. (c) Let the C.P. = Rs 100. Then, M.P. = Rs 120

S.P. of $\frac{3}{4}$th of goods $= \frac{3}{4} \times$ Rs 120 = Rs 90

S.P. of remaining $\frac{1}{4}$th of goods $= \frac{50}{100} \times \frac{1}{4} \times$ Rs 120

$\qquad\qquad = $ Rs 15

∴ Total S.P. = Rs 90 + Rs 15 = Rs 105

∴ Gain = Rs 105 – Rs 100 = Rs 5, *i.e.,*

Gain % = 5%.

23. (c) M.P. of the shirt = Rs 700

S.P. of the shirt 1st shoopkeeper

$\qquad = 70\%$ of 94% of Rs 700

$\qquad = \frac{70}{100} \times \frac{94}{100} \times$ Rs 700 = Rs 460.60

S.P. of the shirt offered by the 2nd shopkeeper

$\qquad = 80\%$ of 84% of Rs 700

$\qquad = \frac{80}{100} \times \frac{84}{100} \times$ Rs 700 = Rs 470.40

∴ 1st shopkeeper offers better discounts and
Required difference = Rs 470.40 – Rs 460.60

$\qquad\qquad\qquad\qquad = $ Rs 9.80.

24. (b) Let the C.P. = Rs 100, M.P. = Rs 110

Loss = 1% $\Rightarrow$ S.P. = Rs $\left(\frac{99}{100} \times 100 \right)$ = Rs 99

∴ Discount = Rs 110 – Rs 99 = Rs 11

∴ Discount per cent $= \left(\frac{11}{110} \times 100 \right)\% = 10\%$

25. (d) Let the original price be Rs 100.
Then C.P. = Rs 80

S.P. = 140% of Rs 80 = Rs $\left(\frac{140}{100} \times 80 \right)$

$\qquad = $ Rs 112

∴ Required % $= \frac{(112-100)}{100} \times 100 = 12\%.$

26. (a) S.P. = Rs 2880, Profit = 20%
Let the labelled price be Rs x. Then,

120% of $x = 2880 \Rightarrow x = \frac{2880 \times 100}{120} = 2400$

∴ C.P. = 85% of Rs 2400 = Rs $\left(\frac{85}{100} \times 2400 \right)$

$\qquad = $ Rs 2040.

27. (d) Let the M.P. of the article = Rs 100
Discount = 5% $\Rightarrow$ S.P. = Rs 95

Profit = 23.5% $\Rightarrow$ C.P. = Rs $\left(\frac{95 \times 100}{123.5} \right)$

$\qquad\qquad\qquad = $ Rs $\frac{95000}{1235}$

Had there been no discount, S.P. = Rs 100

Then, Profit % $= \dfrac{\left(100 - \dfrac{95000}{1235} \right)}{\dfrac{95000}{1235}} \times 100$

$\qquad = \dfrac{\left(\dfrac{123500 - 95000}{1235} \right)}{\dfrac{95000}{1235}} \times 100$

$\qquad = \dfrac{28500}{95000} \times 100 = 30\%$

28. (d) S.P. = Rs 6750, Discount = 10%

∴ M.P. = Rs $\left(\frac{6750 \times 100}{90} \right)$ = Rs 7500

If there was no discount S.P. = Rs 7500
and Profit % = 50%

∴ C.P. = Rs $\left(\frac{7500 \times 100}{150} \right)$ = Rs 5000

∴ Actual profit = Rs 6750 – Rs 5000 = Rs 1750

Actual profit% $= \frac{1750}{6750} \times 100 = 35\%.$

29. (b) Let the M.P. = Rs 100, Discount = 40%

∴ C.P. of the retailer = 60% of Rs 100 = Rs 60
S.P. of the retailer = Rs 100

∴ Profit % $= \frac{40}{60} \times 100 = \frac{200}{3}\% = 66\frac{2}{3}\%$

30. (a) Let the C.P. of the total stock = Rs 100
Then, M.P. of the total stock = Rs 120

∴ S.P. $= \frac{1}{2} \times 120 + \frac{1}{4} \times \frac{80}{100} \times 120 + \frac{1}{4} \times \frac{60}{100} \times 120$

$\qquad = $ Rs (60 + 24 + 18) = Rs 102

∴ Total gain = Rs 102 – Rs 100 = Rs 2, *i.e.,* 2%.

Self Assessment Sheet–16(c)

1. Two dealers offer an article at the same list price. The first allows discount 20%, 10% and 5% and the other of 15%, 12% and 8%. Which is a better offer for the customer?

   (a) 1st offer

   (b) 2nd offer

   (c) Both 1st offer and 2nd offer

   (d) Cannot be determined

2. If a discount of 10% is given to a customer on the marked price of an article, the gain of the trader is 20%. What will be the gain per cent of the trader if the discount is increased to 15%?

   (a) $12\frac{1}{2}\%$

   (b) $13\frac{1}{3}\%$

   (c) $14\frac{1}{4}\%$

   (d) $15\frac{1}{5}\%$

3. A tradesman allows a discount of 15% on the written price. How much above the cost price should he mark his goods to make a profit of 19%?

   (a) 20%   (b) 40%

   (c) 30%   (d) 25%

4. A pen is listed for Rs 12. A discount of 15% is given on it. A second discount is given bringing the price down to Rs 8.16. The rate of second discount is:

   (a) 15%   (b) 18%

   (c) 20%   (d) 25%

5. A shopkeeper claims to sell his articles at a discount of 10%, but marks his articles by increasing the cost of each by 20%. His gain per cent is:

   (a) 6%   (b) 8%

   (c) 10%   (d) 12%

6. A seller allows a discount of 5% on a watch. If he allows a discount of 7%, he earns Rs 15 less in the profit. What is the marked price?

   (a) Rs 697.50   (b) Rs 712.50

   (c) Rs 750   (d) Rs 817.50

7. A shopkeeper sold an air conditioner for Rs 25935 with a discount of 9% and earned a profit of 3.74%. What would have been the percentage of profit earned if no discount was offered?

   (a) 12.3%   (b) 15.6%

   (c) 16%   (d) 14 %

8. On selling an article at a discount of 20%, the profit is 20%. Find the profit per cent if the article is sold at a discount of 10%.

   (a) 25%   (b) 30%

   (c) 15%   (d) 35%

9. A trader bought some goods at a discount of 20% of the list price. He wants to mark them at such a price that he can give a discount of 20% on the marked price and still make a profit of 25%. Find the per cent of the list price at which he should mark the goods.

   (a) 25% above the list price

   (b) 15% below the list price

   (c) 25% below the list price

   (d) 15% above the list price

10. In one shop, on article is marked 75% above the cost price, but the purchaser is allowed a discount of 20% on the marked price. In another shop a similar article is sold for Rs 58 at a gain of 45%. What did the purchased pay for this article in the first shop?

    (a) Rs 60   (b) Rs 56

    (c) Rs 62   (d) Rs 65

Answers

1. (a)	2. (b)	3. (b)	4. (c)	5. (b)	6. (c)	7. (d)	8. (d)	9. (b)

10. (a) [**Hint.** Find the profit % in the 1st step. Then calculate S.P. of the 1st shop using the C.P. of the second shop and the profit % you found]

Chapter 17

AVERAGE

KEY FACTS

1. Average of n observations $= \dfrac{\text{Sum of } n \text{ observations}}{n}$

2. Sum of n observations = Average of n observations $\times n$

Solved Examples

Ex. 1. *What is the average of squares of consecutive odd numbers between 1 and 13 ?*

Sol. The consecutive odd numbers from 1 to 13 = 3, 5, 7, 9, 11

$\therefore$ Required average $= \dfrac{3^2 + 5^2 + 7^2 + 9^2 + 11^2}{5} = \dfrac{9 + 25 + 49 + 81 + 121}{5} = \dfrac{285}{5} = \mathbf{57}$

Ex. 2. *The average age of r boys in a class is a years. If the average age of s of them is b years, then what is the average age of the remaining boys ?*

Sol. Total age of r boys $= r \times a = ra$ years

Total age of s boys $= s \times b = sb$ years

$\therefore$ Total age of remaining boys $= ra - sb$

Total number of remaining boys $= r - s$

$\therefore$ Average age of remaining boys $= \dfrac{ra - sb}{r - s}$ **years.**

Ex. 3. *The average marks of 48 students in a class is 45. The average marks of the boys in the class is 40 and the average marks of the girls in the class is 50. What is the ratio between the numbers of boys and the number of girls in the class ?*

Sol. Total marks of 48 students = 48 × 45 = 2160

Let the number of boys be B and the number of girls be $(48 - B)$. Then,

Total marks of boys = $40B$

Total marks of girls = $50 (48 - B) = 2400 - 50B$

$\Rightarrow 40B + 2400 - 50B = 2160 \Rightarrow 10B = 240 \Rightarrow B = 24$

$\therefore$ Number of girls = 48 − 24 = 24.

$\therefore$ Required ratio = 24 : 24 = **1 : 1.**

Ex. 4. *From a class 24 boys, a boy aged 10 years leaves the class and in his place a new boy is admitted. As a result the average age of the class is increased by 2 months. What is the age of the new boys ?*

Sol. Let the average age of the class be x years.
Then the total age of the class = $24x$ years

New average age of the class = $\left(x - \dfrac{2}{12}\right)$ years = $\left(x - \dfrac{1}{6}\right)$ years

Let the age of the new boy be y years.

Given, $\dfrac{24x - 10 + y}{24} = x - \dfrac{1}{6}$

$\Rightarrow$ $24x - 10 + y = 24x - 4$ $\Rightarrow$ $y = 6$.

$\therefore$ The age of the new boy is **6 years**.

Ex. 5. *The average of 8 numbers is 20. The average of first two numbers is $15\dfrac{1}{2}$ and that of next three is $21\dfrac{1}{3}$. If the sixth number be less than the seventh and eighth numbers by 4 and 7 respectively, then find the eighth number ?*

Sol. Let the sixth number be x. Then,
Seventh number = $x + 4$, Eighth number = $x + 7$
Total of all 8 numbers = $8 \times 20 = 160$

Total of first 2 numbers = $2 \times \dfrac{31}{2} = 31$

Total of next 3 numbers = $3 \times \dfrac{64}{3} = 64$

Given, $31 + 64 + x + x + 4 + x + 7 = 160$

$\Rightarrow$ $3x + 106 = 160$ $\Rightarrow$ $3x = 160 - 106 = 54$ $\Rightarrow$ $x = 18$

$\therefore$ Eigth number = $18 + 7 = $ **25**.

Ex. 6. *The average age of 8 persons in a committee is increased by 2 years when two men aged 35 years and 45 years are substituted by two women. What is the average age of these two women ?*

Sol. Let the average age of 8 persons be x years. Then,
Total age of 8 persons = $8x$
Also let y_1, and y_2 be the ages of the two women who replaced the two men. Then,

$\dfrac{8x - 35 - 45 + (y_1 + y_2)}{8} = x + 2$

$\Rightarrow$ $8x - 80 + y_1 + y_2 = 8x + 16$ $\Rightarrow$ $y_1 + y_2 = 96$

$\therefore$ Average age of the two women = $\dfrac{y_1 + y_2}{2} = \dfrac{96}{2} = $ **48 years**.

Ex. 7. *The average of the age of a husband and wife, five years ago was 25 years. The average of the present age of husband, wife and a child born during the time is 21 years. Determine the present age of the child.*

Sol. Sum of the ages of husband and wife, **5 years ago** = $2 \times 25 = 50$ years
Sum of the ages of husband and wife, **at present** = $50 + 5 + 5 = 60$ years
Sum of the ages of husband, wife and child, at present = $3 \times 21 = 63$ years

$\therefore$ Present age of the child = $(63 - 60)$ years = **3 years**.

Ex. 8. *Of the three numbers, the first is twice the second and the second is twice the third. The average of the reciprocal of the numbers is $\dfrac{7}{72}$. What are the three numbers ?*

Sol. Let the third number be x. Then
Second number = $2x$ and first number = $4x$

Sum of the reciprocals of these 3 numbers = $\dfrac{1}{4x} + \dfrac{1}{2x} + \dfrac{1}{x} = \dfrac{1 + 2 + 4}{4x} = \dfrac{7}{4x}$

Given, $\dfrac{7}{4x} = 3 \times \dfrac{7}{72}$ $\Rightarrow$ $4x = 24$ $\Rightarrow$ $x = 6$

$\therefore$ The three numbers are $4 \times 6, 2 \times 6, 6$, *i.e.*, $24, 12, 6$.

Ex. 9. *The average temperature of Monday to Wednesday was 37°C and of Tuesday to Thursday was 34°C. If the temperature on Thursday was $\frac{4}{5}$ that of Monday, what was the temperature on Thursday ?*

Sol. Let the temperatures on Monday, Tuesday, Wednesday and Thursday be M, T, W and Th respectively. Then,

M + T + W = 3 × 37°C = 111°C ... (i)

T + W + Th = 3 × 34°C = 102°C ... (ii)

⇒ Eq (i) – Eq (ii)

⇒ M – Th = 111°C – 102°C = 9°C

Also given, Th = $\frac{4}{5}$ M ⇒ M – $\frac{4}{5}$ M = 9 ⇒ $\frac{M}{5}$ = 9 ⇒ M = 45°C

∴ Temperature on Thursday = $\frac{4}{5}$ × 45°C = **36°C**.

Ex. 10. *The average age of a class is 40 years. 12 new students with an average age of 32 years join the class, there by decreasing the average by 4 years. What is the original strength of the class ?*

Sol. Let the original strength of the class be x. Then,

Total age of x students = 40x years

Total age of 12 new students = 12 × 32 = 384 years

New average of (x + 12) students = 40 – 4 = 36 years

∴ Total age of (x + 12) students = (x + 12) × 36 = (36x + 432) years

∴ 40x + 384 = 36x + 432

⇒ 4x = 48 ⇒ **x = 12**.

Question Bank–17

1. The average of the squares of the first ten natural numbers is
   (a) 40
   (b) 50
   (c) 47.5
   (d) 38.5

2. The average of the two digit numbers, which remain the same when the digits interchange their positions is
   (a) 33
   (b) 44
   (c) 55
   (d) 66

3. If the average of m numbers is n^2 and that of n numbers is m^2, then the average of (m + n) numbers is
   (a) m – n
   (b) mn
   (c) (m + n)
   (d) m / n

4. If the average of a, b, c is M and ab + bc + ca = 0, then the average of a^2, b^2, c^2 is
   (a) M^2
   (b) $3M^2$
   (c) $6M^2$
   (d) $9M^2$

5. The average of six numbers is x and the average of three of these is y. If the average of the remaining three is z, then
   (a) x = y + z
   (b) 2x = y + z
   (c) x = 2y + 2z
   (d) x = y + 2z

6. The average weight of 120 students in the second year class of college is 56 kg. If the average weight of boys and that of girls in the class are 60 kg and 50 kg respectively, then the number of boys and girls in the class are respectively :
   (a) 72, 64
   (b) 38, 64
   (c) 72, 48
   (d) 62, 58

7. The average price of 10 books is Rs 12, while the average price of 8 of these books is Rs 11.75. Of the remaining two books, if the price of one is 60% more than the price of the other, what is the price of each of these two books ?
   (a) Rs 8; Rs 12
   (b) Rs 10; Rs 16
   (c) Rs 5; Rs 7.50
   (d) Rs 12; Rs 14

8. The average salary of male employees in a firm was Rs 5200 and that of females was Rs 4200. The average salary of all the employees was Rs 5000. What is the percentage of female employees ?
   (a) 80%
   (b) 20%
   (c) 40%
   (d) 30%

9. A batsman has a certain average of runs for 16 innings. In the 17th inning, he makes a score of 85 runs there by increasing the average by 3. What is the average of 17 innings ?
   (a) 38
   (b) 37
   (c) 36
   (d) 35

10. There are 50 boys in a class. One boy weighing 40 kg goes away and at the same time another boy joins the class. If the average weight of the class is thus decreased by 100 g, find the weight of the new boy?
    (a) 32 kg
    (b) 40 kg
    (c) 35 kg
    (d) 37 kg

11. X has twice as much money as that of Y and Y has 50% more money than that of Z. If the average money of all of them is Rs 110, then the money which X has
(a) Rs 55   (b) Rs 60
(c) Rs 90   (d) Rs 180

12. The average score of two sections is 28. The average score of section A is 30, in which there are 25 students. If there are 20 students in section B, then the average score of section B is
(a) 25.5   (b) 27.5
(c) 29   (d) 33

13. The average age of a father and his two sons is 27 years. Five years ago, the average age of the two sons was 12 years. If the difference between the ages of two sons is 4 years, then the present age of the father is
(a) 34 years   (b) 47 years
(c) 64 years   (d) 27 years

14. The average marks obtained by five students A, B, C, D and E is $52\frac{3}{5}$. If the average marks obtained by A, C and D are $53\frac{2}{3}$ and the average marks obtained by B, C and E are $52\frac{2}{3}$, then the marks obtained by C were
(a) 56   (b) 53
(c) 57   (d) 60

15. In an examination, a pupil's average marks were 63 per paper. If he had obtained 20 more marks for his Geography paper and 2 more marks for his History paper, his average per paper would have been 65. How many papers were there in the examination ?
(a) 8   (b) 9
(c) 10   (d) 11

16. The average of marks of 28 students in Mathematics was 50. Eight students left the school and then this average increased by 5. What is the average of marks obtained by the students who left the school ?
(a) 37.5   (b) 42.5
(c) 45   (d) 50.5

17. The average age of 11 players of a cricket team is decreased by 2 months when two of them aged 17 years and 20 years are replaced by two new players. The average age of the new players is
(a) 17 years 1 month   (b) 17 years 7 months
(c) 17 years 11 months   (d) 18 years 3 months

18. The average monthly salary of the workers in a workshop is 8500. If the average monthly salary of 7 technicians is Rs 10,000 and the average monthly salary of the rest is Rs 7800, the total number of workers in the workshop is
(a) 18   (b) 20
(c) 22   (d) 24

19. The average age of 30 boys in a class is 15 years. One boy aged 20 years left the class, but two new boys come in his place whose ages differ by 5 years. If the average age of all the boys now in the class still remains 15 years, the age of the younger new comer is
(a) 20 years   (b) 15 years
(c) 10 years   (d) 8 years

20. 3 years ago, the average age of a family of 5 members was 17 years. A baby having been born, the average age of the family is same today. The present age of the baby is
(a) 3 years   (b) 2 years
(c) $1\frac{1}{2}$ years   (d) 1 year

21. The average annual income (in Rs) of certain agricultural workers is S and that of other workers is T. The number of agricultural workers is 11 times that of other workers. Then, the average annual income (in Rs) of all the workers is
(a) $\dfrac{S+11T}{12}$ (b) $\dfrac{S+T}{2}$
(c) $\dfrac{11S+T}{12}$ (d) $\dfrac{1}{11S}+T$

22. The average monthly income of X and Y is Rs 5050. The average monthly income of Y and Z is Rs 6250 and the average monthly income of X and Z is Rs 5200. The monthly income of X is
(a) Rs 4050   (b) Rs 3500
(c) Rs 4000   (d) Rs 5000

23. The average of four positive integers is 72.5. The highest integer is 117 and the lowest integer is 15. The difference between the remaining two integers is 12. Which is the higher of these two remaining integers ?
(a) 70   (b) 73
(c) 85   (d) 80

24. The average score of a class of boys and girls in an examination is A. The ratio of boys and girls in the class is 3 : 1. If the average score of the boys is $(A + 1)$, the average score of the girls is
(a) $(A - 1)$ (b) $(A - 3)$
(c) $(A + 1)$ (d) $(A + 3)$

25. Of the four numbers, the first is twice the second, the second is one-third of the third and third is 5 times the fourth. The average of the numbers is 24.75. The largest of these numbers is
(a) 9   (b) 25
(c) 30   (d) 45

26. The average temperature of the town in the first four days of a month was 58 degrees. The average for second, third, fourth and fifth days was 60 degrees. If the temperatures if the first and fifth days were in the ratio 7 : 8, then what is the temperature on the fifth day ?
(a) 64 degrees   (b) 62 degrees
(c) 56 degrees   (d) 48 degrees

27. A company produces on an average 4000 items per month for the first three months. How many items it must produce on an average per month over the next 9 months to average 4375 items per month over the whole year ?
(a) 4500   (b) 4600
(c) 4680   (d) 4710

28. The average of 5 consecutive numbers is m. If the next three natural numbers are also included, how much more than m will the average of these 8 numbers be ?
(a) 2   (b) 1
(c) 1.4   (d) 1.5

29. The average weight of three men A, B and C is 84 kg. D joins them and the average weight of the four becomes 80 kg. If E whose weight is 3 kg more than that of D replaces A, the average weight of B, C, D and E becomes 79 kg. The weight of A is
(a) 65 kg   (b) 70 kg
(c) 75 kg   (d) 80 kg

30. The batting average of a cricket player for 40 innings is 50 runs. His highest score in an innings exceeds his lowest score by 172 runs. If these two innings are excluded, the average of the remaining 38 innings is 48 runs. Determine his highest score, scored in one innings.
(a) 175   (b) 180
(c) 174   (d) 185

Answers

1. (d)	**2.** (c)	**3.** (b)	**4.** (b)	**5.** (b)

6. (c) **7.** (b) **8.** (b) **9.** (b) **10.** (c)
11. (d) **12.** (a) **13.** (b) **14.** (a) **15.** (d) **16.** (a) **17.** (b) **18.** (c) **19.** (b) **20.** (b)
21. (c) **22.** (c) **23.** (c) **24.** (b) **25.** (d) **26.** (a) **27.** (a) **28.** (d) **29.** (c) **30.** (c)

Hints and Solutions

1. (d) Average
$$= \frac{1^2+2^2+3^2+4^2+5^2+6^2+7^2+8^2+9^2+10^2}{10}$$
$$= \frac{1+4+9+16+25+36+49+64+81+100}{10}$$
$$= \frac{385}{10} = 38.5$$

2. (c) The required two digit numbers are 11, 22, 33, 44, 55, 66, 77, 88, 99.
∴ Required Average
$$= \frac{11+22+33+44+55+66+77+88+99}{9}$$
$$= \frac{495}{9} = 55$$

3. (b) Average of m numbers = n^2
∴ Sum of m numbers = $m \times n^2 = mn^2$
Average of n numbers = m^2
∴ Sum of n numbers = $n \times m^2 = m^2 n$
∴ Required average $= \frac{mn^2+nm^2}{m+n} = \frac{mn(n+m)}{m+n} = mn$

4. (b) Given, $\frac{a+b+c}{3} = M \Rightarrow a+b+c = 3M$
Now, $(a+b+c)^2 = a^2+b^2+c^2+2(ab+bc+ca)$
$\Rightarrow (3M)^2 = a^2+b^2+c^2+0 \Rightarrow a^2+b^2+c^2 = 9M^2$
Average of a^2, b^2, $c^2 = \frac{a^2+b^2+c^2}{3}$
$$= \frac{9M^2}{3} = 3M^2$$

5. (b) Sum of six numbers = $6x$
Sum of three of these 6 numbers = $3y$
Sum of remaining three numbers = $3z$
∴ $3y + 3z = 6x \Rightarrow 2x = y + z$

6. (c) Let the number of boys be x. Then,
The number of girls = $(120 - x)$
Total weight of 120 students in the class
$= 120 \times 56$ kg = 6720 kg
Total weight of x boys = $60x$ kg
Total weight of $(120-x)$ girls = $(120-x) \times 50$ kg
$= 6000 - 50x$
Given, $60x + 6000 - 50x = 6720$

$\Rightarrow 10x = 720 \Rightarrow x = 72$

$\therefore$ Number of boys = 72,

Number of girls = 120 – 72 = 48

7. (b) Total price of the remaining two books

$= 10 \times 12 - 8 \times 11.75$

$= 120 - 94 = $ Rs 26

Suppose the price of one book = Rs x

Then, price of the other book $= x + 60\%$ of x

$= \dfrac{160}{100}x = \dfrac{8x}{5}$

Given, $x + \dfrac{8x}{5} = 26 \Rightarrow \dfrac{13x}{5} = 26$

$\Rightarrow x = \dfrac{26 \times 5}{13} = $ Rs 10

$\therefore$ Price of the other book = Rs $\dfrac{8 \times 10}{5} = $ **Rs 16**.

8. (b) Let the number of male and female employees be M and F respectively. Then,

Total salary of male employees = Rs 5200 M

Total salary of female employees = Rs 4200 F

Total number of employees = $M + F$

$\therefore$ Total salary of all the employees

$= $ Rs 5000 $(M + F)$

$\therefore 5200M + 4200F = 5000M + 5000F$

$\Rightarrow 200M = 800F \Rightarrow \dfrac{M}{F} = \dfrac{4}{1} \Rightarrow M:F = 4:1$

$\therefore$ Percentage of female employees $= \left(\dfrac{1}{5} \times 100\right)\%$

$= 20\%$.

9. (b) Let the average of runs for 16 innings be A.

Then, average of runs for 17 innings = $A + 3$

Total runs for 16 innings = 16 A

$\Rightarrow$ Total runs for 17 innings = $16A + 85$

Given, $\dfrac{16A + 85}{17} = A + 3$

$\Rightarrow 16A + 85 = 17A + 51 \Rightarrow A = 34$

$\therefore$ Average of 17 innings = 34 + 3 = 37.

10. (c) Let the average weight of the class be x kg.

Total weight of 50 boys = $50x$ kg

If y is weight of the new boy, then

$50x - 40 + y = 50 \times (x - 0.1)$

$\Rightarrow 50x - 40 + y = 50x - 5 \Rightarrow y = 35$ kg.

11. (d) Let Z have Rs X. Then,

Y has 150% of Rs X = Rs $\dfrac{150X}{100} = $ Rs $\dfrac{3X}{2}$

X has $2 \times $ Rs $\dfrac{3X}{2} = $ Rs $3X$

Given, $\dfrac{X + \dfrac{3X}{2} + 3X}{3} = 110$

$\Rightarrow \dfrac{2X + 3X + 6X}{2} = 330$

$\Rightarrow 11X = 660 \Rightarrow X = 60$

$\therefore$ Amount that X has = $3 \times 60 = $ Rs 180.

12. (a) Total number of students in section A and B

$= 25 + 20 = 45$

Average score of both the sections = 28

$\therefore$ Total score of both the sections = $28 \times 45 = 1260$

Total score of section $A = 25 \times 30 = 750$

Let the average score of section $B = x$

Then, total score of section $B = 20x$

Given, $20x + 750 = 1260$

$\Rightarrow 20x = 510 \Rightarrow x = 25.5$.

13. (b) Let the ages of father and his two sons be x, y and z years respectively.

Given, $x + y + z = 3 \times 27 = 81$ ... (i)

and $(y - 5) + (z - 5) = 2 \times 12 \Rightarrow y + z = 34$ ...(ii)

From (i) and (ii)

Age of father $(x) = 81 - 34 = 47$ years.

14. (a) Total marks obtained by

$A + B + C + D + E = 5 \times 52\dfrac{3}{5} = 5 \times \dfrac{263}{5} = 263$ ...(1)

Total marks obtained by

$A + C + D = 3 \times 53\dfrac{2}{3} = 3 \times \dfrac{161}{3} = 161$ ...(2)

Total marks obtained by

$B + C + E = 3 \times 52\dfrac{2}{3} = 3 \times \dfrac{158}{3} = 158$ ...(3)

$\therefore$ (1) — (2)

$\Rightarrow$ Marks obtained by

$B + E = 263 - 161 = 102$ ... (4)

(3) — (4)

$\Rightarrow$ Marks obtained by $C = 158 - 102 = 56$.

15. (d) Let there be n papers in all.

Total marks originally for n papers = $63n$

Total marks with increased average = $65n$

Given, $65n - 63n = 20 + 2 \Rightarrow 2n = 22 \Rightarrow n = 11$.

16. (a) Total marks of 28 students in Mathematics

$= 28 \times 50 = 1400$

New number of students = 28 – 8 = 20

New average of marks in Mathematics
$$= 50 + 5 = 55$$

∴ Total marks of 20 students in Mathematics
$$= 20 \times 55 = 1100$$

∴ Total marks of 8 students who left the school
$$= 1400 - 1100 = 300$$

∴ Average marks of these 8 students $= \dfrac{300}{8}$
$$= 37.5.$$

17. (b) Let the average age of 11 players be x.
Then, total age of 11 players $= 11x$
Total age of 9 players $= 11x - (17 + 20)$
$$= 11x - 37$$
Let y be the total age of the 2 new players.

Then $\dfrac{11x - 37 + y}{11} = x - \dfrac{1}{6} = \dfrac{6x-1}{6}$

$\Rightarrow 66x - 222 + 6y = 66x - 11$

$\Rightarrow 6y = -11 + 222 = 211 \Rightarrow y = \dfrac{211}{6}$ years

∴ Average age of the two players $= \dfrac{y}{2} = \dfrac{211}{12}$
$$= 17 \text{ years 7 months.}$$

18. (c) Let the total number of workers in the workshop be x.
Then, number of other workers (besides technicians) $= (x - 7)$
Given, $(x - 7) \times 7800 + 7 \times 10000 = 8500x$

$\Rightarrow 7800x - 54600 + 70000 = 8500x$

$\Rightarrow 700x = 15400 \Rightarrow x = \dfrac{15400}{700} \Rightarrow x = 22.$

19. (b) Total age of 30 boys $= 30 \times 15 = 450$
Total age of the rest of boys after the boy aged 20 years leaves $= 450 - 20 = 430$
Let x be the age of the younger new comer.

Then, $\dfrac{430 + x + (x+5)}{31} = 15$

$\Rightarrow 435 + 2x = 31 \times 15 = 465$
$\Rightarrow 2x = 30 \Rightarrow x = 15$ years.

20. (b) Total age of the family three years ago
$$= 17 \times 5 = 85 \text{ years.}$$
Let the present age of the child be x years.
Present total age of the family $= 85 + 5 \times 3 + x$
$$= (100 + x) \text{ years.}$$

Given $\dfrac{100 + x}{6} = 17$

$\Rightarrow 100 + x = 102 \Rightarrow x = 2$ years.

21. (c) Let the number of other workers $= x$
Then, the number of agricultural workers $= 11x$
∴ Total income of other workers $= $ Rs Tx
Total income of agricultural workers $= $ Rs $11Sx$

∴ Average income of all the workers $= \dfrac{11Sx + Tx}{11x + x}$

$$= \dfrac{x(11S + T)}{12x} = \dfrac{11S + T}{12}$$

22. (c) Total income of $(X + Y) = 2 \times$ Rs 5050
$$= \text{Rs } 10100$$
Total income of $(Y + Z) = 2 \times$ Rs 6250
$$= \text{Rs } 12500$$
Total income of $(Z + X) = 2 \times$ Rs 5200
$$= \text{Rs } 10400$$

∴ $2(X + Y + Z) = $ Rs 10100 + Rs 12500 + Rs 10400
$$= \text{Rs } 33000$$

$\Rightarrow X + Y + Z = $ Rs 16500
∴ X's monthly income $= (X + Y + Z) - (Y + Z)$
$$= \text{Rs } 16500 - \text{Rs } 12500 = \text{Rs } 4000.$$

23. (c) Let the higher integer be x.
Then the other integer $= x - 12$

∴ $\dfrac{15 + (x-12) + x + 117}{4} = 72.5$

$\Rightarrow 120 + 2x = 72.5 \times 4 = 290$
$\Rightarrow 2x = 170 \Rightarrow x = 85.$

24. (b) Let the number of boys be $3x$ and the number of girls be x.
Then, total number of students $= 4x$
Average score of the class $= A$
∴ Total score of all the students $= 4Ax$
Average score of all the boys $= A + 1$
∴ Total score of all the boys $= 3x(A + 1)$
$$= 3Ax + 3x$$
$\Rightarrow$ Total score of all the girls $= 4Ax - 3Ax - 3x$
$$= Ax - 3x = (A - 3)x$$

$\Rightarrow$ Average score of the girls $= \dfrac{(A-3)x}{x} = (A - 3)$

25. (d) Let the fourth number be x. Then,
Third number $= 5x$
Second number $= \dfrac{5x}{3}$

First number = $\dfrac{10x}{3}$

$\therefore \quad x + 5x + \dfrac{5x}{3} + \dfrac{10x}{3} = (24.75 \times 4) = 99$

$\Rightarrow 3x + 15x + 5x + 10x = 99 \times 3$

$\Rightarrow 33x = 99 \times 3 \quad \Rightarrow \quad x = \dfrac{99 \times 3}{33} = 9$

$\therefore$ The numbers are 9, 45, 15, 30

$\Rightarrow$ Largest number = 45

26. (a) Sum of temperatures on 1st, 2nd, 3rd and 4th days
   = (58×4) = 232 degrees   ... (1)

   Sum of temperatures on 2nd, 3rd, 4th and 5th days
   = (60×4) = 240 degrees   ...(2)

   Subtracting eqn. (1) from eqn. (2), we get

   Temp. on 5th day – Temp. on 1st = 8 degrees

   Given, Temp. on 1st day : Temp. on 5th day = 7 : 8

   $\therefore \quad 8x - 7x = 8 \quad \Rightarrow \quad x = 8$

   $\therefore$ Temperature on 5th day = 64 degrees.

27. (a) Total number of items produced for first three
   months = $4000 \times 3 = 12000$

   Total number of items required to be produced
   over a period of 12 months = $4375 \times 12 = 52500$

   $\Rightarrow$ Number of items to be produced over a period
   of 9 months = $52500 - 12000 = 40500$

   $\therefore$ Average number of articles per month produced

   over 9 months = $\dfrac{40500}{9} = 4500$

28. (d) Let the five consecutive numbers be

   $x, x + 1, x + 2, x + 3, x + 4$

   Then $\dfrac{x + x + 1 + x + 2 + x + 3 + x + 4}{5} = m$

$\Rightarrow 5x + 10 = 5m \Rightarrow 5x = 5(m - 2) \Rightarrow x = m - 2.$

$\therefore$ The 8 consecutive numbers are

$m - 2, m - 1, m, m + 1, m + 2, m + 3, m + 4, m + 5$

Average of these 8 numbers

$= \dfrac{\begin{array}{c} m - 2 + m - 1 + m + m + 1 + m \\ + 2 + m + 3 + m + 4 + m + 5 \end{array}}{8}$

$= \dfrac{8m + 12}{8} = m + \dfrac{3}{2}$

$\therefore$ Required difference $= m + \dfrac{3}{2} - m = \dfrac{3}{2} = 1.5$

29. (c) Total weight of $A + B + C = 84 \times 3 = 252$ kg

   Total weight of $A + B + C + D = 80 \times 4 = 320$ kg

   $\therefore$ Weight of $D = 320 - 252 = 68$ kg

   Given, $E = D + 3 = (68 + 3)$ kg = 71 kg

   Total weight of $B + C + D + E = (79 \times 4)$ kg

   $= 316$ kg

   $\therefore$ Weight of $B + C = 316$ kg $- 71$ kg $- 68$ kg

   $= 177$ kg

   Weight of A only = 320 kg – 177 kg – 68 kg = 75 kg

30. (c) Let the lowest score in one innings be x. Then,

   Highest score in an innings = $x + 172$

   Total runs in 40 innings = $40 \times 50 = 2000$

   Total runs in 38 innings = $38 \times 48 = 1824$

   Given, $2000 - \{x + (x + 172)\} = 1824$

$\Rightarrow 1828 - 2x = 1824 \Rightarrow 2x = 4 \Rightarrow x = 2$

$\therefore$ Highest score $= 172 + 2 = 174.$

Self Assessment Sheet–17

1. The average age of a husband and a wife, who were
   married 4 years ago, was 25 years at the time of
   their marriage. The average age of the family
   consisting of husband, wife and a child born during
   that interval is 20 years today. The age of the child
   is
   (a) 1 year   (b) 2 years
   (c) 2.5 years   (d) 3 years

2. The average of marks obtained by 120 condidates
   was 35. If the average of the passed candidates was
   39 and that of the failed candidates was 15, then the
   number of those candidates who passed the

   examination was
   (a) 100   (b) 110
   (c) 120   (d) 150

3. In a competitive examination, the average marks
   obtained was 45. It was later discovered that there
   was some error in computerisation and the marks
   of 90 candidates had to be changed from 80 to 50
   and the average came down to 40 marks. The total
   number of candidates who appeared in examination
   is
   (a) 520   (b) 550
   (c) 540   (d) 560

4. The average of marks of 28 students in Mathematics was 50. Eight students left the school and then this average increased by 5. What is the average of marks obtained by the students who left the school?

(a) 37.5   (b) 42.5

(c) 45   (d) 50.5

5. The average of five consecutive natural numbers is m. If the next three natural numbers are also included, how much more than m will the average of these 8 numbers be ?

(a) 1   (b) 1.5

(c) 1.4   (d) 2

6. The weight of a body, calculated as the average of seven different experiments is 53.735 g. The average of the first three experiments is 54.005 g. The fourth was greater than the fifth by 0.004 g and the average of sixth and seventh was 0.010 g less than the average of the first three. Find the weight of the body in the third experiment.

(a) 52.071 g   (b) 53.072g

(c) 51.450 g   (d) 53.005 g

7. The average weight of A, B and C is x kg. A and C lose y kg each after dieting and B puts on $y/2$ kg. After this their average weight decreases by 1 kg. Find y.

(a) 1.5 kg   (b) 3 kg

(c) 2 kg   (d) 3.5 kg

8. A goods train in five successive minutes from its starts runs 68 metres, 127 metres, 208 metres, 312 metres and 535 metres and for the next 5 minutes maintains an average speed of 33 km/hr. Find the average speed of the train in km/hr in covering this total distance.

(a) 20 km/hr   (b) 30 km/hr

(c) 28 km/hr   (d) 24 km/hr

9. The age of the captain of a cricket team if 11 players is 25 years and the wicket keeper is 3 years older than the captain. If the ages of these two are excluded, the average age of the remaining players of the team becomes 1 year less than the average age of the whole team. What is the average age of the whole team?

(a) 18 years   (b) 22 years

(c) 25 years   (d) 23 years

10. There are four natural numbers. The average of any three numbers is added in the fourth number and in this way the number 29, 23, 21 and 17 are obtained. One of the number is

(a) 11   (b) 24

(c) 21   (d) 10

Answers

1. (b)	2. (a)	3. (c)	4. (a)	5. (b)	6. (b)	7. (c)	8. (d)	9. (b)	10. (c)

Chapter 18

SIMPLE INTEREST

KEY FACTS

Formulae:

1. $S.I. = \dfrac{P \times R \times T}{100}$

2. Amount $(A) = P + S.I.$

3. $P = \dfrac{S.I. \times 100}{R \times T}$

4. $R = \dfrac{S.I. \times 100}{P \times T}$

5. $T = \dfrac{S.I. \times 100}{P \times R}$

Question Bank–18

Revision Assignment on Simple Interest

1. If the simple interest for 6 years be equal to 30% of the principal, it will be equal to the principal after
   (a) 10 years   (b) 20 years
   (c) 22 years   (d) 30 years

2. Kruti took a loan at simple interest at 6% in the first year with an increase of 0.5% in each subsequent year. She paid Rs 3375 as interest after 4 years. How much loan did she take ?
   (a) Rs 12500   (b) Rs 15800
   (c) Rs 33250   (d) Rs 30,000

3. A sum of Rs 18750 is left by will by a father to be divided between two sons, 12 and 14 years of age, so that when they attain maturity at 18, the amount received by each of them at 5 per cent simple interest will be the same. Find the sum allotted at present to each son ?
   (a) Rs 9500, Rs 9250   (b) Rs 8000, Rs 1750
   (c) Rs 9000, Rs 9750   (d) Rs 10000, Rs 8750

4. The difference between the simple interest received from two different sources on Rs 1500 for 3 years is Rs 13.50. The difference between their rates of interests is
   (a) 0.1%   (b) 0.2%
   (c) 0.3%   (d) 0.4%

5. David invested certain amount in three different schemes A, B and C with the rate of interest 10% p.a., 12% p.a. and 15% p.a. respectively. If the total interest accrued in one year was Rs 3200 and the amount invested in Scheme C was 150% of the amount invested in Scheme A and 240% of the amount invested in Scheme B, what was the amount invested in scheme B ?
   (a) Rs 5000   (b) Rs 6500
   (c) Rs 6000   (d) Rs 8000

6. A person invested in all Rs 2600 at 4%, 6% and 8% per annum simple interest. At the end of the year, he got the same interest in all the three cases. The money invested at 4% is :
   (a) Rs 200   (b) Rs 600
   (c) Rs 800   (d) Rs 1200

7. Divide Rs 2379 into 3 parts, so that their amounts after 2, 3 and 4 years respectively may be equal the rate of interest being 5% per annum at simple interest. The first part is
   (a) Rs 759   (b) Rs 792
   (c) Rs 818   (d) Rs 828

8. A sum of Rs 1440 is lent out in three parts in such a way that the interests on first part at 2% for 3 years, second part at 3% for 4 years and third part at 4% for 5 years are equal. Then the difference between the largest and smallest sum is
   (a) Rs 400   (b) Rs 560
   (c) Rs 460   (d) Rs 200

9. With a given rate of simple interest, the ratio of principal and amount for a certain period of time is 4 : 5. After 3 years, with the same rate of interest, the ratio of the principal and amount becomes 5 : 7. The rate of interest per annum is
   (a) 4%   (b) 5%
   (c) 6%   (d) 7%

10. Rs 1000 invested at 5% p.a. simple interest. If the interest is added to the principal after every 10 years, the amount will become Rs 2000 after

   (a) 15 years   (b) $16\frac{2}{3}$ years
   (c) 18 years   (d) 20 years

11. Two equal sums of money are lent at the same time at 8% and 7% per annum simple interest. The former is recovered 6 months earlier than the later and the amount in each case is Rs 2560. The sum and time for which the sums of money are lent out are
   (a) Rs 1500, 3.5 years and 4 years
   (b) Rs 2000, 3.5 years and 4 years
   (c) Rs 2000, 4 years and 5.5 years
   (d) Rs 3000, 4 years and 4.5 years

12. A man borrowed Rs 40,000 at 8% simple interest per year. At the end of second year he paid back certain amount and at the end of fifth year he paid back Rs 35960 and cleared the debt. What is the amount that he paid back after the second year ?

   (a) Rs 16200   (b) Rs 17400
   (c) Rs 18600   (d) Rs 19200

13. Anand deposited Rs 6000 on simple interest. He with- draw Rs 4000 and its interest from that amount after 2 years. After next 3 years, he withdraw the rest of the amount and its interest accrued till that time. In all he obtained Rs 900 as interest. The rate of interest per annum was
   (a) 3%   (b) 4%
   (c) 5%   (d) 6%

14. Divide Rs 1586 in three parts in such a way that their amounts at the end of 2, 3 and 4 years at 5% per annum simple interest be equal.
   (a) Rs 552, Rs 528, Rs 506
   (b) Rs 560, Rs 520, Rs 506
   (c) Rs 556, Rs 524, Rs 506
   (d) Rs 548, Rs 528, Rs 510

15. A owes B Rs 1573 payable $1\frac{1}{2}$ years hence. Also B owes A Rs 1444.50 payable 6 months hence. If they want to settle the account forth with, keeping 14 % as the rate of interest, then who should pay whom and how much ?
   (a) A to B, Rs 28.50
   (b) B to A, Rs 37.50
   (c) A to B, Rs 50
   (d) B to A, Rs 50

Answers

1. (b)　**2.** (a)　**3.** (c)　**4.** (c)　**5.** (a)　**6.** (d)　**7.** (d)　**8.** (b)　**9.** (b)　**10.** (b)
11. (b)　**12.** (b)　**13.** (c)　**14.** (a)　**15.** (d)

Hints and Solutions

1. (b) Let the principal = Rs x and rate of interest = $R\%$ p.a.

Then, $S.I. = 30\%$ of Rs $x = \frac{30}{100} \times x$

$\therefore \frac{x \times R \times 6}{100} = \frac{30}{100} \times x$

$\Rightarrow R = \frac{30}{6} = 5\%$ p.a.

Let the time in which the principal is equal to simple interest be t years, then

$\frac{x \times 5 \times t}{100} = x \Rightarrow t = \frac{100}{5}$ years = 20 years.

2. (a) Let the loan taken be Rs x. Then,

$\frac{x \times 6 \times 1}{100} + \frac{x \times 6.5 \times 1}{100} + \frac{x \times 7 \times 1}{100} + \frac{x \times 7.5 \times 1}{100} = 3375$

$\Rightarrow \frac{27x}{100} = 3375 \Rightarrow x = \frac{3375 \times 100}{27} = $ Rs 12500.

3. (c) Let sum left for the 12 years old and 14 years old sons be Rs x and Rs $(18750-x)$ respectively. Then,

$\frac{x \times 6 \times 5}{100} + x = \frac{(18750-x) \times 4 \times 5}{100} + (18750-x)$

$\Rightarrow \frac{30x}{100} + x = 3750 - \frac{20x}{100} + 18750 - x$

$$\Rightarrow \frac{50x}{100} + 2x = 22500 \Rightarrow \frac{250x}{100} = 22500$$

$$\Rightarrow x = \text{Rs } \frac{22500 \times 100}{250} = \text{Rs } 9000$$

∴ The sum allotted to the other son is Rs 9750.

4. (c) Let the two rates of simple interests p.a. be x % and y %.

Then, $\dfrac{1500 \times x \times 3}{100} - \dfrac{1500 \times y \times 3}{100} = 13.50$

$\Rightarrow 45x - 45y = 13.50 \Rightarrow 45(x - y) = 13.50$

$\Rightarrow x - y = \dfrac{13.50}{45} = 0.3$ %.

5. (a) Let x, y and z be the amounts invested in schemes A, B and C respectively. Then,

$$\left(\frac{x \times 10 \times 1}{100}\right) + \left(\frac{y \times 12 \times 1}{100}\right) + \left(\frac{z \times 15 \times 1}{100}\right) = 3200$$

$\Rightarrow 10x + 12y + 15z = 320000$ ... (i)

Given, $z = 240\%$ of $y = \dfrac{240}{100} \times y = \dfrac{12}{5} y$ ... (ii)

and $z = 150\%$ of $x = \dfrac{150}{100} \times x = \dfrac{3}{2} x$

$\Rightarrow x = \dfrac{2}{3} z = \dfrac{2}{3} \times \dfrac{12}{5} y = \dfrac{8}{5} y$ ... (iii)

∴ From (i), (ii) and (iii) we have

$$10 \times \frac{8}{5} y + 12y + 15 \times \frac{12}{5} y = 320000$$

$\Rightarrow 16y + 12y + 36y = 320000 \Rightarrow 64y = 320000$

$\Rightarrow y = \dfrac{320000}{64} = 5000.$

∴ Sum invested in scheme B = Rs 5000.

6. (d) Let the three parts be x, y and z. Then,

$$\frac{x \times 4 \times 1}{100} = \frac{y \times 6 \times 1}{100} = \frac{z \times 8 \times 1}{100}$$

$\Rightarrow 0.04 x = 0.06 y = 0.08 z$

$\Rightarrow \dfrac{x}{y} = \dfrac{0.06}{0.04} = \dfrac{3}{2}$ and $\dfrac{y}{z} = \dfrac{0.08}{0.06} = \dfrac{8}{6} = \dfrac{4}{3}$

$\Rightarrow \dfrac{x}{y} = \dfrac{6}{4}$ and $\dfrac{y}{z} = \dfrac{4}{3} \Rightarrow x : y : z = 6 : 4 : 3$

∴ Sum invested at 4 % = $\dfrac{6}{13} \times$ Rs 2600 = Rs 1200.

7. (d) Let the three parts be x, y and z. Then,

$$x + \frac{x \times 2 \times 5}{100} = y + \frac{y \times 3 \times 5}{100} = z + \frac{z \times 4 \times 5}{100}$$

$\Rightarrow x + \dfrac{x}{10} = y + \dfrac{3y}{20} = z + \dfrac{z}{5}$

$\Rightarrow \dfrac{11x}{10} = \dfrac{23y}{20} = \dfrac{6z}{5} = k$ (say)

∴ $x = \dfrac{10k}{11}, y = \dfrac{20k}{23}, z = \dfrac{5k}{6}$

Given, $x + y + z = 2379$

$\Rightarrow \dfrac{10k}{11} + \dfrac{20k}{23} + \dfrac{5k}{6} = 2379$

$\Rightarrow \dfrac{1380k + 1320k + 1265x}{1518} = 2379$

$\Rightarrow 3965k = 2379 \times 1518$

$\Rightarrow k = \dfrac{\overset{3}{\cancel{2379}} \times 1518}{\underset{5}{\cancel{3965}}}$

$\Rightarrow$ First part = Rs $\dfrac{10 \times 3 \times 1518}{11 \times 5}$ = Rs 828.

8. (b) Let the three parts be Rs x, Rs y and Rs z. Then,

$$\frac{x \times 2 \times 3}{100} = \frac{y \times 3 \times 4}{100} = \frac{z \times 4 \times 5}{100}$$

$\Rightarrow \dfrac{6x}{100} = \dfrac{12y}{100} = \dfrac{20z}{100} = k$ (say)

$\Rightarrow x = \dfrac{100k}{6}, y = \dfrac{100k}{12}, z = \dfrac{100k}{20}$

$\Rightarrow x = \dfrac{50k}{3}, y = \dfrac{25k}{3}, z = 5k$

Given, $x + y + z = 1440$

$\Rightarrow \dfrac{50k}{3} + \dfrac{25k}{3} + 5k = 1440$

$\Rightarrow 50k + 25k + 15k = 1440 \times 3$

$\Rightarrow 90k = 1440 \times 3 \Rightarrow k = \dfrac{1440 \times 3}{90} = 48$

∴ Largest share = $x = \dfrac{50 \times 48}{3}$ = Rs 800

Smallest share = $z = 5 \times 48$ = Rs 240
Required difference = Rs 800 – Rs 240 = Rs 560.

9. (b) Suppose after t years, P = Rs $4x$ and Amount = Rs $5x$

$\Rightarrow$ $P + S.I.$ after t years = Rs $5x$ ... (i)

Also, given $P : [P + S.I.$ after $(t + 3)$ years]

$= 5 : 7 = 1 : \dfrac{7}{5} = 4x : \dfrac{7}{5} \times 4x = 4x : \dfrac{28x}{5}$

$\Rightarrow$ $P + S.I.$ after $(t + 3)$ years = Rs $\dfrac{28x}{5}$ ... (ii)

$\therefore$ Eq (ii) – Eq (i) gives

$S.I.$ after 3 years = Rs $\left(\dfrac{28x}{5} - 5x\right)$ = Rs $\dfrac{3x}{5}$

$\therefore$ Rate of interest = $\dfrac{\dfrac{3x}{5} \times 100}{4x \times 3}$ = 5% p.a.

10. (b) $S.I.$ for 10 years = Rs $\left(1000 \times \dfrac{5}{100} \times 10\right)$

$= $ Rs 500

Principal after 10 years becomes

$= $ Rs $(1000 + 500)$

$= $ Rs 1500

Amount on that principal after t years = Rs 2000

$\therefore$ $S.I.$ on it = Rs $(2000 - 1500)$ = Rs 500

$\therefore$ $t = \left(\dfrac{500 \times 100}{1500 \times 5}\right)$ years $= 6\dfrac{2}{3}$ years

$\therefore$ Total time $= \left(10 + 6\dfrac{2}{3}\right)$ years $= 16\dfrac{2}{3}$ years.

11. (b) Let the sum be Rs x.

$\therefore$ Rs x are lent at 8% for t years and Rs x are lent

at 7% for $\left(t + \dfrac{1}{2}\right)$ years.

$\therefore$ $\dfrac{x \times t \times 8}{100} + x = 2560$

$\Rightarrow$ $8tx + 100x = 256000$ ... (i)

and $\dfrac{x \times (2t + 1) \times 7}{2 \times 100} + x = 2560$

$\Rightarrow$ $\dfrac{14tx + 7x + 200x}{200} = 2560$

$\Rightarrow$ $14tx + 207x = 512000$ ... (ii)

Performing (i) × 7 – (ii) × 4, we get

$(56tx + 700x) - (56tx + 828x)$

$= 256000 \times 7 - 512000 \times 4$

$\Rightarrow$ $700x - 828x = 1792000 - 2048000$

$\Rightarrow$ $128x = 256000$ $\Rightarrow$ $x = 2000$

Putting the value of x in (i), we get

$8 \times t \times 2000 + 100 \times 2000 = 256000$

$\Rightarrow$ $16000t = 256000 - 200000$

$\Rightarrow$ $16000t = 56000$ $\Rightarrow$ $t = \dfrac{56000}{16000} = 3.5$ years.

12. (b) For the first two years,

Amount $= 40000 + \dfrac{40000 \times 8 \times 2}{100}$

$= 40000 + 6400 = $ Rs 46400

Let x be the principal for the last 3 years.

Then, $x + \dfrac{x \times 8 \times 3}{100} = 35960$

$\Rightarrow$ $x + \dfrac{6x}{25} = 35960$ $\Rightarrow$ $\dfrac{25x + 6x}{25} = 35960$

$\Rightarrow$ $\dfrac{31x}{25} = 35960$ $\Rightarrow$ $x = \dfrac{35960 \times 25}{31} = $ Rs 29000

$\therefore$ Amount paid back after the Second year

$= $ Rs 46400 – Rs 29000 = Rs 17400.

13. (c) Let the rate of interest p.a. be x %

Then, interest on Rs 6000 after 2 years

$= \dfrac{2 \times 6000 \times x}{100} = 120x = I_1$

After withdrawing Rs 4000, Rs 2000 remained in the account.

$\therefore$ Interest on Rs 2000 for 3 years $= \dfrac{3 \times 2000 \times x}{100}$

$= 60x = I_2.$

Given, $I_1 + I_2 = 900$

$\Rightarrow$ $120x + 60x = 900$ $\Rightarrow$ $180x = 900 \Rightarrow x = 5\%.$

14. (a) Similar to Q. 7.

15. (d) Let the present value of what A owes B be Rs x.

Then, $x + \dfrac{x \times 14 \times 3}{2 \times 100} = 1573$

$\Rightarrow$ $x + \dfrac{21x}{100} = 1573 \Rightarrow \dfrac{121x}{100} = 1573$

$\Rightarrow$ $x = \dfrac{1573 \times 100}{121} = 1300$

Let y be the present value of what B owes A.

Then, $y + y \times \dfrac{1}{2} \times \dfrac{14}{100} = 1444.50$

$\Rightarrow$ $y + \dfrac{7y}{100} = 1444.50$ $\Rightarrow$ $107y = 1444.50 \times 100$

$\Rightarrow$ $y = \dfrac{144450}{107} = $ Rs 1350

Hence B must pay Rs 50 to A.

Self Assessment Sheet–18

1. Divide Rs 2000 into two sums such that, if the first be put out at simple interest for 6 years at $3\frac{1}{2}$ per cent and the second for 3 years at $4\frac{1}{2}$ per cent, the interest on the first sum shall be double that of the second.
   (a) Rs 725, Rs 1275   (b) Rs 1125, Rs 875
   (c) Rs 1500, Rs 500   (d) Rs 635, Rs 1365

2. A sum of money was borrowed at 6% per annum simple interest. At the end of first year Rs 6800 was paid off and the rate of interest on the balance was reduced to 5% per annum. If the interest for the second year was $\frac{11}{20}$ of the interest for the first year what was the original sum borrowed?
   (a) Rs 10000   (b) Rs 12000
   (c) Rs 17000   (d) Rs 15000

3. Divide Rs 7053 into three parts so that the amount after 2, 3 and 4 years respectively may be equal, the rates of interest being 4% per annum.
   (a) Rs 2500, Rs 3500, Rs 1053
   (b) Rs 2436, Rs 2349, Rs 2268
   (c) Rs 2568, Rs 3200, Rs 1285
   (d) Rs 2360, Rs 2289, Rs 2404

4. A sum of money at simple interest amounts to Rs 9440 in 3 years. If the rate of interest is increased by 25% the same sum amounts to Rs 9800 in the same time. The original rate of interest is
   (a) 10% p.a.   (b) 8% p.a.
   (c) 7.5% p.a.   (d) 6 % p.a.

5. Ramesh borrowed Rs 7000 from a bank on simple interest. He returned Rs 3000 to the bank at the end of three years and Rs 5450 at the end of five years and closed the account. Find the rate of interest p.a.
   (a) 4%   (b) 5%
   (c) 6%   (d) 8%

Answers

1. (b)   2. (c)   3. (b) [**Hint.** Let the three parts be X, Y and Z. Then, amount after 4 years $= X + \dfrac{X \times 2 \times 4}{100} = \dfrac{27X}{25}$,

$Y + \dfrac{Y \times 3 \times 4}{100} = \dfrac{28Y}{25}$, $Z + \dfrac{Z \times 4 \times 4}{100} = \dfrac{29Z}{25}$. Given $\dfrac{27X}{25} = \dfrac{28Y}{25} = \dfrac{29Z}{25} = K$. Now find $X : Y : Z$ and hence the shares]

4. (d)   5. (b)

Chapter

19

COMPOUND INTEREST

Section-A
COMPOUND INTEREST

KEY FACTS

1. **Definition:** If the borrower and lender agree to fix up a certain interval of time (say, a year or a half year or a quarter of a year etc.) so that the amount (= principal + Interest) at the end of the interval becomes the principal for the next interval, then the total interest over all the intervals calculated in this way is called **Compound Interest** (C.I.).

   Compound Interest = Amount at the end of the period – Original principal

2. **Conversion Period**

   The fixed interval of time at the end of which the interest is calculated and added to the principal at the beginning of the interval is called the **conversion period**.

   $\therefore$ When the interest is calculated and added to the principal every six months, the conversion period is six months.

   Similarly, the conversion period is 3 months when the interest is calculated and added quarterly.

   If no conversion period is specified, the conversion period is taken to be **1 year.**

3. **Formulae**

   Let P be the principal, $R\%$ per annum be the rate of interest, then the amount A and compound interest C.I. at the end of n years is :

   (i) **When the interest is compounded yearly**

   $$A = P\left(1+\frac{R}{100}\right)^n \quad \text{and} \quad \text{C.I.} = A - P = P\left[\left(1+\frac{R}{100}\right)^n - 1\right]$$

   (ii) **When the interest is compounded half yearly**

   $$A = P\left(1+\frac{R}{200}\right)^{2n} \quad \text{and} \quad \text{C.I.} = P\left[\left(1+\frac{R}{200}\right)^{2n} - 1\right]$$

   (iii) **When the interest is compounded quarterly**

   $$A = P\left(1+\frac{R}{400}\right)^{4n} \quad \text{and} \quad \text{C.I.} = P\left[\left(1+\frac{R}{400}\right)^{4n} - 1\right]$$

(iv) When interest is compounded annually but time is a fraction, say $2\frac{1}{3}$ years.

$$\text{Amount} = P\left(1+\frac{R}{100}\right)^2\left(1+\frac{\frac{R}{3}}{100}\right)$$

(v) When the rates are different for different years say, $R_1\%$, $R_2\%$, $R_3\%$ p.a. for 1st year, IInd year and IIIrd year respectively.

$$\text{Amount} = P\left(1+\frac{R_1}{100}\right)\left(1+\frac{R_2}{100}\right)\left(1+\frac{R_3}{100}\right)$$

Solved Examples

Ex. 1. *What will be the compound interest on a sum of Rs 25000 after 3 years at the rate of 12 per cent p.a. ?*

Sol. $P = $ Rs 25000, $n = 3$ years, $r = 12\%$ p.a.

$$\therefore \text{Amount} = P\left(1+\frac{r}{100}\right)^n = \text{Rs } 25000\times\left(1+\frac{12}{100}\right)^3$$

$$= \text{Rs } 25000\times\left(\frac{112}{100}\right)^3 = \text{Rs } 25000\times\frac{28}{25}\times\frac{28}{25}\times\frac{28}{25} = \text{Rs } 35123.20$$

$$\therefore \text{Compound interest} = \text{Rs }(35123.20 - 25000) = \text{Rs } 10123.20$$

Ex. 2. *At what rate per cent per annum will Rs 3000 amount to Rs 3993 in 3 years, if the interest is compounded annually ?*

Sol. $A = $ Rs 3993, $P = $ Rs 3000, $n = 3$, $r = ?$

$$\therefore A = P\left(1+\frac{r}{100}\right)^n \Rightarrow 3993 = 3000\left(1+\frac{r}{100}\right)^3 \Rightarrow \frac{3993}{3000} = \left(1+\frac{r}{100}\right)^3 \Rightarrow \frac{1331}{1000} = \left(1+\frac{r}{100}\right)^3$$

$$\Rightarrow \left(\frac{11}{10}\right)^3 = \left(1+\frac{r}{100}\right)^3 \quad 1+\frac{r}{100} = \frac{11}{10} \Rightarrow \frac{r}{100} = \frac{11}{10}-1 = \frac{1}{10}$$

$$\therefore r = \frac{100}{10} = 10\% \text{ p.a.}$$

Ex. 3. *In how many years will the compound interest on Rs 1000 at the rate of 10% p.a. be Rs 210 ?*

Sol. $P = $ Rs 1000, $C.I. = $ Rs 210 $\Rightarrow A = P + C.I. = $ Rs 1000 + Rs 210 = Rs 1210
$r = 10\%$ p.a., $n = ?$

$$\therefore 1210 = 1000\left(1+\frac{10}{100}\right)^n \Rightarrow \frac{1210}{1000} = \left(1+\frac{1}{10}\right)^n$$

$$\Rightarrow \frac{121}{100} = \left(\frac{11}{10}\right)^n = \left(\frac{11}{10}\right)^2 = \left(\frac{11}{10}\right)^n \Rightarrow n = 2 \text{ years}.$$

Ex. 4. *Find the compound interest on Rs 5000 for one year at 4% per annum, the interest being compounded half yearly ?*

Sol. $P = $ Rs 5000, $r = 4\%$ p.a. $= 2\%$ per half year
$T = 1$ year $= 2$ half years $\Rightarrow n = 2$

$$\therefore \ C.I. = A - P = 5000\left(1 + \frac{2}{100}\right)^2 - 5000 = 5000 \times \left[\left(\frac{51}{50}\right)^2 - 1\right] = 5000\left[\frac{(51)^2 - (50)^2}{50^2}\right]$$

$$= 5000 \times \frac{(51-50) \times (51+50)}{2500} = 2 \times 101 = \text{Rs } 202.$$

Ex. 5. *At what rate per cent per annum will Rs 32000 yield a compound interest of Rs 5044 in 9 months interest being compounded quarterly ?*

Sol. Principal = Rs 32000

Amount = Rs (32000 + 5044) = Rs 37044

Rate = r% p.a. or $\frac{r}{4}$% per quarter

Time = 9 months = 3 quarters, i.e., $n = 3$

$\therefore$ Applying $A = P\left(1 + \frac{r}{100}\right)^n$, we have

$$37044 = 32000\left(1 + \frac{r}{400}\right)^3 \Rightarrow \frac{37044}{32000} = \left(1 + \frac{r}{400}\right)^3$$

$$\Rightarrow \frac{9261}{8000} = \left(1 + \frac{r}{400}\right)^3 \Rightarrow \left(\frac{21}{20}\right)^3 = \left(1 + \frac{r}{400}\right)^3$$

$$\Rightarrow 1 + \frac{r}{400} = \frac{21}{20} \Rightarrow \frac{r}{400} = \frac{21}{20} - 1 = \frac{1}{20} \Rightarrow r = \frac{400}{20} = 20\% \text{ p.a.}$$

Ex. 6. *What is the difference between the compound interest and simple interest on Rs 2500 for 2 years at 4% p.a.?*

Sol. $S.I. = \dfrac{2500 \times 2 \times 4}{100} = \text{Rs } 200$

$$C.I. = 2500\left(1 + \frac{4}{100}\right)^2 - 2500 = 2500 \times \frac{26}{25} \times \frac{26}{25} - 2500$$

$$= 2704 - 2500 = \text{Rs } 204$$

$C.I. - S.I. = \text{Rs } 204 - \text{Rs } 200 = \text{Rs } 4.$

Ex. 7. *On what sum does the difference between the compound interest and the simple interest for 3 years at 10% is Rs 31 ?*

Sol. Let the sum be Rs 100. Then,

$$C.I. = 100\left(1 + \frac{10}{100}\right)^3 - 100 = 100 \times \left(\frac{11}{10}\right)^3 - 100 = 100 \times \frac{1331}{1000} - 100$$

$$= \text{Rs } 133.10 - \text{Rs } 100 = \text{Rs } 33.10$$

$$S.I. = \frac{100 \times 3 \times 10}{100} = \text{Rs } 30$$

$\therefore$ Difference = Rs 33.10 – Rs 30 = Rs 3.10

Now, Rs 3.10 is the difference when the sum is Rs 100.

$\therefore$ Rs 31 is the difference when the sum is Rs $\left(\dfrac{100}{3.10} \times 31\right)$ = Rs 1000.

Ex. 8. *A sum of money at compound interest (compounded annually) doubles itself in 4 years. In how many years will it amount to eight times of itself ?*

Sol. Given, $2P = P\left(1+\dfrac{R}{100}\right)^4 \Rightarrow \left(1+\dfrac{R}{100}\right)^4 = 2$ ...(i)

Also, let the time in which it amounts to eight times of itself be T years.

Then, $8P = P\left(1+\dfrac{R}{100}\right)^T \Rightarrow \left(1+\dfrac{R}{100}\right)^T = 8 = 2^3 = \left(\left(1+\dfrac{R}{100}\right)^4\right)^3$ (from (i))

$$= \left(1+\dfrac{R}{100}\right)^{12}$$

$\therefore$ $T = 12$ years.

Ex. 9. *A sum of money amounts to Rs 4840 in 2 years and Rs 5324 in 3 years at compound interest compounded annually. What is the rate of interest per annum ?*

Sol. Let the principal be Rs P and rate of interest p.a. = r%. Then,

$$P\left(1+\dfrac{r}{100}\right)^2 = 4840 \quad\quad ...(i) \quad\quad \text{and} \quad\quad P\left(1+\dfrac{r}{100}\right)^3 = 5324 \quad\quad ...(ii)$$

$$\Rightarrow \dfrac{5324}{4840} = \dfrac{(1+r/100)^3}{(1+r/100)^2} \Rightarrow 1+\dfrac{r}{100} = \dfrac{1331}{1210}$$

$$\Rightarrow \dfrac{r}{100} = \dfrac{1331}{1210} - 1 = \dfrac{121}{1210} = \dfrac{1}{10} \Rightarrow r = \dfrac{1}{10} \times 100 = 10\% \text{ p.a.}$$

Ex. 10. *If the compound interest on a certain sum for 2 years at 3% p.a. is Rs 101.50, then what will be the simple interest on the same sum at the same rate and for the same time ?*

Sol. Let the principal be Rs P. Then,

$$101.50 = P\left[\left(1+\dfrac{3}{100}\right)^2 - 1\right] = P\left[\left(\dfrac{103}{100}\right)^2 - 1\right]$$

$$\Rightarrow 101.50 = P\left[\dfrac{(103)^2 - (100)^2}{10000}\right] = P\left[\dfrac{(103+100)(103-100)}{10000}\right]$$

$$\Rightarrow P = \dfrac{101.50 \times 10000}{203 \times 3} = \text{Rs} \dfrac{5000}{3}$$

$$\therefore S.I. = \dfrac{5000 \times 3 \times 2}{3 \times 100} = \text{Rs } 100.$$

Ex. 11. *If the rate of interest be 4% per annum for first year, 5% per annum for second year and 6% per annum for third year, then what is the compound interest on Rs 10000 for 3 years ?*

Sol. $A = P\left(1+\dfrac{r_1}{100}\right)\left(1+\dfrac{r_2}{100}\right)\left(1+\dfrac{r_3}{100}\right) \Rightarrow A = 10000\left(1+\dfrac{4}{100}\right)\left(1+\dfrac{5}{100}\right)\left(1+\dfrac{6}{100}\right)$

$$= 10000 \times \dfrac{104}{100} \times \dfrac{105}{100} \times \dfrac{106}{100} = \text{Rs } 11575.20$$

$\therefore$ $C.I. = \text{Rs } 11575.20 - \text{Rs } 10000 = \text{Rs } 1575.20.$

Ex. 12. *The difference between compound interest and simple interest at the same rate on Rs 5000 for 2 years is Rs 72. What is the rate of interest per annum ?*

Sol. Let the rate per cent p.a. be r. Then,

$$S.I. = Rs \left(5000 \times \frac{r}{100} \times 2\right) = Rs \ 100r$$

$$C.I. = Rs \left[5000\left(1+\frac{r}{100}\right)^2 - 5000\right] = Rs \ 5000\left[\left(1+\frac{r}{100}\right)^2 - 1\right] = Rs \ 5000\left[\left(1+\frac{r^2}{10000}+\frac{2r}{100}\right)-1\right]$$

$$= Rs \ 5000\left(\frac{r^2}{10000}+\frac{r}{50}\right) = Rs \ \frac{5000(r^2+200r)}{10000} = Rs \left(\frac{r^2}{2}+100r\right)$$

$\therefore$ C.I. – S.I. = 72

$$\Rightarrow \frac{r^2}{2}+100r-100r=72 \Rightarrow \frac{r^2}{2}=72 \Rightarrow r^2=144 \Rightarrow r=12\% \ \text{p.a.}$$

Question Bank–19(a)

1. What is the compound interest on an amount of Rs 4800 at the rate of 6 per cent p.a. at the end of 2 years ?
   (a) Rs 544.96   (b) Rs 576
   (c) Rs 593.28   (d) Rs 588

2. The principal that amounts to Rs 4913 in 3 years at $6\frac{1}{4}\%$ per annum compound interest compounded annually is
   (a) Rs 4096   (b) Rs 4085
   (c) Rs 4076   (d) Rs 3096

3. Rs 8000 invested at compound interest gives Rs 1261 as interest after 3 years. The rate of interest per annum is
   (a) 25%   (b) 17.5%
   (c) 10%   (d) 5%

4. The compound interest on Rs. 30,000 at 7% per annum is Rs 4347. The period (in years) is
   (a) 2   (b) $2\frac{1}{2}$
   (c) 3   (d) 4

5. How much should a sum of Rs 16000 approximately amount to in 2 years at 10% p.a. compounded half yearly ?
   (a) Rs 17423   (b) Rs 18973
   (c) Rs 19448   (d) Rs 19880

6. The difference between simple interest and compound interest on Rs 1200 for one year at 10% per annum reckoned half yearly is
   (a) Rs 2.50   (b) Rs 3
   (c) Rs 3.75   (d) Rs 4

7. In how many years will a sum of Rs 800 at 10% per annum compound interest, compounded semi-annually become Rs 926.10 ?
   (a) 1 year   (b) 3 years
   (c) 2 years   (d) $1\frac{1}{2}$ years

8. The compound interest on Rs 16000 for 9 months at 20% p.a. compounded quarterly is
   (a) Rs 2518   (b) Rs 2520
   (c) Rs 2522   (d) Rs 2524

9. A sum of Rs 24000 is borrowed for $1\frac{1}{2}$ years at the rate of interest 10% per annum compounded semi anually. What is the compound interest (x) ?
   (a) $x <$ Rs 3000
   (b) Rs 3000 $< x <$ Rs 4000
   (c) Rs 4000 $< x <$ Rs 5000
   (d) $x >$ Rs 5000

10. An amount of Rs x at compound interest at 20% per annum for 3 years becomes y. What is $y : x$?
    (a) 3 : 1   (b) 36 : 25
    (c) 216 : 125   (d) 125 : 216

11. A man deposited Rs 6000 in a bank at 5% simple interest. Another man deposited Rs 5000 at 8% compound interest. After 2 years, the difference of their interests will be
    (a) Rs 230   (b) Rs 232
    (c) Rs 600   (d) Rs 832

12. The difference between compound interest and simple interest on a sum for 2 years at 8% p.a. is Rs 768. The sum is
    (a) Rs 100000   (b) Rs 110000
    (c) Rs 120000   (d) Rs 170000

13. The compound interest on a sum for two years is Rs 832 and the simple interest on the same sum at the same rate for the same period is Rs 800. What is the rate of interest ?
    (a) 6%   (b) 8%
    (c) 10%   (d) 12%

14. A money lender borrows money at 4% p.a. simple interest and pays interest at the end of the year. He

lends it at 6% p.a. compound interest compounded half-yearly and receives the interest at the end of the year. Thus he gains Rs 104.50 a year. The amount of money he borrows is

(a) Rs 4500   (b) Rs 5000
(c) Rs 5500   (d) Rs 6000

15. A sum of money at compound interest amounts to three times of itself in three years. In how many years will it be nine times of itself ?

(a) 6 years   (b) 5 years
(c) 9 years   (d) 7 years

16. If the compound interest on a certain sum for three years at 10% is Rs 993, what would be the simple interest on the same sum at the same rate for the same time ?

(a) Rs 930   (b) Rs 920
(c) Rs 900   (d) Rs 890

17. If the amount is $2\frac{1}{4}$ times of the sum after 2 years, then the rate of compound interest must be

(a) 60%   (b) 40%
(c) 64%   (d) 50%

18. A sum of money invested at compound interest amounts in 3 years to Rs 2400 and in 4 years to Rs 2520. The interest rate per annum is

(a) 5%   (b) 6%
(c) 10%   (d) 12%

19. The compound interest on Rs 2000 in 2 years if the rate of interest is 4% per annum for the first year and 3% per annum for the second year will be

(a) Rs 142.40   (b) Rs 140.40
(c) Rs 141.40   (d) Rs. 143.40

20. Two partners A and B together lend Rs 84100 at 5% compounded annually. The amount which A gets at

the end of 3 years is the same as what B gets at the end of 5 years. Determine the ratio of shares of A and B.

(a) 21 : 20   (b) 441 : 400
(c) 1 : 4   (d) 5 : 21

21. A principal sum of money is lent out at compound interest compounded annually at the rate of 20% per annum for 2 years. It would give Rs 2410 more if the interest is compounded half yearly. Find the principal sum.

(a) Rs 120000   (b) Rs 125000
(c) Rs 100000   (d) Rs 110000

22. A man borrowed Rs 5000 at 10% per annum compound interest. At the end of each year he has repaid Rs 1500. The amount of money he still owes after the third year is

(a) Rs 1600   (b) Rs 1690
(c) Rs 1700   (d) Rs 1790

23. The compound interest on a certain sum for 2 years at 10% per annum is Rs 525. The simple interest on the same sum for double the time at half the rate per cent per annum is

(a) Rs 400   (b) Rs 600
(c) Rs 500   (d) Rs 800

24. Two friends A and B jointly lent out Rs 81600 at 4% compound interest. After 2 years A gets the same amount as B gets after 3 years. The investment made by B was

(a) Rs 40000   (b) Rs 30000
(c) Rs 45000   (d) Rs 38000

25. Find the compound interest on Rs 24000 at 15% per annum for $2\frac{1}{3}$ years.

(a) Rs 8000   (b) Rs 9237
(c) Rs 9327   (d) Rs 9732

Answers

1. (c)	**2.** (a)	**3.** (d)	**4.** (a)	**5.** (c)	**6.** (b)	**7.** (d)	**8.** (c)	**9.** (b)	**10.** (c)
11. (b)	**12.** (c)	**13.** (b)	**14.** (b)	**15.** (a)	**16.** (c)	**17.** (d)	**18.** (a)	**19.** (a)	**20.** (b)
21. (c)	**22.** (b)	**23.** (c)	**24.** (a)	**25.** (c)					

Hints and Solutions

1. (c) $P = $ Rs 4800, $r = 6\%$ p.a., $n = 2$

$\therefore \quad C.I. = A - P = P\left(1 + \frac{r}{100}\right)^n - P$

$= 4800\left(1 + \frac{6}{100}\right)^2 - 4800$

$= 4800 \times \frac{53 \times 53}{50 \times 50} - 4800$

$= 5393.28 - 4800 = $ Rs 593.28

2. (a) $A = $ Rs 4913, $n = 3$, $r = 6\frac{1}{4}\% = \frac{25}{4}\%$, $P = ?$

$\therefore \quad 4913 = P\left(1 + \frac{25}{400}\right)^3 \Rightarrow 4913 = P\left(1 + \frac{1}{16}\right)^3$

$\Rightarrow 4913 = P\left(\frac{17}{16}\right)^3 \Rightarrow 4913 = P \times \frac{4913}{4096}$

$\Rightarrow P = $ Rs 4096

3. (d) $P = $ Rs 8000, $C.I. = $ Rs 1261

   $\Rightarrow$ Amount $= $ Rs 9261, $n = 3$, $r = ?$

   $\therefore \quad 9261 = 8000\left(1 + \dfrac{r}{100}\right)^3$

   $\Rightarrow \left(1 + \dfrac{r}{100}\right)^3 = \dfrac{9261}{8000} = \left(\dfrac{21}{20}\right)^3$

   $\Rightarrow 1 + \dfrac{r}{100} = \dfrac{21}{20}$

   $\Rightarrow \dfrac{r}{100} = \dfrac{21}{20} - 1 = \dfrac{1}{20}$

   $\Rightarrow r = \dfrac{100}{20}\% = 5\%$ p.a.

4. (a) $P = $ Rs 30000, $r = 7\%$ p.a., $C.I. = $ Rs 4347, $n = ?$

   $\Rightarrow$ Amount $= $ Rs 30000 + Rs 4347 = Rs 34347

   $\therefore \quad 34347 = 30000\left(1 + \dfrac{7}{100}\right)^n$

   $\Rightarrow \left(\dfrac{107}{100}\right)^n = \dfrac{34347}{30000} = \dfrac{11449}{10000}$

   $\Rightarrow \left(\dfrac{107}{100}\right)^n = \left(\dfrac{107}{100}\right)^2 \Rightarrow n = 2$

5. (c) $P = $ Rs 16000, $r = 10\%$ p.a., $= 5\%$ per half year

   $n = 2$ years $= 4$ half years

   $\therefore$ Amount $= 16000\left(1 + \dfrac{5}{100}\right)^4 = 16000 \times \left(\dfrac{21}{20}\right)^4$

   $= \dfrac{16000 \times 194481}{160000}$

   $= 19448.10 = $ Rs 19448 (approx)

6. (b) $P = $ Rs 1200, $R = 10\%$ p.a., $T = 1$ year

   $\therefore \quad S.I. = \dfrac{P \times R \times T}{100} = \dfrac{1200 \times 10 \times 1}{100} = $ Rs 120

   For $C.I.$ reckoned half yearly,
   $P = $ Rs 1200, $r = 5\%$ per half year, $n = 2$ half year

   $\therefore \quad C.I. = $ Rs $\left(1200\left(1 + \dfrac{5}{100}\right)^2 - 1200\right)$

   $= $ Rs $\left(1200 \times \dfrac{21}{20} \times \dfrac{21}{20} - 1200\right)$

   $= $ Rs 1323 - Rs 1200 = Rs 123.

   $\therefore$ Required difference = Rs 123 - Rs 120 = Rs 3.

7. (d) $P = $ Rs 800, $r = 10\%$ p.a. $= 5\%$ per half year,
   $A = $ Rs 926.10, Time $= 2n$

   $\therefore \quad 926.10 = 800\left(1 + \dfrac{5}{100}\right)^{2n}$

   $\Rightarrow \dfrac{9261}{8000} = \left(1 + \dfrac{1}{20}\right)^{2n} \Rightarrow \left(\dfrac{21}{20}\right)^3 = \left(\dfrac{21}{20}\right)^{2n}$

   $\Rightarrow 2n = 3 \Rightarrow n = \dfrac{3}{2} = 1\dfrac{1}{2}$ years.

8. (c) $P = $ Rs 16000, $r = 20\%$ p.a. $= 5\%$ per quarter,
   Time = 9 months = 3 quarters

   $\therefore \quad A = 16000\left(1 + \dfrac{5}{100}\right)^3$

   $= 16000 \times \left(\dfrac{21}{20}\right)^3 = 16000 \times \dfrac{21 \times 21 \times 21}{8000}$

   $= $ Rs 18522

   $\therefore C.I. = $ Rs 18522 - Rs 16000 = Rs 2522.

9. (b) $P = $ Rs 24000, Time $= 1\dfrac{1}{2}$ years = 3 half years,
   $r = 10\%$ p.a. $= 5\%$ per half year

   $\therefore \quad A = 24000\left(1 + \dfrac{5}{100}\right)^3 = 24000 \times \left(\dfrac{21}{20}\right)^3$

   $= 24000 \times \dfrac{9261}{8000} = $ Rs 27783.

   $\therefore \quad C.I. = $ Rs 27783 - Rs 24000 = Rs 3783.

   $\therefore \quad C.I. (x)$ lies between Rs 3000 and Rs 4000.

10. (c) $P = $ Rs x, $A = $ Rs y, $r = 20\%$ p.a., $n = 3$ years

    $\therefore \quad y = x\left(1 + \dfrac{20}{100}\right)^3 \Rightarrow y = x\left(\dfrac{6}{5}\right)^3 \Rightarrow \dfrac{y}{x} = \dfrac{216}{125}$

11. (b) First man's $S.I. = \dfrac{\text{Rs } 6000 \times 5 \times 2}{100} = $ Rs 600

    Second man's $C.I. = $ Rs $5000\left[\left(1 + \dfrac{8}{100}\right)^2 - 1\right]$

    $= $ Rs $5000\left[\left(\dfrac{27}{25}\right)^2 - 1\right]$

    $= $ Rs $5000\left[\dfrac{729 - 625}{625}\right]$

    $= $ Rs $\dfrac{5000 \times 104}{625} = $ Rs 832.

    $\therefore$ Difference in interests = Rs 832 - Rs 600
    $= $ Rs 232.

12. (c) Let the sum be Rs 100. Then,

$$S.I. = Rs\ \frac{100 \times 2 \times 8}{100} = Rs\ 16$$

$$C.I. = Rs\ 100\left[\left(1+\frac{8}{100}\right)^2 - 1\right]$$

$$= Rs\ 100\left[\left(\frac{27}{25}\right)^2 - 1\right] = Rs\ 100 \times \left[\frac{729-625}{625}\right]$$

$$= Rs\left(\frac{100 \times 104}{625}\right) = Rs\ 16.64$$

∴ Difference = Re 0.64

If the difference between C.I. and S.I. is Re 0.64, Sum = Rs 100

If the difference is Rs 768, sum = Rs $\left(\frac{100}{0.64} \times 768\right)$

$$= Rs\ 120000.$$

13. (b) Let the sum be Rs P and rate of interest per annum be $R\%$

Then, $P\left[\left(1+\frac{R}{100}\right)^2 - 1\right] - \frac{2PR}{100}$

$$= Rs\ 832 - Rs\ 800 = Rs\ 32$$

$$\Rightarrow P\left[1+\frac{2R}{100}+\frac{R^2}{10000}-1\right] - \frac{2PR}{100} = 32$$

$$\Rightarrow \frac{PR^2}{10000} = 32 \Rightarrow PR \times R = 320000 \quad \ldots (i)$$

Also, $\frac{2PR}{100} = 800\ (S.I.) \Rightarrow PR = 40000 \quad \ldots (ii)$

∴ From (i) and (ii)

$$40000 \times R = 320000 \Rightarrow R = 8\% \text{ p.a.}$$

14. (b) Let the amount of money he borrows be Rs x. Then,

Interest paid by the money lender $= \frac{x \times 4 \times 1}{100}$

$$= Rs\ \frac{4x}{100}$$

Interest received by the money lender

$$= x\left[\left(1+\frac{3}{100}\right)^2 - 1\right]$$

$$= x\left[\frac{(103)^2}{(100)^2} - 1\right] = x\left[\frac{103^2 - 100^2}{10000}\right]$$

$$= x\left[\frac{(103+100)(103-100)}{10000}\right] = \frac{609x}{10000}$$

Given, $\frac{609x}{10000} - \frac{4x}{100} = 104.50$

$$\Rightarrow \frac{209x}{10000} = 104.50 \Rightarrow 209x = 1045000$$

$$\Rightarrow x = \frac{1045000}{209} = Rs\ 5000.$$

15. (a) Given, $3P = P\left(1+\frac{r}{100}\right)^3$

$$\Rightarrow \left(1+\frac{r}{100}\right)^3 = 3 \qquad \ldots(i)$$

Let t be the time in years in which the sum will be nine times of itself. Then,

$$9P = P\left(1+\frac{r}{100}\right)^t \Rightarrow \left(1+\frac{r}{100}\right)^t = 9 = 3^2$$

$$\Rightarrow \left(1+\frac{r}{100}\right)^t = \left(\left(1+\frac{r}{100}\right)^3\right)^2 \text{ (From (i))}$$

$$\Rightarrow \left(1+\frac{r}{100}\right)^t = \left(1+\frac{r}{100}\right)^6$$

$$\Rightarrow t = 6 \text{ years.}$$

16. (c) Given, $993 = P\left[\left(1+\frac{10}{100}\right)^3 - 1\right]$

$$\Rightarrow 993 = P\left[\left(\frac{11}{10}\right)^3 - 1\right] \Rightarrow 993 = P\left[\frac{1331-1000}{1000}\right]$$

$$\Rightarrow P = \frac{993 \times 1000}{331}$$

$$\Rightarrow P = Rs\ 3000.$$

$$\therefore S.I. = Rs\ \frac{3000 \times 3 \times 10}{100} = Rs\ 900.$$

17. (d) Let the sum be Rs x. Then, Amount = Rs $\frac{9x}{4}$, $n=2, r=?$

$$\therefore \frac{9x}{4} = x\left(1+\frac{r}{100}\right)^2$$

$$\Rightarrow \frac{9}{4} = \left(1+\frac{r}{100}\right)^2 \Rightarrow \left(\frac{3}{2}\right)^2 = \left(1+\frac{r}{100}\right)^2$$

$\Rightarrow \dfrac{r}{100} = \dfrac{3}{2} - 1 = \dfrac{1}{2} \Rightarrow r = 50\%$ p.a.

18. (a) Let the sum of money invested be Rs x and interest rate per annum = $r\%$

Then, $x\left(1 + \dfrac{r}{100}\right)^3 = $ Rs 2400 ... (i)

and $x\left(1 + \dfrac{r}{100}\right)^4 = $ Rs 2520 ... (ii)

Dividing equation (ii) by (i), we get

$\left(1 + \dfrac{r}{100}\right) = \dfrac{2520}{2400} \Rightarrow \dfrac{r}{100} = \dfrac{2520 - 2400}{2400} = \dfrac{120}{2400}$

$\Rightarrow r = \dfrac{1}{20} \times 100 = 5\%$ p.a.

19. (a) $P = $ Rs 2000, $r_1 = 4\%$ p.a. for 1st year, $r_2 = 3\%$ p.a. for 2nd year

$\therefore A = P\left(1 + \dfrac{r_1}{100}\right)\left(1 + \dfrac{r_2}{100}\right)$

$= 2000\left(1 + \dfrac{4}{100}\right)\left(1 + \dfrac{3}{100}\right)$

$= $ Rs $\left(2000 \times \dfrac{26}{25} \times \dfrac{103}{100}\right)$

$= $ Rs 2142.40

$\therefore$ C.I. = Rs 2142.40 – Rs 2000 = Rs 142.40.

20. (b) Let the share of $A = $ Rs x. Then,
Share of $B = $ Rs $(84100 - x)$

$\therefore x\left(1 + \dfrac{5}{100}\right)^3 = (84100 - x)\left(1 + \dfrac{5}{100}\right)^5$

$\Rightarrow$ Ratio of shares of A and B

$= \dfrac{x}{84100 - x} = \left(1 + \dfrac{5}{100}\right)^2 = \left(\dfrac{21}{20}\right)^2 = \dfrac{441}{400}$

21. (c) Let the principal sum be Rs x. Then,
Amount (at interest compounded annually)

$= x\left(1 + \dfrac{20}{100}\right)^2 = \dfrac{36x}{25}$

Amount (at interest compounded half yearly)

$= x\left(1 + \dfrac{10}{100}\right)^4 = \dfrac{14641x}{10000}$

Given, $\dfrac{14641x}{10000} - \dfrac{36x}{25} = 2410$

$\Rightarrow \dfrac{14641x - 14400x}{10000} = 2410$

$\Rightarrow \dfrac{241x}{10000} = 2410 \Rightarrow x = $ Rs 100000

22. (b) Since the man repays Rs 1500 at the end of each year.

$\therefore$ Amount he owes at the end of 1st year

$= 5000\left(1 + \dfrac{10}{100}\right) - 1500$

$= 5500 - 1500 = $ Rs 4000

Amount he owes at the end of 2nd year

$= 4000\left(1 + \dfrac{10}{100}\right) - 1500$

$= 4400 - 1500 = $ Rs 2900

Amount he owes at the end of 3rd year

$= 2900\left(1 + \dfrac{10}{100}\right) - 1500$

$= 3190 - 1500 = $ Rs 1690

23. (c) Let the sum be Rs x. Then,

$x\left[\left(1 + \dfrac{10}{100}\right)^2 - 1\right] = 525 \Rightarrow x\left[\left(\dfrac{11}{10}\right)^2 - 1\right] = 525$

$\Rightarrow x\left[\dfrac{121 - 100}{100}\right] = 525 \Rightarrow x = \dfrac{525 \times 100}{21} = $ Rs 2500

For S.I.
$P = $ Rs 2500, $R = 5\%$ p.a. and $T = 4$ years

Then, $S.I. = $ Rs $\left(\dfrac{2500 \times 5 \times 4}{100}\right) = $ Rs 500

24. (a) Let the investment made by A be Rs x. Then, investment made by $B = $ Rs $(81600 - x)$

Given, $x\left(1 + \dfrac{4}{100}\right)^2 = (81600 - x)\left(1 + \dfrac{4}{100}\right)^3$

$\Rightarrow \dfrac{x}{(81600 - x)} = \dfrac{\left(1 + \dfrac{4}{100}\right)^3}{\left(1 + \dfrac{4}{100}\right)^2}$

$\Rightarrow \dfrac{x}{(81600 - x)} = \left(1 + \dfrac{4}{100}\right) = \dfrac{26}{25}$

$\Rightarrow 25x = (81600 - x) \times 26$

$\Rightarrow 25x + 26x = 81600 \times 26$

$\Rightarrow 51x = 81600 \times 26$

$\Rightarrow x = \dfrac{81600 \times 26}{51} = 41600$

∴   Investment made by B = Rs (81600 – Rs 41600)
        = Rs 40000.

25. (c) P = Rs 24000, R = 15% p.a., $T = 2\dfrac{1}{3}$ years

∴ $A = 24000\left(1 + \dfrac{15}{100}\right)^2 \left(1 + \dfrac{\frac{1}{3} \times 15}{100}\right)$

$= 24000\left(1 + \dfrac{3}{20}\right)^2 \left(1 + \dfrac{1}{20}\right)$

$= 24000\left(\dfrac{23}{20}\right)^2 \times \dfrac{21}{20} = 24000 \times \dfrac{23 \times 23 \times 21}{8000}$

$= $ Rs 33327

∴ $C.I.$ = Rs 33327 – Rs 24000 = Rs 9327.

Section-B
PROBLEMS BASED ON APPRECIATION AND DEPRICIATION

KEY FACTS

1. In our day-to-day life, we observe that there are certain values as population of a city, property, height of a tree, etc., which increase in magnitude over a period of time. The **relative increase in the value** is called **growth** and the **growth per unit of time** is called the **rate of growth**. Thus there is an **appreciation** in the value of the article. Similarly certain values as the value of a car or a machinery etc., decrease in magnitude over a period of time. Such articles are said to **depreciate** in value. We use the compound interest formula to calculate the appreciation and depreciation in values.

2. **Formulae** :
   (i) If P is the population of a town or city at a beginning of a certain year and the population grows at the constant rate of R% per annum, then population after n years $= P\left(1 + \dfrac{R}{100}\right)^n$

   (ii) If V is the value of a car or machinery at the beginning of a certain year and it depreciates at a constant rate of R% per annum, then its value after n years $= V\left(1 - \dfrac{R}{100}\right)^n$

   (iii) If V is the value of an entity at a beginning of a certain year and it appreciates at R_1% for 1st year, at R_2% for 2nd year and depreciates at R_3% for the 3rd year, then its value after 3 years is $= V\left(1 + \dfrac{R_1}{100}\right)\left(1 + \dfrac{R_2}{100}\right)\left(1 - \dfrac{R_3}{100}\right)$.

Solved Examples

Ex. 1. *The annual increase in the population of a town is 10%. If the present population of the town is 180000, then what will be its population after two years ?*

Sol. Population after two years $= 180000\left(1+\dfrac{10}{100}\right)^2$

$$= 180000\times\left(\dfrac{11}{10}\right)^2 = 180000\times\dfrac{121}{100} = 217800.$$

Ex. 2. *The population of a town was 1,60,000 three years ago. If it increased by 3%, 2.5% and 5% respectively in the last three years, then what is the present population ?*

Sol. Present population $= 1,60,000\left(1+\dfrac{3}{100}\right)\left(1+\dfrac{2.5}{100}\right)\left(1+\dfrac{5}{100}\right)$

$$= 160000\times\dfrac{103}{100}\times\dfrac{102.5}{100}\times\dfrac{105}{100} = 177366.$$

Ex. 3. *The population of a town increases by 5% annually. If the population in 2009 is 1,38,915 what was it in 2006 ?*

Sol. Let the population in 2006 be x.

Pop. in 2009 $= x\times\left(1+\dfrac{5}{100}\right)^3 \Rightarrow 1,38,915 = x\times\left(\dfrac{21}{20}\right)^3 \Rightarrow x = \dfrac{138915\times8000}{9261} = 120000.$

Ex. 4. *The population of a village is 10000. If the population increases by 10% in the first year, by 20% in the second year and due to mass exodus it decreases by 5% in the third year, what will be its population after 3 years ?*

Sol. Population after 3 years $= 10000\left(1+\dfrac{10}{100}\right)\left(1+\dfrac{20}{100}\right)\left(1-\dfrac{5}{100}\right)$

$$= 10000\times\dfrac{11}{10}\times\dfrac{6}{5}\times\dfrac{19}{20} = 12540.$$

Ex. 5. *Depreciation applicable to an equipment is 20%. By how much per cent will the value of the equipment 3 years from now be less than the present value ?*

Sol. Let the present value of the equipment be x. Then,

The value of the equipment three years from now $= x\left(1-\dfrac{20}{100}\right)^3 = x\times\left(\dfrac{4}{5}\right)^3 = \dfrac{64x}{125}$

$\therefore$ Required percentage $= \dfrac{\left(x-\dfrac{64x}{125}\right)}{x}\times100 = \dfrac{\left(\dfrac{125x-64x}{125}\right)}{x}\times100$

$$= \dfrac{61}{125}\times100 = 48.8\%.$$

Question Bank–19(b)

1. It is observed that the population of a city increases at the rate of 8% per annum. If the present population of the city is 125000, then the population of the city after 2 years will be
   (a) 145000   (b) 145800
   (c) 154000   (d) 154800

2. The population of a city is 7.26 lakh today. If the population has been increasing at the rate of 10% per year, then two years ago, the population would have been
   (a) 6 lakh   (b) 5.5 lakh
   (c) 5 lakh   (d) 4.5 lakh

3. The population of a village was 20,000 and after 2 years it became 22050. What is the rate of increase per annum ?
   (a) 10%   (b) 8%
   (c) 5%   (d) 6%

4. The population of a city increases at the rate of 4% per annum. There is additional annual increase of 1% due to influx of job seekers. The % increase in the population after 2 years is
   (a) 10 (b) 10.25
   (c) 10.5 (d) 10.75

5. The value of a machine depreciates at the rate of 10% every year. It was purchased 3 years ago. If its present value is Rs 8748, its purchase price was
   (a) Rs 10000 (b) Rs 11372
   (c) Rs 12000 (d) Rs 12500

6. In the month of January, the railway police caught 4000 ticketless travellers. In February, the number rose by 5%. However, due to constant vigil by the police and the railway staff, the number reduced by 5% and in April it further reduced by 10% . The total number of ticketless travellers caught in the month of April was
   (a) 3125 (b) 3255
   (c) 3575 (d) 3591

7. The population of a variety of tiny bush in an experimental field increased by 10% in the first year, increased by 8% in the second year but decreased by 10% in the third year. If the present number of bushes in the experimental field is 26730, then the number of bushes in the beginning was
   (a) 25000 (b) 27000
   (c) 28000 (d) 24600

8. A building worth Rs 133100 is constructed on land worth Rs 72900. After how many years will the value of both be same if the land appreciates at 10% p.a. and building depreciates at 10% p.a. ?
   (a) $1\frac{1}{2}$ years (b) 2 years
   (c) $2\frac{1}{2}$ years (d) 3 years

9. The bacteria in a culture grows by 10% in the first hour, decreases by 10% in the second hour and again increases by 10% in the third hour. If the original count of the bacteria in a sample is 10000, find the bacteria count at the end of 3 hours.
   (a) 13310 (b) 10890
   (c) 10990 (d) 11000

10. Sanjay opened a grocery shop with an initial investment of Rs 40,000. In the first year, he incurred a loss of 5%. However during the second year, he earned a profit of 10% which in the third year rose to $12\frac{1}{2}$ %. Calculate the net profit for the entire period of three years ?
    (a) Rs 6500 (b) Rs 47025
    (c) Rs 7025 (d) Rs 46500

11. The production of a company has ups and downs every year. The production increases for two consecutive years consistently by 15% and in the third year it decreases by 10%. Again in the next two years it increases by 15% each year and decreases by 10% in the third year. If we start counting from the year 1998, approximately what will be the effect on production of the company in 2002 ?
    (a) 37% increase (b) 32% increase
    (c) 42% increase (d) 52% increase

12. The population of a town increases 4% annually but is decreased by emigration annually to the extent of $\frac{1}{2}$% . What will be the increase per cent in 3 years ?
    (a) 9.8 (b) 10
    (c) 10.5 (d) 10.8

13. The current birth rate per thousand is 32, whereas the corresponding death rate is 11 per thousand. The net growth rate in terms of population increase in per cent is given by
    (a) 0.0021% (b) 0.021%
    (c) 2.1% (d) 21%

14. Food grain production in India in 1994 was 1520 lakh tonnes. If the product is increasing at a constant rate of 2.8% per annum, what is the production expected to be in 2010 A.D. on that basis ?
    (Given that $(1.028)^{16} = 1.55557$)
    (a) 2265 lakh tonnes (b) 2300.5 lakh tonnes
    (c) 2364.5 lakh tonnes (d) 3000 lakh tonnes

15. The half life of Uranium - 233 is 160000 years, i.e., Uranium 233 decays at a constant rate in such a way that it reduces to 50% in 160000 years. In how many years will it reduce to 25% ?
    (a) 80000 years (b) 240000 years
    (c) 320000 years (d) 40000 years

Answers

1. (b) 2. (a) 3. (c) 4. (b) 5. (c) 6. (d) 7. (a) 8. (d) 9. (b) 10. (c)
11. (a) 12. (d) 13. (c) 14. (c) 15. (c)

Hints and Solutions

1. (b) Population of the city after 2 years

$$= 125000\left(1 + \frac{8}{100}\right)^2 = 125000 \times \frac{27}{25} \times \frac{27}{25}$$

$$= 145800.$$

2. (a) Let the population 2 years ago be x. Then,

$$7.26 = x\left(1 + \frac{10}{100}\right)^2 \Rightarrow 7.26 = x \times \frac{11}{10} \times \frac{11}{10}$$

$$\Rightarrow x = \frac{7.26 \times 100}{121} = 6 \text{ lakh}$$

3. (c) Let the rate of increase of population per annum be $r\%$. Then,

$$22050 = 20000\left(1 + \frac{r}{100}\right)^2$$

$$\Rightarrow \frac{22050}{20000} = \left(1 + \frac{r}{100}\right)^2$$

$$\Rightarrow \frac{441}{400} = \left(1 + \frac{r}{100}\right)^2 \Rightarrow \left(\frac{21}{20}\right)^2 = \left(1 + \frac{r}{100}\right)^2$$

$$\Rightarrow 1 + \frac{r}{100} = \frac{21}{20} \Rightarrow \frac{r}{100} = \frac{21}{20} - 1 = \frac{1}{20}$$

$$\Rightarrow r = \frac{1}{20} \times 100 = 5\% \text{ p.a.}$$

4. (b) Let the population of the city be 100. Then

Population of city after 2 years $= 100\left(1 + \frac{5}{100}\right)^2$

$$= 100 \times \frac{21}{20} \times \frac{21}{20} = 110.25$$

$\therefore$ Increase per hundred in 2 years $= 110.25 - 100$
$= 10.25\%$.

5. (c) Let the purchase price of the machine be Rs x.

Then, $8748 = x\left(1 - \frac{10}{100}\right)^3 = x \times \frac{9}{10} \times \frac{9}{10} \times \frac{9}{10}$

$$\Rightarrow x = \frac{8748 \times 1000}{729} = \text{Rs } 12000.$$

6. (d) Total number of ticketless travellers in April

$$= 4000 \times \left(1 + \frac{5}{100}\right)\left(1 - \frac{5}{100}\right)\left(1 - \frac{10}{100}\right)$$

$$= 4000 \times \frac{21}{20} \times \frac{19}{20} \times \frac{9}{10}$$

$$= 3591.$$

7. (a) Let the number of bushes in the beginning be x.

Then, $26730 = x\left(1 + \frac{10}{100}\right)\left(1 + \frac{8}{100}\right)\left(1 - \frac{10}{100}\right)$

$$\Rightarrow 26730 = x \times \frac{11}{10} \times \frac{27}{25} \times \frac{9}{10}$$

$$\Rightarrow x = \frac{26730 \times 10 \times 25 \times 10}{11 \times 9 \times 27}$$

$$\Rightarrow x = 25000.$$

8. (d) Let the required time be t years. Then,

$$72900\left(1 + \frac{10}{100}\right)^t = 133100\left(1 - \frac{10}{100}\right)^t$$

$$\Rightarrow \left(\frac{11}{10}\right)^t \times \left(\frac{10}{9}\right)^t = \frac{133100}{72900}$$

$$\Rightarrow \left(\frac{11}{10} \times \frac{10}{9}\right)^t = \left(\frac{11}{9}\right)^3 \Rightarrow \left(\frac{11}{9}\right)^t = \left(\frac{11}{9}\right)^3$$

$$\Rightarrow t = 3 \text{ years.}$$

9. (b) Bacteria count after 3 years

$$= 10000\left(1 + \frac{10}{100}\right)\left(1 - \frac{10}{100}\right)\left(1 + \frac{10}{100}\right)$$

$$= 10000 \times \frac{11}{10} \times \frac{9}{10} \times \frac{11}{10} = 10890.$$

10. (c) At the end of three years, worth of Sanjay's

assets $= 40000\left(1 - \frac{5}{100}\right)\left(1 + \frac{10}{100}\right)\left(1 + \frac{12.5}{100}\right)$

$$= 40000 \times \frac{19}{20} \times \frac{11}{10} \times \frac{9}{8} = \text{Rs } 47025$$

$\therefore$ His net profit $= \text{Rs } 47025 - \text{Rs } 40000$
$= \text{Rs } 7025.$

11. (a) Let the production in 1998 be 100 units. Then, Production in 2002

$$= 100\left(1 + \frac{15}{100}\right)^2\left(1 - \frac{10}{100}\right)\left(1 + \frac{15}{100}\right)$$

$$= 100 \times \frac{23}{20} \times \frac{23}{20} \times \frac{9}{10} \times \frac{23}{20} = 136.88$$

$\therefore$ Increase in production $= (136.88 - 100)\%$
$= 36.88\% = 37\%.$

12. (d) Net increase in population in a year

$$= \left(4 - \frac{1}{2}\right)\% = 3\frac{1}{2}\%$$

Let the original population of the town be 100. Then,

Population after 3 years $= 100\left(1 + \frac{3.5}{100}\right)^3$

$$= \frac{100 \times 207 \times 207 \times 207}{100 \times 100 \times 100} = 110.87$$

∴ % increase $= (110.87 - 100)\% = 10.87\% = 10.8\%$.

13. (c) Net growth in population on 1000 = (32 — 11)
$$= 21$$

∴ Net growth rate $= \left(\frac{21}{1000} \times 100\right)\% = 2.1\%$.

14. (c) Food grain production in 1994 = 1520 lakh tonnes

Rate of increase of production = 2.8% p.a.

Number of years = 16

∴ Food grain production in 2010 A.D.

$$= \left[1520 \times \left(1 + \frac{2.8}{100}\right)^{16}\right] \text{ lakh tonnes}$$

$$= 1520 \times (1.028)^{16} \text{ lakh tonnes}$$

$$= (1520 \times 1.55557) \text{ lakh tonnes}$$

$= 2364.4664$ lakh tonnes

$= 2364.5$ lakh tonnes

15. (c) Let the rate of decay of Uranium be R per cent per year.

Also, let the initial amount of Uranium be 1 unit. Since, the half life of Uranium - 233 is 160000 years, therefore

$$\left(1 - \frac{R}{100}\right)^{160000} = \frac{1}{2} \qquad \ldots (i)$$

Suppose Uranium - 233 reduces to 25% in t years. Then,

$$\left(1 - \frac{R}{100}\right)^{t} = \frac{25}{100} = \frac{1}{4} = \left(\frac{1}{2}\right)^{2}$$

$$= \left(\left(1 - \frac{R}{100}\right)^{160000}\right)^{2} = \left(1 - \frac{R}{100}\right)^{320000}$$

⇒ $t = 320000$ years.

Self Assessment Sheet–19

1. A man borrows money at 3% per annum interest payable yearly and lends it immediately at 5% per annum, interest payable half-yearly and gains there by Rs 165 at the end of a year. What sum of money does he borrow?

(a) Rs 6000   (b) Rs 7500

(c) Rs 8000   (d) Rs 9000

2. Two partners A and B together lend Rs 2523 at 5% compounded annually. The amount A gets at the end of 3 years is the same as B gets at the end of 5 years. Determine the share of A.

(a) Rs 1200   (b) Rs 1323

(c) Rs 1450   (d) Rs 1563

3. A principal sum of money is lent out at compound interest compounded annually at the rate of 20% per annum for 2 years. It would give Rs 2410 more if the interest is compounded half-yearly. The principal sum is :

(a) Rs 1,00,000   (b) Rs 20,000

(c) Rs 10,000   (d) Rs 1,10,000

4. The difference between simple and compound interest on a certain sum for 2 years at 5% per annum compounded annually is Rs 75. Find the sum:

(a) Rs 40000   (b) Rs 36000

(c) Rs 45000   (d) Rs 30000

5. A father divides his property between his two sons

A and B. A invests the amount at compound interest of 8% p.a. B invests the amount at 10% p.a. simple interest. At the end of 2 years, the interest received by B is Rs 1336 more than the interest received by A. Find A's share in the father's property of Rs 25,000.

(a) Rs 12000   (b) Rs 13000

(c) Rs 10000   (d) Rs 12500

6. A sum is invested for 3 years compounded at 5%, 10% and 20% respectively. In three years, if the sum amounts to Rs 16,632, then find the sum.

(a) Rs 11000   (b) Rs 12000

(c) Rs 13000   (d) Rs 14000

7. A tree increases annually by $\frac{1}{8}$th of its height. By how much will it increase after 2 years, if it stands today 64 cm high?

(a) 72 cm   (b) 74 cm

(c) 75 cm   (d) 81 cm

8. The population of a town increases annually by 25%. If the present population is 1 crore, then what is the difference between the population three years ago and 2 years ago?

(a) 25,00,000   (b) 12,80,000

(c) 15,60,000   (d) None of these

9. A finance company declares that, with compound interest rate, a sum of money deposited by anyone will become 8 times in three years. If the same amount is deposited at the same compound-rate of interest then in how many years it will become 16 times?

(a) 4   (b) 5

(c) 6   (d) 7

10. A merchant starts with a certain capital and gains annually at the rate of 25%. At the end of 3 years his capital is worth Rs 10,000. What was his original capital?

(a) Rs 5120   (b) Rs 5220

(c) Rs 5210   (d) Rs 5130

Answers

| 1. (c) | 2. (b) | 3. (a) | 4. (d) | 5. (c) | 6. (b) | 7. (d) | 8. (b) | 9. (a) | 10. (a) |

Chapter 20

TIME AND WORK

KEY FACTS

1. If a person can finish a piece of work in n days, then the work done by the person in 1 day is $\frac{1}{n}$ th part of the work.

2. If a person completes $\frac{1}{n}$ th part of a work in 1 day, then the time taken by the person to finish the work is n days.

3. A cistern is filled with two pipes, one pipe to fill it which is called an **inlet** and the other pipe to empty it, which is called an **outlet**.

4. If an inlet fills a tank in n hours, then it fills $\frac{1}{n}$ th part of the tank in 1 hour, i.e., work done by it in 1 hour is $\frac{1}{n}$.

5. If an outlet empties a full tank in m hours, then it will empty $\frac{1}{m}$ th part of the tank in 1 hour, i.e., work done by it is $-\frac{1}{m}$.

Solved Examples

Ex. 1. *A and B can do a piece of work in 72 days; B and C can do it in 120 days; A and C can do it in 90 days. In what time can A alone do it ?*

Sol. (A + B)'s 1 days' work = $\frac{1}{72}$; (B + C)'s 1 days' work = $\frac{1}{120}$; (A + C)'s 1 days' work = $\frac{1}{90}$

∴ (A + B)'s + (B + C)'s + (A + C)'s 1 days' work $= \frac{1}{72} + \frac{1}{120} + \frac{1}{90} = \frac{5 + 3 + 4}{360} = \frac{12}{360} = \frac{1}{30}$

$\Rightarrow$ 2 (A + B + C)'s 1 days' work $= \frac{1}{30}$ $\Rightarrow$ (A + B + C)'s 1 days' work $= \frac{1}{60}$

∴ A's 1 days' work = (A+B+C)'s 1 days' work – (B + C)'s 1 days' work

$= \frac{1}{60} - \frac{1}{120} = \frac{2-1}{120} = \frac{1}{120}$

∴ A alone will complete the work in 120 days.

Ex. 2. *A can do a piece of work in 80 days. He works at it for 10 days and then B alone finishes the remaining work in 42 days. In how many days can the two of them complete the work together ?*

Sol. A's 1 days' work = $\frac{1}{80}$

A's 10 days' work = $\dfrac{10}{80} = \dfrac{1}{8}$

Remaining work = $1 - \dfrac{1}{8} = \dfrac{7}{8}$

$\dfrac{7}{8}$ of the work is completed by B in 42 days.

∴ The whole work is completed by B in $\dfrac{42 \times 8}{7}$ days = 48 days

∴ B's 1 days' work = $\dfrac{1}{48}$

(A + B)'s 1 days' work = $\dfrac{1}{80} + \dfrac{1}{48} = \dfrac{3+5}{240} = \dfrac{8}{240} = \dfrac{1}{30}$

∴ A and B can complete the whole work together in 30 days.

Ex. 3. *A can do a certain job in 12 days. B is 60% more efficient than A. What is the number of days it takes B to do the same piece of work ?*

Sol. A's 1 days' work = $\dfrac{1}{12}$

∴ B's 1 days' work = $\dfrac{1.6}{12} = \dfrac{16}{120} = \dfrac{2}{15}$

∴ B shall complete the work in $\dfrac{15}{2}$ days = $7\dfrac{1}{2}$ days.

Ex. 4. *Two men undertook to do a job for Rs 1400. One of them can do it alone in 7 days and the other in 8 days. With the assistance of a boy, they together completed the work in 3 days. How much money did the boy get ?*

Sol. First man's 1 days' work = $\dfrac{1}{7}$

Other man's 1 days' work = $\dfrac{1}{8}$

Let the boy complete the whole work in x days. Then

Boys' 1 days' work = $\dfrac{1}{x}$

Given, $3\left(\dfrac{1}{7} + \dfrac{1}{8} + \dfrac{1}{x}\right) = 1$

$\Rightarrow \dfrac{1}{7} + \dfrac{1}{8} + \dfrac{1}{x} = \dfrac{1}{3}$ $\Rightarrow \dfrac{1}{x} = \dfrac{1}{3} - \left(\dfrac{1}{7} + \dfrac{1}{8}\right) = \dfrac{56 - (24 + 21)}{168} = \dfrac{56 - 45}{168} = \dfrac{11}{168}$

∴ The boy can complete the work in $\dfrac{168}{11}$ days.

∴ Ratio of their shares = Ratio of their one days' work

$= \dfrac{1}{7} : \dfrac{1}{8} : \dfrac{11}{168} = 24 : 21 : 11$

∴ The boy's share = $\dfrac{11}{56} \times \text{Rs } 1400 = \text{Rs } 275$.

Ex. 5. *12 children take 16 days to complete a work which can be completed by 8 adults in 12 days. 16 adults started working and after 3 days 10 adults left and 4 children joined them. How many days will it take them to complete the remaining work?*

Sol. 12 children take 16 days to complete a work.

So, 1 child's 1 days' work = $\dfrac{1}{12\times16}=\dfrac{1}{192}$

8 adults take 12 days to complete a work

So, 1 adult's 1 days' work = $\dfrac{1}{8\times12}=\dfrac{1}{96}$

16 adults' 3 days' work = $16\times3\times\dfrac{1}{96}=\dfrac{1}{2}$

Remaining work = $1-\dfrac{1}{2}=\dfrac{1}{2}$

Now (6 adults' + children's) 1 days' work = $\dfrac{6}{96}+\dfrac{4}{192}=\dfrac{16}{192}=\dfrac{1}{12}$

$\therefore$ $\dfrac{1}{12}$th of the work is done by them in 1 day

$\therefore$ $\dfrac{1}{2}$ of the work will be done in $\dfrac{1}{1/12}\times\dfrac{1}{2}$ days = 6 days.

Ex. 6. *If 6 men and 8 boys can do a piece of work in 10 days, while 26 men and 48 boys can do the same work in 2 days, then, what is the time taken by 15 men and 20 boys in doing the same type of work?*

Sol. Let 1 man's 1 days' work = x and

1 boy's 1 days' work = y. Then,

$$6x+8y=\dfrac{1}{10} \qquad \ldots(i)$$

$$26x+48y=\dfrac{1}{2} \qquad \ldots(ii)$$

Performing $(i)\times6-(ii)$, we get

$(36x+48y)-(26x+48y)=\dfrac{6}{10}-\dfrac{1}{2}$

$10x=\dfrac{1}{10}\implies x=\dfrac{1}{100}$

Putting $x=\dfrac{1}{100}$ in (i) we get $\dfrac{6}{100}+8y=\dfrac{1}{10}$

$\implies 8y=\dfrac{1}{10}-\dfrac{6}{100}=\dfrac{4}{100}\implies y=\dfrac{1}{200}$

$\therefore$ (15 men's + 20 boys') 1 days' work = $\dfrac{15}{100}+\dfrac{20}{200}=\dfrac{1}{4}$

$\therefore$ 15 men and 20 boys shall complete the work in 4 days.

Ex. 7. *A, B and C are employed to do a piece of work for Rs 529. A and B together are supposed to do $\dfrac{19}{13}$ of the work and B and C together $\dfrac{8}{23}$ of the work. What amount should A be paid?*

  (a) Rs 315   (b) Rs 345   (c) Rs 355   (d) Rs 375

Sol. $(B + C)$ do $\dfrac{8}{23}$ of the work

$\therefore$ A does $1 - \dfrac{8}{23} = \dfrac{15}{23}$ of the work

$\therefore$ Ratio of A's share to $(B + C)$'s share $= \dfrac{15}{23} : \dfrac{8}{23} = 15 : 8$

$\therefore$ A's share $= \dfrac{15}{23} \times$ Rs 529 = Rs 345.

Ex. 8. *A contractor undertook to complete a project in 90 days and employed 60 men on it. After 60 days, he found that $\dfrac{3}{4}$ of the work has already been completed. How many men can he discharge so that the project may be completed on time ?*

Sol. $\dfrac{3}{4}$th of the work is completed by 60 men in 60 days.

$\dfrac{3}{4}$th of the work is completed by 1 man in (60×60) days.

$\therefore$ The whole work will be completed 1 man in $\left(60 \times 60 \times \dfrac{4}{3}\right)$ days.

Let the number of men who can complete the remaining $\dfrac{1}{4}$ of the work in remaining 30 days be x. Then,

$\dfrac{1}{4}$th of the work will be completed by x men in 30 days.

$\dfrac{1}{4}$th of the work will be completed by 1 man in $(30 \times x)$ days.

$\therefore$ The whole work will be completed by 1 man in $(x \times 30 \times 4)$ days.

$\therefore$ $60 \times 60 \times \dfrac{4}{3} = x \times 30 \times 4 \Rightarrow x = 40$

$\therefore$ Number of men who can be discharged $= 60 - 40 = 20$.

Ex. 9. *12 men can complete a work in 30 days by working 9 hours a day. What is the number of men required to complete 10 times of this work in 24 days by working 5 hours a day ?*

Sol. 12 men can complete a work in 30 days by working 9 hours a day.

$\therefore$ 1 man can complete a work in 1 day by working $(12 \times 30 \times 9)$ hours.

Let the number of men required be x.

Then x men can complete 10 times the work in 24 days by working 5 hours a day.

$\therefore$ 1 man can complete 1 work in 1 day by working $\dfrac{x \times 24 \times 5}{10}$ hours

$\therefore$ $12 \times 30 \times 9 = \dfrac{x \times 24 \times 5}{10} \Rightarrow x = 270$ men.

Ex. 10. *10 men working 6 hours a day can complete a work in 18 days. How many hours a day must 15 men work to complete the same work in 12 days.*

Sol. To complete the work in 18 days, 10 men work 6 hours a day

To complete the work in 18 days, 1 man work (10×6) hours a day

To complete the work in a day, 1 man will work $(6 \times 10 \times 18)$ hours in a day.

(less days, less men, more hours)

To complete the work in 1 day, 15 men will work $\dfrac{(6 \times 10 \times 18)}{15}$ hours a day.

∴ To complete the work in 12 days, 15 men will work $\left(\dfrac{6 \times 10 \times 18}{12 \times 15}\right)$ hours a day.

(more days, more men, less hours)

= 6 hours a day.

Ex. 11. *If 6 persons working 8 hours a day earn Rs 8400 per week, then how much will 9 persons working 6 hours a day each per week?*

Sol. 6 persons working 8 hours a day earn Rs 8400 in a week.

6 persons working 1 hour a day earn Rs $\dfrac{8400}{8}$ in a week.

∴ 1 person working 1 hour a day earns Rs $\dfrac{8400}{6 \times 8}$ in a week.

(less persons, less days, less earnings)

1 person working 6 hours a day earns Rs $\dfrac{8400 \times 6}{6 \times 8}$ in a week.

∴ 9 persons working 6 hours a day each Rs $\dfrac{8400 \times 9 \times 6}{6 \times 8}$ in a week.

(more persons, more hours, more earning)

= Rs 9450.

> **Note.** In Examples 10 and 11, for more you multiply and for less you divide.

Ex. 12. **24 men can complete a piece of work in 16 days and 18 women can complete the same work in 32 days. 12 men and 6 women work together for 16 days. If the remaining work was to be completed in 2 days, how many additional men would be required besides 12 men and 6 women ?**

Sol. 24 men can complete a work in 16 days

∴ 1 man's 1 days' work = $\dfrac{1}{24 \times 16}$ ∴ 12 men's 1 day's work = $\dfrac{12}{24 \times 16} = \dfrac{1}{32}$

18 women can complete the work in 32 days

∴ 1 woman's 1 days' work = $\dfrac{1}{18 \times 32}$ ∴ 6 women's 1 day's work = $\dfrac{6}{18 \times 32} = \dfrac{1}{96}$

∴ (12 men's + 6 women's) 16 days' work = $16 \times \left(\dfrac{1}{32} + \dfrac{1}{96}\right) = 16 \times \left(\dfrac{3+1}{96}\right) = 16 \times \dfrac{4}{96} = \dfrac{2}{3}$

∴ Remaining work $= 1 - \dfrac{2}{3} = \dfrac{1}{3}$

(12 men's + 6 women's) 2 days' work = $2 \times \left(\dfrac{1}{32} + \dfrac{1}{96}\right) = 2 \times \left(\dfrac{4}{96}\right) = \dfrac{1}{12}$

Remaining work = $\dfrac{1}{3} - \dfrac{1}{12} = \dfrac{3}{12} = \dfrac{1}{4}$

∴ $\dfrac{1}{384}$ work is done in 1 day by 1 man

∴ $\dfrac{1}{12}$ work will be done in 2 days by $\left(384 \times \dfrac{1}{12} \times \dfrac{1}{2}\right)$ men = 16 men.

Question Bank–20(a)

1. A can do $\frac{1}{5}$th part of the work in 12 days, B can do 60% of the work in 30 days and C can do $\frac{1}{3}$rd of the work in 15 days. Who will complete the work first ?
   (a) A   (b) B
   (c) C   (d) A and B both

2. Ten women can complete a piece in 15 days. Six men can complete the same piece of work in 10 days. In how many days can 5 women and six men together complete the piece of work ?
   (a) 15 days   (b) 7.5 days
   (c) 9 days   (d) 12.5 days

3. 6 men or 10 boys can complete a piece of work in 15 days. If 7 men and x boys complete the same piece of work in 9 days, then x is equal to
   (a) 4   (b) 5
   (c) 6   (d) 7

4. Ramu can complete a job in 8 days. Shyamu can complete the same job in 12 days. However, Monu takes 12 days to complete $\frac{3}{4}$th of the same job. Ramu works for 2 days and then leaves the job. The number of days that will be taken by Shyamu and Monu to complete the balance together is
   (a) 4.2   (b) 5.1
   (c) 6.2   (d) 6.5

5. If Ajit can do $\frac{1}{4}$ of the work in 3 days and Sujit can do $\frac{1}{6}$ of the same work in 4 days, how much will Ajit get if both work together and are paid Rs 180 in all ?
   (a) Rs 120   (b) Rs 108
   (c) Rs 60   (d) Rs 36

6. The work done by a man, a woman and a child are in the ratio 3 : 2 : 1. If the daily wages of 20 men, 30 women and 36 children amount to Rs 78, what will be the wages of 15 men, 21 women and 30 children for 18 weeks ?
   (a) Rs 7371   (b) Rs 8645
   (c) Rs 9000   (d) Rs 9500

7. 40 men can do a piece of work in a given time. If only 30 men be engaged, 6 more days are needed to complete the work. In how many days can 60 men do the work ?
   (a) 9   (b) 10
   (c) 12   (d) 15

8. A man and a boy working together can complete a work in 24 days. If for the last 6 days, the man alone does the work, then it is completed in 26 days. How long will the boy take to complete the work alone ?
   (a) 72 days   (b) 20 days
   (c) 24 days   (d) 36 days

9. A and B can do a piece of work in 20 days and 12 days respectively. A started the work alone and then after 4 days B joined him till the completion of the work. How long did the work last ?
   (a) 10 days   (b) 20 days
   (c) 15 days   (d) 6 days

10. A and B can do a piece of work in 28 and 35 days respectively. They began to work together but A leaves after some time and B completed the remaining work in 17 days. After how many days did A leave ?
    (a) $14\frac{2}{5}$ days   (b) 9 days
    (c) 8 days   (d) $7\frac{5}{9}$ days

11. A alone can do a piece of work in 6 days and B alone in 8 days. A and B undertook to do it for Rs 3200. With the help of C, they completed the work in 3 days. How much is to be paid to C ?
    (a) Rs 375   (b) Rs 400
    (c) Rs 600   (d) Rs 800

12. If 20 men working 7 hours a day can do a piece of work in 10 days, in how many days will 15 men working 8 hours a day do the same piece of work ?
    (a) $15\frac{5}{21}$ days   (b) $11\frac{2}{3}$ days
    (c) $6\frac{9}{16}$ days   (d) $4\frac{1}{5}$ days

13. 12 cows together eat 756 kg of grass in 7 days. How much grass will be eaten by 15 cows in 10 days ?
    (a) 1500 kg   (b) 1200 kg
    (c) 1350 kg   (d) 1400 kg

14. If 5 spiders can catch 5 flies in 5 minutes, how many flies can 100 spiders catch in 100 minutes ?
    (a) 100   (b) 500
    (c) 1000   (d) 2000

15. 8 men working for 9 hours a day can complete a piece of work in 20 days. In how many days can 7 men working for 10 hours a day complete the same piece of work ?

(a) $20\frac{1}{2}$ days   (b) $20\frac{3}{5}$ days

(c) 21 days   (d) $20\frac{4}{7}$ days

16. A job can be completed by 12 men in 12 days. How many extra days will be needed to complete the job, if 6 men leave after working for 6 days ?
(a) 10 days   (b) 12 days
(c) 8 days   (d) 24 days

17. A contractor undertakes to complete a road 360 m long in 120 days and employs 30 men for the work. After 60 days he finds that only 120 m length of the road has been made. How many more men should he employ so that the work may be completed in time ?
(a) 20   (b) 30
(c) 15   (d) 45

18. A wall of 100 metres can be built by 7 men or 10 women in 10 days. How many days will 14 men and 20 women take to build a wall of 600 metres ?
(a) 15   (b) 20
(c) 25   (d) 30

19. Twenty four men can complete a work in sixteen days. Thirty two women can complete the same work in twenty four days. Sixteen men and sixteen women started working and worked for 12 days. How many more men are to be added to complete the remaining work in 2 days ?
(a) 16   (b) 24
(c) 36   (d) 48

20. 12 men and 16 boys can do a piece of work in 5 days; 13 men and 24 boys can do it in 4 days. The ratio of the daily work done by a man to that of a boy is
(a) 2 : 1   (b) 3 : 1
(c) 3 : 2   (d) 5 : 4

21. Three men, four women and six children can complete a work in seven days. A woman does double the work a man does and a child does half the work a man does. How many women alone can complete the work in 7 days ?
(a) 7   (b) 8
(c) 14   (d) 12

22. A man, a woman and a boy can complete a job in 3, 4 and 12 days respectively. How many boys must assist 1 man and 1 woman to complete the job in $\frac{1}{4}$ of a day ?
(a) 1   (b) 4
(c) 19   (d) 41

23. A machine P can print one lakh books in 8 hours, machine Q can print the same number of books in 10 hours while machine R can print them in 12 hours. All the machines are started at 9 a.m. While machine P is closed at 11 a.m. and the remaining two machines complete the work. Approximately at what time will the work be finished ?
(a) 11 : 30 a.m.   (b) 12 noon
(c) 12 : 30 p.m.   (d) 1 p.m.

24. A takes 5 hours more time than that taken by B to complete a work. If working together they can complete a work in 6 hours, then the number of hours that A takes to complete the work individually is
(a) 15   (b) 12
(c) 10   (d) 9

25. If x men working x hours per day can do x units of a work in x days, then y men working y hours per day would be able to complete how many units of work in y days ?
(a) $\dfrac{x^2}{y^3}$ (b) $\dfrac{x^3}{y^2}$
(c) $\dfrac{y^2}{x^3}$ (d) $\dfrac{y^3}{x^2}$

26. A started a work and left after working for 2 days. Then B was called and he finished the work in 9 days. Had A left the work after working for 3 days, B would have finished the remaining work in 6 days. In how many days can each of them, working alone finish the whole work ?
(a) 5 days, 8.5 days   (b) 2.5 days, 7.5 days
(c) 5 days, 15 days   (d) 3 days, 15 days

27. 15 men can complete a work in 210 days. They started the work but at the end of 10 days, 15 additional men, with double efficiency, were inducted. How many days in all did they take to finish the work ?
(a) $72\frac{1}{2}$ days   (b) $84\frac{3}{4}$ days
(c) $76\frac{2}{3}$ days   (d) 70 days

Answers

1. (c)	**2.** (b)	**3.** (b)	**4.** (b)	**5.** (a)	**6.** (a)	**7.** (c)	**8.** (a)	**9.** (a)	**10.** (c)
11. (b)	**12.** (b)	**13.** (c)	**14.** (d)	**15.** (d)	**16.** (b)	**17.** (b)	**18.** (a)	**19.** (b)	**20.** (a)
21. (a)	**22.** (d)	**23.** (d)	**24.** (c)	**25.** (d)	**26.** (c)	**27.** (c)			

Hints and Solutions

1. (c) A can do $\dfrac{1}{5}$ th part of the work in 12 days

$\Rightarrow A$ can do the whole work in (12×5) days

$\qquad\qquad\qquad\qquad\qquad\qquad = 60$ days

B can do 60% of the work in 30 days

$\Rightarrow B$ can do the whole work in $\left(\dfrac{30 \times 100}{60}\right)$ days

$\qquad\qquad\qquad\qquad\qquad\qquad = 50$ days

C can do $\dfrac{1}{3}$ rd of the work in 15 days

$\Rightarrow C$ can do the whole work in (15×3) days

$\qquad\qquad\qquad\qquad\qquad\qquad = 45$ days.

$\therefore\ C$ will complete the work first.

2. (b) 10 women can complete a piece of work in 15 days

$\therefore$ 1 woman can complete in 1 day, $\dfrac{1}{10 \times 15} = \dfrac{1}{150}$ part of work

6 men can complete the same work in 10 days,

$\therefore$ 1 man can complete in 1 day, $\dfrac{1}{6 \times 10} = \dfrac{1}{60}$ part of the work

$\therefore$ (6 men's + 5 women's) 1 days' work

$= \dfrac{6}{60} + \dfrac{5}{150} = \dfrac{1}{10} + \dfrac{1}{30} = \dfrac{3+1}{30} = \dfrac{4}{30} = \dfrac{2}{15}$

$\therefore$ 6 men and 5 women can complete the whole work in $\dfrac{15}{2} = 7.5$ days.

3. (b) 6 men = 10 boys $\Rightarrow$ 1 man = $\dfrac{10}{6}$ boys

$\Rightarrow$ 7 men = $\left(\dfrac{10}{6} \times 7\right)$ boys = $\dfrac{35}{3}$ boys

10 boys can complete the work in 15 days

$\Rightarrow$ 1 boy can complete the work in (10×15) days

$\left(\dfrac{35}{3} + x\right)$ boys can complete the work in 9 days

$\Rightarrow$ 1 boy can complete the work in $\left(\dfrac{35}{3} + x\right) \times 9$ days.

$\therefore\ 10 \times 15 = \left(\dfrac{35}{3} + x\right) \times 9$

$\Rightarrow 9x = 150 - 105 = 45 \Rightarrow x = 5.$

4. (b) Ramu's one days' job = $\dfrac{1}{8}$

Shyamu's one days' job = $\dfrac{1}{12}$

$\dfrac{3}{4}$ th of the job can be done by Monu in 12 days

$\therefore$ Whole job can be done by Monu in $12 \times \dfrac{4}{3}$ days

$\qquad\qquad\qquad\qquad\qquad\qquad = 16$ days

$\therefore$ Monu's one days' job = $\dfrac{1}{16}$

Ramu's two days' job = $2 \times \dfrac{1}{8} = \dfrac{1}{4}$,

Remaining job = $1 - \dfrac{1}{4} = \dfrac{3}{4}$

(Shyamu's + Monu's) one days' job = $\dfrac{1}{12} + \dfrac{1}{16}$

$\qquad\qquad\qquad\qquad\qquad = \dfrac{4+3}{48} = \dfrac{7}{48}$

$\dfrac{7}{48}$ of the job can be finished by Shyamu and Monu in 1 day.

$\therefore\ \dfrac{3}{4}$ of the job can be finished by Shyamu and

Monu in $\dfrac{48}{7} \times \dfrac{3}{4} = \dfrac{36}{7} = 5.1$ days.

5. (a) Ajit's 3 days' work = $\dfrac{1}{4}$

$\Rightarrow$ Ajit's 1 days' work = $\dfrac{1}{3} \times \dfrac{1}{4} = \dfrac{1}{12}$.

Sujit's 4 days' work = $\dfrac{1}{6}$

$\Rightarrow$ Sujit's 1 days' work = $\dfrac{1}{4} \times \dfrac{1}{6} = \dfrac{1}{24}$

$\therefore$ They are paid in the ratio $\dfrac{1}{12} : \dfrac{1}{24} = 2 : 1$

$\therefore$ Ajit's share = $\dfrac{2}{3} \times$ Rs 180 = Rs 120

6. (a) Given, Man : Woman : Child = 3 : 2 : 1

$\Rightarrow M = 3K$, $W = 2K$, $C = K$, where K is a constant

$\Rightarrow M = 3C$, $W = 2C$

$\therefore\ 20M + 30W + 36C = 60C + 60C + 36C = 156C$

and $15M + 21W + 30C = 45C + 42C + 30C$
$$= 117C$$

It is given that
(20 men + 30 women + 36 children)
$$= 156 \text{ children earn Rs } 78 \text{ per day}$$

$\Rightarrow$ 1 child earns Rs $\dfrac{78}{156}$ per day = Rs $\dfrac{1}{2}$ per day.

$\therefore$ (15 men + 21 women + 30 children) = 117

children should get $\left(\text{Rs } \dfrac{1}{2} \times 117 \right)$ per day,

i.e., Rs $\left(\dfrac{1}{2} \times 117 \times 18 \times 7 \right)$ = Rs 7371 in 18

weeks.

7. (c) Suppose 40 men can do a piece of work in x days. Then,

30 men can do the same work in $\dfrac{x \times 40}{30}$ days
$$= (x + 6) \text{ days}$$
$\Rightarrow 4x = 3x + 18 \Rightarrow x = 18$

$\therefore$ 60 men can do the same work in $\dfrac{18 \times 40}{60}$ days
$$= 12 \text{ days.}$$

8. (a) Let 1 man's 1 days' work = $\dfrac{1}{x}$ and 1 boy's

1 days' work = $\dfrac{1}{y}$, Then, $\dfrac{1}{x} + \dfrac{1}{y} = \dfrac{1}{24}$

Also given , $\dfrac{20}{x} + \dfrac{20}{y} + \dfrac{6}{x} = 1$

$\Rightarrow 20\left(\dfrac{1}{x} + \dfrac{1}{y} \right) + \dfrac{6}{x} = 1 \Rightarrow 20\left(\dfrac{1}{24} \right) + \dfrac{6}{x} = 1$

$\Rightarrow \dfrac{6}{x} = 1 - \dfrac{5}{6} \Rightarrow \dfrac{6}{x} = \dfrac{1}{6} \Rightarrow x = 36$

$\therefore \dfrac{1}{y} = \dfrac{1}{24} - \dfrac{1}{36} = \dfrac{3-2}{72} = \dfrac{1}{72} \Rightarrow y = 72$

$\therefore$ The boy will take 72 days to complete the work alone.

9. (a) A's one days' work = $\dfrac{1}{20}$

B's one days' work = $\dfrac{1}{12}$

Work done by A in 4 days = $\dfrac{4}{20} = \dfrac{1}{5}$

Remaining work = $1 - \dfrac{1}{5} = \dfrac{4}{5}$

$(A + B)$'s 1 days' work = $\dfrac{1}{20} + \dfrac{1}{12} = \dfrac{2}{15}$

$\therefore$ Number of days to do the remaining work
$$= \dfrac{4}{5} \div \dfrac{2}{15} = \dfrac{4}{5} \times \dfrac{15}{2} = 6 \text{ days}$$

$\therefore$ Total number of days taken in the completion of work = 6 + 4 = 10 days

10. (c) Suppose A left after x days. Then,

Work of $(A + B)$ for x days = $\dfrac{x}{28} + \dfrac{x}{35}$

Work of B for 17 days = $\dfrac{17}{35}$

$\therefore \dfrac{x}{28} + \dfrac{x}{35} + \dfrac{17}{35} = 1$

$\Rightarrow \dfrac{5x + 4x}{140} = 1 - \dfrac{17}{35} = \dfrac{18}{35}$

$\Rightarrow \dfrac{9x}{140} = \dfrac{18}{35} \Rightarrow x = \dfrac{18 \times 140}{35 \times 9} = 8 \text{ days.}$

11. (b) Let C does the same work in x days. Then,

$3\left(\dfrac{1}{6} + \dfrac{1}{8} + \dfrac{1}{x} \right) = 1 \Rightarrow \dfrac{7}{24} + \dfrac{1}{x} = \dfrac{1}{3}$

$\Rightarrow \dfrac{1}{x} = \dfrac{1}{3} - \dfrac{7}{24} = \dfrac{1}{24}$

$\therefore C$ will do the same work in 24 days.

The ratio of their one days' work = $\dfrac{1}{6} : \dfrac{1}{8} : \dfrac{1}{24}$
$$= 4 : 3 : 1$$

The amount to be paid to C
$$= \dfrac{1}{(4+3+1)} \times \text{Rs } 3200 = \text{Rs } 400.$$

12. (b) 20 men working 7 hours a day can complete the work in 10 days.

20 men working 1 hour a day can complete the work in (10×7) days

(Less hours $\Rightarrow$ More days)

1 man working 1 hour a day can complete the work in $(10 \times 7 \times 20)$ days

(Less men $\Rightarrow$ More days)

1 man working 8 hours a day can complete the work in $\dfrac{(10 \times 7 \times 20)}{8}$ days

(More hours $\Rightarrow$ Less days)

15 men working 8 hours a day can complete the work in $\left(\dfrac{10 \times 7 \times 20}{8 \times 15} \right)$ days

(More men $\Rightarrow$ Less days) = $\dfrac{35}{3}$ days = $11\dfrac{2}{3}$ days

13. (c) 12 cows in 7 days eat 756 kg of grass

12 cows in 1 day eat $\dfrac{756}{7}$ kg of grass

(Less days $\Rightarrow$ Less grass)

1 cow in 1 day eats $\left(\dfrac{756}{12\times7}\right)$ kg of grass

(Less cows $\Rightarrow$ Less grass)

1 cow in 10 days eats $\left(\dfrac{756\times10}{7\times12}\right)$ kg of grass

(More days $\Rightarrow$ More grass)

15 cows in 10 days eat $\left(\dfrac{756\times10\times15}{7\times12}\right)$ kg of grass = 1350 kg

(More cows $\Rightarrow$ More grass)

14. (d) 5 spiders in 5 minutes catch 5 flies

5 spiders in 1 minute catch $\dfrac{5}{5}$ flies

1 spider in 1 minute catches $\dfrac{5}{5\times5}$ flies

1 spiders in 100 minutes catches $\dfrac{5}{5\times5}\times100$ flies

100 spiders in 100 minutes catch $\dfrac{5}{5\times5}\times100\times100$ flies = 2000 flies.

15. (d) 8 men working for 9 hours a day can complete a piece in 20 days.

8 men working for 1 hour a day can complete the work in (20×9) days

(Less hours, more days)

1 man working for 1 hour a day can complete the work in $(20\times9\times8)$ days

(Less men, more days)

1 man working for 10 hours a day can complete the work in $\left(\dfrac{20\times9\times8}{10}\right)$ days

7 men working for 10 hours a day can complete the work in $\left(\dfrac{20\times9\times8}{7\times10}\right)$ days $=\dfrac{144}{7}$ days

$=20\dfrac{4}{7}$ days .

16. (b) 12 men can complete in 12 days, 1 work

1 man can complete in 1 day $\dfrac{1}{12\times12}$ part of the work

$\therefore$ 6 men can complete in 6 days $\dfrac{6\times6}{12\times12}$ part of the work $=\dfrac{1}{4}$th of the work.

Number of remaining men = 6,

Remaining work $=\dfrac{3}{4}$

$\because$ 12 men can complete 1 work in 12 days

6 men can complete $\dfrac{3}{4}$ work in $\dfrac{12\times12\times3}{6\times4}$ =18 days

$\therefore$ Number of extra days = 18 – 6 = 12 days.

17. (b) Let the number of extra men employed be x.

30 men in 60 days complete 120 m long road.

1 man in 1 day will complete $\dfrac{120}{30\times60}$ m long road.

Also, $(30 + x)$ men in 60 days complete 240 m long road.

$\therefore$ 1 man in 1 day will complete $\dfrac{240}{(30+x)\times60}$ m long road.

$\Rightarrow \dfrac{120}{30\times60}=\dfrac{240}{(30+x)\times60}\Rightarrow\dfrac{1}{15}=\dfrac{4}{(30+x)}$

$\Rightarrow 30+x=60\Rightarrow x=30$

$\therefore$ 30 more men have to be employed.

18. (a) According to question, 10 women = 7 men

$\therefore$ 20 women = 14 men

14 men + 20 women = 14 men + 14 men

$\qquad\qquad\qquad = 28$ men.

A wall of 100 m is built by 7 men in 10 days

A wall of 100 m is built by 1 man in (10×7) days

A wall of 1 m is built by 1 man in $\dfrac{10\times7}{100}$ days

(less length, less days)

A wall of 600 m is built by 1 man in $\dfrac{10\times7\times600}{100}$ days (more length, more days)

A wall of 600 m is built by 28 men in $\dfrac{10\times7\times600}{100\times28}$ days = 15 days.

19. (b) 1 man's 1 days' work $=\dfrac{1}{24\times16}$

$\therefore$ 16 men's 1days' work $=\dfrac{16}{24\times16}=\dfrac{1}{24}$

1 woman's 1 days' work = $\dfrac{1}{32 \times 24}$

16 women's 1 days' work = $\dfrac{16}{32 \times 24} = \dfrac{1}{48}$

∴ 12 days' work of (16 men + 16 women)

$= 12\left(\dfrac{1}{24} + \dfrac{1}{48}\right)$

$= 12\left(\dfrac{2+1}{48}\right) = 12 \times \dfrac{3}{48} = \dfrac{3}{4}$

Remaining work = $1 - \dfrac{3}{4} = \dfrac{1}{4}$

Now (16 men's + 16 women's) 2 days' work

$= 2\left(\dfrac{1}{24} + \dfrac{1}{48}\right) = 2 \times \dfrac{1}{16} = \dfrac{1}{8}$

∴ Remaining work = $\dfrac{1}{4} - \dfrac{1}{8} = \dfrac{1}{8}$

$\dfrac{1}{384}$ work is done in 1 day by 1 man

∴ $\dfrac{1}{8}$ work will be done in 2 days by $\left(384 \times \dfrac{1}{8} \times \dfrac{1}{2}\right)$

= 24 men.

20. (a) Let 1 man's 1 days' work = x and 1 boy's 1 days' work = y

Then, $12x + 16y = \dfrac{1}{5}$ ... (i)

$13x + 24y = \dfrac{1}{4}$ ...(ii)

Multiplying (i) by 3 and (ii) by 2, we get

$36x + 48y = \dfrac{3}{5}$ ... (iii)

$26x + 48y = \dfrac{1}{2}$ ... (iv)

Subtracting eqn (iv) from eqn (iii), we get

$10x = \dfrac{3}{5} - \dfrac{1}{2} = \dfrac{6-5}{10} = \dfrac{1}{10} \Rightarrow x = \dfrac{1}{100}$

Putting $x = \dfrac{1}{100}$ in (i), we get

$\dfrac{12}{100} + 16y = \dfrac{1}{5}$

$\Rightarrow 16y = \dfrac{1}{5} - \dfrac{12}{100} = \dfrac{20-12}{100} = \dfrac{8}{100} \Rightarrow y = \dfrac{1}{200}$

∴ Required ratio = $x : y = \dfrac{1}{100} : \dfrac{1}{200} = 2 : 1$.

21. (a) Let 1 woman's 1 days' work = x. Then

1 man's 1 days' work = $\dfrac{x}{2}$ and 1 child's 1 days' work = $\dfrac{x}{4}$

Given, (3 men's + 4 women's + 6 children's) 1 days' work = $\dfrac{1}{7}$

$\Rightarrow \dfrac{3x}{2} + 4x + \dfrac{6x}{4} = \dfrac{1}{7} \Rightarrow \dfrac{6x + 16x + 6x}{4} = \dfrac{1}{7}$

$\Rightarrow \dfrac{28x}{4} = \dfrac{1}{7} \Rightarrow x = \dfrac{1}{49}$

∴ 1 woman can alone complete the work in 49 days To complete the work in 7 days, number of women required = $\dfrac{49}{7} = 7$.

22. (d) 1 man's 1 days' work = $\dfrac{1}{3}$

1 woman's 1 days' work = $\dfrac{1}{4}$

1 boy's 1 days' work = $\dfrac{1}{12}$

Let the number of boys to assist be x. Then,

$\dfrac{1}{3} + \dfrac{1}{4} + \dfrac{x}{12} = \dfrac{1}{1/4}$

$\Rightarrow \dfrac{4+3+x}{12} = 4 \Rightarrow 7 + x = 48$

$\Rightarrow x = 41$ boys.

23. (d) $(P + Q + R)$'s 1 hours' work $= \left(\dfrac{1}{8} + \dfrac{1}{10} + \dfrac{1}{12}\right)$

$= \dfrac{37}{120}$

∴ Work done by $(P + Q + R)$ in 2 hours $= 2 \times \dfrac{37}{120}$

$= \dfrac{37}{60}$

Remaining work $= \left(1 - \dfrac{37}{60}\right) = \dfrac{23}{60}$

$(Q + R)$'s 1 hours' work $= \left(\dfrac{1}{10} + \dfrac{1}{12}\right) = \dfrac{11}{60}$

$\dfrac{11}{60}$ of the work is done by $(Q + R)$ in 1 hour

∴ $\dfrac{23}{60}$ of the work will be done by $(Q + R)$ in

$\dfrac{23}{60} \div \dfrac{11}{60} = \dfrac{23}{11} = 2\dfrac{1}{11}$ hours ≃ 2 hours.

So, the work will be finished approximately 2 hours after 11 a.m., i.e., around 1 p.m.

24. (c) Let B take x hours to complete a work. Then, A shall take $(x + 5)$ hours to complete the same work.

B's 1 hours' work $= \dfrac{1}{x}$, A's 1 hours' work $= \dfrac{1}{x+5}$

Given, $\dfrac{1}{x} + \dfrac{1}{(x+5)} = \dfrac{1}{6}$

$\Rightarrow \dfrac{x+5+x}{x(x+5)} = \dfrac{1}{6} \Rightarrow \dfrac{2x+5}{x^2+5x} = \dfrac{1}{6}$

$\Rightarrow 12x + 30 = x^2 + 5x \Rightarrow x^2 - 7x - 30 = 0$

$\Rightarrow x^2 - 10x + 3x - 30 = 0$

$\Rightarrow x(x - 10) + 3(x - 10) = 0$

$\Rightarrow (x - 10) = 0 \text{ or } (x + 3) = 0$

$x = 10 \text{ or } -3$

Neglecting negative value

$x = 10$.

25. (d) x men working x hours in x days $= x$ units work

Then, 1 man working 1 hour in 1 day $= \dfrac{x}{x^3}$ units

work y men working y hours in y days

$= \dfrac{xy^3}{x^3} = \dfrac{y^3}{x^2}$

26. (c) Suppose A and B do the work in x and y days respectively.

Now, work done by A in 2 days + work done by B in 9 days $= 1$

$\Rightarrow \dfrac{2}{x} + \dfrac{9}{y} = 1$

Similarly, $\dfrac{3}{x} + \dfrac{6}{y} = 1$

Let $\dfrac{1}{x} = a$ and $\dfrac{1}{y} = b$. Then the equations become

$2a + 9b = 1$ ... (i)

$3a + 6b = 1$ ... (ii)

Performing $(i) \times 2 - (ii) \times 3$, we get

$(6a + 12b) - (6a + 27b) = 2 - 3$

$\Rightarrow -15b = -1 \Rightarrow b = \dfrac{1}{15} \Rightarrow x = 15$

$\therefore$ Putting the value of b in (i)

$2a + 9 \times \dfrac{1}{15} = 1$

$\Rightarrow 2a = 1 - \dfrac{3}{5} = \dfrac{2}{5} \Rightarrow a = \dfrac{1}{5} \Rightarrow y = 5.$

27. (c) 15 men's 1 days' work $= \dfrac{1}{210}$

$\therefore$ 15 men's 10 days' work $= \dfrac{10}{210} = \dfrac{1}{21}$

Remaining work $= 1 - \dfrac{1}{21} = \dfrac{20}{21}$

The additional 15 men's 1 days' work $= \dfrac{2}{210}$

$= \dfrac{1}{105}$

$\therefore$ These 30 men's 1 days' work $= \left(\dfrac{1}{210} + \dfrac{1}{105} \right)$

$= \dfrac{1+2}{210} = \dfrac{3}{210} = \dfrac{1}{70}$

$\dfrac{1}{70}$th of the work can be done in 1 day

$\therefore \dfrac{20}{21}$th of the work can be done in $\left(70 \times \dfrac{20}{21} \right)$ days

$= \dfrac{200}{3}$ days $= 66\dfrac{2}{3}$ days

$\therefore$ Total no. of days $= 66\dfrac{2}{3}$ days + 10 days

$= 76\dfrac{2}{3}$ days.

PIPES AND CISTERNS

Solved Examples

Ex. 1. *Two pipes A and B can fill a tank in 15 hours and 30 hours respectively, while a third pipe C can empty the full tank in 25 hours. All the three pipes are opened in the beginning. After 10 hours, C is closed. In how much time will the tank be full?*

Sol. Part of the tank filled when all the three pipes are opened for 10 hours

$$= 10 \times \left(\frac{1}{15} + \frac{1}{20} - \frac{1}{25} \right) = 10 \times \left(\frac{200 + 150 - 120}{300} \right) = \frac{230}{300} = \frac{23}{30}$$

Remaining empty part $= 1 - \frac{23}{30} = \frac{7}{30}$

Part of the tank filled in 1 hour by pipes (A and B) $= \frac{1}{15} + \frac{1}{20} = \frac{4+3}{60} = \frac{7}{60}$

$\therefore \frac{7}{60}$ part is filled in 1 hour

$\frac{7}{30}$ part is filled in 2 hours.

$\therefore$ Total time to fill the tank $= 10 + 2 = 12$ hours.

Ex. 2. *Two pipes A and B can fill a water tank in 20 minutes and 24 minutes respectively and the third pipe C can empty at the rate of 3 gallons per minute. If A, B and C opened together fill the tank in 15 minutes, the capacity (in gallons) of the tank is*

    (a) 180   (b) 150   (c) 120   (d) 60

Sol. Part of the tank filled when all the three pipes are opened in 1 minute $= \frac{1}{15}$

Part of the tank filled when ($A + B$) are opened in 1 minute $= \frac{1}{20} + \frac{1}{24} = \frac{6+5}{120} = \frac{11}{120}$

$\therefore$ Part of the tank emptied by C in 1 min. $= \frac{11}{120} - \frac{1}{15} = \frac{1}{40}$

$\therefore$ Capacity of $\frac{1}{40}$ part of the tank $= 3$ gallons

$\therefore$ Capacity of the tank $= (3 \times 40)$ gallons $= 120$ gallons.

Ex. 3. *A tap can fill a tank in 16 minutes and another can empty it in 8 minutes. If the tank is already $\frac{1}{2}$ full and both the taps are opened together, will the tank be filled or emptied? How long will it take before the tank is either filled completely or emptied completely, as the case may be?*

Sol. In one minute, $\frac{1}{16} - \frac{1}{8} = -\frac{1}{16}$ of the tank can be filled, *i.e.*, $\frac{1}{16}$ of the whole tank can be emptied.

$\therefore$ The whole tank can be emptied in 16 minutes.

Since the tank is half full, therefore it will be emptied in 8 minutes.

Ex. 4. *Two pipes A and B can fill a cistern in 20 min and 25 min respectively. Both the pipes are opened together but at the end of 5 min, the first is turned off. How long does it take to fill the cistern?*

Sol. Let it take x minutes to fill the cistern, then work done by (Pipe A in 5 min. + Pipe B in x min.) $= 1$

$$\Rightarrow \frac{5}{20} + \frac{x}{25} = 1 \Rightarrow \frac{25 + 4x}{100} = 1 \Rightarrow 25 + 4x = 100 \Rightarrow 4x = 75 \Rightarrow x = 18.75 \text{ min}$$

Ex. 5. *A cistern can be filled by two pipes in 30 and 40 min respectively. Both the pipes are opened at once, when the first pipe must be turned off so that the tank may be just filled in 18 min?*

Sol. Let the first pipe be turned off after x minutes, then work done by (First pipe in x min. + Second pipe in 18 min.) $= 1$

$$\Rightarrow \frac{x}{30} + \frac{18}{40} = 1 \Rightarrow \frac{4x + 54}{120} = 1 \Rightarrow 4x + 54 = 120 \Rightarrow 4x = 66 \Rightarrow x = \frac{66}{4} = 16.5 \text{ min.}$$

Ex. 6. *Two pipes can fill a cistern in 30 and 15h respectively. The pipes are opened simultaneously and it is found that due to leakage in the bottom, 5h extra are taken for cistern to be filled up. If the cistern is full, in what time would the leak empty it?*

Sol. Part of the cistern filled when the two pipes are opened simultaneously $= \dfrac{1}{30} + \dfrac{1}{15} = \dfrac{1+2}{30} = \dfrac{3}{10} = \dfrac{1}{10}$, *i.e.*, The two

pipes can fill the cistern together in 10 hours.

Together with the leak, the cistern can be filled in $(10 + 5)$ hours = 15 hours

Let the leak take x hours to empty the cistern. Then,

$$\dfrac{1}{10} - \dfrac{1}{x} = \dfrac{1}{15} \quad \Rightarrow \quad \dfrac{1}{x} = \dfrac{1}{10} - \dfrac{1}{15} = \dfrac{1}{30}$$

$\therefore$ The leak takes 30 hours to empty the cistern.

Ex. 7. *Pipes A and B running together can fill a cistern in 6 minutes. If B takes 5 minutes more than A to fill the cistern then, what will be the respective times in which A and B will fill the cistern separately.*

Sol. Suppose pipe A fills the cistern in x min.

Therefore pipe B will fill the cistern in $(x + 5)$ min.

$\therefore$ In one minute, pipe A and B can together fill $\left[\dfrac{1}{x} + \dfrac{1}{x+5}\right]$ part of the cistern

Given $\dfrac{1}{x} + \dfrac{1}{x+5} = \dfrac{1}{6}$

$\Rightarrow \dfrac{x+5+x}{x(x+5)} = \dfrac{1}{6} \Rightarrow \dfrac{2x+5}{x^2+5} = \dfrac{1}{6} \Rightarrow 12x + 30 = x^2 + 5 \Rightarrow x^2 - 7x - 30 = 0$

$\Rightarrow x^2 - 10x + 3x - 30 = 0 \Rightarrow x(x-10) + 3(x-10) = 0 \Rightarrow (x+3)(x-10) = 0$

$\Rightarrow x = -3$ or 10

Neglecting negative values $x = 10$

$\therefore$ A and B will the cistern separately in 10 min and 15 min.

Ex. 8. *Three pipes A, B and C can fill a tank in 6 min., 8 min, and 12 min. respectively. The pipe C is closed for 4 min before the tank is filled. In what time would the tank be fill?*

Sol. Let the tank be full in x min. Then,

$$\dfrac{x}{6} + \dfrac{x}{8} + \dfrac{x-4}{12} = 1 \Rightarrow \dfrac{4x + 3x + 2(x-4)}{24} = 1 \Rightarrow 9x - 8 = 24$$

$\Rightarrow 9x = 32 \Rightarrow x = 3\dfrac{5}{9}$ min.

Ex. 9. *Three pipes A, B and C can fill a cistern in 10h, 12h and 15h respectively. First A was opened. After 1 hour, B was opened and after 2 hours from the start of A, C was also opened. Find the time in which the cistern was just fill.*

Sol. Let it take x hours to fill the tank by all the three pipes. Then,

Work done by {A in x hours + B in $(x-1)$ hours + C in $(x-2)$ hours} = 1

$\Rightarrow \dfrac{x}{10} + \dfrac{(x-1)}{12} + \dfrac{(x-2)}{15} = 1 \Rightarrow 6x + 5(x-1) + 4(x-2) = 60$

$\Rightarrow 15x - 13 = 60 \Rightarrow 15x = 73$

$\Rightarrow x = \dfrac{73}{15} = 4\dfrac{13}{15} = 4$ hours $+ \dfrac{13}{15} \times 60$ min = 4 hours 52 min.

Ex. 10. *Two pipes A and B can fill a tank in 4 hours and 5 hours respectively. If the pipes A and B are turned on alternately for 1 hour each, then what is the time taken to fill the tank?*

Sol. When the two pipes are opened alternately for one hour each, part of the tank filled by them in

$$2 \text{ hours} = \frac{1}{4} + \frac{1}{5} = \frac{9}{20}$$

∴ Part of the tank filled in 4 hours $= \dfrac{9}{10}$

Remaining part $= 1 - \dfrac{9}{10} = \dfrac{1}{10}$

This $\dfrac{1}{10}$th of the tank has to be filled by pipe A in $\dfrac{4}{10} = \dfrac{2}{5}$ hours

∴ The whole tank can be filled in $4\dfrac{2}{5}$ hours.

Question Bank–20(b)

1. Two pipes A and B can separately fill a cistern in 60 minutes and 75 minutes respectively. There is a third pipe at the bottom of the cistern to empty it. If all the three pipes are simultaneously opened, then the cistern is full in 50 minutes. In how much time, the third pipe alone can empty the cistern?
   (a) 110 minutes   (b) 100 minutes
   (c) 120 minutes   (d) 90 minutes

2. A cistern has a leak which would empty it in 8 hours. A tap is turned on, which admits 6 litres a minute into the cistern and it is now emptied in 12 hours. How many litres can the cistern hold?
   (a) 8000 litres   (b) 8400 litres
   (c) 8640 litres   (d) 8650 litres

3. A tank can be filled by two pipes A and B separately in 3 hours and 3 hours 45 minutes respectively. A third pipe C can empty the full tank in 1 hour, when the tank was exactly half-filled, all the three pipes are opened. The tank will become empty in
   (a) 1 hour 15 min   (b) 2 hours 30 min
   (c) 3 hours 15 min   (d) 4 hours 10 min

4. A drum can be filled with oil with a supply pipe in 40 minutes. Again the full drum can be made empty by another pipe in 60 minutes. When $\dfrac{2}{3}$ part of the drum was filled, the second pipe was opened and it was stopped after 15 minutes. If the supply pipe is opened now, time required to fill the drum is
   (a) $23\dfrac{1}{3}$ min   (b) $25\dfrac{2}{3}$ min
   (c) $27\dfrac{1}{3}$ min   (d) $28\dfrac{2}{3}$ min

5. Two pipes can fill a tank with water in 15 and 12 hours respectively and the third pipe can empty it in 4 hours. If the pipes are opened in order at 8, 9

and 11 A.M. respectively, the tank will be emptied at
   (a) 11 : 40 a.m.   (b) 12 : 40 p.m.
   (c) 1 : 40 p.m.   (d) 2 : 40 p.m.

6. Two pipes can fill a cistern in 14 hours and 16 hours respectively. The pipes are opened simultaneously and it is found that due to leakage in the bottom, 32 minutes extra are taken for the cistern to be filled up. If the cistern is full, in what time would the leak empty it?
   (a) 96 hours   (b) 102 hours
   (c) 106 hours   (d) 112 hours

7. A pipe can empty a tank in 40 minutes. A second pipe with diameter twice as much as that of the first is also attached with the tank to empty it. The two together can empty the tank in
   (a) 8 min   (b) $13\dfrac{1}{3}$ min
   (c) 30 min   (d) 38 min

8. Three pipes A, B and C can fill a tank in 6 hours. After working at it together for 2 hours, C is closed and A and B can fill the remaining part in 7 hours. The number of hours taken by C alone to fill the tank is
   (a) 10   (b) 12
   (c) 14   (d) 16

9. A large tanker can be filled by two pipes A and B in 60 minutes and 40 minutes respectively. How many minutes will it take to fill the tanker from empty state if B is used for half the time and A and B fill it for the other half?
   (a) 15 minutes   (b) 20 minutes
   (c) 27.5 minutes   (d) 30 minutes

10. Two taps can fill a tank in 20 minutes and 30 minutes respectively. There is an outlet tap at exactly half level of that rectangular tank which can pump out 100 litres of water per minute. If the outlet tap is open, then it takes 24 minutes to fill an empty tank. What is the volume of the tank?

    (a) 1800 litres   (b) 1500 litres
    (c) 1200 litres   (d) 2400 litres

11. Two pipes A and B can fill a cistern in 12 min and 16 min respectively. Both the pipes are opened together for a certain time but due to some obstruction, the flow of water was restricted to $\frac{7}{8}$ of the full flow in pipe A and $\frac{5}{6}$ of the full in pipe B. The obstruction, is removed after some time and the tank is now filled in 3 min from that moment. For how many minutes was the obstruction there?

    (a) 8 min   (b) 3 min
    (c) 5.6 min   (d) 4.5 min

12. Pipes A, B and C are attached to a cistern. A and B can fill it in 20 min and 30 min respectively, while C can empty it in 15 min. If A, B and C are kept in operation successively for 1 minute each, how soon will the cistern be filled?

    (a) 167 min   (b) 160 min
    (c) 166 min   (d) 164 min

13. A pipe A can fill a tank in 27 min. Due to development of a hole in the bottom of the tank $\frac{1}{10}$ of the water filled by pipe A leaks out. Find the time when the tank will be full.

    (a) 32 min   (b) 34 min
    (c) 36 min   (d) 30 min

14. If two pipes are put in operation simultaneously, the tank is filled in 24 min. One pipe fills the tank in 20 min faster than the other. How many hours does the faster pipe take to fill the tank?

    (a) 60 min   (b) 45 min
    (c) 40 min   (d) 30 min

15. Two pipes A and B can fill a cistern in 30 min and 40 min respectively. Both the pipes are turned on simultaneously. When should the second pipe be closed if the cistern is to be filled in 24 min.

    (a) 6 min   (b) 8 min
    (c) 10 min   (d) 12 min

Answers

1. (b)   2. (c)   3. (a)   4. (a)   5. (d)   6. (d)   7. (b)   8. (c)   9. (d)   10. (a)
11. (d)   12. (a)   13. (d)   14. (c)   15. (a)

Hints and Solutions

1. (b) Let the third pipe empty the whole cistern in x minutes.

    $\therefore \frac{1}{60}+\frac{1}{75}-\frac{1}{x}=\frac{1}{50}$

    $\Rightarrow \frac{1}{x}=\frac{1}{60}+\frac{1}{75}-\frac{1}{50}=\frac{5+4-6}{300}=\frac{3}{300}=\frac{1}{100}$

    The third pipe can empty the cistern in 100 minutes.

2. (c) Suppose the cistern holds x litres of water.
    The tap admits 6 litres of water in 1 minute, i.e., it admits 360 litres of water in 1hour.

    $\therefore$ x litres of water can be filled in $\frac{x}{360}$ hours.

    The leak would empty the tank in 8 hours.

    $\therefore$ In one hour, $\frac{1}{8}$ of the tank becomes empty.

    If the tap and the leak both are allowed simultaneously, then the tank becomes empty in 12 hours, i.e.,

    $\frac{1}{8}-\frac{360}{x}=\frac{1}{12} \Rightarrow \frac{360}{x}=\frac{1}{8}-\frac{1}{12}=\frac{3-2}{24}=\frac{1}{24}$

    $\Rightarrow x=(360\times24)$ litres $=8640$ litres.

3. (a) Part of the tank emptied in 1 hour

    $=1-\left(\frac{1}{3}+\frac{4}{15}\right)=1-\frac{9}{15}=\frac{6}{15}=\frac{2}{5}$

    ($\because$ 3 hr 45 min $=3\frac{45}{60}=\frac{15}{4}$ hrs.)

    Time taken to empty the full tank $=\frac{1}{2/5}=\frac{5}{2}$ hours

    $\therefore$ Time taken to empty the exactly half filled tank $=\frac{1}{2}\times\frac{5}{2}$ hours $=\frac{5}{4}$ hours $=1$ hr 15 min.

4. (a) Part of the drum that can be emptied in 15 min.

by the second pipe $= 15 \times \dfrac{1}{60} = \dfrac{1}{4}$

After 15 min, part of the drum filled with

oil $= \dfrac{2}{3} - \dfrac{1}{4} = \dfrac{8-3}{12} = \dfrac{5}{12}$

After 15 min, part of the drum emptied

$= 1 - \dfrac{5}{12} = \dfrac{7}{12}$

Given, the whole drum can be filled in 40 min.

$\therefore \dfrac{7}{12}$ part of the drum can be filled in $\left(40 \times \dfrac{7}{12} \right)$

min. $= \dfrac{70}{3}$ min. $= 23\dfrac{1}{3}$ min.

5. (d) Let the tank be emptied in x hrs after 8 A.M.

$\therefore$ Tank filled by first pipe in x hours $= \dfrac{x}{15}$

Tank filled by second pipe in $(x-1)$ hours

$= \dfrac{(x-1)}{12}$

Tank emptied by third pipe in $(x-3)$ hours

$= \dfrac{(x-3)}{4}$

$\therefore \dfrac{x}{15} + \dfrac{(x-1)}{12} - \dfrac{(x-3)}{4} = 0$

$\Rightarrow \dfrac{4x + 5(x-1) - 15(x-3)}{60} = 0$

$\Rightarrow 4x + 5x - 5 - 15x + 45 = 0 \Rightarrow -6x + 40 = 0$

$\Rightarrow x = \dfrac{40}{6} = \dfrac{20}{3} = 6\dfrac{2}{3}$ hrs. $= 6$ hrs. 40 min.

$\therefore$ The tank will be emptied at 2:40 p.m.

6. (d) Part of the cistern filled by the two pipes in 1

hour $= \left(\dfrac{1}{14} + \dfrac{1}{16} \right) = \dfrac{15}{112}$

$\therefore$ Total time taken by the two pipes and the leak

to fill up the cistern $= \left(\dfrac{112}{15} + \dfrac{32}{60} \right)$ hrs $= 8$ hrs.

$\therefore$ Part of the cistern emptied by the leak in 1 hour

$= \left(\dfrac{1}{8} - \dfrac{15}{112} \right) = \dfrac{16-15}{112} = \dfrac{1}{112}$

$\therefore$ The leak will empty the cistern in 112 hours.

7. (b) A pipe with double the diameter can empty a tank in half the time, *i.e.,* in 20 min.

$\therefore$ Part of the tank emptied by the two pipes

together $= \left(\dfrac{1}{40} + \dfrac{1}{20} \right) = \dfrac{1+2}{40} = \dfrac{3}{40}$

$\therefore$ The tank will be emptied by the two pipes

together in $\dfrac{40}{3}$ min. $= 13\dfrac{1}{3}$ min.

8. (c) Part of the tank filled by $(A + B + C)$ in 2 hours

$= 2 \times \dfrac{1}{6} = \dfrac{1}{3}$

Remaining part $= 1 - \dfrac{1}{3} = \dfrac{2}{3}$

$\therefore \dfrac{2}{3}$rd of the tank is filled by $(A + B)$ in 7 hours.

$\therefore$ The whole tank is filled by $(A + B)$ in $\left(7 \times \dfrac{3}{2} \right)$

hours $= \dfrac{21}{2}$ hours

$\therefore$ Part of the tank filled by C in 1 hour

$=$ Part of the tank filled by $(A + B + C)$ in 1 hour

$-$ Part of the tank filled by $(A + B)$ in 1 hour

$= \dfrac{1}{6} - \dfrac{2}{21} = \dfrac{3}{42} = \dfrac{1}{14}$

$\therefore$ C can fill the whole tank in 14 hours.

9. (d) Let the total time taken to fill the tank be x minutes. Then,

work by B in half hour $+$ work done by $(A + B)$ in half hour $= 1$

$\left(\dfrac{1}{40} \times \dfrac{x}{2} \right) + \left(\dfrac{1}{60} + \dfrac{1}{40} \right) \times \dfrac{x}{2} = 1$

$\Rightarrow \dfrac{x}{80} + \dfrac{x}{48} = 1 \Rightarrow \dfrac{3x + 5x}{240} = 1$

$\Rightarrow 8x = 240 \Rightarrow x = 30$ min.

10. (a) Part of the tank that can be filled by the two

filler taps $= \left(\dfrac{1}{30} + \dfrac{1}{20} \right)$

$= \dfrac{2+3}{60} = \dfrac{5}{60} = \dfrac{1}{12}$

$\therefore$ The two taps can fill the tank in 12 minutes.

$\Rightarrow$ Half of the tank will be filled in 6 minutes.

$\Rightarrow$ The time taken by the remaining half of the tank to be filled when the outlet pipe is open $= (24 - 6)$ min $= 18$ min.

⇒ Part of that half of the tank that can be emptied

in 1 hour $= \dfrac{1}{6} - \dfrac{1}{18} = \dfrac{1}{9}$

∴ Time taken to empty half of the cistern
= 9 minutes

∴ Capacity of the tank $= 100 \times 9 \times 2 = 1800$ litres.

11. (d) Let the obstruction remain for x minutes only.

∴ Part of cistern filled in x minutes + Part of cistern filled in 3 minutes = cistern filled

$\left[\left(\dfrac{7}{8} \times \dfrac{x}{12}\right) + \left(\dfrac{5}{6} \times \dfrac{x}{16}\right)\right] + \left[\dfrac{3}{12} + \dfrac{3}{16}\right] = 1$

$\Rightarrow \dfrac{12x}{96} + \dfrac{7}{16} = 1 \Rightarrow \dfrac{12x}{96} = \dfrac{9}{16}$

$\Rightarrow x = \dfrac{9 \times 96}{16 \times 12} = 4.5$ min.

12. (a) Work done in 3 min. $= \left(\dfrac{1}{20} + \dfrac{1}{30} - \dfrac{1}{15}\right) = \dfrac{1}{60}$

∴ Work done in $3 \times 55 = 165$ min $= \dfrac{55}{60}$

∴ Remaining tank $= \left(1 - \dfrac{55}{60}\right) = \dfrac{5}{60} = \dfrac{1}{12}$

Now its A's turn, $\dfrac{1}{20}$ part of the tank is filled by A in 1 min.

Since there is still $\left(\dfrac{1}{12} - \dfrac{1}{20}\right) = \dfrac{1}{30}$ tank to be filled, which will be filled by B in 1 min. Therefore, required time $= (165 + 2) = 167$ min.

13. (d) In one minute $\dfrac{1}{27}$ part of the tank is filled. Due to leak every minute, $\dfrac{1}{10} \times \dfrac{1}{27} = \dfrac{1}{270}$ part of the tank leaks out. Hence the whole tank will be emptied in 270 min by the leak.

So, resultant work by pipe and leak in one min.

$= \dfrac{1}{27} - \dfrac{1}{270} = \dfrac{9}{270} = \dfrac{1}{30}$

∴ The tank will be filled in 30 min.

14. (c) Let one pipe take x minutes to fill the tank, then the other will take $(x + 20)$ minutes.

Given, $\dfrac{1}{x} + \dfrac{1}{x+20} = \dfrac{1}{24}$

$\Rightarrow \dfrac{x+20+x}{x^2+20x} = \dfrac{1}{24} \Rightarrow (2x+20)24 = x^2 + 20x$

$\Rightarrow x^2 - 28x - 480 = 0 \Rightarrow x^2 - 40x + 12x - 480 = 0$

$\Rightarrow x(x-40) + 12(x-40) = 0$

$\Rightarrow (x-40)(x+12) = 0$

$\Rightarrow x = 40$ or -12

Neglecting negative values $x = 40$ min.

15. (a) Let the second pipe be closed after x minutes.

Then, $\dfrac{24}{30} + \dfrac{x}{40} = 1$

$\Rightarrow \dfrac{96+3x}{120} = 1 \Rightarrow 3x = 120 - 96 = 24 \Rightarrow x = 6$ min.

Self Assessment Sheet–20

1. A and B working separately can do a piece of work in 9 days and 12 days respectively. If they work for 10 days alternately. A beginning, in how many days the work will be completed?

(a) 11 days
(b) $11\dfrac{1}{2}$ days
(c) $10\dfrac{1}{4}$ days
(d) $10\dfrac{1}{2}$ days

2. A, B and C complete a piece of work in 25, 20 and 24 days respectively. All work together for 2 days and then A and B leave the work. C works for next $8\dfrac{3}{5}$ days and then A along with D joins C and all finish the work in next three days. In how many days D alone can complete the whole work?

(a) 22 days
(b) $21\dfrac{1}{2}$ days
(c) 23 days
(d) $22\dfrac{1}{2}$ days

3. 75 boys do a piece of work in 24 days. How many men will finish double the work in 20 days when one days' work of 2 men is equal to one days' work of 3 boys?

(a) 100 men
(b) 150 men
(c) 120 men
(d) 80 men

4. It takes 8, 12 and 16 days for X, Y and Z respectively to complete a work. How many days will it take to complete the work if X works on it for 2 days, and then Y works on it for until 25% of the work is left for Z to do, and then Z complete the work?

(a) 10 days
(b) 12 days
(c) 8 days
(d) 14 days

5. 12 children take 16 days to complete a work which can be completed by 8 adults in 12 days. 16 adults started working and after 3 days 10 adults left and 4 children joined them. How may days will it take them to complete the remaining work?

   (a) 6 (b) 8

   (c) 4 (d) 3

6. 40 men take 8 days to earn Rs 2000. How many men will earn Rs 200 in 2 days?

   (a) 10 (b) 1

   (c) 8 (d) 20

7. A cistern which could be filled in 9 hours takes one hour more to be filled owing to a leak in its bottom. If the cistern is full, in what time will the leak empty it?

   (a) 19 hours (b) 1 hours

   (c) 90 hours (d) $\dfrac{10}{9}$ hours

8. A swimming pool is fitted with three pipes with uniform flow. The first two pipes operating simultaneously fill the pool in the same time as that taken by the third pipe alone. The second pipe fills the pool 5 hours faster than the first pipe and 4 hours slower than the third pipe. Find the time required by the third pipe to fill the pool.

   (a) 10 hours (b) 6 hours

   (c) 16 hours (d) 5 hours

9. Two pipes A and B can fill a cistern is 12 minutes and 15 minutes respectively but a third pipe C can empty the full tank in 6 minutes. A and B are kept open for 5 minutes in the beginning and then C is also opened. In what time will the cistern be emptied?

   (a) 30 min (b) 33 min

   (c) 37.5 min (d) 45 min

10. A tap can fill a bath in 20 minutes and another tap can fill it in 30 minutes. Amit opens both the taps simultaneously. When the both should have been full, he finds that the waste pipe was open. He then closes the waste pipe and in another 4 minutes the bath is full. In what time would the waste pipe empty it?

    (a) 35 min (b) 38 min

    (c) 36 min (d) 39 min

Answers

1. (c)   2. (d) [**Hint.** Total work completed by A, B, C and $D = 1 \Rightarrow$ Work completed by (A in 5 days + B in 2 days + C in $(2 + 8\frac{3}{5} + 3)$ days + D in 3 days) $\Rightarrow \dfrac{5}{25} + \dfrac{2}{20} + \dfrac{68}{5 \times 24} + \dfrac{3}{x} = 1$]

3. (c)   4. (b)   5. (a)   6. (b)   7. (c)   8. (b)   9. (d)

10. (c) [**Hint.** Bath filled by both the taps in 1 min $= \left[\dfrac{1}{20} + \dfrac{1}{30} \right] = \dfrac{1}{12}$

    ∴ The waste pipe empties as much bath in 12 minutes as the other two taps together can fill in 4 minutes.

    Bath filled in 4 min $= \dfrac{1}{12} \times 4 = \dfrac{1}{3}$. Now find the reqd. time.]

Chapter
21
DISTANCE, TIME AND SPEED

Section-A
DISTANCE, TIME AND SPEED

KEY FACTS

1. (i) Speed = $\dfrac{\text{Distance}}{\text{Time}}$ (ii) Distance = Speed × Time   (iii) Time = $\dfrac{\text{Distance}}{\text{Speed}}$

2. (i) To convert a speed in m/s to km/h, multiply by $\dfrac{18}{5}$.

    (ii) To convert a speed in km/h to m/s, multiply by $\dfrac{5}{18}$.

3. Average speed = $\dfrac{\text{Total distance covered}}{\text{Total time taken}}$

Solved Examples

Ex. 1. *A thief seeing a policeman from a distance of 200 metres, starts running with a speed of 8 km/hr. The policeman gives chase immediately with a speed of 9 km/hr and the thief is caught. What is the distance run by the thief?*

Sol. Relative speed of the policeman with respect to that of thief = $(9 - 8)$ km/hr = $1 \times \dfrac{5}{18}$ m/sec

$\Rightarrow$ Time taken by the policeman to catch the thief = $\dfrac{200}{5/18}$ sec = 720 seconds

$\therefore$ The distance run by the thief = $\left(8 \times \dfrac{5}{18} \times 720 \right)$ metres = 1600 metres

Ex. 2. *In covering a distance of 30 km, Abhay takes 2 hours more than Sameer. If Abhay doubles his speed, then he would take 1 hour less than Sameer. What is Abhay's speed?*

Sol. Let Abhay's speed be x km/hr and Sameer's speed be y km/hr.

Then, $\dfrac{30}{x} - \dfrac{30}{y} = 2$ .... (i)   and $\dfrac{30}{y} - \dfrac{30}{2x} = 1$ ... (ii)

Adding equation (i) and (ii), we get $\dfrac{30}{x} - \dfrac{30}{2x} = 3$

$\Rightarrow \dfrac{30}{2x} = 3$ $\Rightarrow$ $2x = 10$ $\Rightarrow$ $x = 5$ km/hr.

Ex. 3. *I started on my bicycle at 7 A.M. to reach a certain place. After going a certain distance my bicycle went out of order. Consequently I rested for 35 minutes and came back to my house walking all the way. I reached my house at 1 P.M. If my cycling speed is 10 km/hr and my walking speed is 1 km/hr, then what distance did I cover on my bicycle?*

Sol. Let the distance covered on the bicycle was x km. Then,

$$\frac{x}{10} + \frac{35}{60} + x = 6 \text{ hrs} \Rightarrow \frac{6x + 35 + 60x}{60} = 6 \Rightarrow 66x + 35 = 360$$

$$\Rightarrow 66x = 325 \Rightarrow x = \frac{325}{66} \text{ km} = 4\frac{61}{66} \text{ km}.$$

Ex. 4. *In a race of 200 metres, B can give a start of 10 metres to A and C can give a start of 20 metres to B. What is the start that C can give to A in the same race?*

Sol. Ratio of the distances covered by A and B = 190 : 200 = 1710 : 1800

Ratio of the distances covered by B and C = 180 : 200 = 1800 : 2000

$\Rightarrow$ Ratio of the distances covered by A and C = 1710 : 2000 = 171 : 200

Hence C will give a start of (200 m – 171 m) = 29 m to A in the same race.

Ex. 5. *A cyclist cycles non-stop from A to B, a distance of 14 km at a certain average speed. If his average speed reduces by 1 km / hr, then he takes $\frac{1}{3}$ hour more to cover the same distance. What was the original average speed of the cyclist?*

Sol. Let the original average speed of the cyclist be x km/hr.

Then, $\dfrac{14}{(x-1)} - \dfrac{14}{x} = \dfrac{1}{3} \Rightarrow \dfrac{14x - 14(x-1)}{x(x-1)} = \dfrac{1}{3}$

$\Rightarrow \dfrac{14}{x^2 - x} = \dfrac{1}{3} \Rightarrow x^2 - x - 42 = 0 \Rightarrow x^2 - 7x + 6x - 42 = 0$

$\Rightarrow x(x - 7) + 6(x - 7) = 0 \Rightarrow (x - 7)(x + 6) = 0$

$\Rightarrow x = 7$ or -6

Since speed cannot be negative, $x = 7$ km/hr.

Ex. 6. *The ratio between the rates of walking of A and B is 2 : 3 and therefore A takes 10 minutes more than the time taken by B to reach the destination. If A had walked at double the speed, then in what time would he have covered that distance?*

Sol. Ratio of speeds = 2 : 3

$\Rightarrow$ Ratio of times taken = 3 : 2

Given $3x - 2x = 10 \Rightarrow x = 10$

$\Rightarrow$ A would have taken 30 minutes.

But if A walks with double the speed, then he takes half the time, *i.e.*, 15 minutes.

Ex. 7. *A train X starts from New Delhi at 4 pm and reaches Ghaziabad at 5 pm. While another train Y starts from Ghaziabad at 4 pm and reaches New Delhi at 5 : 30 pm. At what time will the two trains cross each other?*

Sol. Let Delhi – Ghaziabad distance be a km. Then,

Speed of train $X = a$ km/hr

Speed of train $Y = \dfrac{a}{3/2}$ km/hr $= \dfrac{2a}{3}$ km/hr

Suppose they meet after b hours, then $ab + \dfrac{2ab}{3} = a \Rightarrow \dfrac{5b}{3} = 1 \Rightarrow b = \dfrac{3}{5}$ hrs = 36 minutes

∴ They meet at 4 : 36 pm.

Ex. 8. *A and B walk from X to Y, a distance of 27 km at 5 km/hr and 7 km/hr respectively. B reaches Y and immediately turns back meeting A at Z. What is the distance from X to Z?*

Sol. Time taken by A in covering $(27 - x)$ km is the same as time taken by B in covering $(27 + x)$ km.

$$\therefore \frac{27 - x}{5} = \frac{27 + x}{7}$$

$$\Rightarrow 189 - 7x = 135 + 5x \Rightarrow 12x = 54 \Rightarrow x = \frac{54}{12} = \frac{27}{6} \text{ km.}$$

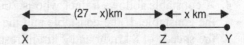

$$\therefore XZ = 27 - \frac{27}{6} = 27 \times \frac{5}{6} = 22.5 \text{ km.}$$

Ex. 9. *A man travels from A to B at a speed of x km/hr. He then rests at B for x hours. He then travels from B to C at a speed of 2x km/hr and rests for 2x hours. He moves further to D at a speed twice as that between B and C. He thus reaches D in 16 hours. If the distance A – B, B – C and C – D are all equal to 12 km, then what could be the time for which he rested at B?*

Sol. Total time taken from A to $D = \dfrac{12}{x} + x + \dfrac{12}{2x} + 2x + \dfrac{12}{4x}$

$$= \frac{48}{4x} + \frac{24}{4x} + \frac{12}{4x} + 3x = \frac{21}{x} + 3x$$

Given, $\dfrac{21}{x} + 3x = 16$

$$\Rightarrow 21 + 3x^2 - 16x = 0 \Rightarrow 3x^2 - 16x + 21 = 0 \Rightarrow 3x^2 - 9x - 7x + 21 = 0$$

$$\Rightarrow 3x(x - 3) - 7(x - 3) = 0 \Rightarrow (x - 3)(3x - 7) = 0$$

$$\Rightarrow x = 3 \text{ or } x = \frac{7}{3}$$

$$\therefore x = 3 \text{ hrs or } \frac{7}{3} \text{ hrs.}$$

Ex. 10. *A man travels three-fifths of a distance AB at a speed of 3a, and the remaining at a speed of 2b. If he goes from B to A and returns at a speed of 5c in the same time, then*

(a) $\dfrac{1}{a} + \dfrac{1}{b} = \dfrac{1}{c}$ (b) $a + b = c$ (c) $3a + 2b = 5c$ (d) $\dfrac{1}{a} + \dfrac{1}{b} = \dfrac{2}{c}$

Sol. Time taken to cover $AC = \dfrac{3x}{5 \times 3a} = \dfrac{x}{5a}$ hr

Time taken to cover $CB = \dfrac{2x}{5 \times 2b} = \dfrac{x}{5b}$ hr

Time taken to cover BA and back $AB = \dfrac{2x}{5c}$

Given, $\dfrac{x}{5a} + \dfrac{x}{5b} = \dfrac{2x}{5c} \Rightarrow \dfrac{1}{a} + \dfrac{1}{b} = \dfrac{2}{c}.$

Question Bank–21(a)

1. A person wants to travel a distance of 50 km by his bicycle. He travels with a speed of 12.5 km/hr. After every 12.5 km, he takes a rest of 20 minutes. How much time will he take to complete the whole distance?

(a) 4 hrs 20 min (b) 5 hrs 20 min
(c) 5 hrs (d) 6 hrs

2. Two cars start at the same time from one point and move along two roads at right angles to each other. Their speeds are 36 km/hr and 48 km/hr respectively. After 15 seconds, the distance between them will be

(a) 400 m (b) 150 m
(c) 300 m (d) 250 m

3. In a race of 800 m, A can beat B by 40 m. In a race of 500 m, B can beat C by 5 m. In a race of 200 m, A will beat C by
   (a) 11.9 m   (b) 1.19 m
   (c) 12.7 m   (d) 1.27 m

4. A constable follows a thief who is 200 m ahead of the constable. If the constable and the thief runs at the speeds of 8 km/hr and 7 km/hr respectively, the constable would catch the thief in
   (a) 10 min   (b) 12 min
   (c) 15 min   (d) 20 min

5. Two buses travel to a place at 45 km/hr and 60 km/hr respectively. If the second bus takes $5\frac{1}{2}$ hrs less than the first for the journey, the length of the journey is
   (a) 900 km   (b) 945 km
   (c) 990 km   (d) 1350 km

6. A cyclist covering a distance of 40 km would have reached 1 hour earlier, if he had run at an increased speed of 2 km/hr. His speed in (km/hr) was
   (a) 6   (b) 8
   (c) 10   (d) 12

7. Ram travels from P to Q at 10 km/hr and returns at 15 km/hr. Shyam travels from P to Q and returns at 12.5 km/hr. If he takes 12 minutes less than Ram, then what is the distance between P and Q?
   (a) 60 km   (b) 45 km
   (c) 36 km   (d) 30 km

8. A student reached his school late by 20 minutes by travelling at a speed of 9 km/hr. Had he travelled at the speed of 12 km/hr, he would have reached his school 20 minutes early. Find the distance between his house and the school?
   (a) 12 km   (b) 6 km
   (c) 3 km   (d) 24 km

9. An aircraft was to take off from a certain airport at 8 a.m., but it was delayed by 30 min. To make up for the lost time, it was to increase its speed by 250 km/hour from the normal speed to reach its destination 1500 km on time. What was the normal speed of the aircraft?
   (a) 650 km/hr   (b) 750 km/hr
   (c) 850 km/hr   (d) 1000 km/hr

10. Robert is travelling on his cycle and has calculated to reach point A at 2 p.m. if he travels at 10 km/hr, he will reach there at 12 noon if he travels at 15 km/hr. At what speed must he travel to reach A at 1 p.m.?

    (a) 8 km/hr   (b) 11 km/hr
    (c) 12 km/hr   (d) 14 km/hr

11. A is faster than B. A and B each walk 24 km. The sum of their speeds is 7 km/hr and sum of times taken by them is 14 hours. Then A's speed is equal to
    (a) 3 km/hr   (b) 4 km/hr
    (c) 5 km/hr   (d) 7 km/hr

12. A car travels the first one-third of a certain distance with a speed of 10 km/hr, the next one-third with a speed of 20 km/hr and the last one-third distance with a speed of 60 km/hr. The average speed of the car for the whole journey is
    (a) 18 km/hr   (b) 24 km/hr
    (c) 30 km/hr   (d) 36 km/hr

13. A motor car starts with a speed of 70 km/hr with its speed increasing every two hours by 10 km/hr. In how many hours will it cover 345 kms?

    (a) $2\frac{1}{4}$ hrs   (b) 4 hrs 5 min

    (c) $4\frac{1}{2}$ hrs   (d) 3 hrs

14. A train can travel 50% faster than a car. Both start from point A at the same time and reach point B 75 km away from A at the same time. On the way, however, the train lost about 12.5 minutes while stopping at the stations. The speed of the car is
    (a) 100 km/hr   (b) 110 km/hr
    (c) 120 km/hr   (d) 130 km/hr

15. Shyam went from Delhi to Shimla via Chandigarh by car. The distance from Delhi to Chandigarh is $\frac{3}{4}$ times the distance from Chandigarh to Shimla. The average speed from Delhi to Chandigarh was one and a half times that from Chandigarh to Shimla. If the average speed for the entire journey was 49 km/hr, what was the average speed from Chandigarh to Shimla?
    (a) 39.2 km/hr   (b) 63 km/hr
    (c) 42 km/hr   (d) 35 km/hr

16. A and B walk round a circular track. They start at 8 a.m. from the same point in opposite directions. A and B walk at a speed of 2 rounds per hour and 3 rounds per hour respectively. How many times shall they cross each other before 9.30 a.m.
    (a) 5   (b) 6
    (c) 7   (d) 8

17. A man covered a certain distance at some speed. Had he moved 3 km/hr faster, he would have taken 40 minutes less. If he had moved 2 km/hr slower, he would have taken 40 minutes more. The distance (in km) is:
    (a) 35
    (b) $36\frac{2}{3}$
    (c) $37\frac{1}{2}$
    (d) 40

18. A and B are 25 km apart. If they travel in opposite directions, they meet after one hour. If they travel in the same direction, they meet after 5 hours. If A travels faster than B, then the speed of A is
    (a) 10 km/hr
    (b) 12.5 km/hr
    (c) 15 km/hr
    (d) 20 km/hr

19. A small aeroplane can travel at 320 km/hr in still air. The wind is blowing at a constant speed of 40 km/hr. The total time for a journey against the wind is 135 minutes. What will be the time, in minutes for the return journey with the wind? (Ignore take off and landing times for the aeroplane.)
    (a) 94.5
    (b) 105
    (c) 108.125
    (d) 120

20. A man reduces his speed to two-third to walk a distance and consequently becomes late by 1 hour. With his usual speed, he covers the same distance in
    (a) $\frac{1}{4}$ hour
    (b) $\frac{1}{2}$ hour
    (c) 2 hours
    (d) $1\frac{1}{2}$ hours

21. If I walk at 3 km/hr, I miss a train by 2 minutes. If, however, I walk at 4 km/hr, then I reach the station 2 minutes before the arrival of the train. How far do I walk to reach the station?
    (a) $\frac{3}{4}$ km
    (b) $\frac{4}{5}$ km
    (c) $\frac{5}{4}$ km
    (d) 1 km

22. A car driver, driving in a fog, passes a pedestrian who was walking at the rate of 2 km/hr in the same direction. The pedestrian could see the car for 6 minutes and it was visible to him upto a distance of 0.6 km. What was the speed of the car?
    (a) 15 km/hr
    (b) 30 km/hr
    (c) 20 km/hr
    (d) 8 km/hr

23. A train increases its normal speed by 12.5% and reaches its destination 20 minutes earlier. What is the actual time taken by the train in the journey?
    (a) 220 min
    (b) 180 min
    (c) 145 min
    (d) 160 min

24. A bike travels a distance of 200 km at a constant speed. If the speed of the bike is increased by 5 km/hr, the journey would have taken 2 hours less. What is the speed of the bike?
    (a) 30 km/hr
    (b) 25 km/hr
    (c) 20 km/hr
    (d) 15 km/hr

25. Two persons P and Q start at the same time from city A to city B, 60 km away. P travels 4 km/hr slower than Q. Q reaches city B and at once turns back meeting P, 12 km from city B. What is the speed of P?
    (a) 8 km/hr
    (b) 12 km/hr
    (c) 16 km/hr
    (d) 20 km/hr

26. A starts from a place P to go to a place Q. At the same time, B starts from Q to P. If after meeting each other A and B took 4 hrs and 9 hrs more respectively to reach their destination, the ratio of their speeds is
    (a) 3 : 2
    (b) 5 : 2
    (c) 9 : 4
    (d) 9 : 13

27. In covering a certain distance, the speeds of A and B are in the ratio 3 : 4. A takes 30 minutes more than B to reach the destination. The time taken by A to reach the destination is
    (a) 1 hour
    (b) $1\frac{1}{2}$ hours
    (c) 2 hours
    (d) $2\frac{1}{2}$ hours

28. A and B run a kilometre and A wins by 25 seconds. A and C run a kilometre and A wins by 275 m. B and C run the same distance and B wins by 30 sec. The time taken by A to run a kilometer is
    (a) 2 min 25 sec
    (b) 2 min 50 sec
    (c) 3 min 20 sec
    (d) 3 min 30 sec

29. A hare sees a dog 200 m away from her and scuds off in the opposite direction at a speed of 24 km/hr. Two minutes later, the dog perceives her and gives chase at a speed of 32 km/hr. How soon will the dog overtake the hare and what is the distance from the spot from where the hare took flight?
    (a) 8 min 2 km
    (b) $7\frac{1}{2}$ min, 2 km
    (c) $7\frac{1}{2}$ min, 3 km
    (d) $7\frac{1}{2}$ min, 1 km

30. A, B and C start from the same place and travel the same directions at speeds of 30, 40 and 60 km/hr respectively. B starts two hours after A. If B and C overtake A at the same instant, how many hours after A did C start .
    (a) 3
    (b) 3.5
    (c) 4
    (d) 4.5

Answers

1. (c)	2. (d)	3. (a)	4. (b)	5. (c)	6. (b)	7. (d)	8. (d)	9. (b)	10. (c)
11. (b)	12. (a)	13. (c)	14. (c)	15. (c)	16. (c)	17. (d)	18. (c)	19. (b)	20. (c)
21. (b)	22. (d)	23. (b)	24. (c)	25. (a)	26. (a)	27. (c)	28. (a)	29. (c)	30. (c)

Hints and Solutions

1. (c) Time taken to travel 50 km $= \dfrac{50}{12.5}$ hrs = 4 hrs.

Number of stoppages $= \dfrac{50}{12.5} - 1 = 3$

∴ Total duration of stoppages $= (3 \times 20)$ min
= 60 min = 1 hrs

∴ Total time to complete the whole distance = 5 hrs.

2. (d) Distance travelled by the 1st car in 15 seconds

$= OA = \left(36 \times \dfrac{5}{18}\right) \times 15$ m = 150 m

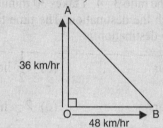

Distance travelled by the 2nd car in 15 seconds

$= OB = \left(48 \times \dfrac{5}{18}\right) \times 15$ m = 200 m

∴ Distance between them after 15 seconds

$= AB = \sqrt{OA^2 + OB^2} = \sqrt{150^2 + 200^2}$

$= \sqrt{22500 + 40000}$

$= \sqrt{62500} = 250$ m.

3. (a) Ratio of distances covered by A and B
= 800 : 760 = 20000 : 190000

Ratio of distances covered by B and C
= 500 : 495 = 190000 : 18810

∴ Ratio of distances covered by A and C
= 20000 : 18810 = 200 : 188.1

Hence A will beat C in a 200 m race by
200 m – 188.1 m = 11.9 m.

4. (b) Relative speed = (8 – 7) = 1 km/hr
(∵ Both are in the same direction)

$= 1 \times \dfrac{5}{18} = \dfrac{5}{18}$ m/sec

∴ Required time $= \dfrac{200}{5/18} = \dfrac{20 \times 18}{5}$ seconds

= 720 seconds = 12 min.

5. (c) Let the length of the journey be x km. Then,

$\dfrac{x}{45} - \dfrac{x}{60} = \dfrac{11}{2} \Rightarrow \dfrac{4x - 3x}{180} = \dfrac{11}{2} \Rightarrow \dfrac{x}{180} = \dfrac{11}{2}$

$\Rightarrow x = \dfrac{11 \times 180}{2} = 990$ km.

6. (b) Let the speed of the cyclist be x km/hr.

∴ $\dfrac{40}{x} - \dfrac{40}{x+2} = 1 \Rightarrow \dfrac{40(x+2) - 40x}{x(x+2)} = 1$

$\Rightarrow 80 = x^2 + 2x \Rightarrow x^2 + 2x - 80 = 0$

$\Rightarrow x^2 + 10x - 8x - 80 = 0$

$\Rightarrow x(x + 10) - 8(x + 10) = 0$

$\Rightarrow (x - 8)(x + 10) = 0$

$\Rightarrow x = 8$ or -10.

Neglecting –ve values, we take x = 8 km/hr.

7. (d) Let the distance between P and Q be x km.
Then, Total time taken by Ram in the round trip
– Total time taken by Shyam in the round trip
= 12 min

$\Rightarrow \left(\dfrac{x}{10} + \dfrac{x}{15}\right) - \left(\dfrac{x}{12.5} + \dfrac{x}{12.5}\right) = \dfrac{12}{60}$

$\Rightarrow \left(\dfrac{3x + 2x}{30}\right) - \dfrac{2x}{12.5} = \dfrac{12}{60}$

$\Rightarrow \dfrac{x}{6} - \dfrac{2x}{12.5} = \dfrac{12}{60} \Rightarrow \dfrac{12.5x - 12x}{75} = \dfrac{12}{60}$

$\Rightarrow 0.5x = 75 \times \dfrac{12}{60} \Rightarrow x = \dfrac{75 \times 12}{0.5 \times 60} = 30$ km.

8. (d) Let the distance of the house from the school
be x km.

Then, $\dfrac{x}{9} - \dfrac{1}{3} = \dfrac{x}{12} + \dfrac{1}{3} \Rightarrow \dfrac{x}{9} - \dfrac{x}{12} = \dfrac{1}{3} + \dfrac{1}{3} = \dfrac{2}{3}$

$\Rightarrow \dfrac{4x - 3x}{36} = \dfrac{2}{3} \Rightarrow \dfrac{x}{36} = \dfrac{2}{3}$

$\Rightarrow x = \left(\dfrac{2}{3} \times 36 \right)$ km = 24 km.

9. (b) Let the normal speed of the aircraft be x km/hr.

$\therefore \quad \dfrac{1500}{x} - \dfrac{1500}{250 + x} = \dfrac{1}{2}$

$\Rightarrow \dfrac{1500(250 + x) - 1500x}{x(x + 250)} = \dfrac{1}{2}$

$\Rightarrow x^2 + 250x = 2 \times 1500 \times 250$

$\Rightarrow x^2 + 250x - 750000 = 0$

$\Rightarrow x^2 + 1000x - 750x - 750000 = 0$

$\Rightarrow x(x + 1000) - 750(x + 1000) = 0$

$\Rightarrow (x + 1000)(x - 750) = 0 \Rightarrow x = -1000$ or 750

Neglecting –ve values, $x = 750$ km/hr.

10. (c) Let the distance travelled be x km. Then,

$\dfrac{x}{10} - \dfrac{x}{15} = 2 \Rightarrow 3x - 2x = 60 \Rightarrow x = 60$ km

Time taken to travel 60 km at 10 km/hr

$= \left(\dfrac{60}{10} \right)$ hrs = 6 hrs.

So Robert started 6 hours before 2 P.M., *i.e.*, at 8 P.M.

$\therefore$ Speed required to reach A at 1 P.M.

$= \left(\dfrac{60}{5} \right)$ km/hr = 12 km/hr

11. (b) Let A's speed = x km/hr. Then B's speed = $(7 - x)$ km/hr

So, $\dfrac{24}{x} + \dfrac{24}{(7 - x)} = 14$

$\Rightarrow 24(7 - x) + 24x = 14x(7 - x)$

$\Rightarrow 168 = 98x - 14x^2 \Rightarrow 14x^2 - 98x + 168 = 0$

$\Rightarrow x^2 - 7x + 12 = 0 \Rightarrow (x - 3)(x - 4) = 0$

$\Rightarrow x = 3$ or 4

Since A is faster than B, A's speed = 4 km/hr and B's speed = 3 km/hr.

12. (a) Let the total distance travelled be x km. Then,

Total time taken $= \dfrac{x/3}{10} + \dfrac{x/3}{20} + \dfrac{x/3}{60}$

$= \dfrac{x}{30} + \dfrac{x}{60} + \dfrac{x}{180}$

$= \dfrac{6x + 3x + x}{180} = \dfrac{10x}{180} = \dfrac{x}{18}$ hrs.

Then, average speed $= \dfrac{\text{Total distance travelled}}{\text{Total time taken}}$

$= \dfrac{x}{x/18}$ km/hr = 18 km/hr.

13. (c) Distance covered in the first 2 hours
$= (70 \times 2)$ km = 140 km

Distance covered in the next 2 hours
$= (80 \times 2)$ km = 160 km

Remaining distance = 345 – (140 + 160) = 45 km

Now, speed in the fifth hour = 90 km/hr

$\therefore$ Time taken to cover 45 km $= \left(\dfrac{45}{90} \right)$ hr $= \dfrac{1}{2}$ hr

$\therefore$ Total time taken $= \left(2 + 2 + \dfrac{1}{2} \right)$ hr $= 4\dfrac{1}{2}$ hrs.

14. (c) Let the speed of the car be x km/hr

Then, speed of the train $= \dfrac{150x}{100}$ km/hr

$= \dfrac{3x}{2}$ km/hr

Time taken by car to reach point $B = \dfrac{75}{x}$ hrs

Time taken by train to reach point $B = \dfrac{75}{\dfrac{3}{2}x}$ hrs

Given, $\dfrac{75}{x} - \dfrac{75}{3/2\,x} = \dfrac{12.5}{60}$

$\Rightarrow \dfrac{75}{x} - \dfrac{50}{x} = \dfrac{125}{10 \times 60} = \dfrac{5}{24} \Rightarrow \dfrac{25}{x} = \dfrac{5}{24}$

$\Rightarrow x = \dfrac{25 \times 24}{5} = 120$ km/hr.

15. (c) Let the distance from Chandigarh to Shimla be x km and the average speed from Chandigarh to Shimla be y km/hr.

Then, average speed from Delhi to Chandigarh $= \dfrac{3}{2}y$ km/hr

Distance from Delhi to Chandigarh $= \dfrac{3}{4}x$ cm

$\therefore$ Average speed for entire journey

$$= \frac{\frac{3}{4}x + x}{\frac{3}{4}x \times \frac{2}{3y} + \frac{x}{y}} \text{ km/hr}$$

$$= \frac{\frac{7x}{4}}{\frac{x}{2y} + \frac{x}{y}} = \frac{\frac{7x}{4}}{\frac{3x}{2y}} = \frac{7x}{4} \times \frac{2y}{3x} = \frac{7y}{6}$$

Given, $\frac{7y}{6} = 49 \Rightarrow y = \frac{49 \times 6}{7} = 42$ km/hr.

16. (c) Relative speed = (2 + 3) km/hr = 5 km/hr
This means that A and B cross each other 5 times in a hour and at least 2 times in half an hour.
$\therefore$ They cross each other 7 times before 9:30 a.m.

17. (d) Let the distance = x km and usual rate = y km/hr. Then,

$$\frac{x}{y} - \frac{x}{y+3} = \frac{40}{60} \Rightarrow \frac{x(y+3) - xy}{y(y+3)} = \frac{2}{3}$$

$$\Rightarrow \frac{3x}{y(y+3)} = \frac{2}{3}$$

$\Rightarrow 2y(y+3) = 9x$ ...(i)

and $\frac{x}{y-2} - \frac{x}{y} = \frac{40}{60} \Rightarrow \frac{xy - x(y-2)}{y(y-2)} = \frac{2}{3}$

$$\Rightarrow \frac{2x}{y(y-2)} = \frac{2}{3}$$

$\Rightarrow y(y-2) = 3x$ ...(ii)

On dividing eqn (i) by eqn (ii), we get

$\frac{2(y+3)}{y-2} = 3 \Rightarrow 2y + 6 = 3y - 6 \Rightarrow y = 12$

$\therefore$ Putting the value of y in (i), we get
$2 \times 12 \times 15 = 9x \Rightarrow x = 40$ km/hr.

18. (c) Let the speeds of A and B be u km/hr and v km/hr respectively.
When they travel in opposite directions,
$1 \times u + 1 \times v = 25 \Rightarrow u + v = 25$ ...(i)
When they travel in same directions,
$5 \times u - 5 \times v = 25 \Rightarrow u - v = 5$ ...(ii)
Adding (i) and (ii), we get $2u = 30$
$\Rightarrow u = 15$ km/hr.

19. (b) Speed of the aeroplane in still air = 320 km/hr
Speed of wind = 40 km/hr
$\therefore$ Aeroplane will travel with the wind at $(320 + 40) = 360$ km/hr and Aeroplane will travel against the wind at $(320 - 40) = 280$ km/hr
Suppose the distance to be travelled = x km
Then,

$$\frac{x}{280} = \frac{135}{60} \text{ hrs} = \frac{9}{4} \text{ hrs}$$

$\Rightarrow x = \frac{280 \times 9}{4} = 630$ km

$\therefore$ Time taken to cover a distance of 630 km at

360 km/hr is $= \left(\frac{630}{360} \times 60\right)$ min = 105 min.

20. (c) Let the distance to be covered be x km at y km/hr.

$\therefore \frac{x}{y} = \frac{x}{\frac{2}{3}y} - 1 \Rightarrow \frac{x}{y} - \frac{3x}{2y} = -1 \Rightarrow \frac{2x - 3x}{2y} = -1$

$\Rightarrow -x = -2y \Rightarrow x = 2y.$

$\therefore$ With his usual speed the time taken to cover a

distance x km. $= \frac{x}{y} = \frac{2y}{y}$ hrs = 2 hrs.

21. (b) Let the distance which I walk to the station be x km. Then,
Time needed to reach the station at 3 km/hr

$= \left(\frac{x}{3} + \frac{2}{60}\right)$ hrs and

Time needed to reach the station at 4 km/hr

$= \left(\frac{x}{4} - \frac{2}{60}\right)$ hrs

But, $\frac{x}{3} + \frac{1}{30} = \frac{x}{4} - \frac{1}{30} \Rightarrow \frac{x}{3} - \frac{x}{4} = \frac{2}{30} = \frac{1}{15}$

$\Rightarrow \frac{4x - 3x}{12} = \frac{1}{15} \Rightarrow \frac{x}{12} = \frac{1}{15} \Rightarrow x = \frac{12}{15} = \frac{4}{5}$ km.

22. (d) Let the speed of the car be x km/hr. Then,
Relative speed of the pedestrian with respect to the car = $(x - 2)$ km/hr ($\because$ Both move in the same direction)

$\therefore \frac{0.6}{(x-2)} = \frac{6}{60} \Rightarrow 6 = x - 2 \Rightarrow x = 8$ km/hr.

23. (b) Let the normal speed of the train be x metres/minute and the actual time taken by the train in the journey be t minutes. Then,

$$x \times t = \frac{112.5x}{100} \times (t - 20)$$

$$\Rightarrow t = \frac{112.5t}{100} - 22.5 \Rightarrow \frac{112.5t}{100} - t = 22.5$$

$$\Rightarrow \frac{12.5t}{100} = 22.5 \Rightarrow t = \frac{22.5 \times 100}{12.5} \text{ minutes} = 180 \text{ min.}$$

24. (c) Let the constant speed of the bike be x km/hr.

Then, $\dfrac{200}{x} - \dfrac{200}{(x+5)} = 2$

$$\Rightarrow \frac{200(x+5) - 200x}{x(x+5)} = 2 \Rightarrow \frac{1000}{x^2 + 5x} = 2$$

$$\Rightarrow x^2 + 5x - 500 = 0 \Rightarrow x^2 + 25x - 20x - 500 = 0$$

$$\Rightarrow x(x+25) - 20(x+25) = 0$$

$$\Rightarrow (x+25)(x-20) = 0$$

$$\Rightarrow x = -25 \text{ or } 20 \Rightarrow x = 20 \text{ km/hr.}$$

$$(\because \text{ speed is not negative})$$

25. (a) Let the speed of P be x km/hr.

Speed of $Q = (x + 4)$ km/hr

Distance travelled by $Q = (60 + 12)$ km $= 72$ km

Distance travelled by $P = (60 - 12)$ km $= 48$ km

$$\Rightarrow \frac{72}{x+4} = \frac{48}{x} \Rightarrow 72x = 48x + 192 \Rightarrow 24x = 192$$

$$\Rightarrow x = \frac{192}{24} = 8 \text{ km/hr.}$$

26. (a) Let the speed of A is x km/hr and speed of B is y km/hr.

Let them meet after t hours. Then,

$xT + yT = $ Total distance

After the meeting, distance left for $A = yT$ and the distance left for $B = xT$

Now, $yT = 4x$ and $xT = 9y$

$$\therefore \frac{yT}{xT} = \frac{4x}{9y} \Rightarrow \frac{x^2}{y^2} = \frac{9}{4} \Rightarrow \frac{x}{y} = \frac{3}{2} \Rightarrow x : y = 3 : 2$$

27. (c) Ratio of speeds $= 3 : 4$

$$\Rightarrow \text{Ratio of times taken} = 4 : 3$$

Let A and B take $4x$ and $3x$ hrs to reach the destination. Then, $4x - 3x = \dfrac{30}{60} = \dfrac{1}{2} \Rightarrow x = \dfrac{1}{2}$

$$\therefore \text{ Time taken by } A = \left(4 \times \frac{1}{2}\right) \text{ hrs} = 2 \text{ hrs.}$$

28. (a) If A covers a distance of 1 km in x seconds, then B covers the distance of 1 km in $(x + 25)$ seconds.

If A covers a distance of 1 km, then in the same time C covers only 725 metres.

If B covers the distance of 1 km in $(x + 25)$ seconds, then C covers the distance of 1 km in $(x + 55)$ seconds. Also C covers a distance of 725 metres in x seconds. Then,

$$\frac{x}{725} \times 1000 = x + 55 \Rightarrow x = 145$$

$\Rightarrow$ A covers a distance of 1 km in 145 seconds, *i.e.*, 2 min 25 sec.

29. (c) Distance covered by hare in 2 min

$$= \frac{24}{60} \times 1000 \times 2 = 800 \text{ m}$$

Now to overtake the hare, the dog has to cover a distance of $(800 + 200) = 1000$ m with a relative speed $= (32 - 24)$ km/hr $= 8$ km/hr

$$\therefore \text{ Time taken} = \frac{1}{8} \text{ hrs} = \left(\frac{1}{8} \times 60\right) \text{ min} = 7\frac{1}{2} \text{ min}$$

$$\therefore \text{ Distance travelled by hare in } \frac{1}{8} \text{ hrs}$$

$$= \left(\frac{1}{8} \times 24\right) \text{ km} = 3 \text{ km.}$$

30. (c) B starts when A has already travelled for 2 hours and covered a distance $= (2 \times 30)$ km $= 60$ km

$\therefore$ Time taken by B to cover A with a relative speed of $(40 - 30)$ km/hr $= 10$ km/hr $= \dfrac{60}{10}$ hrs $= 6$ hrs

$\therefore$ When B overtakes A, A has travelled for 8 hrs and B for 6 hrs

It is given that B and C overtake A at the same instance. Therefore, when C overtakes A, both of them have covered the same distance. Let C take t hours to cover the same distance as covered by A in 8 hours.

$$\therefore \quad 8 \times 30 = t \times 60 \Rightarrow t = 4 \text{ hours}$$

$\therefore$ C started after $(8 - 4 = 4)$ hrs after A started.

Section-B
PROBLEMS ON TRAINS

1. Time taken by a train a metres long to pass a stationary object = Time taken by it to cover a metres.

2. Time taken by a train a metres long to pass a platform or tunnel b metres long = Time taken by it to cover $(a + b)$ metres.

3. **Relative motion**

(i) If two objects are moving in opposite directions towards each other with speeds u km/hr and v km/hr respectively, then their relative speed is $(u + v)$ km/hr.

So, the time taken by two trains of lengths l_1 and l_2 metres running with speeds u km/hr and v km/hr in opposite directions to cross each other is the same as time taken to cover a distance equal to $(l_1 + l_2)$ metres with a relative speed equal to $(u + v)$ km/hr or $\dfrac{5}{18}(u + v)$ m/s.

$\therefore$ Required time = $\dfrac{18(l_1 + l_2)}{5(u + v)}$ seconds

(ii) If two objects are moving in the same direction with speeds u km/hr and v km/hr respectively such that $u > v_1$ then their relative speed is $(u - v)$ km/hr.

So, the time taken by two trains of lengths l_1 and l_2 metres moving in the same direction with speeds u km/hr and v km/hr respectively to cross each other is same as the time taken to cover a distance of $(l_1 + l_2)$ metres with a relative speed of $(u - v)$ km/hr or $\dfrac{5}{18}(u - v)$ m/s.

$\therefore$ Required time = $\dfrac{18(l_1 + l_2)}{5(u - v)}$ seconds

Solved Examples

Ex. 1. *If a train with a speed of 60 km/hr crosses a pole in 30 seconds, what is the length of the train in metres?*

Sol. Speed of train = 60 km/hr = $\left(60 \times \dfrac{5}{18}\right)$ m/s = $\dfrac{50}{3}$ m/s

Time taken = 30 seconds

$\therefore$ Length of train = $\left(\dfrac{50}{3} \times 30\right)$ m = 500 m.

Ex. 2. *Two trains 140 m and 160 m long run at the speed of 60 km/hr and 40 km/hr respectively in opposite directions on parallel tracks. What is the time (in seconds) which they take to cross each other?*

Sol. Required time = $\dfrac{\text{Total distance}}{\text{Relative speed}}$ = $\dfrac{\text{Sum of the lengths of the two trains}}{\text{Sum of speeds}}$

$= \dfrac{(140 + 160) \times 18}{(60 + 40) \times 5}$ seconds = $\dfrac{300 \times 18}{100 \times 5}$ = 10.8 seconds.

Ex. 3. *Two trains travel in the same direction at 50 km/hr and 32 km/hr respectively. A man in the slower train observes that the faster train passes him completely in 15 seconds. What is the length of the faster train in metres?*

Sol. Since the trains travel in the same direction, relative speed of the trains = (50 – 32) km/hr = 18 km/hr

$$= \left(18 \times \frac{5}{18}\right) \text{ m/s} = 5 \text{ m/s}$$

∴ Length of the faster train = Speed × Time
$$= (5 \times 15) \text{ m} = 75 \text{ m}.$$

Ex. 4. *A train 110 m long passes a man running at a speed of 6 km/hr in the direction opposite to the train in 6 seconds. What is the speed of the train?*

Sol. Let the speed of the train be x km/hr.

Speed of man = 6 km/hr

Since the train and the man are in opposite directions
Relative speed = $(x + 6)$ km/hr

$$= \left((x + 6) \times \frac{5}{18}\right) \text{ m/s}$$

∴ $(x+6) \times \dfrac{5}{18} \times 6 = 110$

⇒ $5x + 30 = 330$ ⇒ $5x = 300$ ⇒ $x = 60$ km/hr.

Ex. 5. *A train with a speed of 90 km/hr crosses a bridge in 36 seconds. Another train 100 metres shorter crosses the same bridge at 45 km/hr. What is the time taken by the second train to cross the bridge?*

Sol. Let the length of the bridge be x metres and length of the first train at 90 km/hr be y metres. Then,

$$(x + y) = \left[\left(90 \times \frac{5}{18}\right) \times 36\right] \text{ m} = 900 \text{ m}$$

∴ The second train crosses the bridge by covering a distance of $[x + (y - 100)]$ m at the rate of 45 km/hr or 12.5 m/s, *i.e.*, 800 m at 12.5 m/s ($\because x + y = 900$ m)

⇒ Time taken by the second train to cross the bridge = $\dfrac{800}{12.5}$ sec = 64 seconds.

Ex. 6. *A train 108 m long moving at a speed of 50 km/hr crosses a train 112 m long coming from the opposite direction in 6 seconds. What is the speed of the second train?*

Sol. Let the speed of the second train be x km/hr. Then,
Relative speed of trains = $(x + 50)$ km/hr

$$= (x + 50) \times \frac{5}{18} \text{ m/s}$$

Total time taken $= \dfrac{\text{Sum of lengths of both the trains}}{\text{Relative speed}}$

$$= \frac{108\,\text{m} + 112\,\text{m}}{(x + 50) \times 5/18}$$

⇒ $\dfrac{220 \times 18}{(x + 50) \times 5} = 6$ ⇒ $44 \times 3 = x + 50$ ⇒ $132 = x + 50$ ⇒ $x = 132 - 50 = 82$ km/hr.

Ex. 7. *A train overtakes two persons walking along a railway track. The first one walks at 4.5 km/hr. The other one walks at 5.4 km/hr. The train needs 8.4 and 8.5 seconds respectively to overtake them. What is the speed of the train if both the persons are walking in the same direction as the train?*

Sol. Let the speed of the train be x km/hr. Then,

Relative speed of train w.r.t. first person $= (x - 4.5)$ km/hr $= (x - 4.5) \times \dfrac{5}{18}$ m/s

Relative speed of train w.r.t. second person $= (x - 5.4)$ km/hr $= (x - 5.4) \times \dfrac{5}{18}$ m/s

$\therefore$ Length of the train $= (x - 4.5) \times \dfrac{5}{18} \times 8.4 = (x - 5.4) \times \dfrac{5}{18} \times 8.5$

$\Rightarrow 8.4x - 37.8 = 8.5x - 45.9 \Rightarrow 0.1x = 45.9 - 37.8 = 8.1 \Rightarrow x = 81$ km/hr.

Ex. 8. *A man sitting in a train travelling at the rate of 50 km/hr observes that it takes 9 sec for a goods train travelling in the opposite direction to pass him. If the goods train is 187.5 m long, find its speed ?*

Sol. Let the speed of the goods train be x km/hr.

Relative speed $= (50 + x)$ km/hr

$= (50 + x) \times \dfrac{5}{18}$ m/s

$\therefore \dfrac{187.5}{(50 + x) \times 5/18} = 9 \Rightarrow \dfrac{187.5 \times 18}{250 + 5x} = 9$

$\Rightarrow 250 + 5x = 187.5 \times 2 = 375 \Rightarrow 5x = 375 - 250 = 125$

$\Rightarrow x = 25$ km/hr.

Question Bank–21(b)

1. In what time will a train 100 metres long with a speed of 50 km/hour cross a pillar?
   (a) 7 seconds
   (b) 72 seconds
   (c) 7.2 seconds
   (d) 70 seconds

2. Two trains 160 m and 140 m long are running in opposite directions on parallel rails, the first at 77 km/hr and the other at 67 km/hr. How long will they take to cross each other?
   (a) 7 seconds
   (b) $7\dfrac{1}{2}$ seconds
   (c) 6 seconds
   (d) 10 seconds

3. How much time does a train 50 m long moving at 68 km/hr take to pass another train 75 m long moving at 50 km/hr in the same direction?
   (a) 5 seconds
   (b) 10 seconds
   (c) 20 seconds
   (d) 25 seconds

4. A person standing on a railway platform noticed that a train took 21 seconds to completely pass through the platform which was 84 m long and it took 9 seconds in passing him. The speed of the train was
   (a) 25.2 km/hr
   (b) 32.4 km/hr
   (c) 50.4 km/hr
   (d) 75.6 km/hr

5. A moving train 66 m long overtakes another train 88 m long moving in the same direction in 0.168 min. If the second train is moving at 30 km/hr, at what speed is the first train moving?

   (a) 85 km/hr
   (b) 50 km/hr
   (c) 55 km/hr
   (d) 25 km/hr

6. A train of length 150 m takes 10 sec to pass over another train 100 m long coming from opposite direction. If the speed of the first train be 30 km/hr, the speed of the second train is
   (a) 54 km/hr
   (b) 60 km/hr
   (c) 72 km/hr
   (d) 36 km/hr

7. A train is running at a speed of 45 km/hr and a man is walking at a speed of 5 km/hr in the opposite direction. If the train crosses the man in 18 seconds, then its length is
   (a) 200 m
   (b) 220 m
   (c) 180 m
   (d) 250 m

8. Two trains of equal length take 10 seconds and 15 seconds respectively to cross a telegraph post. If the length of each train be 120 metres, in what time (in seconds) will they cross each other in opposite directions ?
   (a) 16
   (b) 15
   (c) 12
   (d) 10

9. A train passes two bridges of lengths 800 m and 400 m in 100 seconds and 60 seconds respectively. The length of the train is
   (a) 80 m
   (b) 90 m
   (c) 200 m
   (d) 150 m

10. A man standing on a platform finds that a train takes 3 seconds to pass him and another train of same length moving in the opposite direction takes 4 seconds. The time taken by the trains to pass each other will be

(a) $2\frac{3}{7}$ seconds
(b) $3\frac{3}{7}$ seconds
(c) $4\frac{3}{7}$ seconds
(d) $5\frac{3}{7}$ seconds

11. Two trains travel in the same direction at 60 km/hr and 96 km/hr. If the faster train passes a man in the slower train in 20 seconds, then the length of the faster train is
(a) 100 m
(b) 125 m
(c) 150 m
(d) 200 m

12. Two trains each 100 m long moving in opposite directions cross each other in 8 seconds. If one is moving twice as fast as the other, then the speed of the faster train is
(a) 30 km/hr
(b) 45 km/hr
(c) 60 km/hr
(d) 75 km/hr

13. If a train takes 1.5 seconds to cross a telegraph post and 1.75 seconds to overtake a cyclist racing along a road parallel to the track at 10 m/s, then the length of the train is
(a) 135 metres
(b) 125 metres
(c) 115 metres
(d) 105 metres

14. A train passes two persons walking in the same direction at a speed of 3 km/hr and 5 km/hr respectively in 10 seconds and 11 seconds. The speed of the train is
(a) 28 km/hr
(b) 27 km/hr
(c) 25 km/hr
(d) 24 km/hr

15. Two trains are running at 40 km/hr and 20 km/hr respectively in the same direction. Fast train completely passes a man sitting in the slower train in 5 seconds. What is the length of the fast train?
(a) 23 m
(b) $23\frac{2}{9}$ m
(c) 27 m
(d) $27\frac{7}{9}$ m

16. Two trains running in opposite directions cross a man standing on the platform in 27 seconds and 17 seconds respectively and they cross each other in 23 seconds. The ratio of their speeds is
(a) 1 : 3
(b) 3 : 2
(c) 3 : 4
(d) 1 : 2

17. Two trains 130 m and 110 m long are going in the same direction. The faster train takes one minute to

pass the other completely. If they are moving in opposite directions, they pass each other completely in 3 seconds. Find the speed of each train?
(a) 38 m/s, 36 m/s
(b) 42 m/s, 38 m/s
(c) 36 m/s, 42 m/s
(d) 40 m/s, 36 m/s

18. A train 75 m long overtook a person who was walking at the rate of 6 km/hr and passes him in $7\frac{1}{2}$ seconds. Subsequently, it overtook a second person and passed him in $6\frac{3}{4}$ seconds. At what rate was the second person travelling?
(a) 4 km/hr
(b) 1 km/hr
(c) 2 km/hr
(d) 5 km/hr

19. A train travelling at 36 km/hr passes in 12 seconds another train half its length, travelling in the opposite direction at 54 km/hr. If it also passes a railway platform in $1\frac{1}{2}$ minutes, what is the length of the platform?
(a) 800 m
(b) 700 m
(c) 900 m
(d) 1000 m

20. Train A leaves Ludhiana for Delhi at 11 am running at a speed of 60 km/hr. Train B leaves Ludhiana for Delhi by the same route at 2 pm on the same day running at a speed of 72 km/hr. At what time will the two trains meet each other?
(a) 5 am on the next day
(b) 2 am on the next day
(c) 5 pm on the next day
(d) 2 pm on the next day

21. Two men are running in the same direction with a speed of 6 km/hr and $7\frac{1}{2}$ km/hr. A train running in the same direction crosses them in 5 sec and $5\frac{1}{2}$ sec respectively. The length and the speed of the train are
(a) 22.92 m (approx) and 22 km/hr
(b) 22 m (approx) and 22.5 km/hr
(c) 22.90 m (approx) and 20.5 km/hr
(d) 22.92 m (approx) and 22.5 km/hr

22. An express train left Delhi for Howrah at 3 pm at an average speed of 60 km/hr. At 1 pm, a goods train also had left Delhi for Howrah on a parallel track at an average speed of 40 km/hr. How far from Delhi is the express train expected to overtake the goods train?
(a) 200 km
(b) 220 km
(c) 240 km
(d) 280 km

23. Two persons are walking in the same direction at rates 3 km/hr and 6 km/hr. A train comes running from behind and passes them in 9 and 10 seconds. The speed of the train is

(a) 22 km/hr   (b) 40 km/hr

(c) 33 km/hr   (d) 35 km/hr

24. Two trains are 2 km apart and their lengths are 200 m and 300 m. They are approaching towards each other with a speed of 20 m/s and 30 m/s respectively. After how much time will they cross each other.

(a) 50 seconds   (b) 100 seconds

(c) $\dfrac{25}{3}$ seconds   (d) 150 seconds

25. Two trains pass each other on parallel lines. Each train is 100 m long. When they are going in the same direction, the faster one takes 60 seconds to pass the other completely. If they are going in opposite directions, they pass each other completely in 10 seconds. Find the speed of the slower train in km/hr.

(a) 30 km/hr   (b) 42 km/hr

(c) 48 km/hr   (d) 60 km/hr

Answers

1. (c)	2. (b)	3. (d)	4. (a)	5. (a)	6. (b)	7. (d)	8. (c)	9. (c)	10. (b)
11. (d)	12. (c)	13. (d)	14. (c)	15. (d)	16. (b)	17. (b)	18. (c)	19. (b)	20. (a)
21. (d)	22. (c)	23. (c)	24. (a)	25. (a)					

Hints and Solutions

1. (c) Speed of the train = 50 km/hr = $50 \times \dfrac{5}{18}$ m/sec

$$= \dfrac{125}{9} \text{ m/sec}$$

Distance travelled = Length of train = 100 m

∴ Time taken to cross the pillar = $\dfrac{100}{125/9}$ seconds

$$= \dfrac{36}{5} \text{ seconds} = 7.2 \text{ sec.}$$

2. (b) Since the two trains are moving in opposite direction, the relative speed of the faster train w.r.t. the slower train = Sum of their speeds

$$= (77 + 67) \text{ km/hr} = 144 \text{ km/hr} = \left(144 \times \dfrac{5}{18}\right) \text{m/sec}$$

$$= 40 \text{ m/sec}$$

Distance travelled = Sum of the lengths of the two trains = 160 m + 140 m = 300 m

∴ Required time = $\dfrac{300\,\text{m}}{40\,\text{m/sec}} = 7\dfrac{1}{2}$ sec.

3. (d) Since the trains are moving in the same direction the relative speed of the faster train w.r.t. the slower train = Difference of their speeds

$$= (68 - 50) \text{ km/hr} = 18 \text{ km/hr} = \left(18 \times \dfrac{5}{18}\right) \text{m/sec}$$

$$= 5 \text{ m/sec}$$

Distance travelled = Sum of the lengths of the two trains = (50 + 75) m = 125 m

∴ Required time = $\dfrac{125}{5}$ seconds = 25 seconds.

4. (a) Let the length of the train be x metres. Then, Speed of the train in passing through the platform = $\dfrac{x + 84}{21}$ m/sec

and speed of the train in passing the man

$$= \dfrac{x}{9} \text{ m/sec}$$

Since both the speeds are the same,

$$\dfrac{x + 84}{21} = \dfrac{x}{9} \quad \Rightarrow \quad 9x + 756 = 21x$$

$$\Rightarrow 12x = 756 \quad \Rightarrow \quad x = \dfrac{756}{12} = 63 \text{ m}$$

∴ Speed of the train = $\dfrac{(63 + 84)}{21}$ m/sec = 7 m/sec

$$= 7 \times \dfrac{18}{5} \text{ km/hr} = 25.2 \text{ km/hr.}$$

5. (a) Let the speed of the first train be x km/hr. Then, Relative speed of first train w.r.t. second train

$$= (x - 30) \text{ km/hr} = (x - 30) \times \dfrac{5}{18} \text{ m/sec}$$

Total distance travelled = (66 + 88) m = 154 m

Time taken = 0.168 min = (0.168 × 60) sec
= 10.08 sec

$\therefore \quad (x-30) \times \dfrac{5}{18} \times 10.08 = 154$

$\Rightarrow \quad x - 30 = \dfrac{154 \times 18}{5 \times 10.08} = 55 \quad \Rightarrow \quad x = 85$ km/hr.

6. (b) Let the speed of the second train = x km/hr

Since both the trains are running in opposite directions. Their relative speed = $(30 + x)$ km/hr

$$= (30 + x) \times \dfrac{5}{18} \text{ m/sec}$$

$\therefore$ Time taken to pass another train

$$= \dfrac{\text{Sum of the lengths of both the trains in metres}}{\text{Relative speed in m/sec}}$$

$\Rightarrow \quad 10 = \dfrac{(150 + 100)}{(30 + x) \times 5/18}$

$\Rightarrow \quad (30 + x) = \dfrac{250 \times 18}{10 \times 5} = 90$

$\Rightarrow \quad x = (90 - 30)$ km/hr = 60 km/hr.

7. (d) Since the man is walking in the opposite direction of the moving train,

Relative speed of the train = $(45 + 5)$ km/hr

$$= \left(50 \times \dfrac{5}{18}\right) \text{ m/sec} = \dfrac{250}{18} \text{ m/sec}$$

Length of the train

= Relative speed × Time taken in crossing the man

$$= \left(\dfrac{250}{18} \times 18\right) \text{ m} = 250 \text{ m.}$$

8. (c) Speed of the first train = $\dfrac{120}{10} = 12$ m/s

Speed of the second train = $\dfrac{120}{15} = 8$ m/s

$\therefore$ Required time = $\dfrac{\text{Total distance}}{\text{Relative speed}}$

$$= \dfrac{120 + 120}{12 + 8} = \dfrac{240}{20} \text{ seconds}$$

$$= 12 \text{ seconds.}$$

9. (c) Let the length of the train be x metres.

Then, $\dfrac{800 + x}{100} = \dfrac{400 + x}{60}$

$\Rightarrow \quad 2400 + 3x = 2000 + 5x$

$\Rightarrow \quad x = 200$ metres.

10. (b) Let the length of 1st train = length of the 2nd train
= x metres

Then, speed of the 1st train = $\dfrac{x}{3}$ m/s

Speed of the 2nd train = $\dfrac{x}{4}$ m/s

Since both the trains are moving in opposite directions, their relative speed = $\left(\dfrac{x}{3} + \dfrac{x}{4}\right) = \dfrac{7x}{12}$ m/s

$\therefore$ Time taken to pass each other

$$= \dfrac{\text{Distance to be covered}}{\text{Relative speed}}$$

$$= \dfrac{x + x}{7x/12} = \dfrac{2x \times 12}{7x} = \dfrac{24}{7} \text{ seconds}$$

$$= 3\dfrac{3}{7} \text{ sec.}$$

11. (d) Relative speed = $(96 - 60)$ km/hr = 36 km/hr

$$= \left(36 \times \dfrac{5}{18}\right) \text{ m/s} = 10 \text{ m/sec}$$

As we know $D = s \times t$

$\therefore$ Length of faster train = $10 \times 20 = 200$ m.

12. (c) Let the speeds of the two trains be x m/s and $2x$ m/s respectively.

Since both the trains are moving in opposite directions, their relative speed = $(x + 2x)$ m/s
= $3x$ m/s

Given, $\dfrac{100\,\text{m} + 100\,\text{m}}{3x} = 8$

$\Rightarrow \quad x = \dfrac{200}{24} \text{ m/s} = \left(\dfrac{200}{24} \times \dfrac{18}{5}\right) \text{ km/hr} = 30 \text{ km/hr}$

$\therefore$ Speed of the faster train = 60 km/hr.

13. (d) Let the length of train be l metres and its speed be x m/s.

Then, $\dfrac{l}{x} = 1.5 \quad \Rightarrow \quad l = 1.5x \qquad \ldots(i)$

and $\dfrac{l}{(x - 10)} = 1.75 \quad \Rightarrow \quad l = 1.75(x - 10)$

$\Rightarrow \quad l = 1.75x - 17.5 \qquad \ldots(ii)$

From (i) and (ii),

$1.5x = 1.75x - 17.5 \quad \Rightarrow \quad 0.25x = 17.5$

$\Rightarrow x = \dfrac{17.5}{0.25} = 70$ m/s

$\therefore$ Length of the train $= (1.5 \times 70)$ metres
    $= 105$ metres.

14. (c) Distance covered by the first person in

10 seconds $= \left[\left(3 \times \dfrac{5}{18}\right) \times 10\right]$ m $= \dfrac{25}{3}$ m

Distance covered by the second person in

11 seconds $= \left[\left(5 \times \dfrac{5}{18}\right) \times 11\right]$ m $= \dfrac{275}{18}$ m

$\therefore$ The train travels a distance $= \dfrac{275}{18} - \dfrac{25}{3} = \dfrac{125}{18}$ m

in $(11 - 10) = 1$ second.

$\therefore$ Speed of the train $= \left(\dfrac{125}{18} \times \dfrac{18}{5}\right)$ km/hr $= 25$ km/hr.

15. (d) Relative speed $= (40 - 20)$ km/hr

$= \left(20 \times \dfrac{5}{18}\right)$ m/s $= \dfrac{50}{9}$ m/s.

$\therefore$ Length of the faster train $= \left(\dfrac{50}{9} \times 5\right)$ m

$= \dfrac{250}{9}$ m $= 27\dfrac{7}{9}$ m

16. (b) Let the speeds of the two trains be x m/s and y m/s respectively.

$\therefore$ Length of the first train $= 27x$ m and length of the second train $= 17y$ m
    Since the trains are moving in opposite directions.
    Relative speed $= (x + y)$ m/s

$\therefore \dfrac{27x + 17y}{x + y} = 23 \quad \Rightarrow \quad 27x + 17y = 23x + 23y$

$\Rightarrow 4x = 6y \quad \Rightarrow \quad \dfrac{x}{y} = \dfrac{3}{2}.$

17. (b) Let the speeds of the faster train be x m/s and that of the slower train be y m/s. Then,
    When the two trains are moving in the same direction,
    Relative speed $= (x - y)$ m/s

$\therefore \dfrac{130 m + 110 m}{x - y} = 60$

$\Rightarrow \dfrac{240}{x - y} = 60 \quad \Rightarrow \quad x - y = 4 \qquad \qquad \dots(i)$

When the two trains are moving in opposite direction,

Relative speed $= (x + y)$ m/s

$\therefore \dfrac{130 m + 110 m}{x + y} = 3 \quad \Rightarrow \quad x + y = 80 \qquad \dots(ii)$

Adding eqn (i) and (ii), we get $2x = 84$

$\Rightarrow x = 42$ m/s

$\therefore$ Putting in (i), we get $y = 38$ m/s.

18. (c) Let the speed of the train be x m/s.

Speed of the 1st man $= 6$ km/hr $= \left(6 \times \dfrac{5}{18}\right)$ m/s

$= \dfrac{5}{3}$ m/s

$\dfrac{75}{(x - 5/3)} = \dfrac{15}{2}$

$\Rightarrow 150 = 15 \times \dfrac{(3x - 5)}{3} = 15x - 25$

$\Rightarrow 15x = 175 \quad \Rightarrow \quad x = \dfrac{175}{15} = \dfrac{35}{3}$ m/s

Let the speed of the second man be y m/s. Then

$\dfrac{75}{\left(\dfrac{35}{3} - y\right)} = \dfrac{27}{4}$

$\Rightarrow 300 = 27 \times \left(\dfrac{35 - 3y}{3}\right) = 315 - 27y$

$\Rightarrow 27y = 15 \Rightarrow y = \dfrac{15}{27}$ m/s $= \left(\dfrac{15}{27} \times \dfrac{18}{5}\right)$ km/hr

$= 2$ km/hr.

19. (b) Let the length of the first train be x metres.

Then, length of the second train $= \dfrac{x}{2}$ metres

Relative speed $= (36 + 54)$ km/hr $= 90$ km/hr

$= \left(90 \times \dfrac{5}{18}\right)$ m/s $= 25$ m/s

$\therefore \dfrac{x + x/2}{25} = 12 \quad \Rightarrow \quad \dfrac{3x}{2} = 300 \quad \Rightarrow \quad x = 200.$

$\therefore$ Length of the first train $= 200$ m.
    Let the length of the platform be y metres.

Speed of the first train $= \left(36 \times \dfrac{5}{18}\right)$ m/s $= 10$ m/s

$\therefore (200 + y) \times \dfrac{1}{10} = 90$

$\Rightarrow 200 + y = 900 \quad \Rightarrow \quad y = 700$ m.

20. (a) Distance covered by train A before train B leaves Ludhiana = $60 \times 3 = 180$ km

Relative speed = $(72 - 60)$ km/hr = 12 km/hr

$\therefore$ Time taken to cross each other = $\dfrac{180}{12}$ hrs

$= 15$ hrs

$\therefore$ Required time = 2 pm + 15 hrs

$= 5$ am on the next day.

21. (d) Let the length of the train be l m and its speed be x km/hr.

Then, relative speed of train w.r.t.1st man

$= (x - 6)$ km/hr $= (x - 6) \times \dfrac{5}{18}$ m/s

Relative speed of train w.r.t. 2nd man

$= (x - 7.5)$ km/hr

$= (x - 7.5) \times \dfrac{5}{18}$ m/s

Length of the train

= Distance travelled in both the cases

$\Rightarrow (x-6) \times \dfrac{5}{18} \times 5 = (x-7.5) \times \dfrac{5}{18} \times 5.5$

$\Rightarrow (x-6) \times 5 = (x-7.5) \times 5.5$

$\Rightarrow 5x - 30 = 5.5x - 41.25 \Rightarrow 0.5x = 11.25$

$\Rightarrow x = \dfrac{11.25}{0.5} = 22.5$ km/hr

$\therefore$ Length of the train $= \left((22.5-6) \times \dfrac{5}{18} \times 5\right)$ m

$= 22.92$ m (approx)

22. (c) Distance covered by goods train before express train leaves = $40 \times 2 = 80$ km

Relative speed = $(60 - 40)$ km/hr = 20 km/hr

$\therefore$ Time taken to cross each other = $\dfrac{80}{20}$ hrs

$= 4$ hrs

$\therefore$ Distance from Delhi where the express train is expected to overtake the goods train

$= (4 \times 60)$ km = 240 km.

23. (c) Similar to Q. 21.

24. (a) Relative speed of the trains = $(20 + 30)$ m/s

$= 50$ m/s

Total distance covered

$= (2 \times 1000 + 200 + 300)$ m

$= 2500$ m

$\therefore$ The two trains will cross each other after

$\dfrac{2500}{50}$ seconds = 50 seconds.

25. (a) Let the speed of the faster train be x km/hr and that of the slower train be y km/hr.

Relative speed when both move in same direction = $(x - y)$ km/hr

Relative speed when both move in opposite directions = $(x + y)$ km/hr

Total distance travelled

= Sum of lengths of both the trains

$= 200$ m

Given, $\dfrac{200}{(x-y) \times \dfrac{5}{18}} = 60$ and $\dfrac{200}{(x+y) \times \dfrac{5}{18}} = 10$

$\Rightarrow \dfrac{3600}{5(x-y)} = 60$ and $\dfrac{3600}{5(x+y)} = 10$

$\Rightarrow x - y = \dfrac{3600}{300} = 12$ ...(i)

and $x + y = \dfrac{3600}{50} = 72$ ...(ii)

Adding eqn (i) and eqn (ii), we get

$2x = 84 \Rightarrow x = 42$ km/hr

$\therefore$ From (i), $y = 30$ km/hr.

Section-C
BOATS AND STREAMS

KEY FACTS

1. In water, the direction along the stream is called **downstream** and the direction against the stream is called **upstream**.
2. If the speed of a boat in still water is x km/hr and the speed of the stream is y km/hr, then
Speed downstream = $(x + y)$ km/hr
Speed upstream = $(x - y)$ km/hr

3. If the speed downstream is u km/hr and speed upstream is v km/hr, then

Speed in still water $= \frac{1}{2}(u+v)$ km/hr

Rate of stream $= \frac{1}{2}(u-v)$ km/hr

Solved Examples

Ex. 1. *A man rows upstream 13 km and downstream 28 km taking 5 hrs each time. What is the velocity in (km/hr) of the current?*

Sol. Speed upstream $= \frac{13}{5}$ km/hr $= 2.6$ km/hr

Speed downstream $= \frac{28}{5}$ km/hr $= 5.6$ km/hr

$\therefore$ Velocity of current $= \frac{1}{2}(5.6 - 2.6)$ km/hr

$= \left(\frac{1}{2} \times 3\right)$ km/hr $= 1.5$ km/hr.

Ex. 2. *In one hour a boat goes 11 km along the stream and 5 km against the stream. What is the speed of the boat in still water (in km/hr)?*

Sol. Speed downstream $= 11$ km/hr

Speed upstream $= 5$ km/hr

$\therefore$ Speed of boat in still water $= \frac{1}{2}(11 + 5)$ km/hr $= 8$ km/hr.

Ex. 3. *A motor boat whose speed is 15 km/hr in still water goes 30 km downstream and comes back in 4 hours 30 minutes. Determine the speed of the stream.*

Sol. Let the speed of the stream be x km/hr. Then,
Speed of boat downstream $= (15 + x)$ km/hr
and speed of boat upstream $= (15 - x)$ km/hr

$\therefore \frac{30}{15+x} + \frac{30}{15-x} = 4\frac{1}{2} = \frac{9}{2}$

$\Rightarrow \frac{30(15-x) + 30(15+x)}{(15+x)(15-x)} = \frac{9}{2} \Rightarrow \frac{900}{225 - x^2} = \frac{9}{2} \Rightarrow 200 = 225 - x^2$

$\Rightarrow x^2 = 25 \Rightarrow x = 5$ km/hr.

Ex. 4. *A boat takes 4 hours for travelling downstream from point A to point B and coming back to point A upstream. If the velocity of the stream is 2 km/hr and the speed of the boat in still water is 4 km/hr, what is the distance between A and B?*

Sol. Speed of boat downstream $= (4 + 2)$ km/hr $= 6$ km/hr

Speed of boat upstream $= (4 - 2)$ km/hr $= 2$ km/hr

Let the distance between point A and point B be x km. Then,

$\frac{x}{6} + \frac{x}{2} = 4 \Rightarrow \frac{x + 3x}{6} = 4 \Rightarrow 4x = 24 \Rightarrow x = 6$ km.

Ex. 5. *A boat goes 20 km downstream in one hour and the same distance upstream in two hours. What is the speed of the boat in still water?*

Sol. Let the speed of the boat in still water be x km/hr and speed of the stream $= y$ km/hr

Then, speed of boat downstream $= (x + y)$ km/hr

Speed of boat upstream $= (x - y)$ km/hr

Given $(x + y) \times 1 = 20$ and $(x - y) \times 2 = 20$

$\Rightarrow$ $x + y = 20$ ...(i) and $x - y = 10$ ...(ii)

Adding eqn (i) and (ii), we get

$2x = 30$ $\Rightarrow$ $x = 15$ km/hr.

Ex. 6. *A man rows to a place at a distance of 48 km and back in 14 hours. He finds that he can row 4 km with the stream in the same time as 3 km against the stream. What is the rate of the stream?*

Sol. Ratio of the downstream speed to upstream speed = 4 : 3

$\therefore$ Ratio of times taken to cover a certain distance downstream and upstream = 3 : 4

Given $3x + 4x = 14$ hrs $\Rightarrow$ $x = 2$

$\therefore$ The man took 6 hours to row 48 km downstream and 8 hours to row 48 km upstream.

$\therefore$ Speed downstream = $\dfrac{48}{6}$ km/hr = 8 km/hr

Speed upstream = $\dfrac{48}{8}$ km/hr = 6 km/hr

$\therefore$ Rate of stream = $\dfrac{1}{2}(8 - 6)$ km/hr = 1 km/hr.

Ex. 7. *A boat goes 6 km in an hour in still water. It takes thrice as much time in covering the same distance against the current. What is the speed of the current?*

Sol. Speed of boat in still water = 6 km/hr

Time taken by boat to move upstream = 3 hours

$\therefore$ Speed of boat upstream = $\dfrac{6}{3}$ km/hr = 2 km/hr

Let the speed of the current be x km/hr.

Then, $6 - x = 2$ $\Rightarrow$ $x = 4$ km/hr.

Question Bank–21(c)

1. A boat moves downstream at the rate of 1 km in 6 min and upstream at the rate of 1 km in 10 min. The speed of the current (in km/hr) is
   (a) 1 (b) 1.5
   (c) 2 (d) 2.5

2. A boat goes 40 km upstream in 8 hours and 36 km downstream in 6 hours. The speed of the boat in still water is
   (a) 6.5 km/hr (b) 5.5 km/hr
   (c) 6 km/hr (d) 5 km/hr

3. A man can row the boat at 5 km/hr in still water. If the velocity of the current is 1 km/hr and it takes him 1 hour to row to a place and come back, how far is the place?
   (a) 2.5 km (b) 3 km
   (c) 2.4 km (d) 3.6 km

4. A boat covers a distance of 14 km in 4 hours along the flow. What is the speed of the boat in still water, if the speed of the flow of water is 2 km/hr?
   (a) 2 km/hr (b) 3 km/hr
   (c) 2.5 km/hr (d) 1.5 km/hr

5. A river is running at 2 km/hr. It took a man twice as long to row up as to row down the river. The rate (in km/hr) of the man in still water is
   (a) 8 (b) 10
   (c) 4 (d) 6

6. A man rows a boat 18 km in 4 hours downstream and returns upstream in 12 hours. The speed of the stream (in km/hr) is
   (a) 1 (b) 1.5
   (c) 2 (d) 1.75

7. A boat covers a certain distance downstream in 8 hours and comes back upstream in 10 hours. If the speed of the current be 1 km/hr, the distance (in km) of the one way journey is
   (a) 60 (b) 70
   (c) 80 (d) 90

8. A motor in still water travels at a speed of 36 km/hr. It goes 56 km upstream in 1 hour 45 min. The time taken by it to cover the same distance downstream is
(a) 2 hours 25 min   (b) 3 hours
(c) 1 hour 24 min   (d) 2 hours 21 min

9. A streamer goes downstream from one port to another in 4 hours. It covers the same distance upstream in 5 hours. If the speed of the stream is 2 km/hr, the distance between the two ports is
(a) 50 km   (b) 60 km
(c) 70 km   (d) 80 km

10. The speed of a motorboat is to that of the current of water as 36 : 5. The boats goes along with the current in 5 hours 10 min. It will come back in
(a) 5 hrs 50 min   (b) 6 hrs
(c) 6 hrs 50 min   (d) 12 hrs 10 min

11. A man can row at 5 km/hr in still water. If the river is running at 1 km/hr, it takes him 75 minutes to row to a place and back. How far is the place?
(a) 2.5 km   (b) 3 km
(c) 4 km   (d) 5 km

12. A man rows $\frac{3}{4}$ of a km against the stream in $11\frac{1}{4}$ minutes and returns in $7\frac{1}{2}$ min. Find the speed of the man in still water.
(a) 4 km/hr   (b) 3 km/hr
(c) 5 km/hr   (d) 6 km/hr

13. Twice the speed of a boat downstream is equal to thrice the speed upstream. The ratio of its speed in still water to its speed in current is
(a) 1 : 5   (b) 1 : 3
(c) 5 : 1   (d) 2 : 3

14. A boatman goes 2 km against the current of the stream in 1 hour and goes 1 km along the current in 10 minutes. How long will he take to go 5 km in stationary water?
(a) 1 hour   (b) $1\frac{1}{2}$ hours
(c) 1 hour 15 min   (d) 40 min

15. A boat running upstream takes 8 hours 48 min to cover a certain distance, while it takes 4 hours to

cover the same distance running downstream. What is the ratio between the speed of the boat and the speed of the water current respectively?
(a) 2 : 1   (b) 3 : 1
(c) 8 : 3   (d) 4 : 3

16. A boatman rows to a place at a distance of 45 km and back in 20 hours. He finds that he can row 12 km with the stream in the same time as 4 km against the stream. Find the speed of the stream.
(a) 3 km/hr   (b) 2.5 km/hr
(c) 4 km/hr   (d) 3.5 km/hr

17. A boat takes 90 minutes less to travel 36 km downstream than to travel the same distance upstream. If the speed of the boat in still water is 10 km/hr, the speed of the stream is
(a) 4 km/hr   (b) 3 km/hr
(c) 2.5 km/hr   (d) 2 km/hr

18. A boat goes 24 km upstream and 28 km downstream in 6 hours. It goes 30 km upstream and 21 km downstream in 6 hours and 30 minutes. The speed of the boat in still water is
(a) 10 km/hr   (b) 4 km/hr
(c) 14 km/hr   (d) 6 km/hr

19. At his usual rowing rate, Rahul can travel 12 miles downstream in a certain river in 6 hours less than it takes him to travel the same distance upstream. But if he could double his usual rowing rate for his 24 mile round trip, the downstream 12 miles would then take only one hour less than the upstream 12 miles. What is the speed of the current in miles per hour?
(a) $1\frac{1}{3}$ (b) $1\frac{2}{3}$
(c) $2\frac{1}{3}$ (d) $2\frac{2}{3}$

20. A boat takes 11 hours for travelling downstream from point A to point B and coming back to point C midway between A and B. If the velocity of the stream be 3 km/hr and the speed of the boat in still water be 12 km/hr, what is the distance between A and B?
(a) 100 km   (b) 90 km
(c) 110 km   (d) 120 km

Hints and Solutions

1. (c) Speed downstream = $\left(\dfrac{1}{6} \times 60\right)$ = 10 km/hr

   Speed upstream = $\left(\dfrac{1}{10} \times 60\right)$ = 6 km/hr

   ∴ Speed of the current = $\dfrac{1}{2}$ (10 – 6) km/hr
   = 2 km/hr.

2. (b) Speed of boat downstream = $\dfrac{36}{6}$ km/hr
   = 6 km/hr

   Speed of boat upstream = $\dfrac{40}{8}$ km/hr = 5 km/hr

   ∴ Speed of boat in still water = $\dfrac{1}{2}$ (6 + 5) km/hr
   = 5.5 km/hr.

3. (c) Let the distance of the destination be x km.
   Speed of boat downstream = (5 + 1) km/hr
   = 6 km/hr
   Speed of boat upstream = (5 – 1) km/hr
   = 4 km/hr

   Then, $\dfrac{x}{6} + \dfrac{x}{4} = 1$ ⇒ $\dfrac{2x + 3x}{12} = 1$

   ⇒ $x = \dfrac{12}{5}$ = 2.4 km.

4. (d) Speed of the flow of water = 2 km/hr

   Speed of the boat downstream = $3\dfrac{1}{2}$ km/hr

   Let the speed of the boat in still water = x km/hr

   Then, $x + 2 = 3\dfrac{1}{2}$ ⇒ $x = 1\dfrac{1}{2}$ km/hr

5. (d) Let the distance travelled be D km and speed of the man in still water be x km/hr. Then,
   Speed downstream = $(x + 2)$ km/hr
   Speed upstream = $(x – 2)$ km/hr

   Given, $\dfrac{D}{x-2} = 2 \times \dfrac{D}{x+2}$

   ⇒ $x + 2 = 2x - 4$ ⇒ $x = 6$ km/hr.

6. (b) Let the speed of the boat in still water = x km/hr and the speed of the stream y km/hr.

   Then, speed downstream = $x + y = \dfrac{18}{4} = \dfrac{9}{2}$ …(i)

and speed upstream = $x - y = \dfrac{18}{12} = \dfrac{3}{2}$ …(ii)

Subtracting eq (ii) from eq (i), we get,

$2y = \dfrac{6}{2} = 3$ ⇒ $y = 1.5$ km/hr.

7. (c) Speed of boat in still water be u km/hr. Then,
   Distance covered downstream = $(u + 1) \times 8$
   = $8u + 8$
   Distance covered upstream = $(u – 1) \times 10$
   = $10u – 10$
   Since the distance both ways is the same,
   $8u + 8 = 10u – 10 \Rightarrow 2u = 18 \Rightarrow u = 9$ km/hr
   ∴ Distance covered one way = $(9 + 1) \times 8$ km
   = 80 km.

8. (c) Let the speed of the stream be x km/hr

   Then, $\dfrac{56}{7/4} = 36 - x$ ⇒ $\dfrac{56 \times 4}{7} = 36 - x$

   ⇒ $x = 36 – 32 = 4$ km/hr

   ∴ Time taken by it to cover the same distance

   upstream = $\dfrac{56}{36 + 4}$ hrs = $\dfrac{56}{40}$ hrs

   = $1\dfrac{2}{5}$ hours = 1 hour 24 min.

9. (d) Let the distance between the two parts be x km.

   Then, speed downstream = $\dfrac{x}{4}$ km/hr

   and speed upstream = $\dfrac{x}{5}$ km/hr

   ⇒ Speed of stream = $\dfrac{1}{2}\left(\dfrac{x}{4} - \dfrac{x}{5}\right)$

   Given, $\dfrac{1}{2}\left(\dfrac{x}{4} - \dfrac{x}{5}\right) = 2$ ⇒ $\dfrac{x}{4} - \dfrac{x}{5} = 4$

   ⇒ $\dfrac{x}{20} = 4$ ⇒ $x = 80$ km

10. (c) Let the speed of the motorboat be $36x$ km/hr and that of the current of water be $5x$ km/hr.
    Speed downstream = $(36x + 5x)$ km/hr
    = $41x$ km/hr
    Speed upstream = $(36x – 5x)$ km/hr
    = $31x$ km/hr
    Time taken to travel downstream = 5 hrs 10 min

    = $5\dfrac{10}{60}$ hrs = $\dfrac{31}{6}$ hrs

$\therefore$ Distance covered downstream = $\left(41x \times \dfrac{31}{6}\right)$ km

$\Rightarrow$ Distance covered upstream = $\left(41x \times \dfrac{31}{6}\right)$ km

$\therefore$ Time taken to travel upstream = $\left(41x \times \dfrac{31}{6} \times \dfrac{1}{31x}\right)$ hrs

$= \dfrac{41}{6}$ hrs $= 6\dfrac{5}{6}$ hrs = 6 hrs 50 min.

11. (b) Let the distance to the place be x km.
Speed downstream = $(5 + 1)$ km/hr = 6 km/hr
Speed upstream = $(5 - 1)$ km/hr = 4 km/hr

Given, $\dfrac{x}{6} + \dfrac{x}{4} = \dfrac{75}{60}$ hrs $\Rightarrow \dfrac{2x + 3x}{12} = \dfrac{5}{4}$

$\Rightarrow \dfrac{5x}{12} = \dfrac{5}{4} \Rightarrow x = 3$ km.

12. (c) Distance covered upstream in $\dfrac{45}{4}$ min = $\dfrac{3}{4}$ km

$\therefore$ Speed of man upstream = $\left(\dfrac{3}{4} \times \dfrac{4}{45} \times 60\right)$ km/hr
$= 4$ km/hr

Distance covered down stream in $\dfrac{15}{2}$ min = $\dfrac{3}{4}$ km

$\therefore$ Speed of man down stream = $\left(\dfrac{3}{4} \times \dfrac{2}{15} \times 60\right)$ km/hr
$= 6$ km/hr

$\therefore$ Speed of man in still water = $\dfrac{1}{2}$ $(6 + 4)$ km/hr
$= 5$ km/hr.

13. (c) Let the speed downstream be u km/hr and speed upstream be v km/hr.

Given, $2u = 3v \Rightarrow u = \dfrac{3v}{2}$

Required ratio = $\dfrac{1}{2}(u + v) : \dfrac{1}{2}(u - v)$

$= \left(\dfrac{3v}{2} + v\right) : \left(\dfrac{3v}{2} - v\right)$

$= \dfrac{5v}{2} : \dfrac{v}{2} = 5 : 1$.

14. (c) Speed of the boatman upstream

$= \dfrac{2}{1}$ km/hr = 2 km/hr

Speed of the boatman downstream

$= \dfrac{1}{10/60}$ km/hr = 6 km/hr

$\therefore$ Speed of boatman in stationary water

$= \dfrac{1}{2}$ $(6 + 2)$ km/hr = 4 km/hr

$\therefore$ Time taken to cover 5 km in stationary water

$= \dfrac{5}{4}$ hrs = $1\dfrac{1}{4}$ hrs = 1 hr. 15 min.

15. (c) Let the distance covered in one direction be x km.

Then, speed upstream = $\dfrac{x}{44/5}$ km/hr

$\left(8 \text{ hrs } 48 \text{ min} = 8\dfrac{48}{60} = 8\dfrac{4}{5} \text{ hrs}\right)$

$= \dfrac{5x}{44}$ km/hr

Speed downstream = $\dfrac{x}{4}$ km/hr

$\therefore$ Speed of boat in still water = $\dfrac{1}{2}\left(\dfrac{5x}{44} + \dfrac{x}{4}\right)$

$= \dfrac{8x}{44} = \dfrac{2x}{11}$ km/hr

Speed of current = $\dfrac{1}{2}\left(\dfrac{x}{4} - \dfrac{5x}{44}\right) = \dfrac{3x}{44}$ km/hr

$\therefore$ Required ratio = $\dfrac{2x}{11} : \dfrac{3x}{44} = 8 : 3$.

16. (a) Ratio of upstream speed to downstream speed = 1 : 3

$\therefore$ Ratio of times taken to cover a certain distance upstream and downstream = 3 : 1
Given, the man took 20 hours to cover the 45 km distance and back $\Rightarrow$ He took 15 hours to cover 45 km upstream and 5 hours to the cover 45 km downstream.

$\therefore$ Speed downstream = $\dfrac{45}{5} = 9$ km/hr

Speed upstream = $\dfrac{45}{15} = 3$ km/hr

$\therefore$ Speed of stream = $\dfrac{1}{2}$ $(9 - 3)$ km/hr = 3 km/hr.

17. (d) Let the speed of the stream be x km/hr.
Speed of boat in still water = 10 km/hr
$\therefore$ Speed of boat downstream = $(10 + x)$ km/hr
Speed of boat upstream = $(10 - x)$ km/hr

Given, $\dfrac{36}{(10 - x)} - \dfrac{36}{(10 + x)} = \dfrac{90}{60}$

$$\Rightarrow \frac{36(10+x)-36(10-x)}{(10-x)(10+x)}=\frac{3}{2}$$

$$\Rightarrow \frac{72x}{100-x^2}=\frac{3}{2} \quad \Rightarrow \quad 144x=300-3x^2$$

$$\Rightarrow 3x^2+144x-300=0$$

$$\Rightarrow 3x^2+150x-6x-300=0$$

$$\Rightarrow 3x(x+50)-6(x+50)=0$$

$$\Rightarrow (3x-6)(x+50)=0$$

$$\Rightarrow x=2 \ \text{ or } \ -50.$$

Since speed is not negative, $x = 2$ km/hr.

18. (b) Let the speed downstream be x km/hr and speed upstream be y km/hr. Then,

$$\frac{28}{x}+\frac{24}{y}=6 \qquad \qquad \dots(i)$$

and $\dfrac{21}{x}+\dfrac{30}{y}=\dfrac{13}{2}$ $\qquad \dots(ii)$

Multiplying (i) by 3 and (ii) by 4 and then subtracting eqn (i) from eqn (ii), we get

$$4\left(\frac{21}{x}+\frac{30}{y}\right)-3\left(\frac{28}{x}+\frac{24}{y}\right)=4\times\frac{13}{2}-3\times6$$

$$\Rightarrow \frac{84}{x}+\frac{120}{y}-\frac{84}{x}-\frac{72}{y}=26-18$$

$$\Rightarrow \frac{48}{y}=8 \quad \Rightarrow \quad y=6 \text{ km/hr}$$

Putting the value of y in (i), we have

$$\frac{28}{x}+4=6 \quad \Rightarrow \quad \frac{28}{x}=2 \quad \Rightarrow \quad x=14 \text{ km/hr}$$

$\therefore$ Speed of the stream $=\dfrac{1}{2}(14-6)$ km/hr

$$= 4 \text{ km/hr.}$$

19. (d) Let the speed of the boat in still water be x mph and speed of the current be y mph. Then, Speed upstream $=(x-y)$ mph and stream downstream $=(x+y)$ mph

Given, $\dfrac{12}{(x-y)}-\dfrac{12}{(x+y)}=6$

$$\Rightarrow \frac{12(x+y)-12(x-y)}{x^2-y^2}=6$$

$$\Rightarrow 24y=6(x^2-y^2) \Rightarrow x^2-y^2=4y$$

$$\Rightarrow x^2=4y+y^2 \qquad \qquad \dots(i)$$

and $\dfrac{12}{(2x-y)}-\dfrac{12}{(2x+y)}=1$

$$\Rightarrow \frac{12(2x+y)-12(2x-y)}{4x^2-y^2}=1$$

$$\Rightarrow 24y=4x^2-y^2 \Rightarrow 4x^2-y^2=24y$$

$$\Rightarrow x^2=\frac{24y+y^2}{4} \qquad \qquad \dots(ii)$$

From (i) and (ii), $4y+y^2=\dfrac{24y+y^2}{4}$

$$\Rightarrow 16y+4y^2=24y+y^2$$

$$\Rightarrow 3y^2=8y \quad \Rightarrow \quad y=8/3$$

$\therefore$ Speed of the current $= 8/3$ mph $= 2\dfrac{2}{3}$ mph

20. (b) Speed downstream $=(12+3)$ km/hr $=15$ km/hr

Speed upstream $=(12-3)$ km/hr $=9$ km/hr

Let the distance between A and B be x km. Then,

$$\frac{x}{15}+\frac{x/2}{9}=11 \Rightarrow \frac{x}{15}+\frac{x}{18}=11$$

$$\Rightarrow \frac{6x+5x}{90}=11 \Rightarrow \frac{11x}{90}=11 \Rightarrow x=90 \text{ km.}$$

Self Assessment Sheet–21

1. A student walks from his house at $2\dfrac{1}{2}$ km an hour and reaches his school 6 minutes late. The next day he increases his speed by 1 kilometre an hour and reaches 6 minutes early. How far is the school from the house ?

   (a) 2.5 km   (b) 3 km

   (b) 1.75 km   (d) 1 km

2. Distance between two points A and B is 110 km. A motor-cycle rider starts from A towards B at 7 am at a speed of 20 km/hr. Another motor-cycle rider starts from B towards A at 8 am at a speed of 25 km/hr. Find when will they cross each other.

   (a) 11 am   (b) 9 : 30 am

   (c) 8 : 30 am   (d) 10 am

3. A train leaves the station 1 hour before the scheduled time. The driver decreases its speed by 50 km/hr. At the next station 300 km away, the train reached on time. Find the original speed of the train.

   (a) 100 km/hr   (b) 150 km/hr

   (c) 125 km/hr   (d) 200 km/hr

4. A train of length 150 m takes 10 s to cross another train 100 m long coming from opposite direction. If the speed of the first trains is 30 km/hr, what is the speed of the second train?
- (a) 72 km/hr
- (b) 60 km/hr
- (c) 54 km/hr
- (d) 48 km/hr

5. Two trains of equal lengths are running on parallel lines in the same direction at the rate of 46 km/hr and 36 km/hr. The faster train passes the slower train in 36 seconds. The length of each train is :
- (a) 50 m
- (b) 72 m
- (c) 80 m
- (d) 82 m

6. Two persons are walking in the same direction at speeds of 3 km/hr and 6 km/hr. A train comes running from behind and passes them in 9 and 10 seconds. The speed of the train is :
- (a) 33 km/hr
- (b) 40 km/hr
- (c) 22 km/hr
- (d) 35 km/hr

7. A man can row three-quarters of a kilometer against the stream in $11\frac{1}{4}$ minutes and return in $7\frac{1}{2}$ minutes. The speed of the man in still water is :
- (a) 2 km/hr
- (b) 3 km/hr
- (c) 4 km/hr
- (d) 5 km/hr

8. A boat running upstream takes 8 hours 48 minutes to cover a certain distance, while it takes 4 hours to cover the same distance running downsteam. What is the ratio between the speed of the boat and the speed of the water current respectively?
- (a) 2 : 1
- (b) 3 : 1
- (c) 8 : 3
- (d) Cannot be determined

9. A boat covers 24 km upstream and 36 km downstream in 6 hours, while it covers 36 km upstream and 24 km downstream in $6\frac{1}{2}$ hours. The speed of the current is :
- (a) 1 km/hr
- (b) 2 km/hr
- (c) 1.5 km/hr
- (d) 2.5 km/hr

10. The average speed of a train is 20% less on the return journey than on the onward journey. The train halts for half an hour at the destination station before starting on the return journey. If the total time taken for the to and fro journey is 23 hours covering a distance of 1000 km, the speed of the trian on the return journey is :
- (a) 60 km/hr
- (b) 40 km/hr
- (c) 50 km/hr
- (d) 55 km/hr

Answers

1. (c)	2. (d)	3. (b)	4. (b)	5. (a)	6. (a)	7. (d)	8. (c)	9. (b)	10. (b)

Unit Test–3

1. If an amount of Rs 1,50,000 is shared between A, B and C in the ratio 2 : 3 : 5, then A receives the same amount as he would receive if another sum of money is shared between A, B and C in the ratio 5 : 3 : 2. The ratio of 1,50,000 to the second amount of money is :
- (a) 2 : 3
- (b) 3 : 2
- (c) 5 : 3
- (d) 5 : 2

2. In a chemical experiment, two NaOH solution bottles are used. Bottle A contains salt and water in the ratio 7 : 3 and bottle B contains salt and water in the ratio 4 : 3. In what proportion should the quantities be taken from A and B to give the 2 : 1 NaOH solution?
- (a) 2 : 1
- (b) 10 : 7
- (c) 20 : 7
- (d) 1 : 2

3. A man's income is increased by Rs 1200 and at the same time, the rate of tax to be paid is reduced from 12% to 10%. He now pays the same amount of tax as before. What is his increased income, if 20% of his income is exempted from tax in both the cases?
- (a) Rs 6300
- (b) Rs 7200
- (c) Rs 4500
- (d) Rs 6500

4. At a college in New Delhi, 60% of the students are boys and the rest are girls. Further 15% of the boys and 7.5% of the girls are getting a fee waiver. If the number of students getting a fee waiver is 90, find the total number of students getting 50% concession, if it is given that 50% of those not getting a fee waiver are eligible to get half fee concession?
- (a) 360
- (b) 280
- (c) 320
- (d) 330

5. Jai Pal sells a shirt at a profit of 25 per cent. Had he bought it at 25 per cent less and sold it far Rs 25 less, he still would have gained 25 per cent. The cost price of the shirt is :

(a) Rs 50　　　　　　　(b) Rs 75
(c) Rs 80　　　　　　　(d) Rs 100

6. A person bought two clocks. The cost price of one of them exceeds by $\frac{1}{4}$ the price of the other. He sold the dearer one at a gain of 10% and the other at a gain of 7.5% and thus got Rs 98 in all as S.P. Find the total cost price of two clocks.
(a) Rs 150　　　　　　(b) Rs 90
(c) Rs 75　　　　　　　(d) Rs 100

7. An almirah is listed at Rs 1000. A retailer buys it with two successive discounts of 10% and 20% for cash. The other expenses are 10% of the cost of the almirah. At what price should he sell to earn a profit of 15%?
(a) Rs 910.80　　　　(b) Rs 900.50
(c) Rs 910.50　　　　(d) Rs 980.50

8. A merchant buys 40 bicycles and marks them at 25% above the cost price. He allows a discount on the marked price at 10% for cash sales, and at 5% for credit sales. If three-fourth of the stock is sold for cash and the rest for credit, and if the total profit be Rs 2025, what is the cost price of a bicycle?
(a) Rs 350　　　　　　(b) Rs 720
(c) Rs 360　　　　　　(d) Rs 460

9. The average of four consecutive even numbers is one-fourth of the sum of these numbers. What is the difference between the first and last number?
(a) 4　　　　　　　　(b) 6
(c) 2　　　　　　　　(d) 8

10. The average weight of three men A, B and C is 84 kg. D joins them and the average weight of the four becomes 80 kg. If E, whose weight is 3 kg more than that of D, replaces A, the average weight of B, C, D and E becomes 79 kg. Weight of A is :
(a) 65 kg　　　　　　(b) 70 kg
(c) 75 kg　　　　　　(d) 80 kg

11. A person invested some amount at the rate of 12% simple interest and some other amount at the rate of 10% simple interest. He received yearly interest of Rs 130, but if he had interchanged the amounts invested, he would have received Rs 4 more as interest. How much amount did he invest at different rates?
(a) Rs 500 @ 12% ; Rs 700 @ 10%
(b) Rs 700 @ 12% ; Rs 500 @ 10%
(c) Rs 700 @ 12% ; Rs 700 @ 10%
(d) Rs 500 @ 12% ; Rs 500 @ 10%

12. Rohit took a loan of Rs 20,000 to purchase a LCD TV set from a finance company. He promised to make the payment after three years. The company charges compound interest at the rate of 10% per annum for the same. But suddenly, the company announces the rate of interest as 15% per annum for the last one year of the loan period. What extra amount does Rohit have to pay due to this announcement of the new rate of interest?
(a) Rs 7830　　　　　(b) Rs 4410
(c) Rs 1210　　　　　(d) Rs 6620

13. Sanju puts equal amount of money one at 10% per annum compound interest payable half yearly and the second at a certain rate per cent per annum compound interest payable yearly. If he gets equal amounts after 3 years, what is the value of the second rate per cent?
(a) $10\frac{1}{4}\%$　　　　　(b) 10%
(c) $9\frac{1}{2}\%$　　　　　(d) $8\frac{1}{4}\%$

14. A machine depreciates in value each year at the rate of 10% of its previous value. However, every second year there is some maintenance work so that in that particular year, depreciation is only 5% of its previous value. If at the end of the fourth year, the value of the machine stands at Rs 1,46,205, then find the value of the machine at the start at the first year?
(a) Rs 1,90,000　　　(b) Rs 2,00,000
(c) Rs 1,95,000　　　(d) Rs 2,10,000

15. A, B and C can do a piece of work in 36, 54 and 72 days respectively. They started the work but A left 8 days before the completion of the work while B left 12 days before completion. The number of days for which C worked is :
(a) 4　　　　　　　　(b) 8
(c) 12　　　　　　　(d) 24

16. A certain number of men, twice as many women and thrice as many boys earn in 6 days Rs 5100. A woman earns one and a half times as a boy and a man as much as a woman and a boy together per day. How many women were there, if a boy earned Rs 25 daily.
(a) 4　　　　　　　　(b) 8
(c) 12　　　　　　　(d) 36

17. Tap A can fill the tank in two hours while taps B and C can empty it in six and eight hours respectively. All the taps remained open initially. After two hours, tap C was closed and after one more hour, tap B was also closed. In how much time now, would the remaining tank get filled?
(a) 12 min　　　　　(b) 15 min
(c) 30 min　　　　　(d) 20 min

18. At noon ship *A* starts from a point *P* towards a point *Q* and at 1.00 pm ship *B* starts from *Q* towards *P*. If the ship *A* is expected to complete the voyage in 6 hours and ship *B* is moving at a speed $\frac{2}{3}$ that of ship *A,* at what time are the two ships expected to meet one another?

(a) 4 pm (b) 4.30 pm
(c) 3 pm (d) 2.30 pm

19. Two trains are moving in the opposite directions on parallel tracks at the speeds of 64 km/hr and 96 km/hr respectively. The first train passes a telegraph post in 5 seconds whereas the second train passes the post in 6 seconds. Find the time taken by the trains to cross each other completely.

(a) 3 seconds (b) $4\frac{4}{5}$ seconds
(c) $5\frac{3}{5}$ seconds (d) 6 seconds

20. A man rows a boat upstream a certain distance and then returns back to the same place. If the time taken by him in going upstream is twice the time taken in rowing downstream, find the ratio of the speed of the boat in still water and the speed of the stream.

(a) 2 : 1 (b) 3 : 2
(c) 5 : 3 (d) 3 : 1

Answers

1. (d)	2. (c)	3. (b)	4. (d)	5. (c)	6. (b)	7. (a)	8. (c)	9. (b)	10. (c)
11. (a)	12. (c)	13. (a)	14. (b)	15. (d)	16. (b)	17. (c)	18. (a)	19. (c)	20. (d)

GEOMETRY

- *Triangles*
- *Polygons and Quadrilaterals*
- *Circles*

Chapter 22

TRIANGLES

1. A **triangle** is a plane closed figure bounded by three line segments.

   (*i*) Sum of the angles of a triangle is equal to 180°, *i.e.*,
   $$\angle A + \angle B + \angle C = 180°$$

   (*ii*) Exterior angle of a triangle is equal to the sum of its interior opposite angles, *i.e.*,
   $$\angle x = \angle 2 + \angle 3; \ \angle y = \angle 1 + \angle 2 \ ; \ \angle z = \angle 1 + \angle 3$$

   (*iii*) Sum of any two sides of a triangle is always, greater than the third side, *i.e.*,
   $$PQ + QR > PR$$
   $$PQ + PR > QR$$
   $$PR + QR > PQ$$

   (iv) Side opposite to the greatest angle will be greatest in length and vice versa.

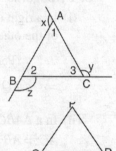

2. **Important Terms of a Triangle**

   (i) **Median and centroid:** A line joining the mid-point of a side of a triangle to its opposite vertex is called the median. D, E, F are the mid-points of the sides QR, PR and PQ respectively of a ΔPQR. Then, PD, QE and RF are the medians of ΔPQR.

   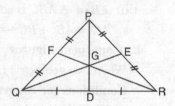

   • The point of concurrency of the three medians of a triangle is called **centroid.**

   • The centroid of a triangle divides each median in the ratio 2:1, *i.e.*,
   $$PG : GD = QG : GE = RG : GF = 2:1$$

   • A median divides a triangle into two parts of equal area.

   (ii) **Perpendicular bisector and circumcentre:** Perpendicular bisector to any side is the line that is perpendicular to that side and passes through its mid-point. Perpendicular bisectors need not pass through the opposite vertex.

   • The point of intersection of the three perpendicular bisectors of a triangle is called its **circumcentre.**

   • The **circumcentre** of a triangle is **equidistant from its three vertices**.

   If we draw a circle with circumcentre as the centre and the distance of any vertex from the circumcentre as radius, the circle passes through all the three vertices and the circle is called **circumcircle.**

   > **Note.** The circumcentre can be inside or outside the circle.

   • Circumcentre of a right angled triangle is the mid-point of the hypotenuse.

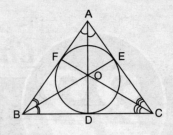

(iii) **Angle bisector and in-centre:**
- The point of intersection of the three angle bisectors of a triangle is called its **in-centre.**
- The **in-centre** always lies inside the triangle.
- It is always **equidistant from the sides of a triangle.**
- The circle drawn with incentre as centre and touching all the three sides of a triangle is called **in-circle.**

(iv) **Altitude and ortho-centre:**

The perpendicular drawn from the vertex of a triangle to the opposite side is called an **altitude.**
- The point of intersection of the three altitudes of a triangle is called **ortho-centre**, which can lie inside or outside the triangle.

> **Note.** • For an isosceles triangle, the median drawn from a vertex to the opposite side is also the perpendicular bisector of that side.
> • In an equilateral triangle, the median, angle bisector, altitude and perpendicular bisector of sides are all represented by the same straight line.
> • The circumcentre, centroid, orthocentre and incentre all coincide in an equilateral triangle.

3. **Pythagoras' theorem:**

(i) In a right angled triangle, the square of the hypotenuse is equal to the sum of the squares of the other two sides.

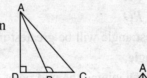

$$\therefore \quad (\text{Perpendicular})^2 + (\text{Base})^2 = (\text{Hypotenuse})^2$$
$$\Rightarrow \quad PQ^2 + QR^2 = PR^2$$

(ii) In a $\triangle ABC$, obtuse angled at B, if $AD \perp CB$, then
$$AC^2 = AB^2 + BC^2 + 2BC.BD$$

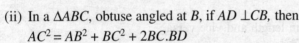

(iii) $\angle B$ of $\triangle ABC$ is acute and $AD \perp BC$, then
$$AC^2 = AB^2 + BC^2 - 2BC.BD$$

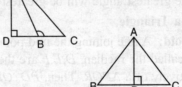

4. **Congruent Figures**

When two geometric figures have the same size and shape, they are said be congruent.

Conditions for congruency of triangles

(i) **SAS axiom (side-angle-side)**

If two triangles have two sides and the included angle of one respectively equal to two sides and the included angle of the other, the triangles are congruent.

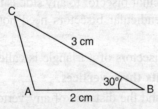
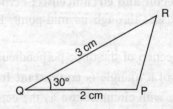

In $\quad \triangle s\ ABC$ and PQR
$$\angle B = \angle Q$$
$$AB = PQ$$
$$BC = QR$$
$$\therefore \quad \triangle ABC \cong \triangle PQR\ (SAS)$$

(ii) *ASA or AAS* **axiom (two angles - one side)**

If the triangles have two angles of one respectively equal to two angles of the other, and also a side of one triangle equal to the corresponding side of the other, the two triangles are congruent.
The side may be the included side of the two angles also.

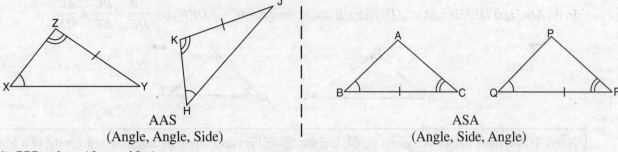

AAS
(Angle, Angle, Side)

ASA
(Angle, Side, Angle)

(iii) **SSS axiom (three sides)**

If two triangles have three sides of one respectively equal to three sides of the other, the triangles are congruent.

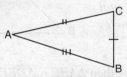 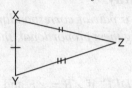

(iv) **RHS axiom (right angle, hypotenuse, sides)**

Two right triangles are congruent, if one side and hypotenuse of one are respectively equal to the corresponding side and hypotenuse of the other.

$$PQ = DE$$
$$\text{Hyp. } PR = \text{Hyp. } DF$$
$$\therefore \ \Delta PQR \cong \Delta DEF$$

5. Basic Proportionality Theorem

If a line is drawn parallel to one side of a triangle intersecting the other two sides, then the other two sides are divided proportionally in the same ratio.

Thus, in ΔABC, if $DE \parallel BC$, then $\dfrac{AD}{DB} = \dfrac{AE}{EC}$

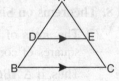

Conversely, *if a straight line divides any two sides of a triangle in the same ratio, then the straight line is parallel to the third side of the triangle.*

Thus, if in ΔABC, a line DE is drawn such that $\dfrac{AD}{DB} = \dfrac{AE}{EC}$, then $DE \parallel BC$.

6. Mid-point Theorem

A straight line drawn through the mid-point of one side, parallel to another side of a triangle bisects the third side.
In ΔPQR, a line ST drawn through the mid-point S of side PQ, $\parallel$ to QR, bisects PR, *i.e.*, T is the mid-point of PR.

Conversely, *the line joining the mid-points of any two sides of a triangle is always parallel to the third side and equal to half of it.*

If S and T are the mid-points of side PQ and PR respectively of ΔPQR, then $ST \parallel QR$ and

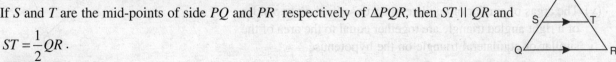

$ST = \dfrac{1}{2} QR$.

7. Similar Triangles

Two triangles (figures) are said to be **similar** *if they have the same shape, but not necessarily the same size, i.e.,*

• *their corresponding angles are equal.*

or

• *their corresponding sides are proportional.*

(i) **AAA-axiom of similarity**

  If two triangles are equiangular, their corresponding sides are proportional.

  If two triangles have two pairs of angles equal, their corresponding sides are proportional.

  In Δs *ABC* and *DEF*, if $\angle A = \angle D$ and $\angle B = \angle E$, then $\Delta ABC \sim \Delta DEF$ and $\dfrac{AB}{DE} = \dfrac{BC}{EF} = \dfrac{AC}{DF}$

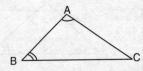 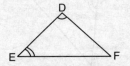

> **Note.** It two pairs angles are given equal, the third pair becomes equal by angle sum property of a triangle.

(ii) **SAS-axiom of similarity**

  If two triangles have a pair of corresponding angles equal and the sides including them proportional, then the triangles are similar.

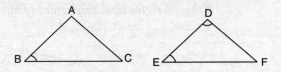

  Thus, in Δs *ABC* and *DEF*, if $\angle B = \angle E$ and $\dfrac{AB}{DE} = \dfrac{BC}{EF}$, then $\Delta ABC \sim \Delta DEF$.

(iii) **SSS -axiom of similarity**

  If two triangles have their three pairs of corresponding sides proportional, then the triangles are similar.

  In Δs *ABC* and *DEF*,

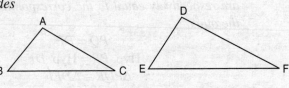

  if $\dfrac{AB}{DE} = \dfrac{BC}{EF} = \dfrac{AC}{DF}$, then $\Delta ABC \sim \Delta DEF$.

8. Theorems on Similar Triangles

  (i) The areas of two similar triangles are proportional to the squares of corresponding sides.

  Thus, if $\Delta ABC \sim \Delta DEF$, then

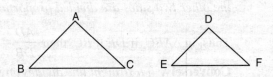

$$\frac{ar\,(\Delta ABC)}{ar\,(\Delta DEF)} = \frac{BC^2}{EF^2} = \frac{AB^2}{DE^2} = \frac{AC^2}{DF^2}$$

  (ii) The ratio of the areas of similar triangles is equal to the ratio of the squares of the corresponding altitudes.

  (iii) The ratio of the areas of similar triangles is equal to the ratio of the squares of the corresponding medians.

  (iv) If the areas of two similar triangles are equal, then the triangles are congruent.

  (v) The areas of similar or equilateral Δs described on two sides of a right angled triangle are together equal to the area of the similar or equilateral triangle on the hypotenuse.

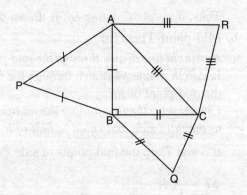

  ar (ΔPAB) + ar(ΔBQC) = ar(ΔARC)

Solved Examples

Ex. 1. *In the given figure, line l is the bisector of an angle A and B is any point on l. BP and BQ are perpendiculars from B to the arms of ∠A. Show that B is equidistant from the arms of ∠A.*

Sol. In Δs *APB* and *ABQ*, we have

∠*APB* = ∠*AQB*	(Each = 90°)
∠*PAB* = ∠*QAB*	(AB bisects ∠ *PAQ*)
AB = *BA*	(common)
∴ Δ *APB* ≅ Δ *ABQ*	(AAS)
⇒ *BP* = *BQ*	(cpct)

⇒   *B* is equidistant from the arms of ∠*A*.

Ex. 2. *ABC is a triangle and D is the mid-point of BC. The perpendiculars from D to AB and AC are equal. Prove that the triangle is isosceles.*

Sol. Let *DE* and *DF* be the perpendiculars from *D* on *AB* and *AC* respectively.

In Δs *BDE* and *CDF*, *DE* = *DF*	(Given)
∠ *BED* = ∠ *CFD* = 90°	
BD = *DC*	(∵ *D* is the mid-point of *BC*)
∴ Δ *BDE* ≅ Δ *CDF*	(RHS)
⇒ ∠*B* = ∠*C*	(cpct)
⇒ *AC* = *AB*	(Sides opp. equal ∠s are equal)
⇒ Δ*ABC* is isosceles.	

Ex. 3. *In the given figure, QA and PB are perpendiculars to AB. If AO = 10 cm, BO = 6 cm and PB= 9 cm, find AQ.*

Sol. In Δs *AOQ* and *BOP*, we have

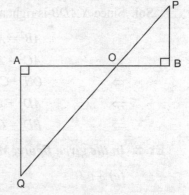

∠ *OAQ* = ∠ *OBP*	(Each equal to 90°)
∠ *AOQ* = ∠ *POB*	(Vertically opp. ∠s)

∴ By AA–similarity,

$$\Delta\, AOQ \sim \Delta\, BOP$$

$$\Rightarrow \quad \frac{AO}{BO} = \frac{OQ}{OP} = \frac{AQ}{BP}$$

$$\Rightarrow \quad \frac{AO}{BO} = \frac{AQ}{BP} \Rightarrow \frac{10}{6} = \frac{AQ}{9} \Rightarrow AQ = \frac{10 \times 9}{6} = \textbf{15 cm}.$$

Ex. 4. *D is a point on the side BC of ΔABC such that ∠ADC = ∠BAC. Prove that* $\dfrac{CA}{CD} = \dfrac{CB}{CA}$ *or CA² = CB × CD*

Sol. In Δ *ABC* and Δ *DAC*, we have

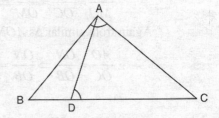

∠*ADC* = ∠*BAC* and ∠*C* = ∠*C*

∴   By AA - axiom of similarity

Δ *ABC* ∼ Δ *DAC*

$$\Rightarrow \quad \frac{AB}{DA} = \frac{BC}{AC} = \frac{AC}{DC}$$

$$\Rightarrow \quad \frac{CB}{CA} = \frac{CA}{CD}$$

Ex. 5. *If △ABC is similar to △DEF such that BC = 3 cm, EF = 4 cm and area of △ABC = 54 cm². Find the area of △DEF.*

Sol. Since the ratio of the areas of two similar triangles is equal to the ratio of the squares of any two corresponding sides,

$$\Rightarrow \quad \frac{ar(\triangle ABC)}{ar(\triangle DEF)} = \frac{BC^2}{EF^2} \quad \Rightarrow \frac{54}{ar(\triangle DEF)} = \frac{3^2}{4^2} \Rightarrow ar(\triangle DEF) = \frac{54 \times 16}{9} = \mathbf{96 \ cm^2}.$$

Ex. 6. *Prove that the area of the equilateral triangle described on the side of a square is half the area of the equilateral triangle described on its diagonal.*

Sol. **Given:** A square *ABCD*,

An equilateral △*BCE* described on side *BC* of the square.

An equilateral △*BDF* described on the diagonal *BD* of the square.

△ *BCE* ~ △ *BDF*   (∵ both are equiangular, each angle = 60°)

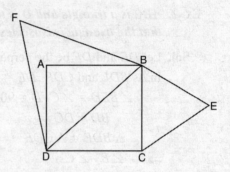

$$\therefore \quad \frac{ar(\triangle BCE)}{ar(\triangle BDF)} = \frac{BC^2}{BD^2} = \frac{BC^2}{(\sqrt{2}BC)^2}$$

(∵ diagonal of a square = $\sqrt{2}$ side)

$$= \frac{BC^2}{2BC^2} = \frac{1}{2}.$$

Ex. 7. *In an isosceles triangle ABC with AB = AC, BD is perpendicular from B to side AC. Prove that BD² – CD² = 2 CD.AD*

Sol. Since △ *ADB* is right angled at *D*,

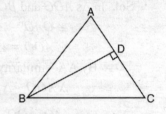

$$AB^2 = AD^2 + BD^2$$
$$\Rightarrow \qquad AC^2 = AD^2 + BD^2 \qquad (\because AB = AC)$$
$$\Rightarrow \qquad (AD+CD)^2 = AD^2 + BD^2$$
$$\Rightarrow \qquad AD^2 + CD^2 + 2AD.CD = AD^2 + BD^2$$
$$\Rightarrow \qquad BD^2 - CD^2 = \mathbf{2AD.CD}$$

Ex. 8. *In the given figure, M is the mid-point of the side CD of the parallelogram ABCD. What is ON:OB?*

(a) 3 : 2   (b) 2 : 1   (c) 3 : 1   (d) 5 : 2

Sol. From similar △s *ABN* and *DMN*

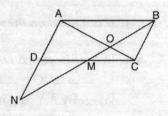

$$\frac{AB}{DM} = \frac{AN}{DN} = \frac{BN}{MN} \Rightarrow \frac{2}{1} = \frac{AN}{DN} \qquad \dots(1)$$

From similar △s *AOB* and *COM*

$$\frac{AB}{MC} = \frac{AO}{OC} = \frac{OB}{OM} \Rightarrow \frac{2}{1} = \frac{AO}{OC} = \frac{OB}{OM} \qquad \dots(2)$$

Again from similar △s *AON* and *BOC*

$$\frac{AO}{OC} = \frac{ON}{OB} \Rightarrow \frac{ON}{OB} = 2 \Rightarrow ON : OB = \mathbf{2:1}.$$

Ex. 9. *In the given triangle, AB is parallel to PQ. AP = c, PC = b, PQ = a, AB = x. What is the value of x?*

Sol. In Δs ABC and PQC

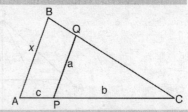

$$\frac{AB}{PQ} = \frac{AC}{PC} = \frac{AP + PC}{PC} = \frac{AP}{PC} + 1$$

$$\Rightarrow \quad \frac{x}{a} = \frac{c}{b} + 1 \Rightarrow x = a + \frac{ac}{b}$$

Ex. 10. *In the given figure, D and E trisect the side BC of a right triangle ABC. Prove that $8AE^2 = 3AC^2 + 5AD^2$*

Sol. Since D and E are the points of trisection of BC, therefore

$$BD = DE = CE$$

Let $BD = DE = CE = x$. Then $BE = 2x$ and $BC = 3x$

In rt Δ ABD,

$$AD^2 = AB^2 + BD^2 = AB^2 + x^2 \qquad \ldots(i)$$

In rt Δ ABE,

$$AE^2 = AB^2 + BE^2 = AB^2 + 4x^2 \qquad \ldots(ii)$$

In rt Δ ABC,

$$AC^2 = AB^2 + BC^2 = AB^2 + 9x^2 \qquad \ldots(iii)$$

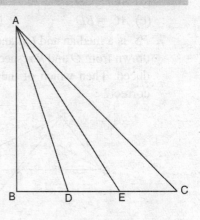

Now, $8AE^2 - 3AC^2 - 5AD^2$

$$= 8(AB^2 + 4x^2) - 3(AB^2 + 9x^2) - 5(AB^2 + x^2)$$

$$= 0$$

$$\Rightarrow \quad \mathbf{8AE^2 = 3AC^2 + 5AD^2}$$

Question Bank–22

Problems on Congruency

1. In the given figure, $\Delta RTQ \cong \Delta PSQ$ by ASA congruency condition. Which of the following pairs does not satisfy the condition.

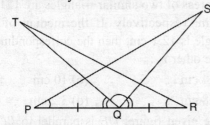

(a) $PQ = QR$ (b) $\angle P = \angle R$

(c) $\angle TQP = \angle SQR$ (d) None of these

2. It is given that $AB = BC$ and $AD = EC$. The $\Delta ABE \cong \Delta CBD$ by ——— congruency.

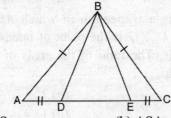

(a) SSS   (b) ASA

(c) SAS   (b) AAS

3. $ABCD$ is a quadrilateral. AM and CN are perpendiculars to BD, $AM = CN$ and diagonals AC and BD intersect at O, then which one of the following is correct?

(a) $AO = OC$ (b) $BO = OD$

(c) $AO = BO$ (d) $CO = DO$

4. Squares $ABDE$ and $ACFH$ are drawn externally on the sides AB and AC respectively of a scalene ΔABC. Which one of the following is correct?

(a) $BH = CE$ (b) $AD = AF$

(c) $BF = CD$ (d) $DF = EH$

5. In the given figure, two sides AB and BC and the median AD drawn to side BC of ΔABC are equal to the two sides PQ and QR and the corresponding median PM of the other ΔPQR. Which of the following is not correct?

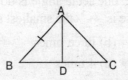

(a) $\Delta ABD \cong \Delta PQM$ (b) $\Delta ABC \cong \Delta PQR$

(c) $\Delta ABD \cong \Delta PMR$ (d) $\Delta ADC \cong \Delta PMR$

6. In the given figure, $OA = OB$, $OC = OD$, $\angle AOB$ $=\angle COD$. Which of the following statements is true?

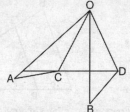

  (a) $AC = CD$ (b) $OA = OD$

  (c) $AC = BD$ (d) $\angle OCA = \angle ODC$

7. PS is a median and QL and RM are perpendiculars drawn from Q and R respectively on PS and PS produced. Then which of the following statements is correct?

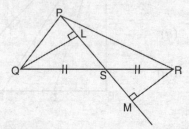

  (a) $PQ = RM$ (b) $QL = RM$

  (c) $PL = SR$ (d) $PS = SM$

8. In the figure, QX and RX are the bisectors of angles Q and R respectively of ΔPQR. If $XS \perp QR$ and $XT \perp PQ$, then $\Delta XTQ \cong \Delta XSQ$ by — congruency.

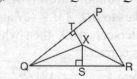

  (a) SAS (b) RHS

  (c) AAS (d) ASA

9. In the given figure, $AD = BC$, $AC = BD$. Then ΔPAB is

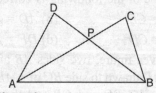

  (a) equilateral   (b) right angled

  (c) scalene   (d) isosceles

10. In a right angled triangle, one acute angle is double the other. The hypotenuse is —— the smallest side.

  (a) $\sqrt{2}$ times   (b) three times

  (c) double   (d) 4 times

Problems on Similar Triangles

11. If $\Delta ABC \sim \Delta EDF$ and ΔABC is not similar to ΔDEF, then which of the following is not true?

  (a) $BC.EF = AC.FD$ (b) $AB.EF = AC.DE$

  (c) $BC.DE = AB.EF$ (d) $BC.DE = AB.FD$

12. In the given figure, x equals

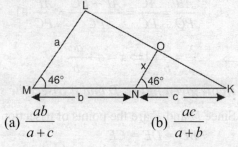

  (a) $\dfrac{ab}{a+c}$ (b) $\dfrac{ac}{a+b}$

  (c) $\dfrac{ac}{b+c}$ (d) $\dfrac{ab}{b+c}$

13. What value of x will make $DE \parallel AB$ in the given figure?

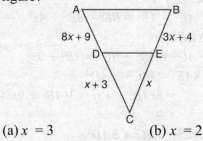

  (a) $x = 3$ (b) $x = 2$

  (c) $x = 1$ (d) $x = 5$

14. If the medians of two equilateral triangles are in the ratio 3 : 2, then what is ratio of the sides?

  (a) 1 : 1   (b) 2 : 3

  (c) 3 : 2   (d) $\sqrt{3} : \sqrt{2}$

15. The areas of two similar triangles are 121 cm² and 64 cm² respectively. If the median of the first triangle is 12.1 cm, then the corresponding median of the other is :

  (a) 6.4 cm   (b) 10 cm

  (c) 8.8 cm   (d) 3.2 cm

16. In the given figure, DE is parallel to BC and the ratio of the areas of ΔADE and trapezium $BDEC$ is 4 : 5. What is $DE : BC$?

  (a) 1 : 2   (b) 2 : 3

  (c) 4 : 5   (d) None of these

17. $ABCD$ is a trapezium in which $AB \parallel DC$ and $AB = 2 DC$. O is the point of intersection of the diagonals. The ratio of the areas of ΔAOB and ΔCOD is:

  (a) 1 : 2   (b) 2 : 1

  (c) 4 : 1   (d) 1 : 4

18. AB, EF and CD are parallel lines. Given that $EG = 5$cm, $GC = 10$ cm, $AB = 15$ cm and $DC = 18$ cm. What is the value of AC?

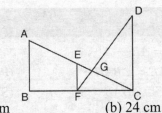

(a) 20 cm   (b) 24 cm
(c) 25 cm   (d) 28 cm

19. In the given figure, if $PA = x$, $RC = y$ and $QB = z$, then which one of the following is correct?

(a) $2y = x + z$

(b) $4y = x + z$

(c) $xy + yz = xz$

(d) $xy + xz = yz$

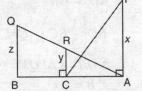

20. In ΔPQR, $QR = 10$, $RP = 11$ and $PQ = 12$. D is the midpoint of PR, DE is drawn parallel to PQ meeting QR in E. EF is drawn parallel to RP meeting PQ in F. What is the length of DF?

(a) $\dfrac{11}{2}$ (b) 6

(c) $\dfrac{33}{4}$ (d) 5

Problems on Pythagoras' Theorem

21. The hypotenuse of a right triangle is 6 m more than twice the shortest side. If the third side is 2 m less than the hypotenuse, find the hypotenuse of the triangle.

(a) 24 m   (b) 34 m
(c) 26 m   (d) 10 m

22. If the distance from the vertex to the centroid of an equilateral triangle is 6 cm, then what is the area of the triangle?

(a) 24 cm^2 (b) $27\sqrt{3}$ cm^2

(c) 12 cm^2 (d) $12\sqrt{3}$ cm^2

23. ΔABC is an equilateral triangle such that $AD \perp BC$, then $AD^2 =$

(a) $\dfrac{3}{2}DC^2$ (b) $2DC^2$

(c) $3CD^2$ (d) $4DC^2$

24. P and Q are points on the sides CA and CB respectively of ΔABC right angled at C. $AQ^2 + BP^2$ equals

(a) $BC^2 + PQ^2$ (b) $AB^2 + PC^2$
(c) $AB^2 + PQ^2$ (d) $BC^2 + AC^2$

25. ABC is a right-angled triangle, right angled at A and AD is the altitude on BC. If $AB : AC = 3 : 4$, what is the ratio $BD : DC$?

(a) $3 : 4$ (b) $9 : 16$
(c) $2 : 3$ (d) $1 : 2$

26. ABC is a right angled triangle, right angled at A. A circle is inscribed in it. The lengths of two sides containing the right angle are 6 cm and 8 cm. Find the radius of the circle?

(a) 3 cm   (b) 2 cm
(c) 5 cm   (d) 4 cm

27. ΔABC is right angled at A and $AD \perp BC$. Then $\dfrac{BD}{DC} =$

(a) $\left(\dfrac{AB}{AC}\right)^2$ (b) $\dfrac{AB}{AC}$

(c) $\left(\dfrac{AB}{AD}\right)^2$ (d) $\dfrac{AB}{AD}$

28. If ΔABC is right angled at B and M, N are the midpoints of AB and BC respectively, then $4(AN^2 + CM^2) =$

(a) $4AC^2$ (b) $5AC^2$

(c) $\dfrac{5}{4}AC^2$ (d) $6AC^2$

29. In ΔPQR, $PD \perp QR$ such that D lies on QR. If $PQ = a$, $PR = b$, $QD = c$ and $DR = d$, then

(a) $(a - d)(a + d) = (b - c)(b + c)$
(b) $(a - c)(b - d) = (a + c)(b + d)$
(c) $(a - b)(a + b) = (c + d)(c - d)$
(d) $(a - b)(c - d) = (a + b)(c + d)$

30. ABC is a triangle right-angled at B and D is a point on BC produced ($BD > BC$), such that $BD = 2DC$. Which one of the following is correct?

(a) $AC^2 = AD^2 - 3CD^2$
(b) $AC^2 = AD^2 - 2CD^2$
(c) $AC^2 = AD^2 - 4CD^2$
(d) $AC^2 = AD^2 - 5CD^2$

Answers

1. (d)	2. (c)	3. (a)	4. (a)	5. (c)	6. (c)	7. (b)	8. (c)	9. (d)	10. (c)
11. (c)	12. (c)	13. (b)	14. (c)	15. (c)	16. (b)	17. (c)	18. (c)	19. (c)	20. (d)
21. (c)	22. (b)	23. (c)	24. (c)	25. (b)	26. (b)	27. (b)	28. (b)	29. (c)	30. (a)

Hints and Solutions

1. (c) In Δs RTQ and PSQ,

   $QR = PQ$ (Given)

   $\angle P = \angle R$ (Given)

   $\angle TQR\ (\angle SQR + \angle SQT)$

   $\qquad\qquad = \angle PQS\ (\angle TQP + \angle SQT)$

   $\therefore\ \Delta RTQ \cong \Delta PSQ$ (ASA)

2. (c) Given, $AD = EC$

   $\Rightarrow AD + DE = DE + EC$

   $\Rightarrow AE = DC$

   Also, $AB = BC$

   $\Rightarrow \angle BCA = \angle BAC$ (isos. Δ property)

   $\Rightarrow \angle BCD = \angle BAE$

   $\therefore$ In Δs ABE and CBD,

   $AB = CB$ (Given)

   $AE = DC$ (Proved above)

   $\angle BAE = \angle BCD$ (Proved above)

   $\therefore \Delta\ ABE \cong \Delta\ CBD$ (SAS)

3. (a) In Δ AMO and Δ CNO

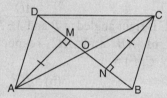

   $AM = CN$ (Given)

   $\angle AMO = \angle CNO = 90°$

   $\angle AOM = \angle CON$ (vert. opp. $\angle s$)

   $\therefore \Delta\ AMO \cong \Delta\ CNO$ (AAS)

   $\Rightarrow AO = OC$ (cpct)

4. (a) Δ ABC is a scalene Δ.

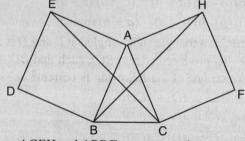

   ACFH and ABDE are squares drawn on sides AC and AB respectively.

   $\angle BAE = \angle CAH = 90°$ (Angle of a square)

   $\therefore \angle BAE + \angle BAC = \angle CAH + \angle BAC$

   $\Rightarrow \angle CAE = \angle BAH$

   In Δs EAC and HAB

   $EA = AB$ { sides of same square}

   $AC = AH$

   $\angle CAE = \angle BAH$ (Proved)

   $\therefore \Delta EAC \cong \Delta HAB$ (SAS)

   $\Rightarrow EC = BH$ (cpct)

5. (c) In Δs ABD and PQM

   $AB = PQ$ $\left.\begin{matrix} \\ \\ \end{matrix}\right\}$ Given

   $AD = PM$

   $BD\left(\dfrac{1}{2}BC\right) = QM\left(\dfrac{1}{2}QR\right)$

   $\therefore \textbf{\textit{ΔABD}} \cong \textbf{\textit{ΔPQM}}$ (SSS)

   $\angle B = \angle Q$ (cpct)

   In Δs ABC and PQR

   $AB = PQ$ (Given)

   $\angle B = \angle Q$ (Proved above)

   $BC = QR$ (Given)

   $\therefore \textbf{\textit{ΔABC}} \cong \textbf{\textit{ΔPQR}}$ (SAS)

   $AC = PR$ (cpct)

   In Δs ADC and PMR

   $AD = PM$ (Given)

   $DC\left(\dfrac{1}{2}BC\right) = MR\left(\dfrac{1}{2}QR\right)$ (Given)

   $AC = PR$ (Proved above)

   $\therefore \textbf{\textit{ΔADC}} \cong \textbf{\textit{ΔPMR}}$ (SSS)

   $\therefore \Delta ABD \not\cong \Delta PMR$

6. (c) In Δs AOC and BOD

   $OA = OB$ (Given)

   $OC = OD$ (Given)

   $\angle AOB - \angle COB = \angle COD - \angle COB,$

   i.e., $\angle AOC = \angle BOD$

   $\therefore \Delta AOC \cong \Delta BOD$ (SAS)

   $\Rightarrow AC = BD$ (cpct)

7. (b) In Δs QLS and RMS,

   $\angle QLS = \angle RMS = 90°$

   $\angle QSL = \angle RSM$ (Vert. opp. $\angle s$)

   $QS = SR$ (PS is the median)

   $\therefore \Delta\ QLS \cong \Delta RMS$ (AAS)

   $\Rightarrow QL = RM$ (cpct)

8. (c) In Δs XTQ and XSQ.

   $XQ = QX$ (Common)

   $\angle XQT = \angle XQS$ (QX bisects $\angle Q$)

   $\angle XTQ = \angle XSQ = 90°$

   $\therefore \Delta XTQ \cong \Delta XSQ$ (AAS)

9. (d) In $\Delta s\ ADB$ and ACB

$AD = BC$	(Given)
$AC = BD$	
$AB = BA$	(Common)

$\therefore\ \Delta ADB \cong \Delta ACB$ (SSS)

$\Rightarrow\ \angle ABD = \angle CAB$ (cpct)

$\Rightarrow\ \angle ABP = \angle PAB$

$\Rightarrow\ PA = PB$ (Sides opp. equal angles are equal)

$\Rightarrow\ \Delta PAB$ is isosceles.

10. (c) Given : A ΔPRQ in which

$\angle Q = 90°$ and $\angle PRQ = 2\angle QPR$

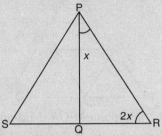

Const. Produce RQ to S such that

$RQ = QS$. Join PS.

In $\Delta s\ PQS$ and PQR

$QS = QR$	(By construction)
$PQ = PQ$	(Common)
$\angle PQS = \angle PQR$	(Each = 90°)

$\therefore\ \Delta PQS \cong \Delta PQR$ (SAS)

$\Rightarrow\ PS = PR$ and $\angle SPQ = \angle RPQ$ (cpct)

Let $\angle SPQ = x$. Then $\angle PRQ = 2x$ (Given)

Then $\angle SPR = \angle SPQ + \angle RPQ$

$\qquad\qquad = x + x = 2x$

$\Rightarrow\ \angle SPR = \angle PRQ \Rightarrow SR = PS$

(Sides opp. equal $\angle s$ are equal)

$\Rightarrow\ 2QR = PS$ $\{ \because SQ = QR \therefore SR = 2QR \}$

$\Rightarrow\ 2QR = PR$ ($\because PS = PR$)

$\Rightarrow$ The hypotenuse PR is **double** the smallest side QR.

11. (c) If $\Delta ABC \sim \Delta EDF$, then

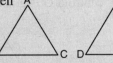

$\dfrac{AB}{ED} = \dfrac{BC}{DF} = \dfrac{AC}{EF}$

$\therefore\ \dfrac{AB}{ED} = \dfrac{BC}{DF} \Rightarrow AB \times DF = BC \times ED$

$\dfrac{BC}{DF} = \dfrac{AC}{EF} \Rightarrow BC \times EF = AC \times DF$

$\dfrac{AB}{ED} = \dfrac{AC}{EF} \Rightarrow AB \times EF = AC \times ED$

$\therefore\ BC \times DE \neq AB \times EF$

12. (c) In $\Delta s\ LMK$ and ONK,

$\angle KML = \angle ONK = 46°$

$\angle K = \angle K$ (Common)

$\therefore\ \Delta LMK \sim \Delta ONK$ (AA similarity)

$\Rightarrow\ \dfrac{KM}{KN} = \dfrac{LM}{ON} \Rightarrow \dfrac{b+c}{c} = \dfrac{a}{x} \Rightarrow x = \dfrac{ac}{b+c}$

13. (b) By the converse of basic proportionality theorem,

if $\dfrac{CD}{DA} = \dfrac{CE}{EB}$, then $DE \parallel AB$

$\Rightarrow\ \dfrac{x+3}{8x+9} = \dfrac{x}{3x+4}$

$\Rightarrow\ (3x+4)(x+3) = x(8x+9)$

$\Rightarrow\ 3x^2 + 13x + 12 = 8x^2 + 9x$

$\Rightarrow\ 5x^2 - 4x - 12 = 0$

$\Rightarrow\ 5x^2 - 10x + 6x - 12 = 0$

$\Rightarrow\ 5x(x-2) + 6(x-2) = 0$

$\Rightarrow\ (5x+6)(x-2) = 0$

$\Rightarrow\ x = \dfrac{-6}{5}$ or 2

Since, x cannot be negative, $x = 2$.

14. (c) Equilateral triangles are similar triangles.

In similar triangles, the ratio of their corresponding sides is the same as the ratio of their medians.

15. (c) The ratio of the areas of two similar triangles is equal to the ratio of the squares of the corresponding medians. Therefore,

$\dfrac{121}{64} = \dfrac{(12.1)^2}{x^2}$, where x is the median of the other Δ.

$\Rightarrow\ x^2 = \dfrac{(12.1)^2 \times 64}{121} \Rightarrow x = \sqrt{\dfrac{121}{100} \times 64}$

$\qquad = \dfrac{11}{10} \times 8 = \mathbf{8.8\ cm}$.

16. (b) In $\Delta s\ ADE$ and ABC,

$\angle A = \angle A$

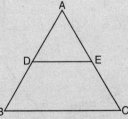

$\angle ADE = \angle ABC$ ($\angle DE \parallel BC$, corr $\angle s$ are equal)

$\therefore\ \Delta ADE \sim \Delta ABC$ (AA similarity)

$\Rightarrow\ \dfrac{ar(\Delta ABC)}{ar(\Delta ADE)} = \dfrac{BC^2}{DE^2}$

$\Rightarrow\ \dfrac{ar(\Delta ADE) + ar(\text{trap.} DEBC)}{ar(\Delta ADE)} = \dfrac{BC^2}{DE^2}$

$$\Rightarrow 1 + \frac{ar \,(trap. \, DEBC)}{ar \,(\Delta ADE)} = \frac{BC^2}{DE^2}$$

$$\Rightarrow 1 + \frac{5}{4} = \frac{BC^2}{DE^2} \Rightarrow \frac{9}{4} = \frac{BC^2}{DE^2} \Rightarrow \frac{BC}{DE} = \frac{3}{2}$$

$$\therefore DE : BC = 2 : 3.$$

17. (c) In $\Delta s \, AOB$ and COD

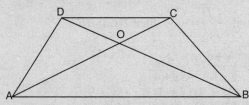

$$\angle AOB = \angle COD \quad (\text{vert. opp.} \angle s)$$
$$\angle OAB = \angle DCO \,(\, DC \parallel AB, \text{ alt.} \angle s \text{ are equal})$$
$$\therefore \Delta \, AOB \sim \Delta \, COD \,(AA \text{ similarity})$$

$$\Rightarrow \frac{ar \,(\Delta \, AOB)}{ar \,(\Delta \, COD)} = \frac{AB^2}{CD^2}$$

{ Ratio of areas of two similar Δs is equal to the ratio of the squares of the corresponding sides}

$$= \frac{(2CD)^2}{CD^2} = \frac{4CD^2}{CD^2} = \frac{4}{1}.$$

18. (c) In ΔEFG and ΔGCD,
$$\angle EFG = \angle GDC \,(EF \parallel CD, \text{ alt.} \angle s \text{ are equal})$$
$$\angle EGF = \angle CGD \,(\text{vert. opp.} \angle s)$$
$$\therefore \Delta EFG \sim \Delta GCD \quad (\text{By } AA \text{ similarity})$$

$$\therefore \frac{EG}{GC} = \frac{EF}{DC} \Rightarrow \frac{EF}{18} = \frac{5}{10} \Rightarrow EF = 9 \text{ cm}$$

Now in $\Delta s \, ABC$ and EFC,
$$\angle ACB = \angle ECF \,(\text{common})$$
$$\angle ABC = \angle EFC \,(AB \parallel EF, \text{ corr.} \angle s \text{ are equal})$$
$$\therefore \Delta ABC \sim \Delta EFC \,(\text{By } AA \text{ similarity})$$

$$\Rightarrow \frac{AC}{EC} = \frac{AB}{EF} \Rightarrow \frac{AC}{(EG+GC)} = \frac{AB}{EF}$$

$$\Rightarrow \frac{AC}{(5+10)} = \frac{15}{9} \Rightarrow AC = \mathbf{25 \text{ cm}}.$$

19. (c) $PA = x, RC = y, QB = z$

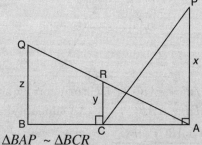

$$\Delta BAP \sim \Delta BCR$$
$$(\because \angle B \text{ is common and } \angle BAP = \angle BCR = 90°)$$

$$\therefore \frac{RC}{PA} = \frac{BC}{AB} \Rightarrow \frac{BC}{AB} = \frac{y}{x} \qquad ...(i)$$

Also, $\Delta ABQ \sim \Delta ACR$
$$(\angle A \text{ common}, \angle ABQ = \angle ACR = 90°)$$

$$\therefore \frac{RC}{BQ} = \frac{AC}{AB} \Rightarrow \frac{AB - BC}{AB} = \frac{y}{z} \Rightarrow 1 - \frac{BC}{AB} = \frac{y}{z}$$
$$...(ii)$$

$\therefore$ From (i) and (ii)

$$1 - \frac{y}{x} = \frac{y}{z} \Rightarrow \frac{x-y}{x} = \frac{y}{z} \Rightarrow xy = xz - yz$$
$$\Rightarrow \mathbf{xy + yz = xz}$$

20. (d) D is the mid-point of PR and $DE \parallel PQ$.

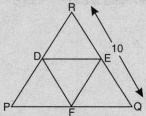

$\therefore$ E is the mid-point of QR (Mid-point Theorem)
$\because$ E is the mid-point of QR and $EF \parallel PR$,
F is the mid-point of PQ (Mid-point Theorem)

$$\therefore DF = \frac{1}{2} \times QR = \frac{1}{2} \times 10 = \mathbf{5}.$$

21. (c) Let the shortest side of the triangle be x m.
Then, hypotenuse $= (2x + 6)$ m
Third side $= (2x + 6) - 2 = (2x + 4)$ m
By Pythagoras' Theorem,
$$(2x + 6)^2 = (2x + 4)^2 + x^2$$
$$\Rightarrow 4x^2 + 24x + 36 = 4x^2 + 16x + 16 + x^2$$
$$\Rightarrow x^2 - 8x - 20 = 0$$
$$\Rightarrow x^2 - 10x + 2x - 20 = 0$$
$$\Rightarrow (x - 10)(x + 2) = 0$$
$$\Rightarrow x = 10 \text{ or } -2$$
$\therefore$ x cannot be negative, $x = 10$
$\therefore$ Hypotenuse $= (2 \times 10 + 6)$ m $= \mathbf{26 \text{ m}}$

22. (b) Let ABC be the equilateral triangle whose centroid G is at a distance 6 cm from vertex A.

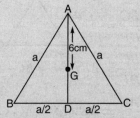

Let each side of ΔABC be a cm.
The median AD is also the perpendicular bisector in case of an equilateral Δ so, $\angle ADB = 90°$ and $BD = DC = a/2$

Now $AG : GD = 2 : 1$

(Centroid divides a median in the ratio 2:1)

$\therefore \quad \dfrac{6}{GD} = \dfrac{2}{1} \Rightarrow GD = 3\,\text{cm}$

$\therefore \quad AD = AG + GD = 6\,\text{cm} + 3\,\text{cm} = 9\,\text{cm}$

Now, $AB^2 = AD^2 + BD^2$ (Pythagoras' Theorem)

$\Rightarrow AB^2 - BD^2 = AD^2$

$\Rightarrow a^2 - \left(\dfrac{a}{2}\right)^2 = 81 \Rightarrow \dfrac{3a^2}{4} = 81 \Rightarrow a^2 = \dfrac{81 \times 4}{3}$

$\Rightarrow a = \sqrt{27 \times 4} = 6\sqrt{3}\ \text{cm}$

$\therefore$ Area of the equilateral triangle

$= \dfrac{\sqrt{3}}{4} a^2 = \dfrac{\sqrt{3}}{4} \times 6\sqrt{3} \times 6\sqrt{3}\ \text{cm}^2$

$= \mathbf{27\sqrt{3}\ cm^2}.$

23. (c) In equilateral $\triangle ABC$, $AD \perp BC$

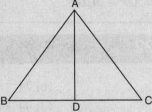

$\Rightarrow AD$ bisects the base BC

$\Rightarrow BD = DC = \dfrac{1}{2}BC = \dfrac{1}{2}AB = \dfrac{1}{2}AC$

Now use Pythagoras' theorem in $\triangle ADB$.

24. (c)

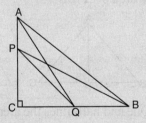

In rt. $\angle d\ \triangle ACQ$,

$\Rightarrow AC^2 + CQ^2 = AQ^2$ ...(i)

In rt. $\angle d\ PCB$,

$PC^2 + CB^2 = PB^2$ ...(ii)

Adding eqn (i) and (ii)

$AC^2 + CQ^2 + PC^2 + CB^2 = AQ^2 + PB^2$

$\Rightarrow (AC^2 + CB^2) + (CQ^2 + PC^2) = AQ^2 + PB^2$

$\underset{(\text{rt} \angle d \triangle ABC)}{AB^2} + \underset{(\text{rt} \angle d \triangle PQC)}{PQ^2} = AQ^2 + PB^2$

$\Rightarrow$

(Pythagoras' Theorem)

25. (b) In $\triangle ABC$ and $\triangle DBA$,

$\angle B$ is common

$\angle CAB = \angle BDA = 90°$

$\Rightarrow \triangle ABC \sim \triangle DBA$

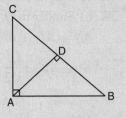

$\Rightarrow \dfrac{AB}{AC} = \dfrac{DB}{DA} \Rightarrow \dfrac{DB}{DA} = \dfrac{3}{4}$

$\Rightarrow AD = \dfrac{4}{3} DB$

...(i)

In $\triangle ABD$ and $\triangle ADC$,

$\angle DAB = \angle ACD$

(Third angles of similar $\triangle$s ABC and DBA)

$\angle ADB = \angle ADC = 90°$

$\therefore \quad \dfrac{AB}{AC} = \dfrac{AD}{CD} = \dfrac{3}{4} \Rightarrow AD = \dfrac{3}{4} CD$ ...(ii)

From (i) and (ii)

$\dfrac{4}{3} DB = \dfrac{3}{4} CD \Rightarrow \dfrac{\boldsymbol{BD}}{\boldsymbol{CD}} = \dfrac{3}{4} \times \dfrac{3}{4} = \dfrac{\boldsymbol{9}}{\boldsymbol{16}}.$

26. (b) Let $\triangle ABC$ be right angled at A.

Since the incentre is equidistant from the sides, let the radius of the incircle be r.

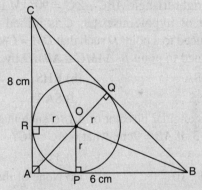

$\therefore \quad OP = OQ = OR = r\ \text{cm}$

By Pythagoras' Theorem

$AC^2 + AB^2 = BC^2$

$\Rightarrow BC^2 = 6^2 + 8^2 = 36 + 64 = 100$

$\Rightarrow BC = 10\ \text{cm}$. Now,

Area of $\triangle ABC$ = Area of $\triangle OAB$

$\qquad\qquad$ + Area of $\triangle OBC$ + Area of $\triangle OCA$

$\Rightarrow \dfrac{1}{2} \times AB \times AC$

$\qquad = \dfrac{1}{2} \times r \times AB + \dfrac{1}{2} \times r \times BC + \dfrac{1}{2} \times r \times CA$

$\Rightarrow \dfrac{1}{2} \times 6 \times 8 = \dfrac{1}{2} \times r \times 6 + \dfrac{1}{2} \times r \times 10 + \dfrac{1}{2} \times r \times 8$

$\Rightarrow 12\,r = 24 \Rightarrow \mathbf{r = 2\ cm}.$

27. (b) Similar to Q. 25.

28. (b) Similar to Q. 24.

$$AN^2 = AB^2 + BN^2$$

$$= AB^2 + \left(\frac{1}{2}BC^2\right)$$

$$CM^2 = \left(\frac{1}{2}AB\right)^2 + BC^2$$

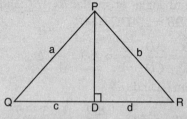

Now solve.

29. (c) Use Pythagoras' Theorem to find PD from ΔPQD and ΔPDR.

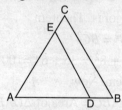

Now equate the two values.

30. (a) Given $BD = 2DC$

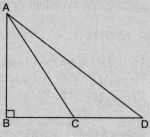

$\Rightarrow BC + CD = 2DC \Rightarrow BC = DC$

In ΔABC,

$$AC^2 = AB^2 + BC^2 \qquad \ldots (i)$$

In ΔABD,

$$AD^2 = AB^2 + BD^2 \qquad \ldots (ii)$$

$$AC^2 - AD^2 = BC^2 - BD^2$$

$$\Rightarrow AC^2 - AD^2 = CD^2 - (2\,CD)^2$$

$$= CD^2 - 4\,CD^2 = -3\,CD^2$$

$$\Rightarrow AC^2 = AD^2 - 3CD^2$$

Self Assessment Sheet–22

1. In a right triangle ABC , $\angle C = 90°$. M is the mid-point of hypotenuse AB. C is joined to M and produced to a point D such that $DM = CM$. Point D is joined to point B. $\Delta AMC \cong \Delta BMD$ by
   - (a) ASA
   - (b) RHS
   - (c) SSS
   - (d) SAS

2. AD is angular bisector of ΔABC such that $BD : DC = 2 : 3$. If $AB = 7$cm, what is $AC : BC$?
   - (a) $2 : 3$
   - (b) $3 : 2$
   - (c) $21 : 10$
   - (d) None of these

3. In the given figure, $DE \parallel BC$.

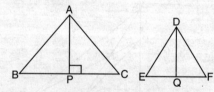

   $AD = x$, $DB = x - 2$
   $AE = x + 2$, $EC = x - 1$
   What is the value of x?
   - (a) 3
   - (b) 4
   - (c) 5
   - (d) 6

4. If in Δs ABC and DEF, $\angle A = \angle E = 37°$, $AB : ED = AC : EF$ and $\angle F = 69°$, then what is the value of $\angle B$?
   - (a) 69°
   - (b) 74°
   - (c) 84°
   - (d) 94°

5. Triangles ABC and DEF are similar. If the length of the perpendicular AP from A on the opposite side BC is 2 cm and the length of the perpendicular DQ from D on the opposite side EF is 1 cm, then what is the area of ΔABC?

   (a) One and half times the area of the triangle DEF.

   (b) Four times the area of triangle DEF.

   (c) Twice the area of the triangle DEF.

   (d) Three times the area of triangle DEF.

6. If $ABCD$ is a parallelogram and E and F are the centroids of triangles ABD and BCD respectively, then EF equals
   - (a) AE
   - (b) BE
   - (c) CE
   - (d) DE

7. In a ΔABC, perpendicular AD from A on BC meets BC at D. If $BD = 8$ cm, $DC = 2$ cm and $AD = 4$ cm, then

(a) ΔABC is isosceles

(b) ΔABC is equilateral

(c) $AC = 2AB$

(d) ΔABC is right angled at A

8. If E is a point on the side CA of an equilateral triangle ABC, such that $BE \perp CA$, then $AB^2 + BC^2 + CA^2 =$

(a) $2 BE^2$ (b) $3BE^2$

(c) $4 BE^2$ (d) $6 BE^2$

9. In a right triangle ABC right angled at C, P and Q are points on the sides CA and CB respectively, which divide these sides in the ratio 2:1. Then, which of the following statements is true?

(a) $9AQ^2 = 9 BC^2 + 4 AC^2$

(b) $9AQ^2 = 9 AC^2 + 4 BC^2$

(c) $9AQ^2 = 9 BC^2 + 4 AC^2$

(d) $9AQ^2 = 9 AB^2 - 4 BP^2$

10. In the given figure ΔABC is a right-angle at B. AD and CE are the two medians drawn from A and C respectively. If $AC = 5$ cm and $AD = \dfrac{3\sqrt{5}}{2}$ cm, then CE equals.

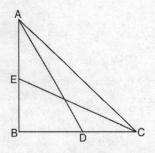

(a) 2 cm   (b) $2\sqrt{5}$ cm

(c) $5\sqrt{2}$ cm   (d) $3\sqrt{2}$ cm

Answers

1. (d)   2. (c)   3. (b)   4. (b)   5. (b)

6. (a) [**Hint.** Centroid divides the median in the ratio 2:1. $OE = \dfrac{1}{3}OA$, $OF = \dfrac{1}{3}OC$]   7. (d)   8. (c)

9. (b) [**Hint.** See Q. No. 24 of Question Bank]   10. (b)

Chapter 23

POLYGONS AND QUADRILATERALS

POLYGONS

1. A simple closed figure formed by three or more line segments is called a **polygon**.
2. If the measure of each interior angle of a polygon is less than 180°, then it is called a **convex polygon**.

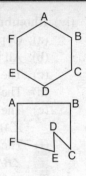

3. If the measure of at least one interior angle of a polygon is greater than 180°, then it is a **concave** or **rentrant polygon**.
4. A polygon with all sides and all angles equal is called a **regular polygon**.
5. **Properties of polygons :** For a polygon of n sides,
   (i) Sum of interior angles $= (2n - 4) \times 90°$
   (ii) Sum of exterior angles $= 360°$ (always)
   (iii) Each interior angle $= \dfrac{(2n-4)}{n} \times 90°$ (regular polygon).
   (iv) Each exterior angle $\dfrac{360°}{n}$ (in regular polygon)
   (v) Interior angle + exterior angle $= 180°$ (always)

QUADRILATERALS

1. **Quadrilaterals** are figures enclosed by four line segments.
2. The sum of the four interior angles of a quadrilateral is 360°.
3. **Types of quadrilaterals and their properties:**
   (i) **Parallelogram:**
      (a) Opposite sides are equal and parallel. $AB\|CD$, $AB = CD$ and $BC \| AD$, $BC = AD$.
      (b) Diagonals bisect each other, $AO = OC$, $BO = OD$
      (c) Each diagonal divides it into two congruent triangles, $\Delta ABC \cong \Delta CDA$ and $\Delta DAB \cong \Delta BCD$
      (d) Sum of any two adjacent angles $= 180°$
         $\angle A + \angle B = 180°$, $\angle B + \angle C = 180°$
         $\angle C + \angle D = 180°$, $\angle D + \angle A = 180°$

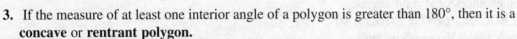

   (ii) **Rectangle:** A parallelogram whose all angles are 90°.
      (a) Opposite sides are parallel and equal. $AB \| DC$, $AB = DC$, $AD \| BC$, $AD = BC$
      (b) All the angles are equal to 90°,
         $\angle A = \angle B = \angle C = \angle D = 90°$
      (c) Diagonals are equal, $AC = BD$
      (d) Diagonals bisect each other, $AO = OC = BO = OD$

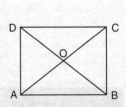

(iii) **Square:** A rectangle whose adjacent sides are equal.

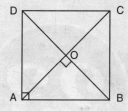

  (a)   All sides are equal

  (b)   Opposite sides are parallel

  (c)   All the angles are equal to 90°

  (d)   The diagonals are equal and bisect each other

  (e)   The diagonals intersect at right angles, *i.e.*, $\angle AOB = \angle BOC = \angle COD = \angle DOA = 90°$

  (f)   The diagonals bisect the opposite angles, *i.e.*

       $\angle DAC = \angle CAB = 45°, \ \angle DCA = \angle BCA = 45°$

       $\angle ABD = \angle DBC = 45°, \angle ADB = \angle BDC = 45°$

(iv) **Rhombus:** A parallelogram with all sides equal.

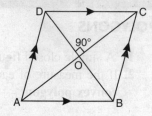

  (a)   Opposite sides are parallel.

  (b)   All the sides are equal.

  (c)   The opposite angles are equal.

  (d)   The diagonals are not equal. $AC \neq BD$.

  (e)   The diagonals intersect at right angles: $\angle AOB = \angle BOC = \angle COD = \angle AOD = 90°$

  (f)   The diagonals bisect the opposite angles: $\angle BAC = \angle DAC, \angle ACB = \angle ACD, \angle DBA = \angle DBC, \angle ADB = \angle BDC.$

(v) **Trapezium:** A quadrilateral in which only one pair of opposite sides is parallel, $AB \parallel CD$,

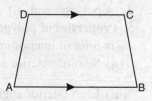

         $\angle A + \angle D = 180°$

         $\angle B + \angle C = 180°$

Isosceles trapezium: A trapezium whose non-parallel sides are equal.

     $AB \parallel DC, \ AD = BC$

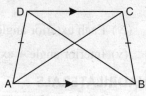

  (a)   The base angles of an isosceles trapezium are equal, *i.e.*, $\angle DAB = \angle ABC$.

  (b)   The diagonals of an isosceles trapezium are equal, *i.e.*, Diagonal $AC =$ Diagonal BD.

(vi) **Kite:** A kite is a quadrilateral in which two pairs of adjacent sides are equal.

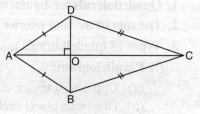

Opposite sides are not equal. $AD = AB, DC = CB$

Diagonals of a kite intersect each other at right angles.

$\angle AOB = \angle BOC = \angle COD = \angle DOA = 90°.$

4. The given chart shows how special quadrilaterals are related,

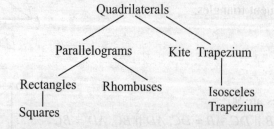

Quadrilaterals
Parallelograms   Kite   Trapezium
Rectangles   Rhombuses
Squares   Isosceles
Trapezium

Solved Examples

Ex. 1. *PQRS is a rhombus with $\angle PQR = 54°$. Determine $\angle PRS$.*

Sol. *PQRS* is a rhombus

     $\angle PSR = \angle PQR = 54°$ (opp. $\angle s$ of a rhombus are equal)

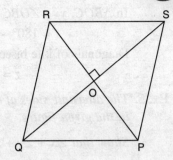

$$\Rightarrow \angle OSR = \frac{1}{2} \times \angle PSR = \frac{1}{2} \times 54° = 27° \qquad (SQ \text{ bisects } \angle PSR)$$

Now in ΔOSR,

$$\angle ORS + \angle ROS + \angle OSR = 180°$$

$$\Rightarrow \angle ORS + 27° + 90° = 180°$$

(Diagonals of a rhombus bisect each other at right $\angle s$)

$$\Rightarrow \angle ORS = 63°$$

$$\Rightarrow \angle PRS = \mathbf{63°.}$$

Ex. 2. *Show that the quadrilateral formed by joining the mid-points of the consecutive sides of a rectangle is a rhombus.*

Sol. A rectangle $ABCD$ in which P,Q,R and S are the mid-points of sides AB, BC, CD and DA respectively. PQ, QR, RS and SP are joined. Join AC.

In ΔABC, P and Q are the mid-points of sides AB and BC respectively.

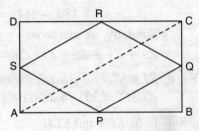

$$\therefore PQ \parallel AC \text{ and } PQ = \frac{1}{2} AC \qquad ...(i) \qquad \text{(Mid pt. Theorem)}$$

In ΔADC, R and S are the mid-points of sides CD and AD respectively.

$$\therefore SR \parallel AC \text{ and } SR = \frac{1}{2} AC \qquad ...(ii) \qquad \text{(Mid. pt. Theorem)}$$

From (i) and (ii),

$$PQ \parallel SR \text{ and } PQ = SR$$

$$\Rightarrow PQRS \text{ is a parallelogram.}$$

Now $ABCD$ is a rectangle. $\therefore AD = BC \Rightarrow \frac{1}{2} AD = \frac{1}{2} BC \Rightarrow AS = BQ \qquad \qquad ...(iii)$

In Δs APS and BPQ, we have

$AP = BP$	(P is the mid-point of AB)
$\angle PAS = \angle PBQ$	(Each is equal to 90°)
and $AS = BQ$	(From (iii))

$$\therefore \Delta APQ \cong \Delta BPQ \qquad \text{(SAS)}$$

$$\Rightarrow PS = PQ$$

$\therefore PQRS$ is a parallelogram whose adjacent sides are equal.

$\Rightarrow PQRS$ is a rhombus.

Ex. 3. *Find x in the given rhombus.*

Sol.

$\angle ODC = 36°$	($AB \parallel CD$, DB is the transversal, alt. $\angle s$ are equal)
$\angle COD = 90°$	(Diagonals of a rhombus bisect each other at right $\angle s$)

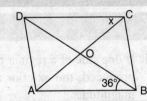

$$\therefore \text{ In } \Delta COD, \ x = \angle OCD = 180° - (\angle ODC + \angle COD)$$

$$= 180° - (36° + 90°) = 180° - 126° = \mathbf{54°.}$$

Ex. 4. *In the given kite, calculate x, y and z.*

Sol. Both pairs of adjacent sides of a kite are equal, therefore,

$$AB = AD \text{ and } BC = CD$$

In $\Delta ABD, \angle ABD = \angle ADB = 40°$ \qquad (isos. Δ property)

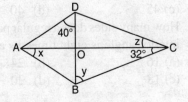

Since the diagonals of a kite intersect at rt. $\angle s$, $\angle AOB = 90°$.

In ΔAOB, $x = \angle OAB = 180° - (90° + 40°)$

$$= 180° - 130° = \mathbf{50°}$$

In ΔBOC, $y = \angle OBC = 180° - (90° + 32°)$
$$= 180° - 122° = 58°$$

Diagonals of kite bisect the angles at the vertices

$\therefore$ $z = 32°$

Ex. 5. *The alternate sides of any regular pentagon are produced to meet so as to form a star shaped figure. shown in the given figure.*

    Show that $\angle x + \angle y + \angle z + \angle t + \angle u = 180°$

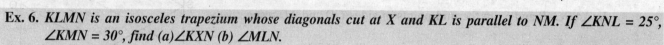

Sol. Each interior angle of a regular pentagon $= \dfrac{(5-2)\times 180°}{5} = 108°$

$\therefore$ Ext. $\angle BAK = $ Ext.$\angle ABK = 180° - 108° = 72°$

$\therefore$ In ΔABK, $x = 180° - (72° + 72°)$
$$= 180° - 144° = 36°$$

Similarly, we can show $u = t = y = z = 36°$

$\therefore$ $\angle x + \angle y + \angle z + \angle t + \angle u = 36° \times 5 = \textbf{180°}$.

Ex. 6. *KLMN is an isosceles trapezium whose diagonals cut at X and KL is parallel to NM. If $\angle KNL = 25°$, $\angle KMN = 30°$, find (a)$\angle KXN$ (b) $\angle MLN$.*

Sol. In Δs KLN and KLM

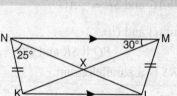

$KN = LM$	(Isos. trap.)
$\angle NKL = \angle KLM$	(Prop. of isos. trap.)
$KL = KL$	(Common)
$\therefore$ $\Delta KLN \cong \Delta KLM$	(SAS)
$\Rightarrow$ $\angle KML = \angle KNL = 25°$	(c.p.c.t.)

$\therefore$ $\angle LMN = 25° + 30° = 55°$

Now $\angle KNM = \angle LMN$

$\Rightarrow \angle KNL + \angle LNM = 55° \Rightarrow \angle LNM = 55° - 25° = 35°$

In ΔNXM, $\angle NXM = 180° - (\angle XNM + \angle XMN)$
$$= 180° - (30° + 30°) = 180° - 60° = 120°$$

(i) Now, $\angle KXN = 180° - \angle NXM = 180° - 120° = \textbf{60°}$ (Linear pair)

Also, $\angle LXM = \angle KXN = 60°$

(ii) Now, in ΔMLX, $\angle MLX (\angle MLN) = 180° - (60° + 25°)$
$$= 180° - 85° = \textbf{95°}$$

Question Bank–23

1. If one side of a regular polygon with seven sides is produced, the exterior angle (in degrees) has the magnitude:
   (a) 60   (b) $51\dfrac{3}{7}$
   (c) 45   (d) 40

2. How many sides does a regular polygon have, whose interior angle is eight times its exterior angle?
   (a) 16   (b) 24
   (c) 18   (d) 20

3. Any cyclic parallelogram having unequal adjacent sides is necessarily a
   (a) square   (b) rectangle
   (c) rhombus   (d) trapezium

4. In the given figure, ABCD is a parallelogram. The quadrilateral PQRS is exactly

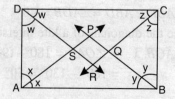

(a) a square (b) a parallelogram
(c) a rectangle (d) a rhombus

5. If angles A, B, C, D of a quadrilateral $ABCD$ taken in order are in the ratio $3 : 7 : 6 : 4$, then $ABCD$ is a
(a) rhombus (b) parallelogram
(c) trapezium (d) kite

6. The figure formed by joining the mid-points of adjacent sides of a rhombus is
(a) a square (b) rectangle
(c) trapezium (d) none of these

7. $ABCD$ is a rectangle. Find x.

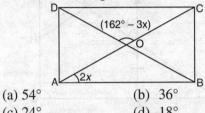

(a) 54° (b) 36°
(c) 24° (d) 18°

8. $PQRS$ is a parallelogram. Then y equals.

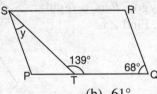

(a) 27° (b) 61°
(c) 41° (d) 28°

9. $ABCD$ is a rhombus.
$\angle DAB = 2x + 15°$,
$\angle DCB = 3x - 30°$,
$\angle BDC$ equals
(a) 45°
(b) 35°
(c) 37.5°
(d) 42.5°

10. $PQRS$ is a kite. $\angle P = 70°$,
$\angle S = 90.5°$, $\angle R$ equals
(a) 99°
(b) 91°
(c) 111°
(d) 109°

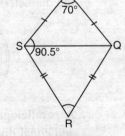

11. If the bisectors of the angles A, B, C and D of a quadrilateral meet at O, then $\angle AOB$ is equal to:

(a) $\angle C + \angle D$ (b) $\frac{1}{2}(\angle C + \angle D)$

(c) $\frac{1}{2}\angle C + \frac{1}{3}\angle D$ (d) $\frac{1}{3}\angle C + \frac{1}{2}\angle D$

12. $ABCD$ is a rectangle with $\angle BAC = 48°$. Then $\angle DBC$ equals
(a) 38° (b) 48°
(c) 132° (d) 42°

13. In a trapezium $ABCD$, $AB \parallel DC$, $AB = AD$, $\angle ADC = 64°$ and $\angle BCD = 54°$. Find $\angle DBC$.

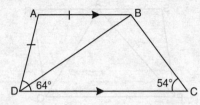

(a) 64° (b) 72°
(c) 94° (d) 116°

14. $ABCD$ is a square, $BA = BQ$, QRC and BPD are straight lines and $\angle PBQ = 21°$. Then, $\angle BAQ$ equals

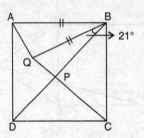

(a) 60° (b) 84°
(c) 78° (d) 74.5°

15. In the diagram, ABD and BCD are isosceles triangles, where $AB = BC = BD$. The special name that is given to quadrilateral $ABCD$ is:

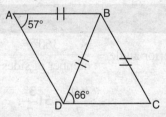

(a) rectangle (b) kite
(c) parallelogram (d) trapezium

16. In the diagram, CDP is a straight line, $\triangle AQD$ is equilateral $\angle BAR = 90°$, $\angle QAR = 135°$, $\angle BCD = 106°$ and $\angle ABC = 100°$. Then, $\angle PDQ$ equals.

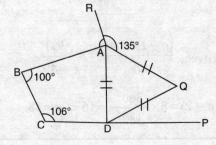

(a) 39°　　　　　　　(b)　21°

(c) 41°　　　　　　　(d)　53°

17. In the given figure, *PQRS* is a parallelogram and ∠*SPQ* = 60°. If the bisectors of ∠*P* and ∠*Q* meet at *A* on *RS*, then which of the following is not correct?

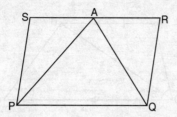

(a) *AS = SP*　　　　　(b)　*AS = AR*

(c) *AR = SP*　　　　　(d)　*AQ = PQ*

18. In the given diagram, *ABCD* is a square and Δ*BCT* is an equilateral triangle. ∠*BTD* equals

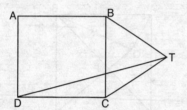

(a) 30°　　　　　　　(b)　15°

(c) 45°　　　　　　　(d)　35°

19. In the given figure, *AD∥BC*, ∠*AFG* = *b* and ∠*BCD* = *c*. Express *b* in terms of *c*.

(a) $\dfrac{c}{2}$　　　　　　(b)　$\dfrac{90° + c}{2}$

(c) $180° - \dfrac{c}{2}$　　　(d)　$90° - \dfrac{c}{2}$

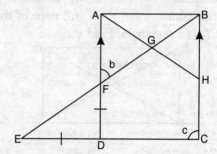

20. Polygon *A* is a regular pentagon and polygon *B* is a regular hexagon. Find ∠*x*.

(a) 108°

(b) 132°

(c) 124°

(d) 96°

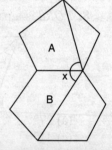

Answers

1. (b)	2. (c)	3. (b)	4. (c)	5. (c)	6. (b)	7. (d)	8. (a)	9. (c)	10. (d)
11. (b)	12. (d)	13. (c)	14. (c)	15. (d)	16. (c)	17. (d)	18. (d)	19. (d)	20. (b)

Hints and Solutions

1. (b) Exterior angle = $\dfrac{360°}{\text{number of sides}}$

$$= \frac{360°}{7} = 51\frac{3}{7}°.$$

2. (c) Interior angle of a regular polygon of '*n*' sides

$$= \frac{(n-2) \times 180°}{n}$$

Exterior angle of a regular polygon of '*n*' sides

$$= \frac{360°}{n}$$

Given, $\dfrac{(n-2) \times 180°}{\not{n}} = 8 \times \dfrac{360°}{\not{n}}$

$$\Rightarrow (n-2) = 8 \times \frac{\overset{2}{\cancel{360°}}}{\underset{1}{\cancel{180°}}} = 16 \Rightarrow n = \mathbf{18}.$$

3. (b) A cyclic parallelogram has diagonals, which are the diameters of the circle.

∴ A parallelogram having unequal adjacent sides and equal diagonals is necessarily a rectangle.

4. (c) In parallelogram *ABCD*, ∠*A* + ∠*D* = 180°

⇒ In Δ*ASD*, ∠*ADS* + ∠*SAD* = $\dfrac{180°}{2}$ = 90°

　(∵ *DS* and *AS* bisect ∠s *D* and *A* respectively)

∴ ∠*ASD* = 180° – (∠*ADS* + ∠*SAD*)

　　　　= 180° – 90° = 90°

⇒ ∠*PSR* = 90°　(vert. opp. ∠s)

Similarly we can show that $\angle SPQ = \angle PQR$ $=\angle QRS = 90°$.

∴ Quadrilateral $PQRS$ is a rectangle.

5. (c) Let the angles of the quadrilateral be $3x$, $7x$, $6x$ and $4x$.

Then, $3x + 7x + 6x + 4x = 360°$

$\Rightarrow 20x = 360° \Rightarrow x = 18°$

∴ The angles A, B, C and D are respectively $3×18°$, $7×18°$, $6×18°$ and $4 × 18°$, *i.e.*, $54°$, $126°$, $108°$ and $72°$.

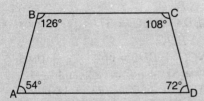

which shows that $\angle A + \angle B = 180°$ and $\angle C + \angle D = 180° \Rightarrow AD \| BC$.

Hence $ABCD$ is a trapezium.

6. (b) To prove, $PQRS$ is a rectangle, it is sufficient to show that it is a parallelogram, whose one angle is a right angle.

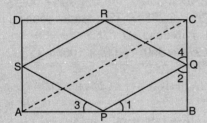

In ΔABC, P and Q are the mid-points of AB and BC respectively,

∴ $PQ \| AC$ and $PQ = \dfrac{1}{2} AC$ ...(*i*)

In ΔADC, R and S are the mid-points of CD and AD respectively.

∴ $RS \| AC$ and $RS = \dfrac{1}{2} AC$ ...(*ii*)

∴ From (*i*) and (*ii*)

$PQ \| RS$ and $PQ = RS$.

$\Rightarrow PQRS$ is a $\|$gm.

Now we prove that one angle of $\|$gm $PQRS$ is a rt. angle.

Since $ABCD$ is a rhombus, $AB = BC$

$\Rightarrow \dfrac{1}{2} AB = \dfrac{1}{2} BC \Rightarrow PB = BQ$ ($\because P$ and Q are mid pts. of AB and BC respectively)

In ΔPBQ, $PB = BQ \Rightarrow \angle 2 = \angle 1$ (isos. Δ prop.)

Also, $ABCD$ is a rhombus, so

$AB = BC = CD = DA$

$\Rightarrow AB = BC$ and $CD = DA$

$\Rightarrow \dfrac{1}{2} AB = \dfrac{1}{2} BC, \dfrac{1}{2} CD = \dfrac{1}{2} DA$

$\Rightarrow AP = CQ, CR = AS$

Now in $\Delta s\ APS$ and CQR, we have

$AP = CQ$

$AS = CR$ (Proved above)

$PS = QR$

$\Rightarrow \Delta APS \cong \Delta CQR$ (SSS congruency)

$\Rightarrow \angle 3 = \angle 4$ (cpct)

Now, $\angle 3 + \angle SPQ + \angle 2 = 180°$ and $\angle 1 + \angle PQR + \angle 4 = 180°$

$\Rightarrow \angle 3 + \angle SPQ + \angle 2 = \angle 1 + \angle PQR + \angle 4$

$\Rightarrow \angle SPQ = \angle PQR$

($\because \angle 3 = \angle 4$ and $\angle 1 = \angle 2$)

Now $PQRS$ being a $\|$gm.

$\angle SPQ + \angle PQR = 180°$ (adj. $\angle$s)

$\Rightarrow \angle SPQ + \angle SPQ = 180°$

$\Rightarrow \angle SPQ = 90°$

Thus $PQRS$ is a $\|$gm with one angle $= 90°$

Hence $PQRS$ is a rectangle.

7. (d) Diagonals of a rectangle are equal and bisect each other.

∴ In ΔAOB, $\angle AOB = 162° - 3x$ (vert. opp. $\angle$s)

$\angle OBA = \angle OAB = 2x$ ($\because OA = OB$)

∴ $\angle AOB + \angle OBA + \angle OAB = 180°$

$\Rightarrow 162° - 3x + 2x + 2x = 180°$

$\Rightarrow x = 18°$

8. (a) $\angle PSR = \angle PQR = 68°$

(opp. $\angle$s of a $\|$gm are equal)

$\angle PTS = 180° - 139° = 41°$

(PTQ is a straight line)

∴ $\angle RST = \angle PTS = 41°$ ($SR \| PQ$ alt. $\angle$s are equal)

∴ $y = \angle PSR - \angle RST = 68° - 41° = \mathbf{27°}$

9. (c) ∴ Opp $\angle$s of a rhombus are equal

$2x + 15° = 3x - 30° \Rightarrow x = 45°$

In ΔBCD, $CD = CB \Rightarrow \angle CBD = \angle BDC$

Also, $\angle DCB = 3 × 45° - 30° = 105°$

∴ $\angle BDC = \dfrac{180° - 105°}{2} = \dfrac{75°}{2} = 37.5°$

10. (d) In a kite, the two pairs of adjacent sides are equal, *i.e.*, $PS = PQ$ and $SR = RQ$

In ΔPSQ,

$\angle PSQ = \angle PQS = \dfrac{180° - 70°}{2} = \dfrac{110°}{2} = 55°$

∴ $\angle QSR = 90.5° - 55° = 35.5°$

$\Rightarrow \angle RQS = \angle QSR = 35.5°$ $(\because RQ = RS)$

In ΔSRQ,

$\angle R + \angle RQS + \angle QSR = 180°$

$\Rightarrow \angle R + 35.5° + 35.5° = 180°$

$\Rightarrow \angle R = 180° - 71° = \mathbf{109°}$.

11. (b) In quad. *ABCD*

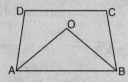

$\angle A + \angle B + \angle C + \angle D = 360°$

$\Rightarrow \angle A + \angle B = 360° - (\angle C + \angle D)$ ...(i)

In ΔAOB,

$\angle AOB + \angle OAB + \angle OBA = 180°$

$\Rightarrow \angle AOB = 180° - (\angle OAB + \angle OBA)$

$= 180° - \frac{1}{2}(\angle A + \angle B)$

$(\because OA$ bisects $\angle A$ and OB bisects $\angle B)$

$\therefore$ From (i),

$\angle AOB = 180° - \frac{1}{2} \{360° - (\angle C + \angle D)\}$

$= 180° - 180° + \frac{1}{2}(\angle C + \angle D)$

$= \frac{1}{2}(\angle C + \angle D)$

12. (d) In ΔAOB,

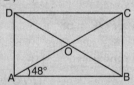

$BO = OA$ (Diagonals of a rectangle are equal and bisects each other)

$\Rightarrow \angle OAB = \angle OBA = 48°$ (isos. Δ property)

$\Rightarrow \angle DBA = \angle OBA = 48°$

$\Rightarrow \angle DBC = \angle CBA - \angle DBA$

$= 90° - 48° = 42°$

13. (c) $\angle DAB = 180° - \angle ADC = 180° - 64° = 116°$

($AB \parallel DC$, co-int. $\angle s$ are supp.)

In ΔDAB, $DA = AB \Rightarrow \angle ABD = \angle ADB$

(isos. Δ prop.)

$\therefore \angle ADB = \frac{180° - 116°}{2} = \frac{64°}{2} = 32°$

$\therefore \angle BDC = 64° - 32° = 32°$

Hence, in ΔDBC,

$\angle DBC = 180° - (\angle BDC + \angle BCD)$

$= 180° - (32° + 54°) = 180° - 86° = \mathbf{94°}$.

14. (c) Since the diagonal of a square bisects the angle at the vertices, $\angle DBC = 45°$.

$\therefore \angle ABQ = 90° - (21° + 45°)$

$= 90° - 66° = 24°$

In ΔBAQ, $BA = BQ$

$\Rightarrow \angle BQA = \angle BAQ$ (isos. Δ property)

$\therefore \angle BAQ = \frac{180° - 24°}{2} = \frac{156°}{2} = \mathbf{78°}$.

15. (d) In ΔBAD,

$\angle BDA = \angle BAD = 57°$ (isos. Δ property)

In ΔBDC,

$\angle BCD = \angle BDC = 66°$ (isos. Δ property)

$\therefore \angle D = 57° + 66° = 123°$ and

$\angle A + \angle D = 57° + 123° = 180°$

Also, $\angle D + C = 123° + 66° = 189°$

Hence, by the property that co-int. angles are supplementary $\Rightarrow$ lines are parallel, we have $AB \parallel DC$ and AD not $\parallel BC$.

Hence $ABCD$ is a trapezium.

16. (c) ΔAQD is an equilateral Δ

$\Rightarrow \angle QAD = \angle QDA = \angle AQD = 60°$

$\angle BAD = 360° - (135° + 90° + 60°)$

$= 360° - 285° = 75°$ (Angles round a pt.)

Also, in quad. $ABCD$,

$\angle CDA = 360° - (100° + 106° + 75°)$

$= 360° - 281° = 79°$.

$\therefore \angle PDQ = 180° - (\angle CDA + \angle QDA)$

(*CDP* is a st. line)

$= 180° - (79° + 60°) = 180° - 139° = \mathbf{41°}$.

17. (d) We have, $\angle SPQ = 60°$

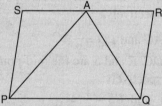

$\angle P + \angle Q = 180°$ (adj. $\angle s$ of a $\parallel$gm are supp.)

$\Rightarrow \angle Q = 180° - 60° = 120°$

Now $PQ \parallel SR$ and AP is the transversal.

$\therefore \angle SAP = \angle APQ = 30°$ $(\because AP$ bisects $\angle SPQ)$

$\angle SPA = \angle SAP \Rightarrow SA = SP$ (isos. Δ property) ...(i)

Also, $\angle RAQ = \angle AQP = 60°$

($PQ \parallel SR$, AQ is transversal, alt. $\angle s$

and $\angle AQP = \frac{1}{2} \angle PQR = 60°$)

$\Rightarrow \angle RQA = \angle RAQ$ (AQ bisects $\angle PQR$)

$\Rightarrow RA = RQ$ (isos. Δ property) ...(ii)

∴ From eqn (*i*) and (*ii*)
AS = AR   (∵ SP = RQ, opp. sides of a ‖gm)
Also, in ΔARQ, ∠ARQ = 60°
⇒ ΔARQ is equilateral
⇒ AR = RQ ⇒ AR = SP.
∴ The incorrect statement is AQ = PQ.

18. (c) ∠DCB = 90° (ABCD is a square)
    ∠TCB = 60° (DCT is an equilateral Δ)
∴ ∠DCT = 90° + 60° = 150°
    DC = CB (Adj sides of a square)
    CB = CT (Sides of an equilateral Δ)
⇒ DC = CT ⇒ ∠CTD = ∠CDT (isos. Δ property)

In ΔDCT, ∠CTD = $\frac{1}{2}$ (180° − ∠DCT)
= $\frac{1}{2}$ (180° − 150°) = 15°
∴ ∠BTD = ∠BTC − ∠CTD = 60° − 15° = **45°**.

19. (d) ∠DFE = ∠AFG = b   (vert. opp. ∠s)
    ∠BEC = ∠DFE = b   (base ∠s of a isos. Δ)
    ∠FDC = b + b = 2b   ...(*i*)
    (ext ∠ = sum of int. opp. ∠s)
    Also, ∠FDC = 180° − ∠BCD
       (∵ AD‖BC, co-int. ∠s)
       = 180° − c   ...(*ii*)
∴ From (*i*) and (*ii*), 2b = 180° − c

⇒ $b = \frac{180° - c}{2} = 90° - \frac{c}{2}$

20. (b) $\angle a = \frac{(5-2) \times 180°}{5} = 108°$

    (Int. ∠ of a pentagon)

$\angle b = \frac{180° - 108°}{2} = 36° (\because \Delta ABC \text{ is isos.})$

$\angle c = \frac{(6-2) \times 180°}{6} = 120°$

    (Int. ∠ of a hexagon)

∠e = 180° − 120° = 60° (∵ SR ‖ QC, co-int. ∠s)
∴ ∠x = ∠y − ∠b − ∠e
   = ∠BCD + ∠DCS − ∠b − ∠e
   = 108° + 120° − 36° − 60°
   = 132°

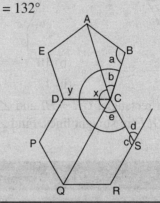

Self Assessment Sheet–23

1. Given a trapezium *ABCD* in which *AB‖CD* and *AD=BC*. If ∠C = 76°, then ∠D equals
   (a) 14°   (b) 104°
   (c) 76°   (b) None of these

2. The figure formed by joining the mid-points of the sides of a quadrilateral *ABCD* taken in order is a square only, if
   (a) *ABCD* is a rhombus.
   (b) diagonals of *ABCD* are equal.
   (c) diagonals of *ABCD* are equal and perpendicular.
   (d) diagonals of *ABCD* are perpendicular.

3. A diagonal of a rectangle is inclined to one side of a rectangle at 34°. The acute angle between the diagonals is :
   (a) 34°   (b) 56°
   (c) 68°   (d) 42°

4. In the given kite, *y* equals

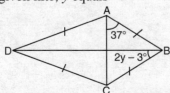

   (a) 37°   (b) 28°
   (c) 20°   (b) 40°

5. In the given figure, *ABCD* is a parallelogram, *AD= AE*, *BAE* is a straight line, ∠ABC= 88°, ∠EAF = 162° and ∠AFC = 48°. Then ∠BCF equals.

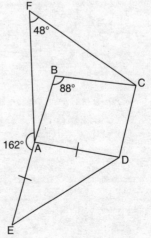

   (a) 18°   (b) 28°
   (c) 22°   (d) 32°

6. Diagonals of a quadrilateral *PQRS* bisect each other at right angles. If *PQ* = 5.5 cm, then the perimeter of *PQRS* is

(a) 11 cm

(b) 22 cm

(c) 16.5 cm

(d) 5.5 (1+ $\sqrt{2}$) cm

7. *ABCDE* is a regular pentagon with sides of length 6 cm. *CD* is also a side of a regular polygon with *n* sides. Given that ∠*EDF* = 90°, find *n*.

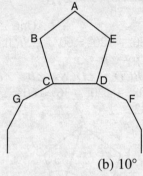

(a) 18°

(b) 10°

(c) 20°

(b) 12°

8. *GHJK* is a rectangle. *GH* = *HI* and ∠*HKJ* = 50°. *HLK* and *GLI* are straight lines. Find ∠*GLK*.

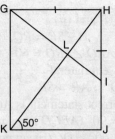

(a) 100°

(b) 130°

(c) 95°

(b) 135°

9. *ABCD* is a rhombus in which altitude from point *D* to the side *AB* bisects *AB*. The angles of the rhombus are:

(a) 150°, 30°, 150°, 30°

(b) 135°, 45°, 135°, 45°

(c) $116\frac{1°}{3}, 63\frac{2°}{3}, 116\frac{1°}{3}, 62\frac{2°}{3}$

(d) 120°, 60°, 120°, 60°

10. Given an equilateral triangle *ABC*, *D*, *E* and *F* are the mid-points of *AB*, *BC* and *AC* respectively, then the quadrilateral *BEFD* is exactly a

(a) square

(b) rectangle

(c) parallelogram

(d) rhombus

Answers

1. (c) **2.** (c) **3.** (c) **4.** (b) **5.** (c) **6.** (b)

7. (c) [**Hint.** Find the measure of the interior angle of the polygon whose sides are unknown, *i.e*, ∠*CDF* and hence the sides.] **8.** (c) **9.** (d) **10.** (d)

CIRCLES

1. A circle is a collection of all such points in a plane which are equidistant from a fixed point. This fixed point is called the **centre** of the circle while the distance of any point on the circle from the centre is called the **radius** of the circle.

2. A line segment joining any two points on a circle is called the **chord** of the circle.

3. The chord passing through the centre of the circle is called the **diameter.** The diameter divides a circle into two equal parts called **semicircles.**

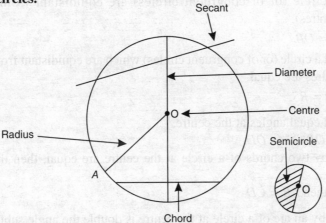

4. A line which intersects a circle in two distinct points is called a **secant** of the circle.

5. (i) A continuous piece of a circle is called an **arc** of the circle.

   • An arc whose length is less than the semicircular arc is called **minor arc** and if greater than semicircular arc is called **major arc.**

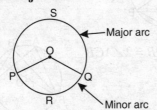   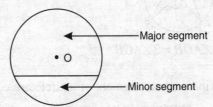

(ii) A chord *AB* of a circle divides the circular region into two parts called **segments** of the circle.

   • The bigger part containing the centre of the circle is called the **major segment** and the smaller part which does not contain the centre is called the **minor segment** of the circle.

6. A line touching a circle at a point is called the **tangent** to the circle.

7. Circles having the same centre are called **concentric circles.**

8. Two circles are said to congruent if and only if they have the same radii.

9. The degree measure of an arc *AB* is denoted by *m* (*AB*). Two arcs of a circle (or of congruent circles) are said to be congruent if and only if they have the same degree measure.

10. **Important Properties of Circles**

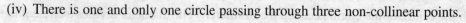

(i) If two arcs of a circle (or of congruent circles) are congruent, then the corresponding chords are equal.

Conversely, if two chords of a circle (or of congruent circles) are equal, then their corresponding arcs are equal.

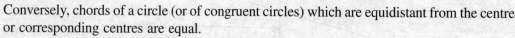

(ii) The angle in a semi-circle is a right angle.

$\angle ACB = 90°$

(iii) The perpendicular from the centre to a chord bisects the chord.

If $OP \perp AB$, then $AP = PB$.

Conversely, the line joining the centre of the circle and the mid-point of the chord is perpendicular to the chord.

If *P* is the mid-point of *AB*, then $OP \perp AB$.

(iv) There is one and only one circle passing through three non-collinear points.

(v) Equal chords of a circle (or of congruent circles) are equidistant from the centre (or corresponding centres).

If *PQ = RS*, then *OA= OB*.

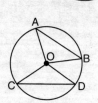

Conversely, chords of a circle (or of congruent circles) which are equidistant from the centre or corresponding centres are equal.

If *OA = OB*, then *PQ = RS*.

(vi) Equal chords subtend equal angles at the centre.

If *AB = CD*, then $\angle AOB = \angle COD$.

If angles subtended by two chords of a circle at the centre are equal, then the two chords are equal.

If $\angle AOB = \angle COD$, then *AB = CD*

(vii) The angle subtended by an arc of a circle at the centre is double the angle subtended by it at any point on the remaining part of the circle.

$\angle AOB = 2\angle ACB$

Reflex $\angle AOB = 2\angle ACB$

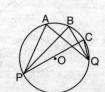

(viii) Angles in the same segment of a circle are equal.

$\angle PAQ = \angle PBQ = \angle PCQ$

(ix) If all the vertices of a quadrilateral lie on a circle, it is called a **cyclic quadrilateral.**

ABCD is a cyclic quadrilateral.

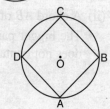

(x) The opposite angles of a cyclic quadrilateral are supplementary.
$\angle A + \angle C = 180°$ and $\angle B + \angle D = 180°$

(xi) If the sum of any pair of opposite angles of a quadrilateral is 180°, then it is cyclic.

(xii) If one side of a cyclic quadrilateral is produced, then the exterior angle is equal
to the interior opposite angle.
$\angle BDE = \angle CAB$

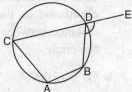

(xiii) The quadrilateral formed by the angle bisectors of a cyclic quadrilateral is also cyclic.

(xiv) If two opposite sides of a cyclic quadrilateral are equal, then the other two sides are parallel.

(xv) An isosceles trapezium is cyclic.

(xvi) Tangent to a circle at a point is perpendicular to the radius through the point of contact.
$OR \perp l$

(xvii) From a point lying outside a circle, two and only two tangents can be drawn to it.

(xviii) The lengths of the two tangents drawn from an external point to a circle are equal
$PA = PB$

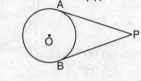

(xix) The angle between two tangents drawn from an external point to a circle is supplementary to the angle subtended by the line segments joining the points of contact at the centre.
$\angle AOB + \angle APB = 180°$

(xx) The tangents are equally inclined to the line joining the external point (from where the tangents are drawn) and the centre of the circle.
$\angle APO = \angle BPO$ and $\angle AOB = \angle BOP$

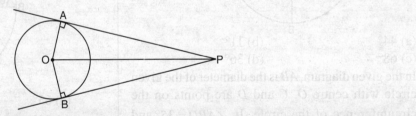

Question Bank–24

1. O is the centre of a circle with radius 5 cm. LM is the diameter of the circle. P is a point on the plane of the circle such that $LP = 6$ cm and $MP = 8$ cm. Then P lies.

(a) on LM
(b) outside the circle
(c) inside the circle
(d) on the circle.

2. If the length of a chord of a circle is equal to its radius, then the angle subtended by it at the minor arc of the circle will be,

(a) 60° (b) 75°
(c) 120° (d) 150°

3. Given a circle with centre O. The smallest chord PQ is of length 4 cm largest chord AB is of length 10 cm and chord EF is of length 7 cm. Then, the radius of the circle is

(a) 3 cm (b) 2 cm
(c) 5 cm (d) 3.5 cm

4. The radius of a circle is 6 cm. The perpendicular distance from the centre of the circle to the chord which is 8 cm in length is

(a) $\sqrt{5}$ cm (b) $2\sqrt{5}$ cm
(c) $2\sqrt{7}$ cm (d) $\sqrt{7}$ cm

5. *PQ* and *RS* are two parallel chords of a circle with centre *C* such that *PQ* = 8 cm and *RS* = 16 cm. If the chords are on the same side of the centre and the distance between them is 4 cm, then the radius of the circle is :

(a) $3\sqrt{2}$ cm   (b) $3\sqrt{5}$ cm

(c) $4\sqrt{5}$ cm   (d) $5\sqrt{5}$ cm

6. In the given figure, *O* is the centre of the circle. The measure of ∠*ADB* is

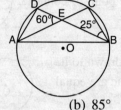

(a) 90°   (b) 85°

(c) 95°   (d) 120°

7. Given that *AOB* is a straight line and *O* is the centre of the circle. Find the value of *y*.

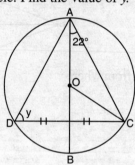

(a) 44°   (b) 11°

(c) 68°   (d) 36°

8. In the given diagram, *AB* is the diameter of the given circle with centre *O*. *C* and *D* are points on the circumference of the circle. If ∠*ABD* = 35° and ∠*CDB* = 15°, then ∠*CBD* equals.

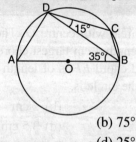

(a) 55°   (b) 75°

(c) 40°   (d) 25°

9. In the diagram, *A, B, C, D, E* are points on the circle. *AD* ∥ *BC*, ∠*ABE* = 39° and ∠*ADC* = 62°. Then the values of *x* and *y* respectively are:

(a) 23°, 51°   (b) 79°, 62°

(c) 62°, 79°   (d) 51°, 23°

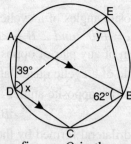

10. In the given figure, *O* is the centre of the circle, ∠*ACB* = 54° and *BCE* is a straight line. Find *x*.

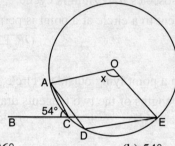

(a) 126°   (b) 54°

(c) 108°   (d) 90°

11. In the given figure, *BOD* is the diameter of the circle with centre *O*. ∠*COD* = 92° and ∠*ABD* = 65°. Then *y* equals

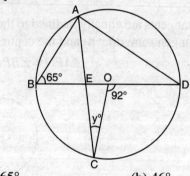

(a) 65°   (b) 46°

(c) 44°   (d) 21°

12. *O* is the centre of the circle *x* and *y* respectively equal.

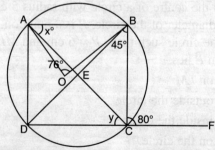

(a) 38°, 45°   (b) 35°, 62°

(c) 62°, 35°   (d) 46°, 38°

13. In the given figure, *AB* and *AC* are tangents to the circle with centre *O*. Given that ∠*BAC* = 70° and *P* is a point on the minor arc *BC*, find ∠*BPC*.

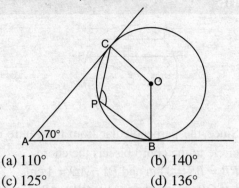

(a) 110°   (b) 140°

(c) 125°   (d) 136°

14. The length of a tangent drawn from a point 10 cm away from the centre of the circle of radius 5 cm is

(a) 5 cm   (b) $5\sqrt{3}$ cm

(c) $2\sqrt{3}$ cm   (d) $\sqrt{15}$ cm

15. In the figure shown here, a circle touches the side *BC* of a triangle *ABC* at *P* and *AB* and *AC* produced at *Q* and *R* respectively. What is *AQ* equal to?

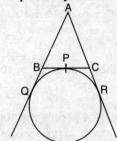

(a) One-third of the perimeter of Δ*ABC*.
(b) Half of the perimeter of Δ*ABC*.
(c) Two-third of the perimeter of Δ *ABC*.
(d) Three-fourth of the perimeter of Δ*ABC*.

16. In the given figure, *AB* and *AC* are tangents to the circle at *B* and *C* respectively and *O* is the centre of the circle, then *x* equals

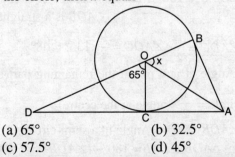

(a) 65°   (b) 32.5°
(c) 57.5°   (d) 45°

17. *ABC* is an isosceles triangle (*AB* = *AC*) circumscribed about a circle. Then, which of the following statements is correct?

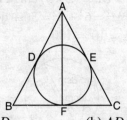

(a) *BD* = *AD*   (b) *AD* = *CF*
(c) *BF* = *CF*   (d) *AE* = *BF*

18. In the figure, *CDE* is a straight line and *A*, *B*, *C* and *D* are points on the circle. ∠*BCD* = 44°, find the value of *x*.

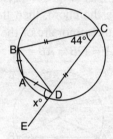

(a) 44°   (b) 68°
(c) 90°   (d) 56°

19. From a point *P* which is at a distance of 13 cm from the centre *O* of a circle of radius 5 cm, the pair of tangents *PQ* and *PR* to the circle are drawn. Then the area of the quadrilateral *PQOR* is :

(a) 60 cm²   (b) 65 cm²
(c) 30 cm²   (d) 32.5 cm²

20. In the given figure, ∠*AFD* = 25°.
∴ ∠*EFC* equals

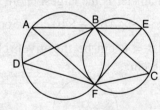

(a) 65°   (b) 155°
(c) 90°   (d) 25°

Answers

1. (d)	2. (d)	3. (c)	4. (b)	5. (c)	6. (c)	7. (c)	8. (c)	9. (b)	10. (c)
11. (d)	12. (b)	13. (c)	14. (b)	15. (b)	16. (c)	17. (c)	18. (c)	19. (a)	20. (d)

Hints and Solutions

1. (d) $LM = 2 \times OL = (2 \times 5)$ cm $= 10$ cm
$LP^2 + PM^2 = 6^2 + 8^2 = 36 + 64 = 100$
and $LM^2 = 10^2 = 100$

$\Rightarrow LP^2 + PM^2 = LM^2 \Rightarrow \Delta LPM$ is right angled
triangle, rt. $\angle d$ at $P \Rightarrow \angle LPM = 90°$
$\Rightarrow P$ lies on the circumference of the circle
($\because$ Angle in a semi-circle is a rt. $\angle$)

2. (d) Let AB be the chord, then the angle subtended
by AB at the minor arc of the circle is $\angle ACB$.
Given, $OA = OB = AB$

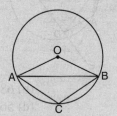

$\Rightarrow \Delta OAB$ is equilateral
$\Rightarrow \angle AOB = 60°$
$\Rightarrow$ Reflex $\angle AOB = 360° - 60° = 300°$
$\therefore \angle ACB = \dfrac{1}{2} \times$ Reflex $\angle AOB = \dfrac{1}{2} \times 300° = 150°$
(Angle at the centre $= 2 \times$ angle at any pt. on
remaining part of the circle)

3. (c) The diameter is the largest chord of a circle, so
radius $= \dfrac{1}{2} \times AB = \dfrac{1}{2} \times 10$ cm $= 5$ cm.

4. (b) Let $AB = 8$ cm be the given chord. Radius
$OA = 6$ cm.

Since the perpendicular from the centre of the
circle to the chord bisects the chord,
$AP = PB = 4$ cm.
In rt. $\angle d \Delta OAP$, $OP^2 = OA^2 - AP^2$
$= 36 - 16 = 20$
$\Rightarrow OP = \sqrt{20} = 2\sqrt{5}$ cm

5. (c) Given, chords $PQ = 8$ cm and $RS = 16$ cm and
$AB = 4$ cm

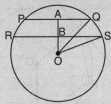

Since, the perpendicular from the centre of the
circle to the chord bisects the chord,
$RB = BS = 8$ cm and $PA = AQ = 4$ cm.
Let $OB = x$ cm, then $OA = OB + AB = (x + 4)$ cm.
In right triangle OBS,
$OS^2 = OB^2 + BS^2 = OB^2 + 64$ ...(i)
In right triangle OAQ,
$OQ^2 = OA^2 + AQ^2 = OA^2 + 16$ ... (ii)
$\therefore$ OS and OQ are the radii of the circle
$\therefore OB^2 + 64 = OA^2 + 16$
$\Rightarrow x^2 + 64 = (x + 4)^2 + 16$
$\Rightarrow x^2 + 64 = x^2 + 8x + 16 + 16$
$\Rightarrow 8x = 32 \Rightarrow x = 4$
$\therefore$ From eqn (i)
Radius of the circle $(OS)^2 = 16 + 64 = 80$
$\Rightarrow OS = 4\sqrt{5}$ cm

6. (c) $\angle CEB = \angle DEA = 60°$ (vert. opp. $\angle s$)
$\therefore$ In ΔCEB, $\angle ECB = 180° - (60° + 25°)$
$= 180° - 85° = 95°$ (Angle sum prop. of a Δ)
$\therefore \angle ADB = \angle BCA (\angle BCE) = \mathbf{95°}$
(Angles in the same segment)

7. (c) $\angle BOC = 2 \times \angle BAC = 2 \times 22° = 44°$
(Angle at the centre $= 2 \times$ angle at any point on
the remaining part of the circle)
$\angle AOC = 180° - \angle BOC = 180° - 44 = 136°$
($\because AOB$ is a straight line)
$\Rightarrow \angle ADC = \dfrac{1}{2} \times \angle AOC = \dfrac{1}{2} \times 136° = 68°$
(Angle at any pt. on the remaining part of the
circle $= \dfrac{1}{2} \times$ angle at the centre)

8. (c) $\angle ADB = 90°$ (Angle in a semi-circle)
In ΔADB, $\angle DAB = 180° - (\angle ADB + \angle DBA)$
$= 180° - (90° + 35°) = 180° - 125° = 55°$
(Angle sum prop. of a Δ)
In cyclic quad $ABCD$,
$\angle A + \angle C = 180° \Rightarrow \angle C = 180° - 55° = 125°$
(Opp. $\angle s$ of a cyclic quad. are supp.)
$\therefore$ In ΔDCB, $\angle CBD = 180° - (15° + 125°) = 40°$

9. (b) Since opp. $\angle s$ of a cyclic quad. are supplementary. $\angle ADC + \angle ABC = 180°$

$\Rightarrow 39° + x + 62° = 180°$

$\Rightarrow x = 180° - 101° = \mathbf{79°}$

Also, $AB \parallel DC \Rightarrow \angle DAB + \angle ADC = 180°$

(co-int $\angle s$)

$\Rightarrow \angle DAB + 39° + 79° = 180°$

$\Rightarrow \angle DAB + 118° = 180°$

$\Rightarrow \angle DAB = 62°$

$\therefore y = \angle DEB = \angle DAB = 62°$

(Angles in the same segment are equal)

10. (c) $\angle ACE = 180° - 54° = 126°$

(BCE is a straight line)

$\Rightarrow \angle ADE = \angle ACE = 126°$

(Angles in the same segement are equal)

$\therefore$ Reflex $\angle AOE = 2 \times \angle ADE = 2 \times 126° = 252°$

(Angle at the centre = 2 × Angle at any other pt. on remaining part of the circle)

$\therefore x = 360° - 252° = \mathbf{108°}.$

11. (d) $\angle BAD = 90°$ (Angle in a semi-circle)

$\angle CAD = \dfrac{1}{2} \times \angle COD = \dfrac{1}{2} \times 92° = 46°$

(Angle at the centre = 2× angle at any other point on the remaining part of the circle)

$\therefore \angle BAC = \angle BAD - \angle CAD = 90° - 46° = 44°$

In $\triangle ABE$, $\angle AEB = 180° - (65° + 44°)$

$= 180° - 109° = 71°$

$\Rightarrow \angle CEO = \angle AEB = 71°$ (Vert. opp. $\angle s$)

Also, $\angle COE = 180° - 92° = 88°$

(BOD is a st. line)

$\therefore$ In $\triangle CEO$,

$y = \angle ECO = 180° - (88° + 71°)$

$= 180° - 159° = \mathbf{21°}.$

12. (b) $\angle DAC = \angle DBC = 45°$

(Angles in same segment)

$\angle DAB = 80°$

(Ext. $\angle$ of a cyclic quad = int. opp. $\angle$)

$\therefore x = \angle DAB - \angle DAC = 80° - 45° = 35°$

Now, $\angle ADB = \dfrac{1}{2} \times \angle AOB = \dfrac{1}{2} \times 76° = 38°$

$\angle ACB = \angle ADB = 38°$

(Angles in the same segment)

$\therefore y = 180° - (38° + 80°) = 180° - 118° = 62°$

($\because DLF$ is a st. line)

13. (c) $\angle AOB + \angle APB = 180°$

(The angle between two tangents drawn from an external pt. to a circle is supplementary to the

angle subtended by the line segments joining the points of contact at the centre)

$\Rightarrow \angle AOB = 180° - 70° = 110°$

Reflex $\angle AOB = 360° - 110° = 250°$

$\therefore \angle BPC = \dfrac{1}{2} \times$ Reflex $\angle AOB = \dfrac{1}{2} \times 250°$

$= \mathbf{125°}$

14. (b) Given, $OP = 10$ cm and $OA = 5$ cm

Since, the tangent to a circle at a point is perpendicular to the radius through the point of contact

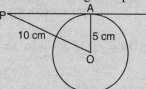

$OA \perp PA \Rightarrow \angle OAP = 90°$

$\therefore$ In $\triangle OAP$, $PA^2 = OP^2 - OA^2$

$\Rightarrow PA^2 = 100 - 25 = 75$

$\Rightarrow PA = \sqrt{75} = 5\sqrt{3}$ cm.

15. (b) Here, $BQ = BP$ ...(i)

$PC = CR$ ...(ii)

$AQ = AR$ ...(iii)

(Lengths of the two tangents drawn from an external point to a circle are equal)

From (iii) we have

$\therefore AB + BQ = AC + CR$

$\Rightarrow AB + BP = AC + CP$ (using (i) and (ii) .... (iv)

Now perimeter of $\triangle ABC = AB + BC + AC$

$= AB + (BP + PC) + AC$

$= (AB + BP) + (PC + AC)$

Perimeter of $\triangle ABC = 2(AB + BP)$ using (iv)

Perimeter of $\triangle ABC = 2(AB + BQ)$

$= 2AQ$

$\therefore AQ = \dfrac{1}{2}$ (Perimeter of $\triangle ABC$)

16. (c) $OC \perp DA$ $\therefore \angle OCD = 90°$

In $\triangle OCD$, $\angle ODC = 180° - (90° + 65°)$

$= 180° - 155° = \mathbf{25°}$

Also $OB \perp AB$, $\therefore \angle OBA = 90°$

In $\triangle DBA$, $\angle DAB = 180° - (\angle BDA + \angle DBA)$

$= 180° - (25° + 90°)$

$= 180° - 115° = 65°$

$\therefore \angle OAB = \dfrac{1}{2} \times \angle CAB$ (or $\angle DAB$) $= \dfrac{1}{2} \times 65° = 32.5°$

($\therefore$ Tangents are equally inclined to the line joining the external pt. and the centre of the circle)

In $\triangle OBA$, $x = 180° - (90° + 32.5°)$

$= 180° - 122.5° = \mathbf{57.5°}$

17. (c) $AB = AC$
$\Rightarrow AD + DB = AE + EC$
$\Rightarrow DB = EC$ ($\because AD = AE$, tangents to a circle from the same external pt. are equal)
$\Rightarrow BF = FC$ ($BD = BF$ and $CE = CF$)

18. (c) $\angle CDB = \dfrac{1}{2}(180° - 44°) = \dfrac{1}{2} \times 136° = 68°$
($\because BCD$ is an isos. Δ)
$\angle BAD = 180° - 44° = 136°$
(opp. $\angle s$ of a cyclic quad. are supp.)
$\therefore \angle ADB = \dfrac{1}{2}(180° - 136°) = \dfrac{1}{2} \times 44° = 22°$
($\because BAD$ is an isos. Δ)
$\therefore \angle ADC = \angle ADB + \angle BDC = 22° + 68° = 90°$
$\Rightarrow x = \angle ADE = 180° - \angle ADC = 180° - 90° = 90°$
($\because EDC$ is a st. line)

19. (a) The radius $\perp$ tangent at the pt. of contact, therefore, $OQ \perp PQ$ and $OR \perp PR$

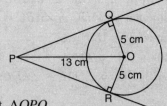

$\therefore$ In rt. ΔOPQ,
$$PQ = \sqrt{OP^2 - OQ^2}$$
$$= \sqrt{169 - 25} = \sqrt{144} = 12 \text{ cm}$$
$\Rightarrow PR = 12$ cm (Two tangents from the same external pt. to a circle are equal)
Now area of quad. $PQOR = 2 \times$ Area of ΔPOQ
$$= \left(2 \times \dfrac{1}{2} \times 12 \times 5\right) \text{cm}^2 = \mathbf{60 cm^2}.$$

20. (d) $\angle ABD = \angle AFD = 25°$
(Angles in the same segment)
$\angle EBC = \angle ABD = 25°$ (Vert. opp. $\angle s$)
$\angle EFC = \angle EBC = 25°$
(Angles in the same segment)

Self Assessment Sheet–24

1. O is the centre of a circle. There is a point P in the region of the circle. If $PA = PB = PC$ where A, B and C are points on the circumference of the circle, then OP must be equal to:
(a) $\dfrac{PA + PB + PC}{3}$
(b) $\dfrac{PA + PB + PC}{2}$
(c) $\dfrac{AB + BC}{2}$
(d) zero

2. In the given figure, O is the centre of the circle. Given that $OD = OE = 3$ cm and $AD = 4$ cm. Find the length of the longest chord.

(a) 6 cm
(b) 8 cm
(c) 10 cm
(d) 9 cm

3. AOD is a diameter of the circle with centre O. Given that $\angle BDA = 18°$ and $\angle BDC = 38°$. $\angle BCD$ equals

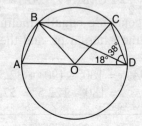

(a) 90°
(b) 108°
(c) 76°
(b) 52°

4. Tangents drawn at the end points of a diameter are
(a) Perpendicular
(b) Parallel
(c) Intersecting
(b) None of these

5. In the given figure, AB is a chord of the circle with centre O and PQ is a tangent at point B of the circle. If $\angle AOB = 110°$, then $\angle ABQ$ is

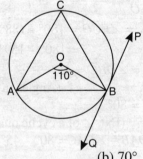

(a) 45°
(b) 70°
(c) 55°
(b) 35°

6. In the given figure, if PA and PB are tangents to the circle with centre O such that $\angle APB = 54°$, then $\angle OAB$ equals
(a) 36°
(b) 18°
(c) 27°
(d) 36°

7. If two tangents inclined at an angle of 60° are drawn to a circle of radius 4 cm, then the length of each tangent is equal to:

(a) $2\sqrt{3}$ cm   (b) 8 cm

(c) 4 cm   (d) $4\sqrt{3}$ cm

8. In the given figure, *RST* is the tangent to the circle with centre *O*, at *S*. *AOS* is a straight line *BO* ∥ *RT* and ∠*ORS* = 46°. Then ∠*BAC* equals

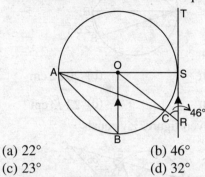

(a) 22°   (b) 46°

(c) 23°   (d) 32°

9. In the diagram, *CB* and *CD* are tangents to the circle with centre *O*. *AOC* is a straight line and ∠*OCB* = 34°. ∠*ABO* equals.

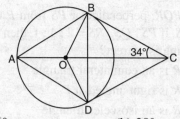

(a) 56°   (b) 28°
(c) 34°   (d) 32°

10. *ED* is the tangent to the circle with centre *O*. ∠*BCD* = 52°. Then, ∠*CAB* equals

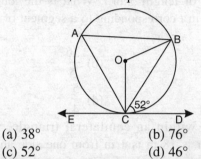

(a) 38°   (b) 76°
(c) 52°   (d) 46°

Answers

1. (d)	2. (c)	3. (b)	4. (b)	5. (c)	6. (c)	7. (d) (Use trigonometrical ratios)
8. (c)	9. (b)	10. (c)				

Unit Test–4

1. *ABC* is a triangle whose altitudes *BE* and *CF* to sides *AC* and *AB* respectively are equal. Which of these conditions is not required to prove Δ*ABE* ≅ Δ*ACF*?
(a) ∠*B* = ∠*C*   (b) ∠*BAE* =∠*FAC*
(c) ∠*AFC* =∠*AEB*   (d) *BE* = *CF*

2. *ABCD* is a square and Δ*DEC* is an equilateral triangle. Δ*ADE* ≅ Δ*BCE* by

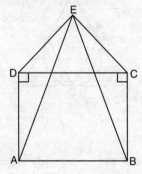

(a) RHS   (b) SSS
(c) AAS   (d) SAS

3. The centroid and the orthocentre are coincident for which one of the following triangles?
(a) Scalene triangle   (b) Isosceles triangle
(c) Equilateral triangle   (d) Right angled triangle

4. *ABC* is an isosceles triangle right angled at *B*. Similar triangles *ACD* and *ABE* are constructed on sides *AC* and *AB*. The ratio between the areas of Δ*ABE* and Δ*ACD* is

(a) $\sqrt{2}$:1   (b) 1 : 2

(c) 2 : 1   (d) $\sqrt{2}$:1

5. In Δ*ABC* and Δ*DEF*, it is given that *AB* = 5 cm, *BC*= 4 cm, *CA* = 4.2 cm, *DE* = 10 cm, *EF* = 8 cm and *FD* = 8.4 cm. If *AL* is perpendicular to *BC* and *DM* is perpendicular to *EF*, then what is the ratio of *AL* to *DM*.

(a) $\dfrac{1}{2}$ (b) $\dfrac{1}{3}$

(c) $\dfrac{1}{4}$ (d) 1

6. In Δ*PQR*, *PQ* = 4 cm, *QR* = 3cm, and *RP* = 3.5 cm. Δ*DEF* is similar to Δ*PQR*. If *EF* = 9 cm, then what is the perimeter of Δ*DEF*?
(a) 10.5 cm
(b) 21 cm
(c) 31.5 cm
(d) Cannot be determined as data is insufficient

7. In a ΔPQR, perpendicular PS from P to QR meets QR at S. If $PS : QS : RS = 2 : 4 : 1$, then which of the following is correct?
   (a) PQR is an equilateral triangle
   (b) PQR is right angled at P
   (c) PQR is an isosceles triangle
   (d) $PQ = 3PR$

8. *s* at *t* are transversals cutting a set of parallel lines such that a segment of length 3 in *s* corresponds to a segment of length 5 in *t*. What is the length of segment in *t* corresponding to a segment of length 12 in *s*?
   (a) 20
   (b) $\dfrac{36}{5}$
   (c) 14
   (d) $\dfrac{5}{4}$

9. A point within an equilateral triangle, where perimeter is 18 m is 1 m from one side and 2 m from another side. Its distance from the third side is:
   (a) $3\sqrt{3}+3$
   (b) $3\sqrt{3}-3$
   (c) $3-\sqrt{3}$
   (d) $3+\sqrt{3}$

10. The perimeter of two similar triangles are 24 cm and 16 cm respectively. If one side of the first triangle is 10 cm, then the corresponding side of the second triangle is
    (a) 9 cm
    (b) 20/3 cm
    (c) 16/3 cm
    (d) 5 cm

11. In a circle of radius 10 cm, a chord is drawn 6 cm from the centre. If a chord half the length of the original chord were drawn, its distance in centimeters from the centre would be
    (a) $\sqrt{84}$
    (b) 9
    (c) 8
    (d) 3π

12. The number of tangents that can be drawn to two non-intersecting circles is
    (a) 4
    (b) 3
    (c) 2
    (d) 1

13. $ABCD$ is a parallelogram. A circle passes through A and D and cuts AB at E and DC at F. Given $\angle BEF = 80°$, find $\angle ABC$.

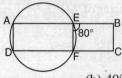

    (a) 100°
    (b) 40°
    (c) 80°
    (d) 104°

14. In the given figure, PR is the diameter of the circle. $PQ = 7$ cm, $QR = 6$ cm and $RS = 2$ cm. The perimeter of the cyclic quadrilateral $PQRS$ is

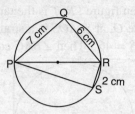

    (a) 18 cm
    (b) $20\sqrt{2}$ cm
    (c) 24 cm
    (d) $22\sqrt{3}$ cm

15. In the given figure, $\angle y$ equals

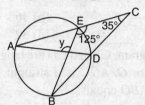

    (a) 75°
    (b) 145°
    (c) 90°
    (d) 105°

16. TP and TQ are tangents from T to the circle with centre O. Then is it possible to draw a circle through the points P, O, Q and T ?

    (a) No
    (b) Yes
    (c) Cannot say
    (d) Data insufficient

17. BC, AB and AC are tangents to the circle at D, E and F respectively. $\angle EBD = x°$, $\angle FCD = y°$. Then $\angle EDF$ equals

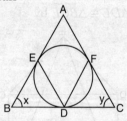

    (a) $x+y$
    (b) $\dfrac{x}{2}-y$
    (c) $90°-(x+y)$
    (d) $\dfrac{x+y}{2}$

18. Find $\angle y$.

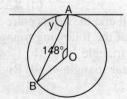

(a) 32°   (b) 72°

(c) 64°   (d) 44°

19. *TP* and *TQ* are the tangents to a circle, with centre *O*. Find *x*.

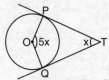

(a) 15°   (b) 60°

(c) 30°   (d) 45°

20. *AB* and *AC* are tangents to the circle with centre *O*. Then *x* equals.

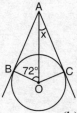

(a) 22°   (b) 18°

(c) 20°   (d) 36°

21. Diagonals of a quadrilateral bisect each other. If $\angle A = 45°$, then $\angle B$ equals

(a) 45°   (b) 55°

(c) 135°   (d) 115°

22. If *APB* and *CQD* are two parallel lines, then the bisectors of the angles *APQ, BPQ, CQP* and *PQD* form:

(a) square   (b) a rhombus

(c) a rectangle   (d) kite

23. The interior angle of a regular polygon with *n* sides is 6 times that of an exterior angle of a regular

polygon with $\frac{3}{2}n$ sides. Then *n* equals

(a) 12   (b) 20

(c) 10   (d) 18

24. In the given figure, *ABCD* is a square. *M* is the mid-point of *AB* and $PQ \perp CM$. Which of the following statements is not true?

(a) *AM = MB*   (b) *CP = CQ*

(c) *CP = CB*   (d) *PM = MQ*

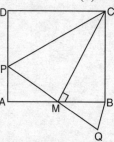

25. In the diagram, *ACDE* is a trapezium with *AC* ∥ *ED*. Given that $\angle EAB = 52°$, $\angle CDR = 126°$ and $\angle PBC = 90°$ and *EQ* ∥ *DR*. Then $\angle BCD$ equals

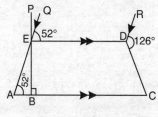

(a) 36°   (b) 74°

(c) 54°   (d) 38°

Answers

1. (a)   **2.** (d)   **3.** (c)   **4.** (b)   **5.** (a)   **6.** (c)

7. (b) [**Hint.** Find sides *PQ* and *PR*, given *PS* = 2*x*, *QS* = 4*x* and *RS* = *x*].   **8.** (a)   **9.** (b)

10. (b) [**Hint.** Ratio of perimeters of two Δ*s* = Ratio of corresponding sides.]   **11.** (a)   **12.** (a)

13. (c)   **14.** (c)   **15.** (d)   **16.** (b)

17. (d) [**Hint.** Show Δ*EBD* and Δ*FDC* are isosceles and find $\angle EDB$ and $\angle FDC$.]

18. (c)   **19.** (c)   **20.** (b)   **21.** (c)   **22.** (c)

23. (c) [**Hint.** Size of an exterior angle of a regular polygon with $\frac{3}{2}n$ sides = $\dfrac{360°}{\frac{3n}{2}}$.]   **24.** (c)   **25.** (b)

UNIT-5

MENSURATION

- *Area and Perimeter of Rectangle, Square, Parallelogram, Triangle and Circle*
- *Area and Perimeter of Rhombus, Trapezium and Polygons*
- *Volume and Surface Area*

Chapter
25
AREA AND PERIMETER OF RECTANGLE, SQUARE, PARALLELO-GRAM, TRIANGLE AND CIRCLE

KEY FACTS

1. RECTANGLE
Area = length × breadth

Perimeter = 2 (length + breadth)

Diagonal = $\sqrt{\text{length}^2 + \text{breadth}^2}$

2. SQUARE
Area = side^2

Perimeter = 4 × side

Diagonal = side $\sqrt{2}$

3. PARALLELOGRAM
Area = base × height

4. TRIANGLE
Area = $\dfrac{1}{2}$ × base × height

Area of an equilateral $\Delta = \dfrac{\sqrt{3}}{4}$ × side^2

Area of a triangle with three sides given = $\sqrt{s(s-a)(s-b)(s-c)}$,

where $s = \dfrac{a+b+c}{2}$ and a, b, c are the three sides.

5. CIRCLE
Area = π × radius^2

Circumference = $2\pi r$ or πd

(where radius = r and diameter = d)

Area of ring = $\pi(r_2^2 - r_1^2)$, where r_2 and r_1 are the outer and inner radius respectively.

6. RHOMBUS
A rhombus is a parallelogram whose all sides are equal. The diagonals of a rhombus bisect each other at right angles.

Area of a rhombus = $\dfrac{1}{2}$ × product of diagonals = $\dfrac{1}{2}d_1 d_2$

Each side of the rhombus = $\dfrac{1}{2}\sqrt{d_1^2 + d_2^2}$

Perimeter = 4 × side = $2 \times \sqrt{d_1^2 + d_2^2}$

Area of rhombus = side × height

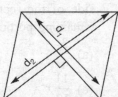

7. TRAPEZIUM

It is a quadrilateral with one pair of opposite sides parallel.

Area $= \frac{1}{2} \times$ (sum of parallel sides) $\times$ height $= \frac{1}{2}(a+b)h$

Perimeter $= a + b + c + d$

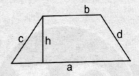

8. QUADRILATERAL

Its area can be found by dividing it into two triangles by a diagonal.

Area $= \frac{1}{2}AC \times h_1 + \frac{1}{2}AC \times h_2$.

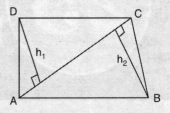

Question Bank–25

(Revision Exercise)

1. The sides of a triangle are 5, 12 and 13 units. A rectangle of width 10 units is constructed equal in area to the area of the triangle. Then, the perimeter of the rectangle is
   (a) 30 units   (b) 26 units
   (c) 13 units   (d) 15 units

2. A square whose side is 2 metres has its corners cut away so as to form an octagon with all sides equal. Then, the length of each side of the octagon in metres is:
   (a) $\dfrac{\sqrt{2}}{\sqrt{2}+1}$ (b) $\dfrac{2}{\sqrt{2}+1}$
   (c) $\dfrac{2}{\sqrt{2}-1}$ (d) $\dfrac{\sqrt{2}}{\sqrt{2}-1}$

3. The perimeter of a square S_1 is 12 m more than the perimeter of the square S_2. If the area of S_1 equals three times the area of S_2 minus 11, then what is the perimeter of S_1?
   (a) 24 m   (b) 32 m
   (c) 36 m   (d) 40 m

4. A square of side x is taken. A rectangle is cut out from this square such that one side of the rectangle is half that of the square and the other is $\frac{1}{3}$ the first side of the rectangle. What is the area of the remaining portion?
   (a) $\dfrac{3}{4}x^2$ (b) $\dfrac{7}{8}x^2$
   (c) $\dfrac{11}{12}x^2$ (d) $\dfrac{15}{16}x^2$

5. A circle is inscribed in an equilateral triangle of side a. What is the area of any square inscribed in the circle?
   (a) $\dfrac{a^2}{3}$ (b) $\dfrac{a^2}{4}$

(c) $\dfrac{a^2}{6}$ (d) $\dfrac{a^2}{8}$

6. If the distance from the vertex to the centroid of an equilateral triangle is 6 cm, then what is the area of the triangle?
   (a) 24 cm^2 (b) $27\sqrt{3}$ cm^2
   (c) 12 cm^2 (d) $12\sqrt{3}$ cm^2

7. What is the area of the region of the circle which is situated outside the inscribed square of side x?
   (a) $(\pi-2)x^2$ (b) $(\pi-2)x^2/2$
   (c) $2(\pi-2)x^2$ (d) $(\pi-2)x^2/4$

8. A rectangle field is half as wide as it is long and is completely enclosed by x metre of fencing. What is the area of the field?
   (a) $\dfrac{x^2}{2}$ m^2 (b) $2x^2$ m^2
   (c) $\dfrac{2x^2}{9}$ m^2 (d) $\dfrac{x^2}{18}$ m^2

9. A square circumscribes a circle and another square is inscribed in this circle with one vertex at the point of contact. The ratio of the areas of the circumscribed and inscribed squares is
   (a) 1 : 2   (b) 2 : 1
   (c) 3 : 1   (d) 4 : 1

10. $ABCD$ is a square inscribed in a circle of radius 14 cm. E, F, G and H are the midpoints of the sides DA, AB, BC and CD respectively. The area of the square $EFGH$ will be

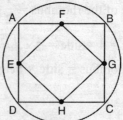

(a) 89 cm^2 (b) 196 cm^2

(c) 98 cm^2 (d) 392 cm^2

11. If an area enclosed by a circle or a square or an equilateral triangle is the same, then the maximum perimeter is possessed by:

   (a) circle

   (b) square

   (c) equilateral triangle

   (d) triangle and square have equal perimeters greater than that of circle.

12. A rectangular field is 80 m long and 60 m wide. If fence posts are placed at the corners and are 10 m apart along the 4 sides of the field, how many posts are needed to completely fence the field?

   (a) 24   (b) 27

   (c) 28   (d) 29

13. The difference between the area of a square and that of an equilateral triangle on the same base is $\frac{1}{4}$ m^2. What is the length of the side of the triangle?

(a) $(4-\sqrt{3})^{1/2}$ cm   (b) $(4+\sqrt{3})^{1/2}$ cm

(c) $(4-\sqrt{3})^{-1/2}$ cm   (d) $(4+\sqrt{3})^{-1/2}$ cm

14. A rectangle and a parallelogram of equal area and equal base are given. If r and p denote their respective perimeters, then

   (a) $r = p$ (b) $r = 2p$

   (c) $r > 2p$ (d) $r < p$

15. Two cardboard pieces in the form of equilateral triangles having a side of 3 cm each are symmetrically glued to form a regular star. The area of the star is

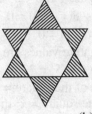

(a) 4 cm^2 (b) 6 cm^2

(c) $3\sqrt{3}$ cm^2 (d) $\dfrac{3\sqrt{3}}{2}$ cm^2

Answers

1. (b)	2. (a)	3. (b)	4. (c)	5. (c)	6. (b)	7. (b)	8. (d)	9. (b)	10. (b)
11. (c)	12. (c)	13. (c)	14. (d)	15. (c)					

Hints and Solutions

1. (b) By Pythagoras theorem, we find that the given triangle is a right angled triangle with 12 as height and 5 as base.

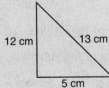

∴ Area of the triangle = $\frac{1}{2} \times 12 \times 5$ sq. units

= 30 sq. units

∴ Area of the rectangle = length × breadth = 30

⇒ Length = $\dfrac{30}{\text{breadth}} = \dfrac{30}{10} = 3$ units

∴ Perimeter of the rectangle = 2 × (10 + 3) units
= 26 units.

2. (a) Let the length of the edge cut away from the corners of the square be x.

Since the resulting figure is a regular octagon.

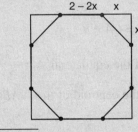

$\sqrt{x^2 + x^2} = 2 - 2x$ ⇒ $\sqrt{2}x = 2 - 2x$

⇒ $\sqrt{2}x + 2x = 2$ ⇒ $\sqrt{2}x(1+\sqrt{2}) = 2$

⇒ $x = \dfrac{2}{\sqrt{2}(1+\sqrt{2})} = \dfrac{\sqrt{2}}{1+\sqrt{2}}$

3. (b) Let the perimeter of the square $S_2 = x$ m, then perimeter of $S_1 = (x + 12)$ m

Length of one side of square $S_2 = \dfrac{x}{4}$ m and

Length of one side of the square $S_1 = \left(\dfrac{x}{4} + 3\right)$ m

Given, $\left(\dfrac{x}{4} + 3\right)^2 = 3\left(\dfrac{x}{4}\right)^2 - 11$

$\Rightarrow \dfrac{x^2}{16} + 9 + \dfrac{3}{2}x = \dfrac{3x^2}{16} - 11$

$\Rightarrow \dfrac{x^2}{8} - \dfrac{3x}{2} - 20 = 0$

$\Rightarrow x^2 - 12x - 160 = 0$

$\Rightarrow x^2 - 20x + 8x - 160 = 0$

$\Rightarrow x(x - 20) + 8(x - 20) = 0$

$\Rightarrow (x - 20)(x + 8) = 0 \Rightarrow x = 20$ or -8

Since negative values are impossible, $x = 20$

∴ $S_2 = 20$ m and $S_1 = 32$ m.

4. (c) Each side of the square = x m
$\Rightarrow$ Area of the square = x^2 m^2

Given, one side of the rectangle = $\dfrac{x}{2}$ m and

other side of the square = $\dfrac{1}{3} \times \dfrac{x}{2}$ m = $\dfrac{x}{6}$ m

∴ Area of the rectangle = $\dfrac{x}{2}$ m $\times \dfrac{x}{6}$ m

$= \dfrac{x^2}{12}$ m^2

∴ Area of the remaining portion = $x^2 - \dfrac{x^2}{12}$

$= \dfrac{11x^2}{12}$ m^2

5. (c) $AB = BC = CA = a$

∴ Height of the equilateral $\Delta = \dfrac{\sqrt{3}}{2} \times a$

O being the centroid of the ΔABC,

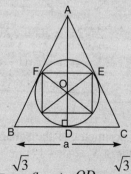

$3OD = \dfrac{\sqrt{3}}{2}a \Rightarrow OD = \dfrac{\sqrt{3}}{6}a$

Diagonal of the square = $2 \times OD$

$= 2 \times \dfrac{\sqrt{3}}{6}a = \dfrac{\sqrt{3}}{3}a$

$= \dfrac{1}{\sqrt{3}}a$

∴ Area of the square = $\dfrac{1}{2}$ (diagonal)^2

$= \dfrac{1}{2} \times \left(\dfrac{1}{\sqrt{3}}a\right)^2 = \dfrac{1}{6}a^2$.

6. (b) Here $OA = 6$ cm ∴ $OD = \dfrac{OA}{2} = 3$ cm

$AD = 6$ cm + 3 cm = 9 cm

Let each side of ΔABC be x cm.

From right angled ΔADB,

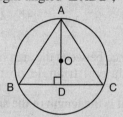

$AB^2 - BD^2 = AD^2$

$\Rightarrow x^2 - \left(\dfrac{x}{2}\right)^2 = 9^2 \Rightarrow \dfrac{3x^2}{4} = 81$

$\Rightarrow x^2 = \dfrac{4 \times 81}{3} = 4 \times 27$

$\Rightarrow x = 6\sqrt{3}$ cm

Area of equilateral $\Delta ABC = \dfrac{\sqrt{3}}{4} \times (6\sqrt{3})^2$

$= \dfrac{\sqrt{3}}{4} \times 108$ cm$^2 = 27\sqrt{3}$ cm^2

7. (b) Diameter of the circle = AC

= diagonal of the square = $\sqrt{2}x$ cm

∴ Radius of the circle = $\dfrac{\sqrt{2}\,x}{2}$ cm = $\dfrac{x}{\sqrt{2}}$ cm

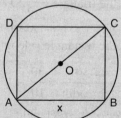

Hence, area of the circle = $\pi\left(\dfrac{x}{\sqrt{2}}\right)^2$ cm^2

$= \dfrac{\pi x^2}{2}$ cm^2

Area of the square $= x^2$

$\therefore$ Required area $= \dfrac{\pi x^2}{2} - x^2 = \dfrac{x^2}{2}(\pi - 2)$ cm^2.

8. (d) Let the breadth of the rectangular field be a m. Then, its length $= 2a$ m

Given, $2(2a + a) = x \Rightarrow 6a = x \Rightarrow a = \dfrac{x}{6}$ m

$\therefore$ Length $= 2a = 2 \times \dfrac{x}{6} = \dfrac{x}{3}$ m

$\therefore$ Area of the rectangular field $= \dfrac{x}{3} \times \dfrac{x}{6} = \dfrac{x^2}{18}$ m^2.

9. (b) Let each edge of square $ABCD = x$ units

Then, area of square $ABCD = x^2$

For square $EFGH$, diag. $HF = x$

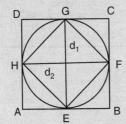

$\Rightarrow$ Side of square $EFGH = x/\sqrt{2}$

$\therefore$ Area of square $EFGH = \left(\dfrac{x}{\sqrt{2}}\right)^2 = \dfrac{x^2}{2}$

$\therefore$ Required ratio $= x^2 : \dfrac{x^2}{2} = 2 : 1$.

10. (b) Given, $OA = 14$ cm $\Rightarrow AC = 28$ cm

Let $AB = BC = a \Rightarrow a^2 + a^2 = 28^2$

$\Rightarrow 2a^2 = 28 \times 28 \Rightarrow a^2 = \dfrac{28 \times 28}{2} = 392$

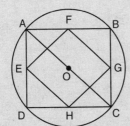

$\Rightarrow$ Area of square $ABCD = 392$ cm^2

Area of square $EFGH = \dfrac{1}{2} \times 392$ cm$^2 = 196$ cm^2

11. (c) Let the radius of the circle $= r$ units,

Length of each side of the square $= a$ units

Length of each side of the equilateral triangle

$= x$ units

Then, $\pi r^2 = A \Rightarrow r = \sqrt{\dfrac{A}{\pi}}$

$a^2 = A \Rightarrow a = \sqrt{A}$

$\dfrac{\sqrt{3}}{4}x^2 = A \Rightarrow x = \sqrt{\dfrac{A \times 4}{3}}$

$\therefore$ Perimeter of the circle $= 2\pi r$

$= 2 \times \pi \times \sqrt{\dfrac{A}{\pi}} = 2\sqrt{\pi A} = 2 \times \sqrt{3.14} \times \sqrt{a}$

$= 2 \times 1.77 \times \sqrt{a} = 3.54\sqrt{a}$

Perimeter of the square $= 4a = 4 \times \sqrt{A}$

Perimeter of the equilateral triangle

$= 3 \times \sqrt{\dfrac{4A}{1.732}} = 3 \times \sqrt{2.31A}$

$= 3 \times 1.52 \times \sqrt{A} = 4.56\sqrt{A}$

Thus we can see that the perimeter of the triangle is the greatest.

12. (c) Let four posts be there at the corners A, B, C, D of the rectangular field.

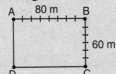

Then, along the length AB, we have 7 posts and along the breadth BC, we have 5 posts.

$\therefore$ Total number of posts $= 2\,(7 + 5) + 4$

$= 28$.

13. (c) Let the base of the square and the equilateral triangle be x units. Then, given

Area of square $-$ Area of equilateral triangle $= \dfrac{1}{4}$

$\Rightarrow x^2 - \dfrac{\sqrt{3}}{4}x^2 = \dfrac{1}{4} \Rightarrow \dfrac{4x^2 - \sqrt{3}x^2}{4} = \dfrac{1}{4}$

$\Rightarrow x^2(4 - \sqrt{3}) = 1$

$\Rightarrow x = \sqrt{\dfrac{1}{4 - \sqrt{3}}} \Rightarrow x = (4 - \sqrt{3})^{-\frac{1}{2}}$

14. (d) Perimeter of the rectangle $ABCD$

$= 2\,(AB + AD) + (DC + CB)$ ... (i)

Perimeter of parallelogram $ABFE$

$= 2\,(AB + AE) + (EF + FB)$ ... (ii)

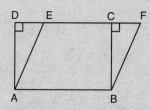

In right angled ΔADE, AE is the hypotenuse and the hypotenuse being the longest side in a Δ,

$AE > AD \quad \Rightarrow \quad BF > BC$

Also, $AB = DC$ and $AB = EF \quad \Rightarrow \quad DC = EF$.

$\therefore$ From (i) and (ii)

Perimeter of rectangle < Perimeter of parallelogram.

15. (c) Area of the star

= Area of ΔABC + Area of ΔEFG
        – Area of hexagon $HIJKLM$

The regular hexagon $HIJKLM$ is made of 6 equilateral triangle of side 1cm.

$AB = BC = CA = EF = FG = GE = 3$ cm

$\therefore$ Required area $= 2 \times \dfrac{\sqrt{3}}{4} \times (3)^2 - 6 \times \dfrac{\sqrt{3}}{4} \times (1)^2$

$= \dfrac{9\sqrt{3}}{2} - \dfrac{3\sqrt{3}}{2} = \dfrac{6\sqrt{3}}{2}$ cm^2

$= 3\sqrt{3}$ cm^2.

Chapter

26

AREA AND PERIMETER OF RHOMBUS, TRAPEZIUM AND POLYGONS

Solved Examples

Ex 1. *The diagonals of a rhombus are 24 cm and 10 cm. What is the perimeter of the rhombus ?*

Sol. Each side $= \sqrt{\left(\dfrac{24}{2}\right)^2 + \left(\dfrac{10}{2}\right)^2} = \sqrt{144+25} = \sqrt{169} = 13$ cm

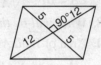

$\therefore$ Perimeter $= 4 \times 13$ cm $= 52$ cm

Ex. 2. *The area of a rhombus is 150 cm^2. The length of one of its diagonals is 10 cm. What is the length of the other diagonal ?*

Sol. Let the length of second diagonal $= d_1$ cm

Then, $\dfrac{1}{2} \times (d_1 \times 10) = 150 \Rightarrow d_1 = \dfrac{150 \times 2}{10} = 30$ cm.

Ex. 3. *The perimeter of a rhombus is 40 cm. If the length of one of its diagonals is 12 cm, what is the length of the other diagonal ?*

Sol. Each side of the rhombus $= \dfrac{40\text{ cm}}{4} = 10$ cm

Let one of the diagonals $BD = 12$ cm. Then $BO = 6$ cm

$\therefore$ In right angled $\triangle BOC$

$OC^2 = \sqrt{BC^2 - BO^2} = \sqrt{100 - 36} = \sqrt{64} = 8$ cm

$\therefore$ Diagonal $AC = 2 \times OC = 2 \times 8$ cm $= 16$ cm.

Ex. 4. *The length of one side of a rhombus is 6.5 cm and its altitude is 10 cm. If the length of one of its diagonals is 26 cm, what will be the length of the other diagonal ?*

Sol. $\dfrac{1}{2} \times 26 \times$ other diagonal $= 6.5 \times 10$

$\Rightarrow$ Other diagonal $= \dfrac{6.5 \times 10 \times 2}{26} = 5$ cm.

Ex. 5. *The area of a field in the shape of a trapezium measures 1440 m^2. The perpendicular distance between its parallel sides is 24 m. If the ratio of the parallel sides is 5 : 3, what is the length of the longer parallel side ?*

Sol. Let the parallel sides be $5x$ and $3x$ respectively.

Then, $\dfrac{1}{2} \times 24 \times (5x + 3x) = 1440 \Rightarrow 8x = \dfrac{1440 \times 2}{24} \Rightarrow x = 15$ m

$\therefore$ The longer parallel side $= 5 \times 15$ m $= 75$ m.

Ex. 6. *The difference between the two parallel sides of a trapezium is 8 m. The perpendicular distance between them is 24 m. If the area of the trapezium is 312 m², then what are the lengths of the two parallel sides ?*

Sol. Let x and $x + 8$ be the two parallel sides of the trapezium. Then,

$$\frac{1}{2} \times (x + x + 8) \times 24 = 312 \Rightarrow 2x + 8 = 26 \Rightarrow 2x = 18 \Rightarrow x = 9$$

∴ The two parallel sides are 9 m and 17 m.

Ex. 7. *If the lengths of the parallel sides of an isosceles trapezium are 20 cm and 30 cm and the area is 100 cm², then what is the length of the non-parallel sides ?*

Sol. Area of the trapezium = $\frac{1}{2}(AB + EF) \times \text{height}$

$$\Rightarrow 100 = \frac{1}{2} \times (20 + 30) \times h \Rightarrow h = 4 \Rightarrow AC = BD = 4 \text{ cm}$$

$$\therefore EA = BF = \sqrt{AC^2 + EC^2} = \sqrt{16 + 25} = \sqrt{41} \text{ cm.}$$

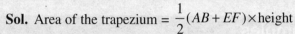

Ex. 8. *What will be the area of the field ABGFEA ?*

Sol. Area of field *ABGFEA*

= Area of △ *ABC* + Area of trapezium *BCHG* + Area of △ *GHF*

+ Area of △ *FDE* + Area of △ *DEA*

$$= \frac{1}{2} \times 40 \times 35 + \frac{1}{2} \times (50 + 35) \times 50 + \frac{1}{2} \times 50 \times 50 + \frac{1}{2} \times 90 \times 45 + \frac{1}{2} \times 45 \times 50$$

$$= 700 \text{ m}^2 + 2125 \text{ m}^2 + 1250 \text{ m}^2 + 2025 \text{ m}^2 + 1125 \text{ m}^2$$

$$= 7225 \text{ m}^2$$

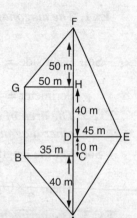

Question Bank–26

1. The side of a rhombus is 10 cm and one diagonal is 16 cm. The area of the rhombus is
   (a) 96 cm²
   (b) 95 cm²
   (c) 94 cm²
   (d) 93 cm²

2. If the diagonals of a rhombus are 24 cm and 10 cm, then the area and perimeter of the rhombus are respectively
   (a) 120 sq cm, 52 cm
   (b) 240 sq cm, 52 cm
   (c) 120 sq cm, 64 cm
   (d) 240 sq cm, 64 cm

3. The diagonals of a rhombus are 32 cm and 24 cm respectively. The perimeter of the rhombus is
   (a) 80 cm
   (b) 72 cm
   (c) 68 cm
   (d) 88 cm

4. The perimeter of a rhombus is 40 cm. If the length of one of its diagonals be 12 m, then the length of the other diagonal is
   (a) 14 cm
   (b) 15 cm
   (c) 16 cm
   (d) 12 cm

5. If the perimeter of a rhombus is $4a$ and the lengths of the diagonals are x and y, then its area is
   (a) $a(x + y)$
   (b) $x^2 + y^2$
   (c) xy
   (d) $\frac{1}{2}xy$

6. If the side of a rhombus is 20 metres and its shorter diagonal is three - fourths of its longer diagonal, then the area of the rhombus must be
   (a) 375 m²
   (b) 380 m²
   (c) 384 m²
   (d) 395 m²

7. A sheet is in the form of a rhombus whose diagonals are 10 m and 8 m. The cost of painting both of its surfaces at the rate of Rs 70 per m² is
   (a) Rs 5600
   (b) Rs 4000
   (c) Rs 2800
   (d) Rs 2000

8. A rhombus and a square have the same base. If the diagonals of the rhombus measure 30 cm and 16 cm respectively, find the area of the square.
   (a) 225 cm²
   (b) 200 cm²
   (c) 240 cm²
   (d) 289 cm²

9. If one of the diagonals of a rhombus is equal to its side, then the diagonals of the rhombus are in the ratio
   (a) $\sqrt{3}:1$
   (b) $\sqrt{2}:1$
   (c) 3 : 1
   (d) 2 : 1

10. A rhombus *OABC* is drawn inside a circle whose centre is at *O* in such a way that the vertices *A, B* and *C* of the rhombus are on the circle. If the area of

the rhombus is $32\sqrt{3}$ m², then the radius of the circle is

(a) 64 m   (b) 8 m

(c) 32 m   (d) 46 m

11. The measure of each of the two opposite angles of a rhombus is 60° and the measure of one of its sides is 10 cm. The length of its smaller diagonal is

(a) 10 cm   (b) $10\sqrt{3}$ cm

(c) $10\sqrt{2}$ cm   (d) $\dfrac{5}{2}\sqrt{2}$ cm

12. If the sum of the lengths of the diagonals of a rhombus of side 4 cm is 10 cm, what is its area ?

(a) 8 cm²   (b) 9 cm²

(c) 10 cm²   (d) 12 cm²

13. In the given figure ABCD is trapezium in which the parallel sides AB and CD both are perpendicular to BC. If AB = 16 m, CD = 8 m and AD = 17 m. What is the area of the trapezium ?

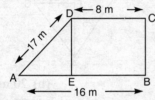

(a) 140 m²   (b) 168 m²

(c) 180 m²   (d) 156.4 m²

14. The lengths of the shorter and longer parallel sides of a trapezium are x cm and y cm respectively. If the area of the trapezium is $(x^2 - y^2)$, then the height of the trapezium is

(a) x   (b) $(x + y)$

(c) y   (d) $2(x - y)$

15. In the given figure, the side of the square is 10 cm. EF = 2.5 cm and C and D are half way between the top and bottom sides of the figure. The area of the shaded portion of the figure is

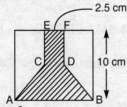

(a) 43.75 cm²   (b) 56.25 cm²

(c) 55.25 cm²   (d) 50.25 cm²

16. The cross-section of a canal is in the shape of trapezium. The canal is 15 m wide at the top and 9 m wide at the bottom. If the area of the cross-section is 720 m², then the depth of the canal is

(a) 58.4 m   (b) 58.6 m

(c) 58.8 m   (d) 60 m

17. The parallel sides of a field in the shape of a trapezium are 20 m and 41 m and the remaining two sides are 10 m and 17 m. Find the cost of levelling the field at the rate of Rs 30 per square metre ?

(a) Rs 6400   (b) Rs 7320

(c) Rs 7500   (d) Rs 7000

18. Top surface of a raised platform is in the shape of a regular octagon as shown in the figure. Find the area of the octagonal surface.

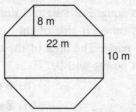

(a) 400 m²   (b) 348 m²

(c) 256 m²   (d) 476 m²

19. The parallel sides of a trapezium are 20 m and 30 m and its non-parallel sides are 6 m and 8 m. Find the area of the trapezium.

(a) 96 m²   (b) 82 m²

(c) 100 m²   (d) 120 m²

20. What is the area of the plot shown in the figure ?

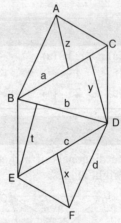

(a) $\dfrac{1}{2}(az + by + ct + dx)$

(b) $\dfrac{1}{2}(bt + cx + ay + az)$

(c) $\dfrac{1}{2}(cx + bt + by + az)$

(d) $\dfrac{1}{2}(d + t)(c + x) + \dfrac{1}{2}(a + b)(y + z)$

21. The area of the figure ABCEFGA is 84 m².

AH = HC = AG = 6 m and CE = HF = 4 m. If the angles marked in the figure are 90°, then the length of DB will be

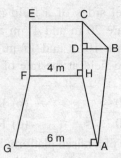

(a) 2.5 m   (b) 5 m

(c) 6 m   (d) 12 m

22. Consider a square of length 3 units. Also consider two points on each side of the square trisecting it into equal parts. The area of the octagon made by these eight points will be

(a) 4 unit^2 (b) 6 unit^2

(c) 7 unit^2 (d) 8 unit^2

23. The area of a regular hexagon of side $2\sqrt{3}$ cm is

(a) $18\sqrt{3}$ cm^2 (b) $12\sqrt{3}$ cm^2

(c) $36\sqrt{3}$ cm^2 (d) $27\sqrt{3}$ cm^2

24. Two sides of a plot measure 32 m and 24 m and the angle between them a perfect right angle. The other two sides measure 25 m each and the other three angles are not right angles. What is the area of the plot (in m^2) ?

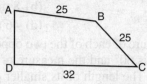

(a) 768   (b) 534

(c) 696.5   (d) 684

25. The area of the given field is 3500 m^2. $AF = 25$ m, $AG = 50$ m, $AH = 75$ m and $AB = 100$ m. The rest of the dimensions are shown in the figure. Find the value of x.

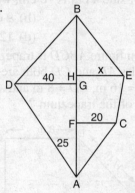

(a) 17 m   (b) 20 m

(c) 22 m   (d) 25 m

Answers

1. (a)	2. (a)	3. (a)	4. (c)	5. (d)	6. (c)	7. (a)	8. (d)	9. (a)	10. (b)
11. (a)	12. (b)	13. (c)	14. (c)	15. (a)	16. (d)	17. (b)	18. (d)	19. (a)	20. (b)
21. (b)	22. (c)	23. (a)	24. (d)	25. (b)					

Hints and Solutions

1. (a) Since the diagonals of a rhombus bisect each other at right angles.

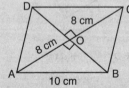

In ΔAOB, $BO^2 = \sqrt{AB^2 - AO^2}$

$= \sqrt{100 - 64}$ cm

$= \sqrt{36}$ cm $= 6$ cm

∴ The other diagonal $= 2 \times 6$ cm $= 12$ cm

∴ Area of the rhombus $= \dfrac{1}{2} \times 16$ cm $\times 12$ cm

$= 96$ cm^2.

2. (a) Each side of the rhombus $= \sqrt{\left(\dfrac{24}{2}\right)^2 + \left(\dfrac{10}{2}\right)^2}$

$= \sqrt{12^2 + 5^2} = \sqrt{144 + 25} = \sqrt{169} = 13$ cm

∴ Area of the rhombus $= \dfrac{1}{2} \times 24$ cm $\times 10$ cm

$= 120$ cm^2

Perimeter $= 4 \times 13$ cm $= 52$ cm.

3. (a) Similar to Q. No. 2.

4. (c) Each side of the rhombus $= \dfrac{40\,\text{cm}}{4} = 10$ cm

Length of the other diagonal $= 2 \times \sqrt{10^2 - \left(\dfrac{12}{2}\right)^2}$

$$= 2 \times \sqrt{100 - 36} = 2 \times \sqrt{64} \text{ cm} = 16 \text{ cm}.$$

5. (d) Area $= \dfrac{1}{2} \times$ Product of diagonals $= \dfrac{1}{2} xy$

6. (c) Let $AC = x$. Then $BD = \dfrac{3x}{4}$

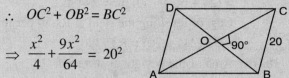

$$\therefore \quad OC^2 + OB^2 = BC^2$$

$$\Rightarrow \quad \frac{x^2}{4} + \frac{9x^2}{64} = 20^2$$

$$\Rightarrow \quad \frac{25x^2}{64} = 400 \quad \Rightarrow \quad x^2 = \frac{400 \times 64}{25} = 1024$$

$$\Rightarrow \quad x = 32 \quad \Rightarrow \quad AC = 32 \text{ cm and } BD = 24 \text{ cm}.$$

$$\therefore \quad \text{Area of the rhombus} = \frac{1}{2} \times AC \times BD$$

$$= \frac{1}{2} \times 32 \times 24 \text{ cm}^2 = 384 \text{ cm}^2 .$$

7. (a) Cost of painting both its surfaces

$$= 2 \times 70 \times \text{Area of one surface}$$

$$= \text{Rs} \left(2 \times 70 \times \frac{1}{2} \times 10 \times 8 \right) = \text{Rs } 5600.$$

8. (d) Each side of the square = each side of

the rhombus $= \sqrt{\left(\dfrac{30}{2}\right)^2 + \left(\dfrac{16}{2}\right)^2}$

$$= \sqrt{225 + 64} = \sqrt{289} = 17 \text{ cm}$$

$$\therefore \quad \text{Area of the square} = (17)^2 \text{ cm}^2 = 289 \text{ cm}^2.$$

9. (a) Let side AB = diagonal $AC = a$ units

Then, the other diagonal

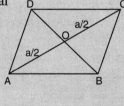

$$= 2 \times \sqrt{a^2 - \left(\frac{a}{2}\right)^2}$$

$$= 2 \times \sqrt{a^2 - \frac{a^2}{4}}$$

$$= 2 \times \sqrt{\frac{3a^2}{4}} = \sqrt{3} a$$

$$\therefore \quad \text{Required ratio} = \frac{\sqrt{3}a}{a} = \sqrt{3} : 1 .$$

10. (b) Let $AC = 2x$, $OB = 2y$.

$$\therefore \quad \text{Radius} = OC = 2y = OB$$

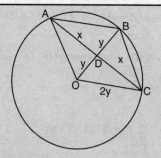

From ΔODC,

$$OC^2 = OD^2 + CD^2$$

$$\Rightarrow \quad 4y^2 = y^2 + x^2 \Rightarrow x^2 = 3y^2 \Rightarrow x = \sqrt{3} y \quad \dots (1)$$

Also,

Area of $\Delta ODC = \dfrac{1}{4} \times$ Area of rhombus $OABC$

$$\Rightarrow \quad \frac{1}{2} \times x \times y = \frac{1}{4} \times 32\sqrt{3} \quad \Rightarrow \quad xy = 16\sqrt{3} \quad \dots (2)$$

From (1) and (2) we have, $y = 4$ and $x = 4\sqrt{3}$

$$\therefore \quad \text{Radius of the circle} = 2y = 8 \text{ m}.$$

11. (a) Let $\angle B = \angle D = 60°$

In ΔADC,

$$AD = DC \qquad \text{(adjacent sides of a rhombus)}$$

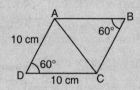

$$\Rightarrow \quad \angle ACD = \angle DAC = \frac{180° - 60°}{2} = 60°$$

(Isosceles Δ property)

$$\therefore \quad \Delta ADC \text{ is an equilateral triangle}$$

$$\Rightarrow \quad AC = AD = DC = 10 \text{ cm}.$$

12. (b) Let d_1 and d_2 be the lengths of the diagonals of the rhombus. Then,

$$\left(\frac{d_1}{2}\right)^2 + \left(\frac{d_2}{2}\right)^2 = 4^2 \Rightarrow \frac{d_1^2}{4} + \frac{d_2^2}{4} = 16$$

$$\Rightarrow \quad d_1^2 + d_2^2 = 64$$

$$\Rightarrow \quad (d_1 + d_2)^2 - 2d_1 d_2 = 64$$

$$\Rightarrow \quad 10^2 - 2d_1 d_2 = 64 \qquad (\because d_1 + d_2 = 10)$$

$$\Rightarrow \quad 2d_1 d_2 = 100 - 64 = 36$$

$$\Rightarrow \quad d_1 d_2 = 18 \quad \Rightarrow \quad \frac{1}{2} d_1 d_2 = 9 \text{ cm}^2$$

$$\therefore \quad \text{Area of the rhombus} = 9 \text{ cm}^2.$$

13. (c) $AE = AB - EB = AB - DC = 16 \text{ m} - 8 \text{ m} = 8 \text{ m}$

$$\therefore \quad \text{In } \Delta AED,$$

$DE^2 = AD^2 - AE^2 = 17^2 - 8^2 = 289 - 64 = 225$

$\Rightarrow DE = 15$ m

$\therefore$ Area of the trapezium $ABCD$

$= \dfrac{1}{2} \times (AB + DC) \times DE$

$= \dfrac{1}{2} \times (16 \text{ m} + 8 \text{ m}) \times 15 \text{ m}$

$= \dfrac{1}{2} \times 24 \text{ m} \times 15 \text{ m} = 180 \text{ m}^2$.

14. (c) Area of trapezium

$= \dfrac{1}{2} \times$ (Sum of parallel sides) $\times$ height

$\Rightarrow \dfrac{1}{2} \times (x + y) \times$ height $= (x^2 - y^2)$

$\Rightarrow$ height $= \dfrac{2(x + y)(x - y)}{(x + y)} = 2(x - y)$

15. (a) Area of the shaded part = Area of rectangle $CDEF$ + Area of trapezium $ABCD$

$= 2.5 \text{ cm} \times 5 \text{ cm} + \dfrac{1}{2} \times (2.5 \text{ cm} + 10 \text{ cm}) \times 5 \text{ cm}$

$= 12.5 \text{ cm}^2 + 31.25 \text{ cm}^2 = 43.75 \text{ cm}^2$.

16. (d) Area of cross-section $= 720 \text{ m}^2$

$\Rightarrow \dfrac{1}{2} \times (9 \text{ m} + 15 \text{ m}) \times d = 720$

$\Rightarrow d = \dfrac{720 \times 2}{24} \text{ m} = 60 \text{ m}.$

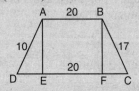

17. (b) $DC = 41$ m (Given)

$\therefore DE + FC = 41 - 20 = 21$ m

Let $DE = x$, therefore

$FC = (21 - x)$

From $\triangle ADE, AE^2 = (10)^2 - x^2 = 100 - x^2$

From $\triangle BCF, BF^2 = (17)^2 - (21 - x)^2$

$= 289 - (441 + x^2 - 42x)$

$= -x^2 + 42x - 152$

Since $AE = BF$, $100 - x^2 = -x^2 + 42x - 152$

$\Rightarrow 42x = 252 \Rightarrow x = 6$

$\therefore AE = BF = \sqrt{100 - 6^2} = 8$

$\therefore$ Area of the trapezium $ABCD$

$=$ Area of $\triangle AED$ + Area of $\triangle BCF$

$\qquad\qquad$ + Area of rectangle $ABEF$

$= \left(\dfrac{1}{2} \times 8 \times 6 + \dfrac{1}{2} \times 15 \times 8 + 20 \times 8 \right) \text{ m}^2$

$= (24 + 60 + 160) \text{ m}^2 = 244 \text{ m}^2$

$\therefore$ Cost of levelling the field at Rs 30 per m^2

$=$ Rs $(244 \times 30) =$ Rs 7320.

18. (d) Area of the octagonal surface

$= 2 \times \dfrac{1}{2} \times (22 \text{ m} + 10 \text{ m}) \times 8 \text{ m} + 22 \text{ m} \times 10 \text{ m}$

$= 32 \text{ m} \times 8 \text{ m} + 220 \text{ m}^2 = 256 \text{ m}^2 + 220 \text{ m}^2 = 476 \text{ m}^2$

19. (a) Similar to Q. 17.

20. (b) Area of the plot

$=$ Area of $\triangle ABC$ + Area of $\triangle BCD$

$\qquad\qquad$ + Area of $\triangle BED$ + Area of $\triangle EFD$

$= \dfrac{1}{2} \times a \times z + \dfrac{1}{2} \times a \times y \times \dfrac{1}{2} \times b \times t + \dfrac{1}{2} \times c \times x$

$= \dfrac{1}{2}(az + ay + bt + cx)$

21. (b) Area of rectangle $CEFH$

$\qquad\qquad$ + Area of trapezium $FHAG$

$= (4 \times 6) \text{ m}^2 + \left(\dfrac{1}{2} \times (6 + 4) \times 6 \right) \text{ m}^2$

$= 24 \text{ m}^2 + 30 \text{ m}^2 = 54 \text{ m}^2$

$\therefore$ Area of right angled $\triangle ABC = 84 \text{ m}^2 - 54 \text{ m}^2$

$\qquad\qquad\qquad\qquad = 30 \text{ m}^2$

$\Rightarrow \dfrac{1}{2} \times 12 \times BD = 30 \Rightarrow BD = 5$ m.

22. (c) Area of octagon $ABHGDCFE$

$=$ Area of rectangle $ABDC$ + Area of trapezium $BHGD$ + Area of trapezium $ACFE$

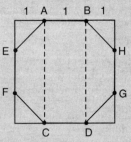

$= (1 \times 3) + \dfrac{1}{2} \times (1 + 3) \times 1 + \dfrac{1}{2} \times (1 + 3) \times 1$

$= 7$ sq. units.

23. (a) A regular hexagon consists of 6 equilateral triangles of equal areas.

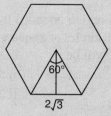

Area of a triangle $= \frac{\sqrt{3}}{4}(2\sqrt{3})^2 = 3\sqrt{3}$ cm^2

$\therefore$ Required area $= 6 \times 3\sqrt{3}$ cm$^2 = 18\sqrt{3}$ cm^2.

24. (d) In right angle $\triangle ADC$,
$AC^2 = AD^2 + DC^2$

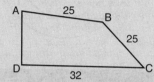

$\Rightarrow AC = \sqrt{24^2 + 32^2} = \sqrt{576 + 1024} = \sqrt{1600}$
$= 40$ m

$\therefore$ Area of the plot
$= $ Area of $\triangle ADC +$ Area of $\triangle ABC$

For $\triangle ABC$, $S = \dfrac{25 + 25 + 40}{2}$ m $= 45$ m

$\therefore$ Required area
$= \dfrac{1}{2} \times 24$ m $\times 32$ m
$\qquad + \sqrt{45(45-25)(45-25)(45-40)}$ m^2
$= 384$ m$^2 + \sqrt{45 \times 20 \times 20 \times 5}$ m^2
$= 384$ m$^2 + 300$ m$^2 = 684$ m^2.

25. (b) Total area $=$ Area of $\triangle AFC +$ Area of $\triangle AGD$
$\qquad +$ Area of trapezium $FCEH +$ Area of $\triangle BGE$
$\qquad +$ Area of $\triangle DGB$
$= \dfrac{1}{2} \times AF \times FC + \dfrac{1}{2} \times AG \times DG$
$\qquad + \dfrac{1}{2} \times (CF + EH) \times HF$
$\qquad + \dfrac{1}{2} \times BH \times HE + \dfrac{1}{2} \times BG \times DG$
$\Rightarrow \dfrac{1}{2} \times 25 \times 20 + \dfrac{1}{2} \times 50 \times 40 + \dfrac{1}{2} \times (20 + x) \times 50$
$\qquad + \dfrac{1}{2} \times 25 \times x + \dfrac{1}{2} \times 50 \times 40 = 3500$
$\Rightarrow 250 + 1000 + (20 + x)25 + 12.5x + 1000 = 3500$
$\Rightarrow 2250 + 500 + 25x + 12.5x = 3500$
$\Rightarrow 37.5x = 3500 - 2750 = 750$
$\Rightarrow x = \dfrac{750}{37.5} = 20$ m

Self Assessment Sheet–25 (Chapters 25 and 26)

1. A table cover 4 m × 2 m is spread on a meeting table. If 25 cm of the table cover hanging all around the table, find the cost of polishing the top of the table at Rs 2.25 per square metres.
   (a) 16
   (b) 12
   (c) 24
   (d) 18

2. Find the ratio of the area of a square inscribed in a semi-circle of radius r to the area of another square inscribed in the entire circle of radius r.
   (a) 2 : 1
   (b) 3 : 2
   (c) 2 : 5
   (d) 3 : 5

3. Find the percentage increase in the area of a triangle if its each side is doubled.
   (a) 50%
   (b) 100%
   (c) 300%
   (d) 150%

4. Perpendiculars are drawn on the sides of an equilateral triangle from any point within the triangle. If the lengths of these perpendiculars be 6 cm, 7 cm and 9 cm, then the length of a side of the triangle is :
   (a) $\dfrac{44}{3}\sqrt{3}$ cm
   (b) $\dfrac{11}{3}\sqrt{3}$ cm
   (c) $\dfrac{22}{3}\sqrt{3}$ cm
   (d) $11\sqrt{3}$ cm

5. A rhombus has an area equal to half the sum of the areas of the squares built on its four sides. The ratio of the long diagonal to the short diagonal is:
   (a) $2 + \sqrt{3}$
   (b) $2 - \sqrt{3}$
   (c) $\dfrac{1}{2}$
   (d) 2

6. In a rectangle, the difference between the sum of the adjacent sides and the diagonal is half the length of the longer side. What is the ratio of the shorter side to the longer side?
   (a) $\sqrt{3} : \sqrt{2}$
   (b) $1 : \sqrt{3}$
   (c) 2 : 5
   (d) 3 : 4

7. The parallel sides of a trapezium are 20 cm and 10 cm. Its non-parallel sides are both equal, each being 13 cm. Find the area of the trapezium?
   (a) 120 cm^2
   (b) 180 cm^2
   (c) 210 cm^2
   (d) 150 cm^2

8. A square and a regular hexagon have equal perimeters. Their areas are in the ratio:

(a) $2:1$

(b) $2\sqrt{3}:1$

(c) $\sqrt{3}:2$

(d) $3:2$

9. If *BC* passes through the centre of the circle, then area of the shaded region in the given figure is :

(a) $\dfrac{a^2}{2}(3-x)$

(b) $a^2(\pi/2-1)$

(c) $2a^2(\pi-1)$

(d) $\dfrac{a^2}{2}(\pi/2-1)$

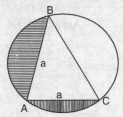

10. If the area of region bounded by the inscribed and circumscribed circles of a square is 9π, then the area of the square will be :

(a) 6π

(b) 5π

(c) 25

(d) 36

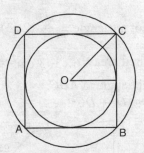

Answers

1. (b)	2. (c)	3. (c)	4. (a)	5. (d)	6. (d)	7. (b)	8. (c)	9. (d)	10. (d)

Chapter 27

VOLUME AND SURFACE AREA

Section-A
CUBES AND CUBOIDS

KEY FACTS

1. A **cuboid** is a solid having six rectangular faces. It has 12 edges and 8 vertices.
2. A cuboid in which length, breadth and height are all equal is called a **cube**. All the six faces of a cube are congruent squares.
3. With the usual meaning for V, l, b and h (Volume, length, breadth and height respectively),
   (i) Surface area of a cuboid = $2(lb + bh + lh)$
   (ii) Surface area of a cube = $6l^2$
   (iii) Lateral surface of a cuboid = Area of four walls of a room
   $$= 2(l+b) \times h$$
   (iv) Volume of a cuboid $(V) = l \times b \times h$
   (v) Volume of a cube $(V) = l^3$
4. **Standard units of Volume**
   (i) $1\ cm^3 = 1000\ mm^3$ (ii) $1\ m^3 = 1,000,000\ cm^3$ (iii) $1\ cm^3 = 1 mL$
   (iv) $1000\ cm^3 = 1$ litre   (v) $1\ m^3 = 1000$ litres = 1 kilo litre.

Question Bank–27

(Revision Exercise on Cubes and Cuboids)

1. The volume of a wall, 5 times as high as it is breadth and 8 times as long as it is high is 18225 m³. Find the breadth of the wall.
   (a) 32.5 m   (b) 5 m
   (c) 4.5 m   (d) 3.5 m

2. The volume of a rectangular block of stone is 10368 dm². Its dimensions are in the ratio of 3 : 2 : 1. If its entire surface is polished at 2 paise per dm², then the total cost will be
   (a) Rs 31.50   (b) Rs 63.00
   (c) Rs 63.36   (d) Rs 31.68

3. One cubic metre of copper is melted and recast into a square cross-section bar 36 m long. An exact cube is cut off from this bar. If 1 cubic metre of copper costs Rs 108, then the cost of this cube is
   (a) 50 paise   (b) 75 paise
   (c) One rupee   (d) 1.50 rupee

4. A water tank is 30 m long, 20 m wide and 12 m deep. It is made up of iron sheet which is 3 m wide. The tank is open at the top. If the cost of the iron sheet is Rs 10 per metre, then the total cost of the iron sheet required to build the tank is
   (a) Rs 6000   (b) Rs 8000
   (c) Rs 9000   (d) Rs 10000

5. The sum of the length, breadth and depth of a cuboid is 19 cm and the diagonal is $5\sqrt{5}$. Its surface area is
   (a) 361 cm²
   (b) 125 cm²
   (c) 236 cm²
   (d) 256 cm²

6. A solid cube with an edge 10 cm is melted to form two equal cubes. The ratio of the edge of the smaller cube to the edge of the bigger cube is
   (a) $\left(\dfrac{1}{3}\right)^{\frac{1}{3}}$
   (b) $\dfrac{1}{2}$
   (c) $\left(\dfrac{1}{2}\right)^{\frac{1}{3}}$
   (d) $\left(\dfrac{1}{4}\right)^{\frac{1}{3}}$

7. A rectangular tank 25 cm long and 20 cm wide contains 4.5 litres of water. When a metal cube is lowered in the tank, the water level rises to a height of 11 cm. Find the length of each edge of the cube ?
   (a) 15 cm
   (b) 5 cm
   (c) 11 cm
   (d) 10 cm

8. An icecream company makes a popular brand of icecream in a rectangular shaped bar 6 cm long, 5 cm wide and 2 cm thick. To cut costs, the company has decided to reduce the volume of the box by 20%. The thickness will remain the same, but the length and width will be decreased by the same percentage amount. Which condition given below will the new length l satisfy ?
   (a) $5.5 < l < 6$
   (b) $5 < l < 5.5$
   (c) $4.5 < l < 5$
   (d) $4 < l < 4.5$

9. Three cubes with sides in the ratio 3 : 4 : 5 are melted to form a single cube whose diagonal is $12\sqrt{3}$ cm. The sides of the cube are
   (a) 6 cm, 8 cm, 10 cm
   (b) 3 cm, 4 cm, 5 cm
   (c) 9 cm, 12 cm, 15 cm
   (d) 12 cm, 16 cm, 20 cm

10. A, B, C denote the areas of three coterminus faces of a cuboid. If P and S denote the product and sum of dimensions of the cuboid respectively, which one of the following is correct ?
    (a) $PS = A^2 + B^2 + C^2$
    (b) $PS = AB + BC + CA$
    (c) $P = S(A + B + C)$
    (d) $SP = (A + B + C)^2$

11. A cuboid has edges of x cm, 1 cm and 2 cm. The total surface area of the cuboid has a numerical value which is some integral multiple of the numerical value of its volume. What is the value of x for minimum possible positive integral multiple ?
    (a) 5 cm
    (b) 2 cm
    (c) 3 cm
    (d) 4 cm

12. A metallic sheet is of rectangular shape with dimensions 48 m × 36 m. From each of its corners, a square is cut off so as to make an open box. The volume of the box is x m³, when the length of the square is 8 m. The value of x is
    (a) 5120
    (b) 8960
    (c) 4830
    (d) 6400

13. Except for one face of given cube, identical cubes are glued through their faces to all the other faces of the given cube. If each side of the given cube measures 3 cm, then what is the total surface area of the solid thus formed ?
    (a) 225 cm²
    (b) 234 cm²
    (c) 270 cm²
    (d) 279 cm²

14. A cube having each side of unit length is cut into two parts by a plane through two diagonals of two opposite faces. What is the total surface area of each of these parts ?
    (a) $3 + \sqrt{2}$ sq. units
    (b) $2 + \sqrt{3}$ sq. units
    (c) $3\sqrt{2}$ sq. units
    (d) 3 sq. units

15. A cuboid is formed of 3 edges measuring 3, 4 and 5 cm. It is sliced into two identical solids by a plane through a diagonal of the smallest of the faces. The surface area of the sliced section is
    (a) 12 cm²
    (b) 15 cm²
    (c) 20 cm²
    (d) 25 cm²

16. A square hole of cross-sectional area 4 cm² is drilled across a cube with its length parallel to a side of the cube. If an edge of the cube measures 5 cm, what is the total surface area of the body so formed ?
    (a) 140 cm²
    (b) 142 cm²
    (c) 162 cm²
    (d) 182 cm²

17. The area of a side of a box is 120 sq. cm. The area of the other side of the box is 72 sq. cm. If the area of the upper surface of the box is 60 sq. cm., then find the volume of the box ?
    (a) 259200 cm³
    (b) 84000 cm³
    (c) 86400 cm³
    (d) 720 cm³

18. A rectangular tank is 225 m by 162 m at the base. With what speed must the water flow into it through an aperture 60 cm by 45 cm that the level may be raised 20 cm in 5 hours.
    (a) 5000 m/hr
    (b) 5200 m/hr
    (c) 5400 m/hr
    (d) 5600 m/hr

19. The water in a rectangular reservoir having a base 80 m by 60 m is 6.5 m deep. In what time can the water be emptied by a pipe of which the cross-

section is a square of side 20 cm, if the water runs through the pipe at the rate of 15 km per hour ?

(a) 26 hours   (b) 42 hours

(c) 52 hours   (d) 65 hours

20. A cake as shown has three layers, each of which is a cuboid as shown. The bases of these cuboids are squares whose sides are of lengths 15 cm, 30 cm and 45 cm. The height of each layer is 5 cm. Find the total area of the cake that must be frosted. Assume that there will be frosting between the layers, but not at the bottom of the cake.

(a) 5000 cm^2 (b) 4000 cm^2

(c) 4950 cm^2 (d) 4500 cm^2

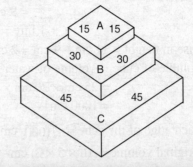

Answers

1. (c)	2. (c)	3. (a)	4. (a)	5. (c)	6. (c)	7. (d)	8. (b)	9. (a)	10. (b)
11. (b)	12. (a)	13. (b)	14. (a)	15. (d)	16. (d)	17. (d)	18. (c)	19. (c)	20. (c)

Hints and Solutions

1. (c) Let the breadth of the wall = h metres

∴ Height of the wall = $5h$ metres

Length of the wall = $40h$ metres

∴ $40h \times 5h \times h = 18225 \Rightarrow h^3 = 91.125$

$\Rightarrow h = 4.5$

2. (c) Let the dimensions of the rectangular block of the stone be $3x$, $2x$ and x respectively.

Volume of the stone = $6x^3 = 10368$

$\Rightarrow x^3 = 1728 \Rightarrow x = 12$

∴ Dimensions of the block of stone are 36 dm, 24 dm and 12 dm.

∴ Total surface area

$= 2 [36 \times 24 + 24 \times 12 + 36 \times 12]$ dm^2

$= 2 (864 + 288 + 432]$ dm$^2 = 3168$ dm^2

∴ Cost of polishing = $3168 \times 2 = 6336$ paise

$= $ Rs 63.36

3. (a) Let a^2 be the area of the square cross-section of the bar.

Therefore, $36 \times a^2 = 1$

$\Rightarrow a^2 = \dfrac{1}{36} \Rightarrow a = \dfrac{1}{6}$

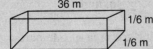

Now a cube of side $\dfrac{1}{6}$ m can be cut from this bar. Total number of such cubes cut off

from a bar of 36 m length $= \dfrac{36}{\frac{1}{6}} = 126$

∴ Cost of a cube $= \dfrac{108}{216} = $ Re $\dfrac{1}{2} = 50$ paise.

4. (a) Surface area of the open tank

$= [2 \times (20 \times 12 + 30 \times 12) + 30 \times 20]$ m^2

$= [2 \times (240 + 360) + 600]$ m^2

$= 1800$ m^2

∴ Length of the iron sheet $= \dfrac{1800}{3} = 600$ m

Cost of the iron sheet = Rs (600×10)

$= $ Rs 6000.

5. (c) Let l, b, h be the length, breadth and height of the cuboid.

∴ $l + b + h = 19$ and $\sqrt{l^2 + b^2 + h^2} = 5\sqrt{5}$

$\Rightarrow l^2 + b^2 + h^2 = \left(5\sqrt{5}\right)^2 = 125$

Surface area of the cuboid = $2 (lb + bh + lh)$

$= (l + b + h)^2 - (l^2 + b^2 + h^2)$

$= 19^2 - 125 = 361 - 125 = 236$ cm^2.

6. (c) Volume of the bigger cube = $(10)^3$ cm^3

$= 1000$ cm^3

Volume of the smaller cube = 500 cm^3

Required ratio $= \dfrac{\text{Edge of smaller cube}}{\text{Edge of bigger cube}}$

$= \dfrac{(500)^{1/3}}{(1000)^{1/3}} = \left(\dfrac{500}{1000}\right)^{1/3} = \left(\dfrac{1}{2}\right)^{1/3}$

7. (d) Height of the water in the tank $= \dfrac{45 \times 1000}{25 \times 20}$ cm

$= 9$ cm

$(\because 1 \text{ litre} = 1000 \text{ cm}^3)$

$\therefore$ Rise in height $= 11$ cm $- 9$ cm $= 2$ cm

$\therefore$ Volume of cube = Volume of water displaced

$= 25$ cm $\times 20$ cm $\times 2$ cm

$= 1000$ cm^3

$\Rightarrow$ Each edge of the cube $= \sqrt[3]{1000}$ cm $= 10$ cm.

8. (b) Original volume $= (6 \times 5 \times 2)$ cm$^3 = 60$ cm^3

New volume $= (60 - 20\%$ of $60)$ cm$^3 = 48$ cm^3

Since the thickness remains the same, $h = 2$ cm

$\Rightarrow l \times b \times 2 = 48 \Rightarrow l \times b = 24$

New length $= (6 - 10\%$ of $6)$ cm $= 5.4$ cm

New breadth $= (5 - 10\%$ of $5)$ cm $= 4.5$ cm

Here $l \times b = 5.4 \times 4.5 = 24.3$ cm^2

$\therefore$ l should clearly lie between 5 and 5.5, *i.e.*, $5 < l < 5.5$.

9. (a) Let the sides of the three cubes be $3k$, $4k$ and $5k$ respectively.

$\therefore$ Volume of the new cube formed by melting the three cubes $= (3k)^3 + (4k)^3 + (5k)^3$

$= 27k^3 + 64 k^3 + 125k^3 = 216k^3$

$\Rightarrow$ Each edge of the new cube $= \sqrt{216k^3} = 6 k$

$\therefore$ Diagonal of the new cube $= 6k\sqrt{3}$

Given, $6k\sqrt{3} = 12\sqrt{3} \Rightarrow 6k = 12 \Rightarrow k = 2$

$\therefore$ Sides of the three cubes are 6 cm, 8 cm and 10 cm respectively.

10. (b) Let l, b and h be the length, breadth and height of the cuboid respectively. Then,

$P = lbh, S = l + b + h, A = lh, B = lb, C = bh$

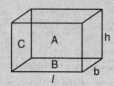

$\therefore$ $PS = lbh (l + b + h)$

$= l^2bh + lb^2h + lbh^2$

$= (lb)(lh) + (lb)(bh) + (lh) (bh)$

$= BA + BC + AC = AB + BC + CA$

11. (b) Total surface of the cuboid

$= 2 (lb + bh + lh)$

$= 2 (x + 2 + 2x) = 6x + 4$

Volume of the cuboid $= l \times b \times h = x \times 1 \times 2 = 2x$

Given, $6x + 4 = n \times 2x$, where n is an integer

$\Rightarrow (2n - 6)x = 4$

Since x is a length of the cuboid; x must be positive.

For positive volume of x, $(2n - 6) > 0$ or $2n > 6$ or $n > 3$

$\therefore$ n is a positive integer, minimum value of $n = 4$

$\therefore$ $6x + 4 = 2 \times 4 \times x \Rightarrow x = 2$ cm.

12. (a) After cutting the squares, the dimensions of the box are $(48 - 8 - 8)$ cm, *i.e.*, 32 m, $(36 - 8 - 8)$ cm, *i.e.*, 20 m and 8 m.

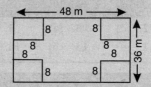

$\therefore$ Volume of the box $= (32 \times 20 \times 8)$ m

$= 5120$ m^3.

13. (b) When one cube is pasted on each of 5 faces of a cube, then

Surface area of each cube pasted $= 5x^2$

$= 5 \times (3)^2$ cm^2

$= 45$ cm^2

For 5 faces, total area $= 5 \times 45$ cm$^2 = 225$ cm^2

But surface area of the bottom face where the cube is not pasted $= x^2 = 3^2 = 9$ cm^2

$\therefore$ Total area $= 225$ cm$^2 + 9$ cm$^2 = 234$ cm^2.

14. (a) Total surface area of the cube $= 6a^2 = 6 \times 1^2$

$= 6$ sq. units

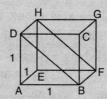

$DB = \sqrt{AD^2 + AB^2} = \sqrt{1 + 1} = \sqrt{2}$

Surface area of the section $DBFH$

$= DB \times BF = \sqrt{2} \times 1$ sq. units

$= \sqrt{2}$ sq. units

$\therefore$ Total surface area of one part $= \dfrac{6}{2} + \sqrt{2}$

$= (3 + \sqrt{2})$ sq. units.

15. (d) Area of sliced section $DBFH = DB \times BF$

$DB^2 = AD^2 + AB^2 = 4^2 + 3^2 = 16 + 9 = 25$

$\Rightarrow DB = 5$

∴   Required area = (5 × 5) cm² = 25 cm².

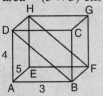

16. (d) Total surface area of the cube have a hole of cross-sectional area 4 cm².

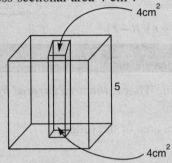

= (6(5)² – 2 × 4) cm²
= (150 – 8) cm² = 142 cm².

Area of the four walls of the hole made parallel to the side of the cube = 4 × 2 × 5 = 40 cm²

∴   Total surface area of the body so formed
= (142 + 40) cm² = 182 cm².

17. (d) Let the length, breadth and height of the cuboid be l, b and h respectively.

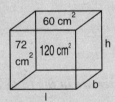

$l \times b = 60$, $b \times h = 72$, $l \times h = 120$

∴ $l \times b \times b \times h \times l \times h = 60 \times 72 \times 120$

⟹ $l^2 b^2 h^2 = 60 \times 36 \times 2 \times 60 \times 2$

∴   Volume of the cuboid = lbh

= $\sqrt{l^2 b^2 h^2}$ = $\sqrt{60 \times 36 \times 2 \times 60 \times 2}$ cm³

= 60 × 6 × 2 cm³ = 720 cm³.

18. (c) Volume of water in the tank in 5 hours

= $\left(225 \times 162 \times \dfrac{20}{100}\right)$ m³ = 7290 m³

Let x m/ hour be the speed of the water flowing through the aperture. Then,

Volume of water flown through the aperture in

5 hours = $\left(\dfrac{60}{100} \times \dfrac{45}{100} \times 5 \times x\right)$ m³ = $\dfrac{27x}{20}$ m³

∴ $\dfrac{27x}{20} = 7290 \Rightarrow x = 7290 \times \dfrac{20}{27}$ m / hour

= 5400 m/hr

19. (c) Volume of water in the reservoir

= $\left(80 \times 60 \times \dfrac{13}{2}\right)$ m³ = 31200 m³

Volume of water flown in 1 hour

= $\left(15 \times 1000 \times \dfrac{20}{100} \times \dfrac{20}{100}\right)$ m³ = 600 m³

Time taken = $\dfrac{31200}{600}$ hours = 52 hours.

20. (c) Area of the cake to be frosted

= Area of tops of
                    (cuboid A + cuboid B + cuboid C)
+ Area of four sides of
                    (cuboid A + cuboid B + cuboid C)

= (15² + 30² + 45²) cm² + [2(15 × 5 + 15 × 5)]
+ [2 (30 × 5 + 30 × 5)] + [2(45 × 5 + 45 × 5)]

= (225 + 900 + 2025) cm²
                                       + (300 + 600 + 900) cm²

= 4950 cm².

Section-B
CYLINDERS

KEY FACTS

Cylinder: A cylinder has a curved (lateral) surface with congruent circular ends.

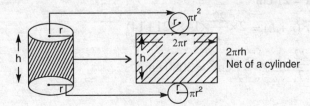

Net of a cylinder

Curved surface area = Area of rectangle = **$2\pi rh$**

Total surface area = Curved surface area + sum of the areas of the two bases

$$= 2\pi rh + 2\pi r^2 = 2\pi r\,(h + r)$$

Total surface area of a hollow circular cylinder = $2\pi Rh + 2\pi rh + 2\pi(R^2 - r^2)$

where R is the external radius and r is the internal radius.

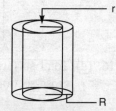

Volume of a right circular cylinder = (Area of base) × height = $\pi r^2 \times h = \pi r^2 h$

Volume of a hollow circular cylinder = $\pi R^2 h - \pi r^2 h = \pi h (R + r)(R - r)$

Solved Examples

Ex. 1. *A roller of diameter 70 cm and length 2 m is rolling on the ground. What is the area covered by the roller in 50 revolutions?*

Sol. Area covered by the roller in 50 revolutions = 50 × Curved surface area of the roller

$$= 50 \times 2 \times \frac{22}{7} \times 35 \times 200 \ \text{cm}^2$$

$$= 2200000 \ \text{cm}^2 = 220 \ \text{m}^2.$$

Ex. 2. *The circumference of the base of a circular cylinder is 6π cm. The height of the cylinder is equal to the diameter of the base. How many litres of water can it hold?*

Sol. Radius $= \dfrac{\text{Circumference}}{2\pi} = \dfrac{6\pi}{2\pi} = 3$ cm

$\Rightarrow$ Height = 6 cm.

$\therefore$ Volume of the cylinder $= \pi r^2 h = \pi \times (3)^2 \times 6 \ \text{cm}^3 = 54\pi \ \text{cm}^3$

$$= \frac{54\pi}{1000} \ \text{litres} = 0.54\pi \ \text{litres}.$$

Ex. 3. *Two cylindrical pots contain the same amount of water. If their diameter are in the ratio 2 : 3, then what is the ratio of their heights?*

Sol. Let r_1, h_1 and r_2, h_2 be the radii and height of the two cylinders respectively. Then

$$\pi r_1^2 h_1 = \pi r_2^2 h_2 \ \Rightarrow \ \frac{h_1}{h_2} = \frac{r_2^2}{r_1^2} = \left(\frac{r_2}{r_1}\right)^2 = \left(\frac{3/2}{2/2}\right)^2 = \frac{9}{4}.$$

Ex. 4. *The ratio between the radius of the base and height of the cylinder is 2 : 3. If the volume is 12936 cm³, what is the total surface area of the cylinder?*

Sol. Let the radius $= 2x$ cm and height $= 3x$ cm. Then,

$$\frac{22}{7} \times (2x)^2 \times 3x = 12936 \ \Rightarrow \ 12x^3 = \frac{12936 \times 7}{22}$$

$$\Rightarrow x^3 = \frac{12936 \times 7}{12 \times 22} = 343 \ \Rightarrow \ x = 7$$

$\therefore$ Radius = 14 cm, Height = 21 cm

Total surface area $= 2\pi r\,(r + h) = 2 \times \dfrac{22}{7} \times 14 \times (21 + 14)$

$$= \left(2 \times \frac{22}{7} \times 14 \times 35\right) \ \text{cm}^2 = 3080 \ \text{cm}^2.$$

Ex. 5. *The ratio of the total surface area to the lateral surface area of the cylinder is 5 : 3. Find the height of the cylinder if the radius of the cylinder is 12 cm?*

Sol. Let the height and radius of the cylinder be h and r respectively. Then,

$$\frac{\text{Total surface area}}{\text{Lateral surface area}} = \frac{2\pi rh + 2\pi r^2}{2\pi rh} = \frac{2\pi r(r+h)}{2\pi rh} = \frac{r+h}{h}$$

Given, $\dfrac{r+h}{h} = \dfrac{5}{3} \Rightarrow \dfrac{12+h}{h} = \dfrac{5}{3} \Rightarrow 36 + 3h = 5h \Rightarrow 2h = 36 \Rightarrow h = 18$ cm.

Ex. 6. *A slab of ice 8 inches in length, 11 inches in breadth and 2 inches thick was melted and resolidified in the form of a rod of 8 inches diameter. What is the length of such a rod in inches?*

Sol. Volume of cylindrical rod = Volume of the slab of ice

$$\Rightarrow \pi \times (4)^2 \times l = 8 \times 11 \times 2 \ (l = \text{length of slab})$$

$$\Rightarrow l = \frac{8 \times 11 \times 2 \times 7}{4 \times 4 \times 22} \text{ inches} = 3.5 \text{ inches}.$$

Ex. 7. *If a solid right circular cylinder made of iron is heated to increase its radius and height by 1% each, then by how much per cent is the volume of the solid increased?*

Sol. Let r and h be the radius and height of the cylinder respectively. Then, after 1% increase,

New radius = 1.1 r and new height = 1.1 h

Original volume = $\pi r^2 h$,

New volume = $\pi(1.01r)^2 (1.01h) = 1.030301 \ \pi r^2 h$

$\therefore$ % increase = $\left(\dfrac{1.030301\pi r^2 h - \pi r^2 h}{\pi r^2 h} \times 100\right)\% = \left(\dfrac{0.030301\pi r^2 h}{\pi r^2 h} \times 100\right)\%$

$$= (0.030301 \times 100)\% = 3.03\%$$

Ex. 8. *A lead pencil is in the shape of a cylinder. The pencil is 21 cm long with radius 0.4 cm and its lead is of radius 0.1 cm. What is the volume of wood in the pencil?*

Sol. Volume of wood = Volume of pencil – Volume of lead

$$= \pi (0.4)^2 \times 21 - \pi \times (0.1)^2 \times 21$$

$$= \frac{22}{7} \times \left[(0.4)^2 - (0.1)^2\right] \times 21$$

$$= \frac{22}{7} \times [0.16 - 0.01] \times 21$$

$$= \frac{22}{7} \times 0.15 \times 21 = 9.9 \text{ cm}^3.$$

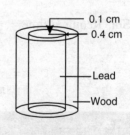

Ex. 9. *The lateral surface area of a hollow cylinder is 5632 cm². It is cut along its height and a rectangular sheet of width 44 cm is formed. Find the perimeter of the rectangular sheet?*

Sol. Since the cylinder is cut along its height, the circumference of its base (or top) = width of rectangular sheet, *i.e.*, $2\pi r = 44$ cm

Given $2\pi rh = 5632$

$$\Rightarrow h = \frac{5632}{44} \text{ cm} = 128 \text{ cm}$$

$\therefore$ Required perimeter = $2(l + b)$

$$= 2(128 + 44) \text{ cm} = 344 \text{ cm}.$$

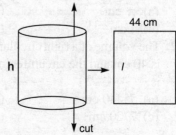

Ex. 10. *Water flows through a cylindrical pipe of internal diameter 7 cm at 2 m per second. If the pipe is always half full, then what is the volume of water (in litres) discharged in 10 minutes?*

Sol. Volume of water flown in 1 second $= \left(\dfrac{22}{7} \times \dfrac{7}{2} \times \dfrac{7}{2} \times 200\right)$ cm^3 = 7700 cm^3.

∴ Volume of water flown in 10 min = (7700 × 10 × 60) cm^3

$$= 4620000 \text{ cm}^3 = \dfrac{4620000}{1000} \text{ litres} = 4620 \text{ litres.}$$

Ex. 11. *A soft drink can has a circular base with diameter 7 cm and height 12 cm. A powder tin has a square base with side 7 cm and height 12 cm. What is the difference in their capacities?*

Sol. Capacity of the soft drink can $= \pi \times \left(\dfrac{7}{2}\right)^2 \times 12$ cm$^3 = \dfrac{22}{7} \times 3.5 \times 3.5 \times 12$ cm^3 = 462 cm^3

Capacity of the powder tin = (7 × 7 × 12) cm^3 = 588 cm^3

∴ Powder tin has a greater capacity by (588 – 462) cm^3 = 126 cm^3.

Ex. 12. *The trunk of a tree is a right cylinder 1.5 m in radius and 10 m high. What is the volume of the timber which remains when the trunk is trimmed just enough to reduce it to a rectangular parallelogram on a square base?*

Sol. Required volume = Area of square base × Height of trunk

Diagonals (AC = BD) of the square base = Diameter of circular base

∴ In △ AOB, AB $= \sqrt{(1.5)^2 + (1.5)^2} = \sqrt{2.25 + 2.25} = \sqrt{4.50}$

∴ Area of the square base of the tree = (AB)^2 = $(\sqrt{4.5})^2$ m^2 = 4.5 m^2

∴ Required volume = 4.5 m^2 × 10 m = 45 m^2.

Ex. 13. *The radius of a cylindrical cistern is 10 metres and its height is 15 metres. Initially the cistern is empty. We start filling the cistern with water through a pipe whose diameter is 50 cm. Water is coming out of the pipe at a velocity of 5 m/s. How many minutes will it take to fill the cistern with water?*

Sol. Required time $= \dfrac{\text{Volume of the cistern}}{\text{Volume of water coming out of the pipe in 1 second}}$

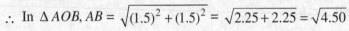

$$= \dfrac{\pi \times 10 \times 10 \times 15}{\pi \times \dfrac{0.5}{2} \times \dfrac{0.5}{2} \times 5} \text{ seconds} = \dfrac{200 \times 200 \times 15}{5 \times 5 \times 5} = 4800 \text{ seconds} = \dfrac{4800}{60} \text{ min} = 80 \text{ min.}$$

Question Bank–27

1. If the volume of a right circular cylinder with its height equal to the radius is $25\dfrac{1}{7}$ cm^3, then the radius of the cylinder is equal to
(a) π cm   (b) 3 cm
(c) 4 cm   (d) 2 cm

2. The volume of a right circular cylinder whose height is 40 cm and the circumference of its base is 66 cm is
(a) 55440 cm^3 (b) 34650 cm^3
(c) 7720 cm^3 (d) 13860 cm^3

3. If π cm^3 of metal is stretched to a wire of length 3600 m, then the diameter of the wire will be
(a) $\dfrac{1}{600}$ cm   (b) $\dfrac{1}{300}$ cm
(c) $\dfrac{1}{200}$ cm   (d) $\dfrac{1}{100}$ cm

4. The number of iron rods, each of length 7 m and diameter 2 cm that can be made out of 0.88 cubic metres of iron is $\left(\pi = \dfrac{22}{7}\right)$
(a) 300   (b) 400
(c) 500   (d) 600

5. The curved surface area of a cylindrical pillar is 264 m² and its volume is 924 m³. Find the ratio of its diameter to its height.

   (a) 3 : 7 (b) 7 : 3
   (c) 6 : 7 (d) 7 : 6

6. The volume of a right circular cylinder, 14 cm in height is equal to that of a cube whose edge is 11 cm. The radius of the base of the cylinder is

   (a) 5.2 cm (b) 5.5 cm
   (c) 11 cm (d) 22 cm

7. The number of coins of radius 0.75 cm and thickness 0.2 cm to be melted to make a right circular cylinder of height 8 cm and base radius 3 cm is

   (a) 460 (b) 500
   (c) 600 (d) 640

8. A hollow iron pipe of 21 cm long and its external diameter is 8 cm. If the thickness of the pipe is 1 cm and iron weighs 8 g/cm³, then the weight of the pipe is

   (a) 3.6 kg (b) 3.696 kg
   (c) 36 kg (d) 36.9 kg

9. A well 20 m in diameter is dug 14 m deep and the earth taken out is spread all round it to a width of 5 m to form an embankment. The height of the embankment is

   (a) 10 m (b) 11 m
   (c) 11.2 m (d) 11.5 m

10. The height of a right circular cylinder is 6 m. If three times the sum of the areas of its two circular faces is twice the area of the curved surface, then the radius of the base is

    (a) 4 m (b) 3 m
    (c) 2 m (d) 1 m

11. The ratio of the radii of two cylinders is $1 : \sqrt{3}$ and their heights are in the ratio 2 : 3. The ratio of their volumes is

    (a) 1 : 9 (b) 2 : 9
    (c) 4 : 9 (d) 5 : 9

12. If the radius of a right circular cylinder is decreased by 50% and its height is increased by 60%, its volume will be decreased by

    (a) 10% (b) 60%
    (c) 40% (d) 20%

13. The radii of the bases of two cylinders are in the ratio 3 : 5 and their heights are in the ratio 2 : 3. The ratio of their curved surfaces will be

    (a) 2 : 5 (b) 2 : 3
    (c) 3 : 5 (d) 5 : 3

14. The diameter of the base of a cylindrical drum is 35 dm and its height is 24 dm. It is full of kerosene. How many tins each of size 25 cm × 22 cm × 35 cm can be filled with kerosene from the drum $\left(\pi = \dfrac{22}{7}\right)$

    (a) 1200 (b) 1020
    (c) 600 (d) 120

15. If the diameter of a wire is decreased by 10%, by how much per cent approximately will the length be increased to keep the volume constant?

    (a) 5% (b) 17%
    (c) 20% (d) 23%

16. The radius of a cylindrical cistern is 10 metres and its height is 15 metres. Initially the cistern is empty. We start filling the cistern with water through a pipe whose diameter is 50 cm. Water is coming out of the pipe with a velocity 5 m/s. How many minutes will it take in filling the cistern with water?

    (a) 20 (b) 40
    (c) 60 (d) 80

17. The volume of a right circular cylinder can be obtained from its curved surface area by multiplying it by its

    (a) $\dfrac{radius}{2}$ (b) $\dfrac{2}{radius}$
    (c) height (d) 2 × height

18. If the radius of a right circular cylinder open at both the ends is decreased by 25% and the height of the cylinder is increased by 25%, then the surface area of the cylinder thus formed.

    (a) Remains unaltered
    (b) Increases by 25%
    (c) Decreases by 25%
    (d) Decreases by 6.25%

19. It is required to fix a pipe such that water flowing through it at a speed of 7 metres per minute fills a tank of capacity 440 cubic metres in 10 minutes. The inner radius of the pipe should be:

    (a) $\sqrt{2}$ m (b) 2 m
    (c) $\dfrac{1}{2}$ m (d) $\dfrac{1}{\sqrt{2}}$ m

20. A rectangular piece of paper is 24 cm long and 22 cm wide. A cylinder is formed by rolling the paper along its length. The volume of the cylinder is

    (a) 924 cm³ (b) 462 cm³
    (c) 264 cm³ (d) 528 cm³

21. A hollow right circular cylinder with height 8 cm and base radius 7 cm is opened out into a rectangle. What are the length and breadth of the rectangle respectively?
(a) 22 cm, 16 cm
(b) 44 cm, 8 cm
(c) 22 cm, 8 cm
(d) 44 cm, 16 cm

22. The magnitude of the volume of a closed right circular cylinder of unit height divided by the magnitude of the total surface area of the cylinder (r being the radius of the cylinder) is equal to
(a) $\dfrac{1}{2}\left(1+\dfrac{1}{r}\right)$
(b) $\dfrac{1}{2}\left(1+\dfrac{1}{r+1}\right)$
(c) $\dfrac{1}{2}\left(1-\dfrac{1}{r}\right)$
(d) $\dfrac{1}{2}\left(1-\dfrac{1}{r+1}\right)$

23. A hollow cylindrical tube 20 cm long is made of iron and its external diameter is 8 cm. The volume of iron used in making the tube is 440 cm³. What is the thickness of the tube?
(a) 1 cm
(b) 0.5 cm
(c) 2 cm
(d) 1.5 cm

24. What length of solid cylinder 2 cm in diameter must be taken to cast into a hollow cylinder of external diameter 12 cm, 0.25 cm thick and 15 cm long?
(a) 44.0123 cm
(b) 42.3215 cm
(c) 44.0625 cm
(d) 44.6023 cm

25. The volume of two cylinders are as $a : b$ and their heights are $c : d$. Find the ratio of their diameters?
(a) $\dfrac{ab}{bc}$
(b) $\dfrac{ad^2}{ac^2}$
(c) $\sqrt{\dfrac{ad}{bc}}$
(d) $\sqrt{\dfrac{a}{b}\times\dfrac{c}{d}}$

26. The inner diameter of a circular building is 54 cm and the base of the wall occupies a space of 2464 cm². The thickness of the wall is
(a) 1 cm
(b) 2 cm
(c) 4 cm
(d) 5 cm

27. The sum of the radius of the base and the height of a solid cylinder is 37 m. If the total surface area of the cylinder be 1628 sq m, its volume is
(a) 4620 m³
(b) 4630 m³
(c) 4520 m³
(d) 4830 m³

28. Water flows out through a circular pipe whose internal diameter is 2 cm, at the rate of 6 metres per second into a cylindrical tank, the radius of whose base is 60 cm. By how much will the level of water rise in 30 minutes?
(a) 2 m
(b) 3 m
(c) 4 m
(d) 5 m

Answers

1. (d)	**2.** (d)	**3.** (b)	**4.** (b)	**5.** (b)	**6.** (b)	**7.** (d)	**8.** (b)	**9.** (c)	**10.** (a)
11. (b)	**12.** (b)	**13.** (a)	**14.** (a)	**15.** (d)	**16.** (d)	**17.** (a)	**18.** (d)	**19.** (a)	**20.** (a)
21. (b)	**22.** (d)	**23.** (a)	**24.** (c)	**25.** (c)	**26.** (a)	**27.** (a)	**28.** (b)		

Hints and Solutions

1. (d) Let the height (h) and radius (r) of the cylinder = x cm

Then, $\pi(x)^2 \times (x) = 25\dfrac{1}{7} \Rightarrow \dfrac{22x^3}{7} = \dfrac{176}{7}$

$\Rightarrow x^3 = \dfrac{176}{22} = 8 \Rightarrow x = 2$ cm.

2. (d) Given, $2\pi r = 66 \Rightarrow r = \dfrac{66\times7}{2\times22} = \dfrac{21}{2}$ cm

$\therefore V = \pi r^2 h = \dfrac{22}{7}\times\dfrac{21}{2}\times\dfrac{21}{2}\times40 = 13860$ cm³.

3. (b) Length of wire = 3600 m = 360000 cm
= h(height of cylinder)

$\therefore$ Vol. of wire = $\pi r^2 \times 360000 = \pi$

$\Rightarrow r^2 = \dfrac{1}{360000} \Rightarrow r = \dfrac{1}{600}$

$\therefore$ Diameter = $2\times\dfrac{1}{600}$ cm = $\dfrac{1}{300}$ cm.

4. (b) Number of iron rods = $\dfrac{\text{Volume of iron}}{\text{Volume of one iron rod}}$

$= \dfrac{0.88}{\dfrac{22}{7}\times(0.01)^2\times7} = \dfrac{0.88}{22\times0.0001} = 400$
($\because$ Radius = 1 cm = 0.01 m)

5. (b) $\dfrac{\text{Volume}}{\text{Curved Surface Area}} = \dfrac{\pi r^2 h}{2\pi rh} = \dfrac{924}{264}$

$\Rightarrow \dfrac{r}{2} = \dfrac{7}{2} \Rightarrow r = 7$ m

and $2\pi rh = 264 \Rightarrow h = 264 \times \dfrac{7}{22} \times \dfrac{1}{2} \times \dfrac{1}{7} = 6$ m

$\therefore$ Required ratio $= \dfrac{2r}{h} = \dfrac{14}{6} = 7:3.$

6. (b) Let the radius of the base of the cylinder be r cm. Then,

$$\dfrac{22}{7} \times r^2 \times 14 = (11)^3$$

$\Rightarrow r^2 = \dfrac{11 \times 11 \times 11 \times 7}{22 \times 14} = \dfrac{121}{4}$

$\Rightarrow r = \sqrt{\dfrac{121}{4}} = \dfrac{11}{2} = 5.5$ cm.

7. (d) Number of coins $= \dfrac{\text{Volume of the cylinder}}{\text{Volume of one coin}}$

$= \dfrac{\pi \times 3 \times 3 \times 8}{\pi \times 0.75 \times 0.75 \times 0.2} = 640.$

8. (b) External radius $= \dfrac{8}{2}$ cm $= 4$ cm

$\therefore$ Internal radius $= 4$ cm $- 1$ cm $= 3$ cm

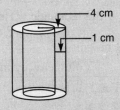

Volume of metal used in the pipe $= \pi(R^2 - r^2)h$

$= \dfrac{22}{7} \times (4^2 - 3^2) \times 21$

$= \dfrac{22}{7} \times 7 \times 21 = 462$ cm^3

$\therefore$ Weight of the pipe $= 462 \times 8$ g $= 3696$ g

$= 3.696$ kg.

9. (c) Volume of earth taken out from the well

$= \pi \times 10 \times 10 \times 14 = 1400\pi$ m^3

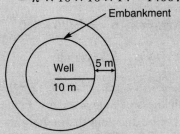

Area of embankment $= \pi \times (15^2 - 10^2)$

$= \pi \times (15 - 10)(15 + 10) = 125\pi$

where 10 m and 15 m are the inner and outer radii of the embankment.

$\therefore$ Height of the embankment $= \dfrac{\text{Volume}}{\text{Area}} = \dfrac{1400\pi}{12\pi}$

$= 11.2$ m.

10. (a) Let the radius and height of the cylinder be r m and h m respectively. Then,

$$3(\pi r^2 + \pi r^2) = 2 \times 2\pi r \times 6$$

$\Rightarrow 6\pi r^2 = 24\pi r \Rightarrow r = 4$ m.

11. (b) Required ratio $= \dfrac{\pi r_1^2 h_1}{\pi r_2^2 h_2} = \left(\dfrac{r_1}{r_2}\right)^2 \times \dfrac{h_1}{h_2}$

$= \left(\dfrac{1}{\sqrt{3}}\right)^2 \times \dfrac{2}{3} = \dfrac{2}{9} = 2:9.$

12. (b) Let the radius of the original cylinder $= r$ cm and height $= h$ cm

$\therefore$ Original volume $= \pi r^2 h$
Then, Radius after decrease $= r/2$ and height after increase $= \dfrac{160}{100}h = \dfrac{8}{5}h$

$\therefore$ New volume $= \pi \left(\dfrac{r}{2}\right)^2 \times \dfrac{8}{5}h$

$= \pi \times \dfrac{r^2}{4} \times \dfrac{8h}{5} = \dfrac{2\pi r^2 h}{5}$

$\therefore$ % decrease $= \left(\dfrac{\pi r^2 h - 2/5\pi r^2 h}{\pi r^2 h} \times 100\right)\%$

$= \left(\dfrac{\frac{3}{5}\pi r^2 h}{\pi r^2 h} \times 100\right)\% = \left(\dfrac{3}{5} \times 100\right)\%$

$= 60\%.$

13. (a) Ratio of curved surfaces $= \dfrac{2\pi r_1 h_1}{2\pi r_2 h_2} = \left(\dfrac{r_1}{r_2}\right) \times \left(\dfrac{h_1}{h_2}\right)$

$= \dfrac{3}{5} \times \dfrac{2}{3} = \dfrac{2}{5} = 2:5.$

14. (a) Required number of tins $= \dfrac{\text{Volume of drum}}{\text{Volume of 1 tin}}$

$= \dfrac{\dfrac{22}{7} \times \dfrac{350}{2} \times \dfrac{350}{2} \times 240}{25 \times 22 \times 35} = 1200.$

15. (d) Since the wire is cylindrical in shape, its

volume $= \frac{1}{4}\pi d^2 l$

where d is the diameter, l is the length. Also

$r^2 = \left(\frac{d}{2}\right)^2 = \frac{d^2}{4}$

Decrease in diameter $= d - 10\%$ of d

$= d\left(1-\frac{10}{100}\right) = \frac{9d}{10}$, new length $= l$ m. Then,

New volume = Original volume

$\Rightarrow \frac{1}{4}\pi\left(\frac{9d}{10}\right)^2 l_n = \frac{1}{4}\pi d^2 l$

$\Rightarrow \frac{l_n}{l} = \frac{10^2}{9^2} = \frac{100}{81} \Rightarrow l_n = \frac{100}{81}l$

$\therefore$ % increase in length $= \left(\frac{\frac{100}{81}l - l}{l}\times100\right)\%$

$= \frac{19}{81}\times100\% = 23\%$ (approx.)

16. (d) Volume of the tank $= \pi r^2 h$

$= \pi\times(10)^2\times15 = 1500\pi$ m³

Area of the pipe $= \frac{\pi D^2}{4} = \frac{\pi\times(0.5)^2}{4} = \frac{\pi}{16}$ m²

Rate of flow of water in the pipe

$=$ Area $\times$ Velocity $= \frac{\pi}{16}\times5 = \frac{5\pi}{16}$ m³/s

$\therefore$ Time required to fill the tank $= \frac{1500\pi}{\frac{5\pi}{16}}$ seconds

$= 4800$ seconds $= \frac{4800}{60}$ min $= 80$ min.

17. (a) $V = \pi r^2 h$ and Curved surface area $= 2\pi rh$

$\therefore V = \frac{2\pi rh}{2}\times r = CSA\times\frac{r}{2}$

18. (d) Original surface area $= 2\pi rh$

New surface area $= 2\pi\times\frac{75^3 r}{100_4}\times\frac{125^5 h}{100_4}$

$= 1.875\,\pi rh$

$\therefore$ % decrease $= \left(\frac{2\pi rh - 1.875\pi rh}{2\pi rh}\times100\right)\%$

$= \left(\frac{0.125\pi rh}{2\pi rh}\times100\right)\% = 6.25\%.$

19. (a) Let the inner radius of pipe be r m.

Then, $440 = \frac{22}{7}\times r^2\times7\times10$

$\Rightarrow r^2 = \frac{440}{22\times10} = 2 \Rightarrow r = \sqrt{2}$ m.

20. (a) Let the radius of the cylinder be R.

Since the cylinder is formed by rolling the rectangular sheet along its length, circumference of base of cylinder $= 2\pi R =$ width of the sheet, i.e.,

$2\times\frac{22}{7}\times R = 22$ cm

$\Rightarrow R = \frac{7}{2}$ cm.

Height of cylinder = length = 24 cm

$\therefore$ Volume of the cylinder $= \pi R^2 h$

$= \frac{22}{7}\times\frac{7}{2}\times\frac{7}{2}\times24$ cm³ $= 924$ cm³.

21. (b) Length of the rectangle $= 2\pi r = 2\times\frac{22}{7}\times7$ cm

$= 44$ cm

Breadth of the rectangle = Height of cylinder $= 8$ cm.

22. (d) Required magnitude

$= \dfrac{\text{Volume of cylinder}}{\text{Total surface area of cylinder}}$

$= \frac{\pi r^2 h}{2\pi r(h+r)} = \frac{\pi r^2}{2\pi r(1+r)}$ ($\because h = 1$)

$= \frac{1}{2}\frac{r}{(1+r)} = \frac{1}{2}\left(1-\frac{1}{r+1}\right)$

23. (a) Let the internal radius of the tube be r cm.

External radius $= \frac{8}{2}$ cm $= 4$ cm,

Height $= 20$ cm

$\therefore$ Volume of iron used in the tube

$= \pi(4^2 - r^2)\times20$ cm

Given, $\pi\times(16-r^2)\times20 = 440$

$\Rightarrow \frac{22}{7}\times20\times(16-r^2) = 440$

$\Rightarrow 16-r^2 = 7 \Rightarrow r^2 = 16-7 = 9 \Rightarrow r = 3$

$\therefore$ Thickness of tube $= 4$ cm $- 3$ cm $= 1$ cm.

24. (c) Let h be the length (or height) of the solid cylinder 2 cm in diameter.

$\therefore$ Volume of solid cylinder $= \pi\,(1)^2 h = \pi h$

External radius of hollow cylinder = 6 cm
Internal radius of hollow cylinder = 6 cm − 0.25 cm
$\qquad\qquad = 5.75$ cm

$\therefore$ Volume of the hollow cylinder

$\quad = \pi(6^2 - 5.75^2)\times 15$

Given, $\pi(6^2 - 5.75^2)\times 15 = \pi h$

$\Rightarrow (6+5.75)(6-5.75)\times 15 = h$
$\Rightarrow h = 11.75 \times 0.25 \times 15 = 44.0625$ cm.

25. (c) Let r_1 and r_2 be radii of the two cylinders. Given

$$h_1 : h_2 = c : d \quad \text{or} \quad \frac{h_1}{h_2} = \frac{c}{d}$$

$$\therefore \frac{\pi r_1^2 h_1}{\pi r_2^2 h_2} = \frac{a}{b} \quad \text{or} \quad \left(\frac{r_1}{r_2}\right)^2 \times \left(\frac{h_1}{h_2}\right) = \frac{a}{b}$$

$$\Rightarrow \left(\frac{r_1}{r_2}\right)^2 \times \frac{c}{d} = \frac{a}{b} \Rightarrow \left(\frac{r_1}{r_2}\right)^2 = \frac{ad}{bc}$$

$$\Rightarrow \frac{r_1}{r_2} = \sqrt{\frac{ad}{bc}}$$

26. (a) Let the outer radius of the wall be R cm. Then,

$$\pi R^2 = 2464 \quad \Rightarrow \quad R^2 = \frac{2464\times 7}{22} = 784$$

$$\Rightarrow R = \sqrt{784} = 28 \text{ cm}$$

Inner diameter = 54 cm
$\Rightarrow$ Inner radius = 27 cm
$\therefore$ Thickness = 28 cm − 27 cm = 1 cm.

27. (a) Given, $r + h = 37$ m.

Also, total surface area $= 2\pi r(r+h)$

$\therefore 2\pi r(r+h) = 1628 \Rightarrow 2\pi r \times 37 = 1628$

$$r = \frac{1628\times 7}{2\times 22\times 37} = 7 \text{ m}$$

$\therefore$ Height $(h) = 37$ m − 7 m = 30 m

Its volume $= \pi r^2 h = \left(\dfrac{22}{7}\times 7\times 7\times 30\right)$ m³ $= 4620$ m³.

28. (b) Length of water flown in 30 minutes
$\qquad\qquad = (6\times 60\times 30)$ m = 10800 m

Radius of pipe $= \dfrac{1}{100}$ m, $h = 10800$ m

$\therefore$ Volume of water flown through the pipe

$$= \left(\pi\times\frac{1}{100}\times\frac{1}{100}\times 10800\right) \text{ m}^3$$

Let h be the rise in the level of water in the tank in 30 minutes.

Then, $\pi\times\dfrac{60}{100}\times\dfrac{60}{100}\times h = \pi\times\dfrac{1}{100}\times\dfrac{1}{100}\times 10800$

$$\Rightarrow h = \frac{108}{100}\times\frac{5}{3}\times\frac{5}{3} = 3 \text{ m}.$$

Self Assessment Sheet–26

1. The four walls of a room can be fully covered by 70 square wall papers of 2 m × 2 m size. The length of the room is 18 m and its breadth is twice that of its height. If the cost of carpeting is Rs 20 per square metre, what will be the total expenditure in carpeting the room?
(a) Rs 1800   (b) Rs 5040
(c) Rs 3600   (d) Rs 1400

2. The weight of a cubic metre of a certain metal is 480 kg. It is melted and then rolled into a square bar 4 m long. Now an exact cube is cut from it. Find the weight of the cube.
(a) 240 kg   (b) 80 kg
(c) 120 kg   (d) 60 kg

3. The length, breadth and height of a cuboid are in the ratio 1 : 2 : 3. The length, breadth and height of the cuboid are increased by 100%, 200% and 200% respectively. Then, the increase in the volume of the cuboid is

(a) 5 times   (b) 6 times
(c) 12 times   (d) 17 times

4. The sum of length, breadth and depth of a cuboid is 19 cm and the length of its diagonal is 11 cm. Find the total surface area of the cuboid.
(a) 361 cm²   (b) 240 cm²
(c) 209 cm²   (d) 121 cm²

5. The external length, breadth and height of a closed rectangular wooden box are 18 cm, 10 cm and 6 cm respectively and the thickness of wood is $\dfrac{1}{2}$ cm. When the box is empty, it weighs 15 kg and when filled with sand it weighs 100 kg. The weight of a cubic cm of wood and a cubic cm of sand are respectively:

(a) $\dfrac{1}{7}$ kg, $\dfrac{1}{9}$ kg   (b) $\dfrac{1}{21}$ kg, $\dfrac{1}{9}$ kg

(c) $\dfrac{1}{21}$ kg, $\dfrac{1}{7}$ kg   (d) $\dfrac{1}{9}$ kg, $\dfrac{1}{7}$ kg

Ch 27–14 **IIT Foundation Mathematics Class – VIII**

6. The radius of the base and height of a right circular cylinder are each increased by 20%. The volume of the cylinder will be increased by:
   (a) 40%   (b) 60%
   (c) 72.8%   (d) 96%

7. The curved surface of a cylinder is 1000 cm². A wire of diameter 5 mm is wound around it, so as to cover it completely. What is the length of the wire used?
   (a) 22 m   (b) 20 m
   (c) 18 m   (d) None of these

8. Three rectangles A_1, A_2 and A_3 have the same area. Their lengths a_1, a_2 and a_3 respectively are such that $a_1 < a_2 < a_3$. Cylinders C_1, C_2 and C_3 are formed from A_1, A_2 and A_3 respectively by joining the parallel sides along the breadth. Then,

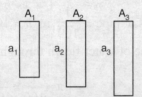

(a) C_1 will enclose maximum volume.
(b) C_2 will enclose maximum volume.
(c) C_3 will enclose maximum volume
(d) Each of C_1, C_2 and C_3 will enclose equal volume.

9. The difference between the outside and inside surfaces of a cylindrical metal pipe 14 cm long is 44 cm². If the pipe is made of 99 cm³ of metal, find the sum of the inner and outer radii of the pipe?
   (a) 4.5 cm   (b) 3.5 cm
   (c) 2 cm   (d) 5.5 cm

10. Water is flowing at the rate of 3 km/hr through a circular pipe of 20 cm internal diameter into a circular cistern of diameter 10 m and depth 2 m. In how much time will the cistern be filled?
   (a) 2 hours   (b) 1 hour 40 min
   (c) 1 hour 15 min   (d) 2 hour 30 min

Answers

1. (c)	2. (d)	3. (d)	4. (b)	5. (b)	6. (c)	7. (b)	8. (c)	9. (a)	10. (b)

Unit Test–5

1. *ABCD* is a rectangle of dimensions 8 units and 6 units. *AEFC* is a rectangle drawn in such a way that diagonal *AC* of the first rectangle is one side and side opposite to it is touching the first rectangle at *D* as shown in the figure. What is the ratio of the area of rectangle *ABCD* to that of *AEFC*?

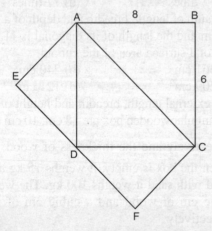

   (a) 2   (b) 3/2
   (c) 1   (d) 8/9

2. The difference between the area of a square and that of an equilateral triangle on the same base is

1/4 cm². What is the length of the side of the triangle?
   (a) $(4-\sqrt{3})^{1/2}$ cm   (b) $(4+\sqrt{3})^{1/2}$ cm
   (c) $(4-\sqrt{3})^{-1/2}$ cm   (d) $(4+\sqrt{3})^{-1/2}$ cm

3. If x and y are respectively the areas of a square and a rhombus of sides of same length. Then what is $x : y$?
   (a) 1 : 1   (b) $2:\sqrt{3}$
   (c) $4:\sqrt{3}$ (d) 3 : 2

4. If the area of a circle inscribed in an equilateral triangle is 4π cm² then what is the area of the triangle?
   (a) $12\sqrt{3}$ cm²   (b) $9\sqrt{3}$ cm²
   (c) $8\sqrt{3}$ cm²   (d) 18 cm²

5. The length of a rectangle is twice the diameter of a circle. The circumference of the circle is equal to the area of a square of side 22 cm. What is the breadth of the rectangle if its perimeter is 668 cm.
   (a) 24 cm   (b) 26 cm
   (c) 52 cm   (d) cannot be determined

6. The two diagonals of a rhombus are of lengths 55 cm and 48 cm. If p is the perpendicular height of the rhombus, then which one of the following is correct?
(a) 36 cm < p < 37 cm   (b) 35 cm < p < 36 cm
(c) 34 cm < p < 35 cm   (d) 33 cm < p < 34 cm

7. In the given figure, AC is a diameter of a circle with radius 5 cm. if $AB = BC$ and $CD = 8$ cm, the area of the shaded region to the nearest whole number is

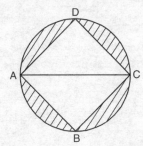

(a) 28 cm^2 (b) 29 cm^2
(c) 30 cm^2 (d) 45 cm^2

8. The radius of a circle is 20 cm. Three more concentric circles are drawn inside it in such a manner that it is divided into 4 equal parts. Find the radius of the smallest circle?
(a) 5 cm   (b) 4 cm
(c) 10 cm   (d) 8 cm

9. $ABCD$ is a trapezium in which $AB \parallel CD$ and $AB = 2\ CD$. If its diagonals intersect each other at O then ratio of the area of triangle AOB and COD is
(a) 1 : 2   (b) 2 : 1
(c) 1 : 4   (d) 4 : 1

10. If the perimeter of an isosceles right triangle is $(6 + 3\sqrt{2})$m, then the area of the triangle is
(a) 4.5 m^2 (b) 5.4 m^2
(c) 9 m^2 (d) 81 m^2

11. Sixteen cylindrical cans each with a radius of 1 unit are placed inside a cardboard box four in a row. If the cans touch the adjacent cans and or the walls of the box, then which of the following could be the interior area of the bottom of the box in square units?
(a) 16   (b) 32
(c) 64   (d) 128

12. Find the number of coins 1.5 cm in diameter and 0.2 cm thick to be melted to form a right circular cylinder of height 10 cm and diameter 4.5 cm?
(a) 430   (b) 440
(c) 460   (d) 450

13. A solid metallic cube of edge 4 cm is melted and recast into solid cubes of edge 1 cm. If x is the surface area of the melted cube and y is the total surface area of the cubes recast, then what is $x : y$?
(a) 2 : 1   (b) 1 : 2
(c) 1 : 4   (d) 4 : 1

14. A cylindrical vessel of base radius 14 cm is filled with water to some height. If a rectangular solid of dimensions 22 cm × 7 cm × 5 cm is immersed in it, what is the rise in the water level?
(a) 0.5 cm   (b) 1.25 cm
(c) 1 cm   (d) 1.5 cm

15. Increasing the radius of the base of a cylinder by 6 units increase the volume by y cubic units. Increasing the altitude of the cylinder by 6 units also increases the volume by y cubic units. If the original altitude is 2 units, find the original radius ?
(a) 8 units   (b) 4 units
(c) 6 units   (d) 5 units

16. A metallic cube of edge 2.5 cm is melted and recasted into the form of a cuboid of base 1.25 cm × 0.25 cm. Find the increase in the surface area.
(a) 123.325 cm^2 (b) 150.625 cm^2
(c) 113.125 cm^2 (d) 37.5 cm^2

17. The value of a metallic cylindrical pipe is 748 cm^3. Its length is 14 cm and its external radius is 9 cm. Find its thickness.
(a) 1.5 cm   (b) 0.5 cm
(c) 2 cm   (d) 1 cm

18. A rectangular paper 11 cm by 8 cm can be exactly wrapped to cover the curved surface of a cylinder of height 8 cm. The volume of the cylinder is
(a) 66 cm^3 (b) 77 cm^3
(c) 88 cm^3 (d) 121 cm^3

19. A solid cylinder has a total surface area of 462 cm^2. Its curved surface area in one-third of its total surface area. The volume of the cylinder is : $\left(\pi = \dfrac{22}{7} \right)$
(a) 792 cm^3 (b) 539 cm^3
(c) 495 cm^3 (d) 676 cm^3

20. The barrel of a fountain pen, cylindrical in shape, is 7 cm long and 5 mm in diameter. A full barrel of ink in the pen will be used up on writing 330 words on an average. How many words would use up a bottle of ink containing one fifth of a litre?
(a) 60000   (b) 66000
(c) 48000   (d) 50000

Answers

1. (c)	2. (c)	3. (a)	4. (a)	5. (b)	6. (a)	7. (c)	8. (c)	9. (d)	10. (a)
11. (c)	12. (d)	13. (c)	14. (b)	15. (c)	16. (c)	17. (d)	18. (b)	19. (b)	20. (c)

UNIT–6
STATISTICS

- *Data Handling*
- *Probability*

Chapter

28

DATA HANDLING

KEY FACTS

1. A collection of numerical facts about objects or events is called **data.**
2. **Statistics** is the science of collecting, organising and analyzing sets of data in order to reveal information.
3. Arranging raw data in ascending or descending order of magnitude is called an **array.**
4. The difference between the greatest and the least values of observation in a set of data is called the **range** of that set.
5. The number of times an observation occurs in a given data set is called its **frequency.**
6. **Grouped frequency distribution:** If the data extends over a wide range, data is usually condensed into groups or classes called the **class intervals.**

    (i) Each class interval is bounded by two figures which are called **class limits.**

    Ex. $0 - 5, 5 - 10, 10 - 15$ are class intervals with class limits as 0, 5, 10, 15 etc.

    For a particular class interval, the number on the left hand side is the lower limit and the number on the right hand side is the upper limit.

    Ex. In the interval $10 - 15$, 10 is the lower limit and 15 is the upper limit.

    There are two types of class intervals:

    (a) **Exclusive form:** Here the upper limit of one class coincides with the lower limit of the next class.

    Ex. $0 - 10, 10 - 20, 20 - 30, \ldots$ etc.

    Note. 5–10 means 5 and less than 10. 10 is included in the class interval $10 - 15$.

    (b) **Inclusive form:** Here the upper limit of a class interval does not coincide with the lower limit of the next class.

    Ex. $1 - 10, 11 - 20, 21 - 30, \ldots$ etc.

    To convert the class interval from an inclusive form to exclusive form, we find the difference between the upper limit of a class interval and the lower limit of the next class interval and divide it by 2.

    Ex. In the above example, diff $= \dfrac{11 - 10}{2} = \dfrac{1}{2} = 0.5$.

    Thus, the class limits are converted to exclusive form by subtracting 0.5 from the lower limit and adding 0.5 to the upper limit.

    Hence, the class boundaries are $0.5 - 10.5, 10.5 - 20.5, \ldots$

    (ii) The difference between the upper and lower class limits is called the **size** or **width** of the **class interval.**

    Ex. Size of the class interval $0 - 10 = 10 - 0 = 10$.

    (iii) The mid-value of a class-interval is called its **class mark.**

    Class mark = upper limit + lower limit

    Ex. Class mark of the interval $10 - 15 = \dfrac{10 + 15}{2} = 12.5$

(iv) The frequency corresponding to a particular class interval is called its **class frequency.**

7. Arithmetic mean

(i) Arithmetic mean for n observations $x_1, x_2, ...x_n$ is given by

$$\frac{x_1 + x_2 + ... + x_n}{n}$$

(ii) Arithmetic mean for a discrete series $= \dfrac{\Sigma (f_i x_i)}{\Sigma f_i}$, where x_i is the observation and f_i is the corresponding frequency.

(iii) Arithmetic mean of grouped data $= \dfrac{\Sigma f_i x_i}{\Sigma f_i}$, where x_i is the class mark of interval corresponding to the class frequency f_i.

(iv) If a particular number is multiplied, divided, added or subtracted to each observation, then the mean also gets multiplied, divided, added or subtracted to that number.

8. Median of ungrouped data:

To find the median of an individual series, arrange the given numbers in ascending or descending order. If the number of observations is:

(a) odd, the median is size of the $\left(\dfrac{n+1}{2}\right)$th term.

(b) even, the median is size of $\left[\dfrac{\left(\dfrac{n}{2}\right)\text{th term} + \left(\dfrac{n+1}{2}\right)\text{th term}}{2}\right]$.

9. Mode: The most occurring observation of a given set of data is called its mode. The mode is the term that has the largest frequency.

10. Empirical formula: Mode = 3 Median – 2 Mean

11. Bar graphs: Bar Graphs are used to represent unclassified frequency distributions. A bar graph consists of axes and a series of labelled horizontal or vertical bars that give you the information.

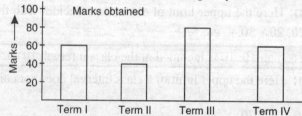

12. Pie graph: A pie-graph is a graphical representation of the given numerical data in the form of sectors of a circle.

$$\text{Central angle for a sector} = \left(\frac{\text{Value of the component}}{\text{Total value}}\right) \times 360°$$

Expenditure of a school on various sports in 2011

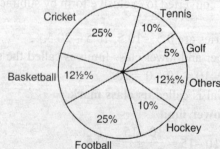

13. A histogram is a graphical representation of grouped data in which class intervals are taken along the horizontal axis and the frequencies along the vertical axis. For each class, a rectangle is constructed with class interval as the base and the height proportional to the corresponding frequency.

Ex. The histogram corresponding to the given grouped frequency table is

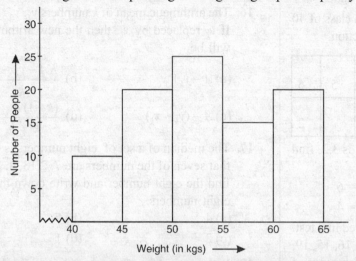

Weight (in kg)	Frequency
40 – 45	10
45 – 50	20
50 – 55	25
55 – 60	15
60 – 65	20

Question Bank–28

1. Given class intervals 0 – 8, 8 – 16, 16 – 24, 24 – 32, ..., then 24 is considered in the class
   (a) 16 – 24   (b) 24 – 32
   (c) 8 – 24   (d) 24 – 38

2. The class mark of a particular class is 17.5 and the class size is 5. The class interval is :
   (a) 14 – 19   (b) 15 –20
   (c) 14.5 – 19.5   (d) 17.5 – 22.5

3. The class marks of a distribution are 105, 115, 125, 135, 145, 155, 165. The class size is :
   (a) 5   (b) 10
   (c) 15   (d) 20

4. The population of four towns *A, B, C* and *D* as on 2011 are as follows:

Town	Population
A	6863
B	519
C	12185
D	1755

What is the most appropriate diagram to present the above data?
   (a) Pie chart   (b) Bar chart
   (c) Histogram   (d) Line graph

5. Let $\overline{x}$ be the mean of *n* observations $x_1, x_2, x_3, ... x_n$. If $(a - b)$ is added to each observation, then what is the mean of the new set of observations?
   (a) zero   (b) $\overline{x}$
   (c) $\overline{x} - (a - b)$ (d) $\overline{x} + (a - b)$

6. Calculate the mean of weekly wages from the following frequency distribution:

Wages (in Rs)	No. of workers
30 – 40	10
40 – 50	20
50 – 60	40
60 – 70	16
70 – 80	8
80 – 90	6

   (a) 52   (b) 43
   (c) 48   (d) 56

7. The observations 29, 32, 48, 50, *x*, *x* + 2, 72, 78, 84, 95 are arranged in ascending order. 'What' is the value of *x* if the median of the data is 63?
   (a) 61   (b) 62
   (c) 62.5   (d) 63

8. The mean weight of the students in a certain class is 60 kg. The mean weight of the boys from the class is 70 kg and that of the girls is 55 kg. What is the ratio of the number of boys to that of girls?
   (a) 2 : 1   (b) 1 : 2
   (c) 1 : 4   (d) 4 : 1

9. If the monthly expenditure pattern of a person who earns a monthly salary of Rs 15,000 is represented in a pie graph, then the sector angle of an item on transport expenses measures 15°. What is his monthly expenditure on transport?
   (a) Rs 450
   (b) Rs 625

(c) Rs 675

(d) Cannot be computed from the given data.

10. The table shows the number of fillings a class of 40 pupils had at the time of a dental inspection.

Number of fillings	0	1	2	3	4	5	6
Number of pupils	1	4	8	x	9	y	2

If the mean number of fillings per pupil is 3.2, find the values of x and y.

(a) $x = 5, y = 4$　　　　(b) $x = 10, y = 6$

(c) $x = 9, y = 6$　　　　(b) $x = 12, y = 4$

11. In a class of 19 students, seven boys failed in a test. Those who passed scored 12, 15,17, 15, 16, 15, 19, 19, 17, 18, 18 and 19 marks. The median score of the 19 students in the class is

(a) 15　　　　(b) 16

(c) 17　　　　(d) 18

12. The mean of 1, 7, 5, 3, 4 and 4 is m. The observations 3, 2, 4, 2, 3, 3 and p have the mean $(m - 1)$. Find the median of this set of data.

(a) 4　　　　(b) 2.5

(c) 3　　　　(d) 5

13. A data set of n observations has mean $2\overline{x}$ while another data set of $2n$ observations has mean $\overline{x}$. The mean of the combined data set of $3n$ observations will be equal to:

(a) $\overline{x}$　　　　(b) $\dfrac{3}{2}\overline{x}$

(c) $\dfrac{2}{3}\overline{x}$　　　　(d) $\dfrac{4}{3}\overline{x}$

14. The pie chart given below shows the expenses incurred and saving by a family in a month. What is the percentage of expenses incurred on account of recreation?

(a) $\dfrac{800}{17}\%$

(b) 20%

(c) 35%

(d) 40%

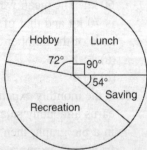

15. The mean age of a class is 16 years. If the class teacher aged 40 years old is also included, the mean age increases to 17 years. The number of students in the class are:

(a) 23　　　　(b) 33

(c) 44　　　　(d) 16

16. The arithmetic mean of k numbers $y_1, y_2, ..., y_k$ is A. If y_k replaced by x_k, then the new arithmetic mean will be

(a) $A - y_k + x_k$　　　　(b) $\dfrac{KA - y_k + x_k}{k}$

(c) $A - (y_k - x_k)$　　　　(d) $\dfrac{(k-1)A}{k} - y_k + x_k$

17. The median of a set of eight numbers is 4.5. Given that seven of the numbers are 7, 2,13, 4, 8, 2 and 1, find the eight number and write down the mode of eight numbers.

(a) 4　　　　(b) 7

(c) 2　　　　(d) 1

18. If the ratio of mode and median is 7 : 4, then find the ratio of mean and mode is:

(a) 7 : 11　　　　(b) 2 : 3

(c) 5 : 14　　　　(d) 8 : 9

19. If the median of $\dfrac{x}{6}, \dfrac{x}{2}, \dfrac{x}{4}, \dfrac{3x}{5}$ and $\dfrac{7x}{10}$ is 3, then the mean of the given observations is :

(a) 2.5　　　　(b) 3.06

(c) 2.16　　　　(d) 2.66

20.

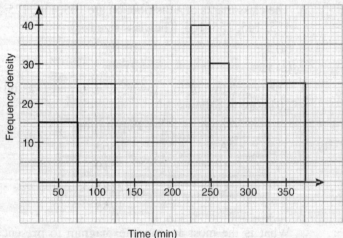

The diagram shows the histogram of the amount of time spent watching TV in a day for a group of students. Remember that the area of each rectangle is proportional to its frequency. Find the total number of students. (Take the most common base of all rectangles as 1 unit)

(a) 165　　　　(b) 140

(c) 130　　　　(d) 110

Answers

1. (b)	2. (b)	3. (b)	4. (b)	5. (d)	6. (d)	7. (b)	8. (b)	9. (b)	10. (b)
11. (a)	12. (c)	13. (d)	14. (d)	15. (a)	16. (b)	17. (c)	18. (c)	19. (d)	20. (b)

Hints and Solutions

2. (b) Class mark $= \dfrac{\text{lower limit} + \text{upper limit}}{2}$

∴ Considering all the options,

$\dfrac{15+20}{2} = \dfrac{35}{2} = \mathbf{17.5}$

5. (d) Given, $\dfrac{x_1+x_2+x_3+....+x_n}{n} = \bar{x}$

∴ $\dfrac{x_1+(a-b)+x_2+(a-b)+x_3+(a-b)+ ....+x_n+(a-b)}{n}$

$= \dfrac{x_1+x_2+x_3+....x_n}{n} + \dfrac{n(a-b)}{n}$

$= \bar{x}+(a-b)$

6. (d)

Wages (in Rs)	Class mark (x)	Frequency (No. of workers) (f)	fx
30 – 40	35	10	350
40 – 50	45	20	900
50 – 60	55	40	2200
60 – 70	65	16	1040
70 – 80	75	8	600
80 – 90	85	6	510
		$\Sigma f = 100$	$\Sigma fx = 5600$

∴ Mean $= \dfrac{\Sigma fx}{\Sigma f} = \dfrac{5600}{100} = \mathbf{56}$

7. (b) Since there are 10 observations,
Median value

$= \dfrac{\left(\dfrac{10}{2}\right)\text{th value} + \left(\dfrac{10}{2}+1\right)\text{th value}}{2}$

$= \dfrac{\text{5th value + 6th value}}{2}$

$= \dfrac{x+x+2}{2} = x+1$

Given, $x+1 = 63 \Rightarrow x = \mathbf{62.}$

8. (b) Let the number of boys be x and the number of girls be y.

Then, $\dfrac{70 \times x + 55 \times y}{(x+y)} = 60$

$\Rightarrow 70x + 55y = 60x + 60y$

$\Rightarrow 10x = 5y \Rightarrow \dfrac{x}{y} = \dfrac{5}{10} = \dfrac{1}{2} \Rightarrow \mathbf{x : y = 1 : 2}.$

9. (b) Monthly expenditure on transport

$= \dfrac{15°}{360°} \times \text{Rs } 15000 = \mathbf{Rs\ 625.}$

10. (b) Given, total frequency = 40

$\Rightarrow 1+4+8+x+9+y+2 = 40$

$\Rightarrow 24+x+y = 40 \Rightarrow x+y = 16$ ...(i)

Also, given mean = 3.2

$\Rightarrow \dfrac{\Sigma fx}{\Sigma f} = 3.2$

$\Rightarrow \dfrac{0\times1+1\times4+2\times8+3\times x+4\times9+5\times y+6\times2}{40} = 3.2$

$\Rightarrow 4+16+3x+36+5y+12 = (3.2\times40) = 128$

$\Rightarrow 3x+5y = 128-68 = 60$ ... (ii)

Now solve the simultaneous equations (i) and (ii) to get the values of x and y.

11. (a) The seven boys who failed the test scored the marks less than these given marks. Let the marks be $x_1, x_2, x_3, x_4, x_5, x_6, x_7$

Then the marks arranged in ascending order are

$x_1, x_2, x_3, x_4, x_5, x_6, x_7$, 12, 15, 15, 15, 16, 17, 17, 18, 18, 19, 19, 19

∴ Median score $= \left(\dfrac{19+1}{2}\right)$th item

$= $ 10th item $= \mathbf{15.}$

12. (c) $m = \dfrac{1+7+5+3+4+4}{6} = \dfrac{24}{6} = \mathbf{4}$

Now mean of 3, 2, 4, 2, 3, 3 and p is $(4-1) = 3$

∴ $\dfrac{3+2+4+2+3+3+p}{7} = 3$

$\Rightarrow \dfrac{17+p}{7} = 3 \Rightarrow 17+p = 21 \Rightarrow p = 4$

∴ The observations are 3, 2, 4, 2, 3, 3 and 4.

Arranged in ascending order, the numbers are 2, 2, 3, 3, 3, 4, 4.

$$\text{Median} = \left(\frac{7+1}{2}\right)\text{th value} = \text{4th value} = 3$$

13. (d) Mean of the combined data $= \dfrac{n \times 2\bar{x} + 2n \times \bar{x}}{n + 2n}$

$$= \frac{2n\bar{x} + 2n\bar{x}}{3n} = \frac{4n\bar{x}}{3n} = \frac{4}{3}\bar{x}$$

14. (d) Sector angle corresponding to recreation
$$= 360° - (72° + 90° + 54°)$$
$$= 360° - 216°$$
$$= 144°$$

∴ % age expenses incurred on account of

recreation $= \left(\dfrac{144°}{360°} \times 100\right)\% = \mathbf{40\%}$

15. (a) Let the number of students $= x$

Mean age = 16 years

∴ Total age of all students = $16x$

New mean age = 17 years

∴ Total age of all the students + teacher
$= 17(x+1)$

Given, $17(x+1) - 16x = 40$
$$x + 17 = 40$$
$$x = \mathbf{23}$$

16. (b) Given, $\dfrac{y_1 + y_2 + y_3 + ... + y_k}{k} = A$

$$y_1 + y_2 + y_3 + ... + y_k = kA$$
If y_k is replaced by x_k. Then,
New sum is $= KA - y_k + x_k$

∴ New mean $= \dfrac{kA - y_k + x_k}{k}$

17. (c) Let the eighth number be x. Then the numbers arranged in ascending order:
$$1, 2, 2, 4, 7, 8, 13, x$$
Since these are 8 in number,

Median value $= \dfrac{\left(\frac{8}{2}\right)\text{th value} + \left(\frac{8}{2}+1\right)\text{th value}}{2}$

$$= \frac{\text{4th value} + \text{5th value}}{2}$$

$$= \frac{4+7}{2} = \frac{11}{2} = 5.5$$

∴ Mode = **2**

18. (c) Mode = 3 Median – 2 Mean

∴ Mode : Median = 7 : 4

∴ Let, Mode = $7x$ and Median = $4x$

∴ $7x = 3 \times 4x - 2$ Mean

$\Rightarrow 7x = 12x - 2$ Mean

$\Rightarrow 2$ Mean $= 5x \Rightarrow$ Mean $= \dfrac{5}{2}x$

∴ Mean : Mode $= \dfrac{5}{2}x : 7x = 5x : 14x$

$$= \mathbf{5 : 14}$$

19. (d) Arranged in ascending order, the values are

$$\frac{x}{6}, \frac{x}{4}, \frac{x}{2}, \frac{3x}{5} \text{ and } \frac{7x}{10}$$

∴ Median value $= \left(\dfrac{5+1}{2}\right)$th value

$$= \text{3rd value} = \frac{x}{2}$$

Given $\dfrac{x}{2} = 3 \Rightarrow x = 6$

∴ The values are $\dfrac{6}{6}, \dfrac{6}{4}, \dfrac{6}{2}, \dfrac{3\times6}{5}, \dfrac{7\times6}{10},$

i.e., 1, 1.5, 3, 3.6 and 4.2

∴ Mean $= \dfrac{1 + 1.5 + 3 + 3.6 + 4.2}{5}$

$$= \frac{13.3}{5} = \mathbf{2.66}$$

20. (b) The most common base is 50, so it is taken as 1 unit. For the class 100 – 200, the base is 2 units. For the class 200 – 225 and 225 – 250, the base

$$= \frac{1}{2} \text{ unit}$$

∴ For the class 100 – 200,
The actual frequency = $10 \times 2 = 20$
For the class 200 – 225,

The actual frequency = $40 \times \dfrac{1}{2} = 20$
For the class 225 – 250,

The actual frequency = $30 \times \dfrac{1}{2} = 15$

∴ Total number of students
$$= 15 + 25 + 20 + 20 + 15 + 20 + 25$$
$$= \mathbf{140}$$

Self Assessment Sheet–27

1. Let m be the mid-point and l the upper class limit of a class in a continuous frequency distribution. The lower limit of the class is
   - (a) $2m + l$
   - (b) $2m - l$
   - (c) $m - l$
   - (d) $m - 2l$

2. When rectangles are drawn with the areas proportional to the frequencies of respective class intervals, what is the diagram called?
   - (a) Bar graph
   - (b) Frequency polygon
   - (c) Pie graph
   - (d) Histogram

3. If a frequency distribution for the number of persons x in a household is prepared with class intervals (1– 4), (5–8), (9–12) etc, then the number of persons x belonging to class interval (5 – 8) satisfies.
   - (a) $5 < x < 8$
   - (b) $5 \leq x < 8$
   - (c) $5 < x \leq 8$
   - (d) $5 \leq x \leq 8$

4. The arithmetic mean of a set of 10 numbers is 20. If each number is first multiplied by 2 and then increased by 5, then what is the mean of the new numbers?
   - (a) 20
   - (b) 25
   - (c) 40
   - (d) 45

5. The mean of twenty observations $x_1, x_2, x_3, x_4, \ldots x_{20}$ is m. If each of the first ten observations is increased by 8 and each of the next 10 observations is decreased by 8, then the new mean is n. The difference $m - n$ equals
   - (a) 16
   - (b) 20
   - (c) 0
   - (d) 8

6. The empirical relationship between mean, median and mode is :
   - (a) Mean > Median > Mode
   - (b) Mean = Median = Mode
   - (c) Mode – Mean = 3 (Median – Mean)
   - (d) Mean – Mode = 3 (Mean – Median)

7.

Class	0–10	10–20	20–30	30–40	40–50
Frequency	5	x	15	16	6

The missing frequency marked x of the above distribution whose mean is 27 is :
   - (a) 7
   - (b) 8
   - (c) 9
   - (d) 10

8. Seven of the eight numbers in a distribution are
   11, 16, 6, 10, 13, 11, 13
   Given that the median of the distribution is 12, find the mean of the distribution.
   - (a) 12
   - (b) 11
   - (c) 11.6
   - (d) 12.2

9. The mean of a, b and c is x. If $ab + bc + ca = 0$, what is the mean of a^2, b^2 and c^2?
   - (a) $\dfrac{x^2}{3}$
   - (b) x^2
   - (c) $3x^2$
   - (d) $9x^2$

10. The median of a set of 11 distinct observations is 15.5. If each of the smallest 5 observations of the set are decreased by 3, then the median of the new set is:
    - (a) Is increased by 3
    - (b) Is decreased by 3
    - (c) Is three times the original median
    - (d) Remains the same as that of the original set

Answers

| 1. (b) | 2. (d) | 3. (d) | 4. (d) | 5. (c) | 6. (c) | 7. (b) | 8. (c) | 9. (c) | 10. (d) |

Chapter 29

PROBABILITY

KEY FACTS

1. **Probability** is the chance of occurring of a certain event when measured quantitatively.
2. **Trial** is the performance of an experiment, such as throwing a dice or tossing a coin.
3. An **event** is anything whose probability we want to measure, as getting an even number on throwing a dice.
4. An **outcome** is any way in which an event can happen.
5. **Sample space** is the set of all possible outcomes of an experiment, generally denoted by S.
   Ex. When a dice is thrown, $S = \{ 1, 2, 3, 4, 5, 6\}$
6. **Equally likely events:** When two or more events have an equal chance of happening, then they are called **equally likely.**
   Ex. In a throw of a dice, all the six faces (1, 2, 3, 4, 5, 6) are equally likely to occur.
7. The probability of a **certain event** is **1** and the probability of an **impossible event** is **0**. Probability is **never greater than 1 or less than 0.**
8. Generally probability examples involve coins, dice and pack of cards. Here is a remainder of their outcomes.
   (i) A toss of a coin has **two outcomes: head or tail.**
   (ii) A throw of a dice has **six outcomes: 1, 2, 3, 4, 5, 6.**
   (iii) A pack of cards consists of **52 cards**, divided into **4 suits**, **each suit** containing **13 cards**.
      Hearts (Red), Spades (Black), Diamonds (Red) and Clubs (Black). Each of the suits has 13 cards bearing the values 2, 3, 4, 5, 6, 7, 8, 9, 10, Jack, Queen, King and Ace. The Jack, Queen and King are called 'picture cards'.
      Hence there are 12 picture cards in all.
      Total number of possible outcomes = 52
9. Probability of an event = $P(E) = \dfrac{\text{Number of ways the event can happen}}{\text{Total number of possible outcomes}}$
   Ex. P(drawing a picture card) = $\dfrac{12}{52} = \dfrac{3}{13}$ ($\therefore$ There are 12 picture cards)
10. **Mutually exclusive and exhaustive events:**
    Events such as tossing a head or a tail with a coin, drawing a Queen or a Jack from a pack of cards, throwing an even or a odd number with a dice are all **mutually exclusive events**. *Here, the occurrence of an event rules out the happening of all the other events in the same experiment, i.e.,*
    If we toss a coin, we can never get a head or a tail in the same toss.

Probability (head) $= \dfrac{1}{2}$, Probability (tail) $= \dfrac{1}{2}$

Also, Probability (head) + Probability (tail) $= \dfrac{1}{2} + \dfrac{1}{2} = 1$

Such events are also called **exhaustive events,** because there are no other possibilities and their probabilities always add up to **1.**

An example of events that are **not mutually exclusive** would be throwing a prime number or an odd number with a dice. There are two prime number 3 and 5 which are also odd numbers.

11. **Complementary event:** The complementary event of A = Event A not happening.

Thus, *an event and its complementary event are both mutually exclusive and exhaustive.*

Hence, **Probability (event A not happening) = 1 – Probability (event A happening)**

Thus, $P(A) + P(\overline{A}) = 1$, where $\overline{A}$ denotes the complementary event of A.

Ex. Probability of drawing a blue ball from a bag of 4 blue, 6 red and 2 yellow balls

$$= \dfrac{4}{4+6+2} = \dfrac{4}{12} = \dfrac{1}{3}.$$

$\therefore\ P$ (not drawing a blue ball) $= 1 - \dfrac{1}{3} = \dfrac{2}{3}$

12. **Addition rule for events:**

When two events are mutually exclusive, we can find the probability of either of them occurring by adding together the separate probabilities.

Ex. The probability of throwing a 3 or a 5 with a dice is

$$P(3) + P(5) = \dfrac{1}{6} + \dfrac{1}{6} = \dfrac{2}{6} = \dfrac{1}{3}.$$

Note. *Addition rule in case of events which are not mutually exclusive.*

Ex. From a well shuffled pack of 52 cards, a card is drawn at random. Find the probability that it is either a spade or a queen.

Sol. Let A be the event of getting a spade and B be the event of getting a queen.

A and B are not mutually exclusive as there is a queen of spades also, so

P(either a spade or a queen) = P(spade) + P(queen) – P(queen of spade)

$$= \dfrac{13}{52} + \dfrac{4}{52} - \dfrac{1}{52} = \dfrac{16}{52} = \dfrac{4}{13}$$

---

13. **Independent events:** If the outcome of event A does not affect the outcome of event B, then the events A and B are called independent events.

$$P\ (A \text{ and } B) = P(A).\ P(B)$$

Ex. Two dice are thrown. Find the probability of getting a prime number on one dice and even number on the other dice.

Sol. Here both the events are independent, so

P (a prime number) $= \dfrac{3}{6} = \dfrac{1}{2}$ ($\because$ There are 3 prime numbers 2, 3, 5)

P (an even number) $= \dfrac{3}{6} = \dfrac{1}{2}$ ($\because$ There are 3 even numbers 2, 4, 6)

$\therefore$ Reqd. probability $= \dfrac{1}{2} \times \dfrac{1}{2} = \dfrac{1}{4}$.

14. **Combined events:** There are many times when two or more events occur together. Some common examples are:

    (i) **Tossing two coins together**

    There are four equally likely outcomes: {HH, HT, TH, TT}

    (ii) **Tossing three coins together**

    There are eight equally likely outcomes: {HHH, HTT, THH, TTH, HHT, HTH, THT, TTT}

    Ex. $P(\text{3 heads or 3 tails}) = \frac{1}{8} + \frac{1}{8} = \frac{2}{8} = \frac{1}{4}$.

    (iii) **Throwing two dice together**

    There are 36 equally likely outcomes:

    {(1,1), (1,2), (1,3), (1,4), (1,5), (1,6), (2,1), (2,2), (2,3), (2,4), (2,5), (2,6), (3,1), (3,2), (3,3), (3,4), (3,5), (3,6), (4,1), (4,2), (4,3), (4,4), (4,5), (4,6), (5,1), (5,2), (5,3), (5,4), (5,5), (5,6), (6,1), (6,2), (6,2), (6,3), (6,4), (6,5), (6,6)}

    Ex. $P(\text{a sum of 5}) = \frac{4}{36} = \frac{1}{9}$ ($\therefore$ There are four combinations to get a sum of 5. (1,4), (2,3), (3,2) and (4,1)).

15. **Odds of an event**

    If there are p outcomes favourable to a certain event and q outcomes unfavourable to the event in a sample space

    then, odds in favour of the event $= \dfrac{\text{Number of favourable outcomes}}{\text{Number of unfavourable outcomes}} = \dfrac{p}{q}$

    Odds against the event $= \dfrac{\text{Number of unfavourable outcomes}}{\text{Number of favourable outcomes}} = \dfrac{q}{p}$

    If odds in favour of an event A are $m : n$, then

    Probability of happening of event $A = P(A) = \dfrac{m}{m+n}$

    Probability of not happening of event $A = P(\bar{A}) = \dfrac{n}{m+n}$

Solved Examples

Ex. 1. *A bag contains 27 balls. Ten are red, 2 are green and the rest are white. Annie takes out a ball from the bag at random. What is the probability that she takes*

    (i) a white ball   *(ii) a ball that is red or green*

Sol. (i) Number of white balls $= 27 - (10 + 2) = 15$

$$P(\text{white ball}) = \frac{\text{Number of white balls}}{\text{Total number of balls}} = \frac{15}{27} = \frac{5}{9}$$

(ii) $P(\text{red ball or green ball}) = P(\text{red}) + P(\text{green})$

$$= \frac{10}{27} + \frac{2}{27} = \frac{12}{27} = \frac{4}{9}.$$

Ex. 2. *Each morning I walk to work or take a taxi to work. The probability that I walk to work is $\frac{3}{5}$. What is the probability that I take a taxi?*

Sol. The events "walk to work" and "take a taxi to work" are mutually exclusive events, therefore are complementary events.

Hence, $P(\text{take a taxi to work}) = 1 - P(\text{walk to work}) = 1 - \frac{3}{5} = \frac{2}{5}$.

Ex. 3. *When two coins are tossed together, what is the probability of getting at least one tail?*

Sol. When two coins are tossed together, there are four equally likely outcomes, *i.e.,* $S = \{ HH, HT, TH, TT \}$
The outcomes having at least one tail are *HT, TH* and *TT*, *i.e.,* 3.

$$\therefore P \text{ (at least one tail)} = \frac{3}{4}.$$

Ex. 4. *Sid draws a card from a pack of cards, replaces it, shuffles the pack and then draws another card. What is the probability that the cards are both aces?*

Sol. $P(\text{ace}) = \frac{4}{52} = \frac{1}{13}$ ($\because$ there are 4 aces in a pack of 52 cards)

$$\therefore P \text{ (both aces)} = P \text{ (first ace \textbf{and} second ace)} = \frac{1}{13} \times \frac{1}{13} = \frac{1}{169}.$$

Ex. 5. *There are seven white and one brown eggs in an egg box. Ruchira decides to make a two-egg omelette. She takes out each egg from the box without looking at its colour. What is the probability that Ruchira gets an omelette made from*

     (i) two white eggs   *(ii) one white and one brown egg*   *(iii) two brown eggs*

Sol. (i) $P \text{ (first white egg)} = \frac{7}{8}$ (7 white eggs, 8 total eggs)

$P \text{ (second white egg)} = \frac{6}{7}$ (6 white eggs, 7 total eggs as 1 egg has been taken out)

$$\therefore P \text{ (omelette of two white eggs)} = \frac{7}{8} \times \frac{6}{7} = \frac{6}{8} = \frac{3}{4}$$

(ii) $P \text{ (first white egg)} = \frac{7}{8}$ $P \text{ (first brown egg)} = \frac{1}{8}$

     $P \text{ (second brown egg)} = \frac{1}{7}$ $P \text{ (second white egg)} = \frac{7}{7} = 1$

$$\therefore P \text{ (omelette of one white and one brown egg)} = \frac{7}{8} \times \frac{1}{7} + \frac{1}{8} \times \frac{7}{7}$$

$$= \frac{1}{8} + \frac{1}{8} = \frac{1}{4}.$$

Note. The eggs can be taken out in any order, so we consider both the cases, it is either first case or second case.

(iii) $P \text{ (first brown egg)} = \frac{1}{8}$

     $P \text{ (second brown egg)} = \frac{0}{7}$

$$\therefore P \text{ (omelette of two brown eggs)} = \frac{1}{8} \times 0 = \mathbf{0}.$$

Question Bank–29

1. In a test, the marks obtained by 15 students are 34, 37, 44, 39, 45, 46, 35, 42, 48, 40, 39, 33, 43, 47, 44. The probability that a pupil chosen at random passed the test, if the passing marks are 40 is:

   (a) $\dfrac{8}{15}$ (b) $\dfrac{3}{5}$

   (c) $\dfrac{7}{15}$ (d) $\dfrac{11}{15}$

2. Which of the following pairs of events is not mutually exclusive?
   (a) Throwing a number greater than 4 with a dice/ Throwing a number less than 4 with a dice.
   (b) Drawing a red card from a pack of cards /Draw a club from a pack of cards.
   (c) Drawing a diamond from a pack of cards/ Drawing an ace from a pack of cards.
   (d) Drawing a vowel card from a set of alphabet cards/Drawing a consonant card from a set of alphabet cards.

3. An electronic machine choses random numbers from 1 to 30. What is the probability that the number chosen is a triangular number?
   (a) $\frac{11}{30}$
   (b) $\frac{1}{10}$
   (c) $\frac{1}{6}$
   (d) $\frac{7}{30}$

4. In a simultaneous toss of two coins, find the probability of getting two tails.
   (a) $\frac{1}{2}$
   (b) $\frac{1}{4}$
   (c) $\frac{3}{4}$
   (d) $\frac{1}{3}$

5. Three coins are tossed simultaneously. Find the probability of at least one head and one tail.
   (a) $\frac{1}{2}$
   (b) $\frac{1}{4}$
   (c) $\frac{3}{4}$
   (d) None of these

6. In a single throw of two dice, what is the probability of getting a total of 11.
   (a) $\frac{1}{9}$
   (b) $\frac{1}{18}$
   (c) $\frac{1}{12}$
   (d) $\frac{35}{36}$

7. A bag contains 4 blue, 5 red and 7 green balls. If 4 balls are drawn one by one with replacement, what is the probability that all are blue?
   (a) $\frac{1}{16}$
   (b) $\frac{1}{4}$
   (c) $\frac{1}{256}$
   (d) $\frac{1}{64}$

8. The set $S = \{n : n$ is an integer, $1 \leq n \leq 50\}$. If an element of S is selected at random, find the probability that it does not contain the digit '2' at all.
   (a) $\frac{7}{25}$
   (b) $\frac{18}{25}$

   (c) $\frac{17}{50}$
   (d) $\frac{7}{10}$

9. In a single throw of two dice, find the probability of getting a doublet of odd numbers.
   (a) $\frac{1}{9}$
   (b) $\frac{1}{18}$
   (c) $\frac{1}{36}$
   (d) $\frac{1}{12}$

10. What is the probability that a card drawn at random from a pack of playing cards is either a king or a jack?
    (a) $\frac{1}{13}$
    (b) $\frac{2}{13}$
    (c) $\frac{3}{13}$
    (d) $\frac{4}{9}$

11. Two dice are thrown. Find the odds in favour of getting the sum 4.
    (a) 1 : 11
    (b) 11 : 1
    (c) 4 : 11
    (d) 11: 4

12. A machine generates a two-digit number randomly. Find the probability that the number generated is either less than 25 or greater than 85.
    (a) $\frac{27}{89}$
    (b) $\frac{28}{89}$
    (c) $\frac{28}{90}$
    (d) $\frac{29}{90}$

13. From a pack of cards, two are drawn, the first being replaced before the second is drawn. Find the probability that the first is a club and the second is a red card.
    (a) $\frac{1}{52}$
    (b) $\frac{1}{26}$
    (c) $\frac{1}{8}$
    (d) $\frac{4}{13}$

14. If A and B are two mutually exclusive and exhaustive events with $P(B) = 3\, P(A)$, then what is the value of $P(\bar{B})$?
    (a) $\frac{3}{4}$
    (b) $\frac{1}{4}$
    (c) $\frac{1}{3}$
    (d) $\frac{2}{3}$

15. The probability that a student passes in Mathematics is $\frac{4}{9}$ and he passes in Physics is $\frac{2}{5}$. Assuming that passing in Mathematics and Physics are independent of each other, what is

the probability that he passes in Mathematics and fails in Physics?

(a) $\dfrac{4}{15}$ (b) $\dfrac{8}{45}$

(c) $\dfrac{26}{45}$ (d) $\dfrac{19}{45}$

16. An aircraft has three engines A, B and C. The aircraft crashes if all the three engines fail. The probabilities of failure are 0.03, 0.02 and 0.05 for engines A, B and C respectively. What is the probability that the aircraft will not crash?

(a) 0.00003   (b) 0.90

(c) 0.99997   (d) 0.90307

17. A card is drawn from an ordinary pack of 52 cards and a gambler bets that it is a spade or an ace. What are the odds against his winning the bet?

(a) 9 : 4   (b) 4 : 9

(c) 5 : 9   (d) 9 : 5

18. A problem in statistics is given to four students A, B, C and D. Their chances of solving it are $\dfrac{1}{3}, \dfrac{1}{4}, \dfrac{1}{5}$

and $\dfrac{1}{6}$ respectively. What is the probability that the problem will be solved?

(a) $\dfrac{1}{3}$ (b) $\dfrac{2}{3}$

(c) $\dfrac{4}{5}$ (d) None of these

19. A man A speaks truth in 80% of the cases and another man B in 90% of the cases. While stating the same fact, what is the probability that they contradict?

(a) $\dfrac{37}{50}$ (b) $\dfrac{13}{50}$

(c) $\dfrac{16}{50}$ (d) None of these

20. From a bag containing 60 standard and 40 substandard articles, two articles are chosen at random. What is the probability that one of them is standard and the other substandard?

(a) $\dfrac{60}{100} \times \dfrac{40}{100}$ (b) $\dfrac{60}{100} \times \dfrac{39}{100}$

(c) $\dfrac{16}{33}$ (d) 24%

Answers

1. (b)	2. (c)	3. (d)	4. (b)	5. (c)	6. (b)	7. (c)	8. (b)	9. (d)	10. (b)
11. (a)	12. (d)	13. (b)	14. (b)	15. (a)	16. (c)	17. (a)	18. (b)	19. (b)	20. (c)

Hints and Solutions

1. (b) Number of students getting marks equal to or more than 40 = 9 (44, 45, 46, 42, 48, 40, 43, 47, 44)

∴ $P(\text{pass}) = \dfrac{\text{Number of students who passed}}{\text{Total number of students}}$

$= \dfrac{9}{15} = \dfrac{3}{5}$.

2. (c) As we have an ace of diamonds, so drawing a diamond from a pack of cards also includes the possibility of drawing an ace. Hence the events are not mutually exclusive.

3. (d) Triangular numbers between 1 and 30 are 1, 3, 6, 10, 15, 21, 27, *i.e.*, 7 in number.

∴ Reqd. probability = $\dfrac{7}{30}$

4. (b) The sample space for a simultaneous toss of two coins is

$S = \{ HH, HT, TH, TT\}$

Favourable cases = $\{TT\}$

∴ $P(\text{getting two tails}) = \dfrac{1}{4}$.

5. (c) The sample space for a simultaneous toss of three coins

$S = \{HHH, HHT, HTT, THH, TTH, THT, HTH, HHH\}$

Favourable cases = $\{HHT, HTT, THH, TTH, THT, HTH\}$

∴ $P(\text{getting at least one head, one tail}) = \dfrac{6}{8} = \dfrac{3}{4}$.

6. (b) Total number of exhaustive cases = 6 × 6 = 36

A total of 11 may be obtained in 2 ways as (5, 6), (6, 5).

∴ $P(\text{total of 11}) = \dfrac{2}{36} = \dfrac{1}{18}$.

7. (c) Probability of drawing a blue ball

$= \dfrac{4}{4+5+7} = \dfrac{4}{16} = \dfrac{1}{4}$

∴ Required probability $= \dfrac{1}{4} \times \dfrac{1}{4} \times \dfrac{1}{4} \times \dfrac{1}{4} = \dfrac{1}{256}$.

8. (b) $S = \{1, 2, 3, 4, 5, ...49, 50\}$

Numbers containing digit 2 are 2, 12, 20, 21, 22, 23, 24, 25, 26, 27, 28, 29, 32, 42, *i.e.,* 14 in number.

Probability (Number contains 2) $= \dfrac{14}{50} = \dfrac{7}{25}$

∴ Probability (Number does not contain 2)

$= 1 - \dfrac{7}{25} = \dfrac{18}{25}.$

9. (d) Total number of exhaustive cases in a single throw of two dice $= 6 \times 6 = 36$
Doublets are obtained as (1, 1), (2, 2), (3, 3), (4, 4), (5, 5), (6, 6)
Number of doublets of odd numbers $= 3$

∴ Required probability $= \dfrac{3}{36} = \dfrac{1}{12}.$

10. (b) P(a king or a jack) $= P$(King) $+ P$(Jack)

$= \dfrac{4}{52} + \dfrac{4}{52} = \dfrac{2}{13}.$

11. (a) Total number of exhaustive cases $= 6 \times 6 = 36$
A sum of 4 can be obtained as (1, 3) (2, 2) (3, 1)
Therefore, there are 3 favourable outcomes and $(36 - 3) = 33$ unfavourable outcomes.

∴ Odds in favour of sum of $4 = \dfrac{3}{33} = \dfrac{1}{11}.$

12. (d) Two digit numbers are $\{10, 11, 12, ..., 98, 99\}$ *i.e.,* 90 in number.
Numbers less than 25 $= \{10, 11, 12, 13, ..., 24\}$, *i.e.,* 15
Numbers greater than 85 $= \{86, 87, 88, ..., 99\}$, *i.e.,* 14

∴ P (number less than 25 or greater than 85)

$= \dfrac{15}{90} + \dfrac{14}{90} = \dfrac{29}{90}.$

13. (c) Total number of cards $= 52$
Clubs $= 13$

∴ P (club) $= \dfrac{13}{52} = \dfrac{1}{4}$
Red cards $= 26$

∴ P(red) $= \dfrac{26}{52} = \dfrac{1}{2}$

∴ P(club and red card) $= \dfrac{1}{4} \times \dfrac{1}{2} = \dfrac{1}{8}.$

14. (b) Since $P(A)$ and $P(B)$ are two mutually exclusive and exhaustive events, $P(A) + P(B) = 1$
$\Rightarrow P(A) + 3 P(A) = 1 \qquad \because P(B) = 3 P(A)$
$\Rightarrow 4 P(A) = 1 \Rightarrow P(A) = \dfrac{1}{4}$

$\Rightarrow P(B) = 1 - \dfrac{1}{4} = \dfrac{3}{4}$
$\Rightarrow P(\bar{B}) = 1 - \dfrac{3}{4} = \dfrac{1}{4}.$

15. (a) P (pass in Maths) $= \dfrac{4}{9}$

P (pass in Phy) $= \dfrac{2}{5}$

∴ P (fail in Phy) $= 1 - \dfrac{2}{5} = \dfrac{3}{5}$

∴ P(pass in Maths and fail in Phy) $= \dfrac{4}{\cancel{9}_3} \times \dfrac{\cancel{3}^1}{5} = \dfrac{4}{15}.$

16. (c) P (aircraft will crash) $= 0.03 \times 0.02 \times 0.05$
$= 0.00003$
P (aircraft will not crash) $= 1 - 0.00003$
$= \mathbf{0.99997.}$

17. (a) Let event A : a spade is drawn and event B : an ace is drawn.
Probability of winning the bet $= P$ (A or B)
P (A or B) $= P(A) + P(B) - P(A \cap B)$

> **Note.** Here A and B are not mutually exclusive events, hence the common part has to be taken into consideration.

$= \dfrac{13}{52} + \dfrac{4}{52} - \dfrac{1}{52}$ (There is **one** ace of spades)
$= \dfrac{16}{52} = \dfrac{4}{13}$

∴ Probability of losing the bet $= 1 - \dfrac{4}{13} = \dfrac{9}{13}$

∴ Odds against winning the bet $= \dfrac{9}{13} : \dfrac{4}{13} = \mathbf{9 : 4}.$

18. (b) $P(A \text{ solving}) = \dfrac{1}{3} \Rightarrow P(A \text{ not solving}) = 1 - \dfrac{1}{3} = \dfrac{2}{3}$

$P(B \text{ solving}) = \dfrac{1}{4} \Rightarrow P(B \text{ not solving}) = 1 - \dfrac{1}{4} = \dfrac{3}{4}$

$P(C \text{ solving}) = \dfrac{1}{5} \Rightarrow P(C \text{ not solving}) = 1 - \dfrac{1}{5} = \dfrac{4}{5}$

$P(D \text{ solving}) = \dfrac{1}{6} \Rightarrow P(D \text{ not solving}) = 1 - \dfrac{1}{6} = \dfrac{5}{6}$

∴ Probability (problem not solved)

$= \dfrac{2}{3} \times \dfrac{3}{4} \times \dfrac{4}{5} \times \dfrac{5}{6} = \dfrac{1}{3}$

$\Rightarrow$ Probability (problem solved) $= 1 - \dfrac{1}{3} = \dfrac{2}{3}.$

19. (b) Probability (contradiction)
$= P(A \text{ speaks truth}) \times P(B \text{ does not speak truth})$
$+ P(A \text{ does not speak truth}) \times P(B \text{ speaks truth})$
$= \dfrac{80}{100} \times \dfrac{10}{100} + \dfrac{20}{100} \times \dfrac{90}{100}$

$$= \frac{8}{100} + \frac{18}{100} = \frac{26}{100} = \frac{13}{50}.$$

20. (c) Here two articles are chosen together without replacement

P(one of the articles is substandard)

= P (first article is standard) × P(second article is substandard) + P (first article is substandard) × P (second article is standard)

$$= \frac{60}{100} \times \frac{40}{99} + \frac{40}{100} \times \frac{60}{99} = \frac{4800}{9900} = \frac{48}{99} = \frac{16}{33}.$$

Self Assessment Sheet–28

1. Each letter of the word "INDEPENDENT" is written on individual cards. The cards are placed in a box and mingled thoroughly. A card with letter 'N' is removed from the box. Now find the probability of picking a card with a consonant?

(a) $\frac{7}{11}$ (b) $\frac{7}{10}$

(c) $\frac{3}{5}$ (d) $\frac{2}{5}$

2. In a simultaneous throw of two dice, what is the number of exhaustive events?

(a) 6   (b) 12

(c) 36   (d) 18

3. Three coins are tossed simultaneously. What is the probability that head and tail show alternately. (*i.e., HTH* or *THT*)?

(a) $\frac{3}{8}$ (b) $\frac{1}{4}$

(c) $\frac{1}{8}$ (d) $\frac{1}{2}$

4. Two cards are drawn from a well shuffled pack of 52 cards without replacement. The probability of drawing a queen and a jack is :

(a) $\frac{16}{663}$ (b) $\frac{2}{663}$

(c) $\frac{4}{663}$ (d) $\frac{8}{663}$

5. In a single throw of two dice, find the probability of getting a total of 3 or 5.

(a) $\frac{1}{3}$ (b) $\frac{5}{6}$

(c) $\frac{1}{9}$ (d) $\frac{1}{6}$

6. A husband and a wife appear in a interview for two vacancies in the same post. The probability of husband's selection is $\frac{1}{7}$ and that of wife's is $\frac{1}{6}$. What is the probability that none of them will be selected.

(a) $\frac{2}{13}$ (b) $\frac{5}{7}$

(c) $\frac{1}{42}$ (d) $\frac{41}{42}$

7. A bag contains x red balls, $(x + 5)$ blue balls and $(3x + 10)$ white balls. If the probability of drawing a blue ball is $\frac{2}{9}$, what is the number of white balls?

(a) 15   (b) 20

(c) 35   (d) 55

8. A letter is chosen at random from the letters in the word "PROBABILITY". What is the probability that the letter will be a B or a vowel?

(a) $\frac{5}{11}$ (b) $\frac{6}{11}$

(c) $\frac{2}{11}$ (d) $\frac{7}{11}$

9. There are three events one of which must and only happen. The odds are 8 : 3 against A, 5 to 2 against B. Find the odds against C?

(a) 43 : 34   (b) 34 : 43

(c) 43 : 77   (d) 77 : 43

10. A can solve 80% of the problems given in a book and B can solve 60%. What is the probability that at least one of them will solve a problem selected at random from the book?

(a) $\frac{12}{25}$ (b) $\frac{97}{100}$

(c) $\frac{23}{25}$ (d) $\frac{11}{25}$

Answers

1. (c)   **2.** (c)   **3.** (b)   **4.** (c)   **5.** (d)   **6.** (d)   **7.** (d)   **8.** (b)

9. (a) [**Hint.** All three are exhaustive events, $P(A) + P(B) + P(C) = 1$]

10. (c) [**Hint.** P (solved) = $1 - P$ (not solved)]

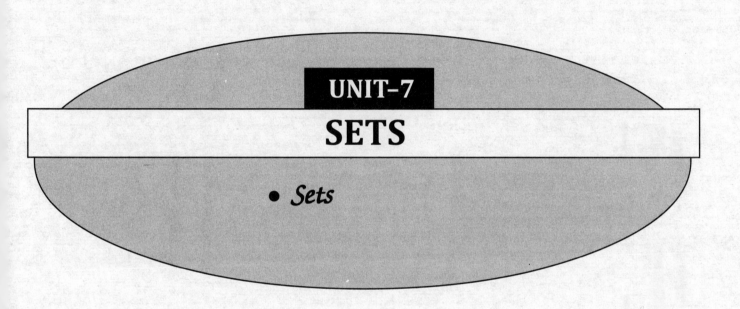

UNIT–7

SETS

- *Sets*

Chapter
30

SETS

1. A **set** is a well defined collection of objects.

2. Sets are usually denoted by capital letters *A, B, C,* etc. and their elements by small letters *a, b, c,* etc.

3. If an element '*a*' is a member of set *A*, then we write $a \in A$.

   If an element '*a*' is not a member of set *A*, then we write $a \notin A$.

4. There are two main ways of expressing a set:

   (i) **Tabular form or Roster form:** In this form we list all the members of the set separating them by commas and enclosing them in only brackets.

   Ex. The set of the first ten perfect square numbers is written as

   $$S = \{1, 4, 9, 16, 25, 36, 49, 64, 81, 100\}$$

   > **Note :** • The elements of a set can be written in any order.
   > • An element of a set is not written more than once.

   (ii) **Set-builder form or Rule method:** In this method, instead of listing all elements of a set, we write the set by some special property or properties satisfied by all the elements and write it as:

   $$A = \{ x : P(x)\} \text{ or } A = \{ x \mid x \text{ has the property } P(x)\}$$

   Ex. The set of the first ten perfect square numbers is written as:

   $$S = \{x \mid x = n^2, 1 \leq n \leq 10, n \in N \}$$

5. **Types of Sets:**

   (i) **Finite set:** A set having no elements or a definite number of elements is called a finite set.

   Ex. *V* = The set of vowels of english alphabet = { *a, e, i, o, u*}

   (ii) **Infinite set:** A set having unlimited number of elements is called an infinite set.

   Ex. *N* = The set of natural numbers = { 1, 2, 3, 4 , ...}

   (iii) **Empty set:** A set containing no element is called the **empty** or **null** or **void** set. The symbol for empty set is ϕ.

   $$\phi = \{ \ \}$$

   (iv) **Singleton set:** A set containing only one element is called a singleton set.

   Ex. {2}, {*a*}, {0}

   > **Note :** { } is the empty set whereas {0} is a singleton set.

6. **Cardinal number of a finite set:** The number of distinct elements in a finite set A is called the cardinal number of A and is denoted by $n(A)$

   Ex. If $A = \{ a, e, i, o, u \}$, then $n(A) = 5$.

7. **Equal sets:** Two sets P and Q containing the same elements are called equal sets.

   Here, $P = Q$ iff $a \in P \Rightarrow a \in Q$ and $a \in Q \Rightarrow a \in P$.

   Ex. $P = \{$letters of the word 'ramp'$\}$

   $Q = \{$letters of the word 'pram '$\}$

   $\Rightarrow$ $P = Q$

8. **Equivalent sets:** Two sets A and B containing equal number of elements, which are not necessarily the same are called equivalent sets.

   Ex. $A = \{$ letters of the word 'flower'$\}$

   $= \{ f, l, o, w, e, r \}$

   $B = \{2, 4, 6, 8, 10, 12\}$

   Here, $n(A) = 6$ and $n(B) = 6$, then $A \sim B$, where $\sim$ is the symbol for equivalence.

   > **Note :** All equal sets are equivalent but all equivalent sets are not necessarily equal.

9. **Subset and superset of a set:**

   If every element of set A is also an element of set B, then A is a subset of B.

   We write it symbolically as $A \subseteq B$ where '$\subseteq$' denotes **'is a subset of'**.

   Ex. $A = \{4, 8, 12\}$, $B = \{ 2, 4, 6, 8, 10, 12\}$, then $A \subseteq B$.

   $\Rightarrow$ A is contained in B

   $\Rightarrow$ B contains A

   $\Rightarrow$ B is a **superset** of A

   $\Rightarrow$ $B \supseteq A$

   > **Note :** (i) Every set is a subset of itself.
   >
   > (ii) Null set is a subset of every set.

10. **Proper subset of a set:**

    All the subsets of a set, other than the set itself are known as proper subsets.

    Ex. The subsets of set $A = \{ 1, 2, 3 \}$ are $\{ \}$, $\{1\}$, $\{2\}$, $\{3\}$, $\{1, 2\}$, $\{2, 3\}$, $\{1, 3\}$ and $\{1, 2, 3\}$

    Here all the subsets except the set $\{1, 2, 3\}$ are proper subsets of A. The symbol '$\subseteq$' denotes ' **is a proper subset of** '. Thus, $\{ \} \subset A$, $\{1\} \subset A$, etc.

    > **Note :** (i) $\not\subseteq$ denotes 'is not a subset of '.
    >
    > (ii) $\not\subset$ denotes 'is not a proper subset of '.

11. If there are n elements in a set P,

    (i) the total number of subsets of P is 2^n.

    (ii) the total number of proper subsets of P is $2^n - 1$.

    Ex. If $A = \{ a, b, c, d \}$, then

    number of subsets of $A = 2^4 = 16$

    number of proper subsets of $A = 2^4 - 1 = 16 - 1 = 15$

12. **Power set:**

    The set of all the subsets of a given set Q is called the **power set** of Q and is denoted by $P(Q)$

    Ex. If $S = \{1\}$, then $P(S) = \{ \phi, s \}$

    If $T = \{a, b\}$, then $P(T) = \{ \phi, \{a\}, \{b\}, T \}$

13. Comparable sets:

Two sets A and B are said to be comparable, if one of them is a subset of the other, *i.e.,* A and B are comparable if either $A \subseteq B$ or $B \subseteq A$.

Ex. $A = \{$set of vowels$\}$, $B = \{$letters of english alphabet$\}$ are comparable as $A \subseteq B$.

14. Universal set:

The superset of all the sets for a particular discussion is called the **universal set.** It is denoted by $\cup$ or ξ.

Ex. The set of integers is the universal set for the set of positive numbers and negative numbers, also for prime numbers and composite numbers.

15. Complement of a set:

The set of elements of universal set, which are not in a given set (say P) is the complement of P, denoted by P'.

$$P' = \{x \in \xi \, ; \, x \notin P)$$

Ex. $\xi = \{$set of natural numbers$\}$, $P = \{$ set of even numbers$\}$.

Then $P' = \{$set of odd numbers$\}$

16. Venn diagrams:

A British mathematician venn pictorially represented universal sets, subsets, properties of sets and operations on sets by diagrams called venn diagrams.

The universal set is represented by a rectangle, the subsets by circles, ovals etc within the rectangle.

The elements of the sets are written inside the curve.

Ex. If $\xi = \{1, 2, 3, 4, 5\}$, $A = \{1,2\}$ and $B = \{4, 5\}$, then the venn diagram representing this information is :

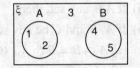

17. (i) Overlapping sets:

Two sets are called overlapping if they have at least one element in common.

Ex. The sets $\{2, 3, 4\}$ and $\{3, 6, 9\}$ are overlapping as the element 3 is common to both of them.

(ii) **Disjoint sets:** If two sets A and B have no elements in common, they are called disjoint sets.

Ex. $A = \{$ set of vowels$\}$ $B = \{$ set of consonants$\}$ are disjoint as they have no letter in common.

18. Operations on Sets.

(i) **Union of sets :** The **union of two sets** P and Q is the set of all the elements which belong to either P or Q or both.

The symbol for 'union of ' is '$\cup$'.

Thus $P \cup Q = \{x :$ either $x \in P$ or $x \in Q\}$

Ex. $P = \{ 4, 6, 8, 16\}$, $Q = \{ 1, 4, 9, 16\}$

Then $P \cup Q = \{ 1, 4, 6, 8, 9, 16\}$

Properties of Union of Sets

(i) If A is any set, then   (a) $A \cup \phi = A$ (b) $A \cup \xi = \xi$ (c) $A \cup A = A$ (d) $A \cup A' = \xi$

(ii) If A, B and C are any sets, then

    (a) $A \cup B = B \cup A$ (Commutative law)   (b) $(A \cup B) \cup C = A \cup (B \cup C)$ (Associative law)

    (c) $A \subseteq A \cup B$ and $B \subseteq A \cup B$ (d) If $A \subseteq B$, then $A \cup B = B$

(i) **Venn diagrams illustrating union of sets**

The shaded portions in the following diagrams illustrate the union of given sets.

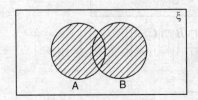

$A \cup B$
(Overlapping sets)

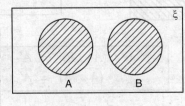

$A \cup B$
(Disjoint sets)

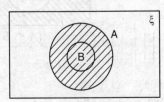

$A \cup B$
(B is a subset of A)

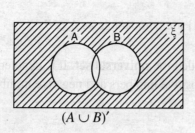

$(A \cup B)'$

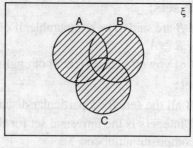

$A \cup B \cup C$

(ii) Intersection of Sets

The intersection of two sets A and B is a set that contains elements that are in both A and B.

The symbol for 'intersection of' is '$\cap$'.

$$A \cap B = \{ x | x \in A \text{ and } x \in B \}$$

Ex. $A = \{$ letters of the word 'fun'$\}$

$B = \{$ letters of the word 'son'$\}$

Then, $A \cap B = \{ n \}$

Properties of Intersection of Sets

(i) If A is any set, then $(a) A \cap \phi = \phi$ $(b) A \cap \xi = A$ $(c) A \cap A = A$ $(d) A \cap A' = \phi$

(ii) If A, B and C be any sets, then

   (a) $A \cap B = B \cap A$ (Commutative law)

   (b) $(A \cap B) \cap C = A \cap (B \cap C)$ (Associative law)

   (c) If $A \subseteq B$, then $A \cap B = A$

   (d) For any sets A and B, we have $A \cap B \subseteq A$ and $A \cap B \subseteq B$

Venn diagrams illustrating the intersection of sets

The shaded portions in the following diagrams illustrate the intersection of the given sets.

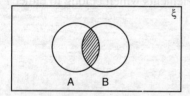

$A \cap B$ (Overlapping sets)

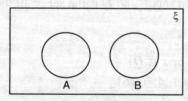

$A \cap B = \phi$ (Disjoint sets)

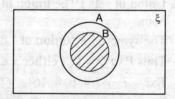

$A \cap B = B$ (B is a proper subset of A)

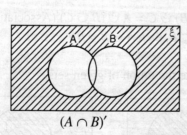

$(A \cap B)'$

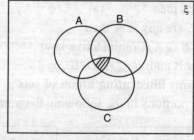

$A \cap B \cap C$

> **Note :** $(A \cup B)' = A' \cap B'$
>
> $(A \cap B)' = A' \cup B'$

(iii) Difference of two sets:

Let A and B be two sets, then $A - B$ is the sets of elements which belong to A but do not belong to B.

Thus, $A - B = \{ x | x \in A \text{ and } x \notin B \}$

$B - A = \{x \mid x \in B \text{ and } x \notin B\}$

Ex. $A = \{2, 3, 5, 6\}$, $B = \{5, 6, 7, 8\}$

Then, $A - B = \{2, 3\}$

$B - A = \{7, 8\}$

> **Note :**   In general, $A - B \neq B - A$

Venn diagrams illustrating the difference of sets.

The shaded portions in the following diagrams show the difference of the given sets.

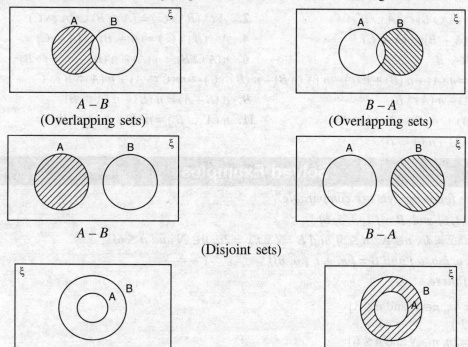

$A - B$
(Overlapping sets)

$B - A$
(Overlapping sets)

$A - B$

$B - A$

(Disjoint sets)

$A - B = \phi$

$B - A$

(When A is a proper subset of B)

(iv) **Symmetric difference of two sets:**

The symmetric difference of two sets A and B denoted by $A \Delta B$ is the set $(A - B) \cup (B - A)$

$$A \Delta B = (A - B) \cup (B - A)$$
$$= \{x \mid x \in A \cap B\}$$

Represented diagramatically, it is shown by the shaded part as

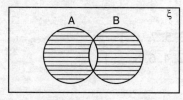

Ex. If $A = \{1, 2, 3, 4\}$ and $B = \{3, 4, 5, 6\}$

Then $A - B = \{1, 2\}$ and $B - A = \{5, 6\}$

$\therefore$ $A \Delta B = (A - B) \cup (B - A)$

$= \{1, 2\} \cup \{5, 6\} = \{1, 2, 5, 6\}$

Cartesian Product of Sets

Let A and B be non-empty sets. The cartesian product of A and B is denoted by $A \times B$ (read 'A cross B') and is defined as the set of all ordered pairs (a, b) where $a \in A$ and $b \in B$. symbolically.

$$A \times B = \{(a, b) : a \in A \text{ and } b \in B\}$$

Ex. $A = \{3, 5, 7\}$ and $B = \{a, b\}$

Then, $A \times B = \{(3, a), (5, a), (7, a), (3, b), (5, b), (7, b)\}$

          $B \times A = \{(a, 3), (a, 5), (a, 7), (b, 3), (b, 5), (b, 7)\}.$

Note : $A \times B \times C = \{(a, b, c) : a \in A,\ b \in B,\ c \in C\}$

SOME IMPORTANT RESULTS

1. $A \cup (B \cap B') = (A \cup B) \cap (A \cup B')$ **2.** $A \cap (B \cup C) = (A \cap B) \cup (A \cap C)$

3. $A - (B \cup C) = (A - B) \cap (A - C)$ **4.** $A - (B \cap C) = (A - B) \cup (A - C)$

5. $A - B = (A \cup B) - B$ **6.** $n(A \cup B) = n(A) + n(B) - n(A \cap B)$

7. $n(A \cup B \cup C) = n(A) + n(B) + n(c) - n(A \cap B) - n(B \cap C) - n(C \cap A) + n(A \cap B \cap C)$

8. $n(A - B) = n(A) - n(A \cap B)$ **9.** $n(B - A) = n(B) - n(A \cap B)$

10. $n(A \triangle B) = n(A) + n(B) - 2n(A \cap B)$ **11.** $n(A' \cup B') = n(\xi) - n(A \cap B)$

12. $n(A' \cap B') = n(\xi) - n(A \cup B)$

Solved Examples

Ex. 1. *Which of the following sets are comparable?*

    (a) $A = \{2, 4, 6\}$ and $B = \{1, 3, 5, 6\}$

    (b) $A = \{x : x = 4n, n \in N,\ n \le 3\}$ and $B = \{x : x = 2n, n \in N$ and $n \le 6\}$

    (c) $A = \{a, e, i, o, u\}$ and $B = \{a, e, i, \{o, u\}\}$

    (d)   *None of these*

Sol. $A = \{x : x = 4n, n \in N$ and $n \le 3\}$

     $= \{4, 8, 12\}$

    $B = \{x : x = 2n, n \in N$ and $n \le 6\}$

     $= \{2, 4, 6, 8, 10, 12\}$

    Here $A \subset B$

$\Rightarrow$ A and B are comparable.

Ex. 2. *Given* $\xi = \{x : x$ *is a natural number*$\}$

       $A = \{x : x$ *is an even number,* $x \in N\}$

       $B = \{x : x$ *is an odd number,* $x \in N\}$

    Then $(B \cap A) - (x - A) = \dots$

Sol. $A = \{2, 4, 6, 8, \dots\}$

    $B = \{1, 3, 5, 7, \dots\}$

    $B \cap A = \{2, 4, 6, 8, \dots\} \cap \{1, 3, 5, 7, \dots\} = \phi$

    $\xi - A = \{1, 2, 3, 4, 5, 6, \dots\} - \{2, 4, 6, 8, \dots\}$

       $= \{1, 3, 5, 7, \dots\}$

$\therefore$ $B \cap A - (\xi - A) = \phi - \{1, 3, 5, 7, \dots\}$

        $= \phi$

Ex. 3. *If* $\xi = \{2, 3, 4, 5, 6, 7, 8, 9, 10, 11\}$

    $A = \{3, 5, 7, 9, 11\}$

    $B = \{7, 8, 9, 10, 11\}$ *then find* $(A - B)'$.

Sol. $A - B = \{3, 5\}$

    $(A - B)' = \{2, 4, 6, 7, 8, 9, 10, 11\}$

Ex. 4. *If P and Q any two sets, then Q − P =*

(a) $Q \cup P'$ (b) $Q \cap P'$ (c) $Q' \cap P'$ (d) $Q' \cap P'$

Sol. For all $x \in Q - P \Rightarrow x \in Q$ and $x \in P$

$\Rightarrow x \in Q$ and $x \in P'$

$\Rightarrow x \in Q \cap P'$

$\therefore \qquad Q - P = Q \cap P'$

Ex. 5. *Let P and Q be two sets, then what is $(P \cap Q') \cup (P \cup Q)'$ equal to ?*

Sol. $(P \cap Q') \cup (P \cup Q)' = (P \cap Q') \cap (P' \cap Q')$

$= (P \cup P') \cap (P \cup Q') \cap (Q' \cup P') \cap (Q' \cup Q')$

$= \xi \cap \{ Q' \cup (P \cap P') \} \cap Q'$

$= \xi \cap \{ Q' \cup \xi \} \cap Q'$

$= \xi \cap Q' \cap Q' = \xi \cap Q' = \xi$

Ex. 6. *If n (A) = 120, n (B) = 250 and n (A − B) = 52 , then find n (A ∪ B).*

Sol. $n (A - B) = n (A) - n (A \cap B)$

$\Rightarrow 52 = 120 - n (A \cap B)$

$\Rightarrow n (A \cap B) = 120 - 52 = 68$

Now, $n (A \cup B) = n (A) + n (B) - n (A \cap B)$

$= 120 + 250 - 68$

$= 302$

Ex. 7. *If A = {x, y} and B = { 3, 4, 5, 7, 9} and C = { 4, 5, 6, 7 } , find A × (B∩C).*

Sol. $B \cap C = \{ 3, 4, 5, 7, 9 \} \cap \{ 4, 5, 6, 7 \}$

$= \{ 4, 5, 7 \}$

$\therefore A \times (B \cap C) = \{ x, y \} \times \{ 4, 5, 7 \}$

$= \{ (x, 4), (x, 5), (x, 7), (y, 4), (y, 5), (y, 7) \}$

Ex. 8. *In a certain group of 36 people, only 18 are wearing hats and only 24 are wearing sweaters. If six people are wearing neither a hat nor a sweater, then how many people are wearing both a hat and a sweater?*

Sol. Number of people wearing a hat or a sweater or both = 36 − 6 = 30

$n (H \cup S) = n (H) + n (S) - n (H \cap S)$

$30 = 18 + 24 - n (H \cap S)$

$\Rightarrow n (H \cap S) + 42 - 30 = 12.$

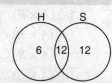

Question Bank–30

1. If A, B and C are three finite sets, then what is $[(A \cup B) \cap C]'$ equal to ?

(a) $(A' \cup B') \cap C'$ (b) $A' \cap (B' \cap C')$

(c) $(A' \cap B') \cup C'$ (d) $(A \cap B) \cap C$

2. If X and Y are any two non-empty sets, then what is $(X - Y)'$ equal to?

(a) $X' - Y'$ (b) $X' \cap Y$

(c) $X' \cup Y$ (d) $X - Y'$

3. Out of 32 persons, 30 invest in National Savings Certificates and 17 invest in Shares. What is the number of persons who invest in both?

(a) 13   (b) 15

(c) 17   (d) 19

4. If $A = P (\{1, 2\})$, where P denotes the power set, then which one of the following statements is correct?

(a) $\{1, 2\} \subset A$ (b) $1 \in A$

(c) $\phi \notin A$ (d) $\{1, 2\} \in A$

5. Which one of the following statements is correct?

(a) $A \cup P(A) = P(A)$ (b) $A \cap P(A) = A$

(c) $A - P(A) = A$ (d) $P(A) - \{A\} = P(A)$

where $P(A)$ denotes the power set of A.

6. If ξ is the universal set and P is a subset of ξ, then what is $P \cap \{ (P- \xi) \cup (\xi - P)\}$ equal to
   (a) ϕ (b) P'
   (c) ξ (d) P

7. If $F(n)$ denotes the set of all divisions of n except 1, what is the least value of y satisfying $[F(20) \cap F(16) \subseteq F(y)$?
   (a) 1   (b) 2
   (c) 4   (d) 8

8. Consider the following for any three non empty sets A, B and C.
   1. $A - (B \cup C) = (A - B) \cup (A - C)$
   2. $A - B = A - (A \cap B)$
   3. $A = (A \cap B) \cup (A - B)$
   which of the above is /are correct?
   (a) Only 1   (b) 2 and 3
   (c) 1 and 2   (d) 1 and 3

9. Let A and B be two non-empty subsets of a set X. If $(A - B) \cup (B - A) = A \cup B$, then which one of the following is correct?
   (a) $A \subset B$ (b) $A \subset (X - B)$
   (c) $A = B$ (d) $B \subset A$

10. Which one of the following is correct?
    (a) $A \cup (B - C) = A \cup (B \cap C')$
    (b) $A - (B \cup C) = (A \cup B') \cap C'$
    (c) $A - (B \cap C) = (A \cap B') \cap C$
    (d) $A \cap (B - C) = (A \cap B) \cap C$

11. If $A = \{ 1, 2, 3\}$, $B = \{1, 2\}$ and $C = \{2, 3\}$, which one of the following is correct?
    (a) $(A \times B) \cap (B \times A) = (A \times C) \cap (B \times C)$
    (b) $(A \times B) \cap (B \times A) = (C \times A) \cap (C \times B)$
    (c) $(A \times B) \cup (B \times A) = (A \times B) \cup (B \times C)$
    (d) $(A \times B) \cup (B \times A) = (A \times B) \cup (A \times C)$

12. If A and B are finite sets. which of the following is the correct statement.
    (a) $n(A - B) = n(A) - n(B)$
    (b) $n(A - B) = n(B - A)$
    (c) $n(A - B) = n(A) - n(A \cap B)$
    (d) $n(A - B) = n(B) - n(A \cap B)$

13. Out of 40 children, 30 can swim, 27 can play chess and 5 can do neither. How many children can swim only?
    (a) 30   (b) 22
    (c) 12   (d) 8

14. Which one of the following is correct?
    (a) $A \times (B - C) = (A - B) \times (A - C)$

(b) $A \times (B - C) = (A \times B) - (A \times C)$
(c) $A \cap (B \cup C) = (A \cap B) \cup C$
(d) $A \cup (B \cap C) = (A \cup B) \cap C$

15. Consider the following statements for non-empty sets A, B and C.
    1. $A - (B - C) = (A - B) \cup C$
    2. $A - (B \cup C) = (A - B) - C$
    which of the statements given above is /are correct?
    (a) 1 only   (b) 2 only
    (c) Both 1 and 2   (d) Neither 1 nor 2

16. Out of 600 students in a school, 125 played cricket, 220 played football and 300 played hockey of the total, 28 played both hockey and football, 70 played cricket and football and 32 played cricket and hockey, 26 played all the three games. What is the number of students who did not play any game?
    (a) 240   (b) 169
    (c) 259   (d) 171

17. If $n(A) = 65$, $n(B) = 32$ and $n(A \cap B) = 14$, then $n(A \Delta B)$ equals
    (a) 65   (b) 47
    (c) 97   (d) 69

18. Two finite sets have m and n elements. The total number of subsets of the first set is 56 more than the total number of subsets of the second set. The values of m and n are
    (a) 7, 6   (b) 6, 3
    (c) 5, 1   (d) 8, 7

19. U is a universal set and $n(U) = 160$. A, B and C are subset of U. If $n(A) = 50$, $n(B) = 70$, $n(B \cup C) = \phi$, $n(B \cap C) = 15$ and $A \cup B \cup C = U$, then $n(C)$ equals.
    (a) 40   (b) 50
    (c) 55   (d) 60

20. Match List-I with List-II and select the correct answer using the codes given below for the lists:

List-I	List-II
(A) $(E - A) \cup (E - A')$	1. ϕ
(B) $(E - [(A \cup A') - (A \cap A')]$	2. A
(C) $(E \cap (A - A') \cup A$	3. A'
(D) $[(E - \phi) \cup (\phi - E)] - A$	4. E

Here A' is the complement set of A, E is the universal set and ϕ is an empty set.

Codes:

	A	B	C	D
(a)	4	1	2	3
(b)	4	3	2	1
(c)	2	3	4	1
(d)	2	1	4	3

Answers

1. (c)	2. (c)	3. (b)	4. (d)	5. (b)	6. (a)	7. (c)	8. (b)	9. (b)	10. (b)
11. (c)	12. (c)	13. (d)	14. (b)	15. (b)	16. (c)	17. (d)	18. (b)	19. (c)	20. (a)

Hints and Solutions

1. (c) Given, A, B and C are three finite sets, then
$$[(A \cup B) \cap C]' = (A \cup B)' \cup C'$$
$$= (A' \cup B') \cup C'$$

2. (c) $X - Y = \{ x : x \in x, x \notin y \}$
$$= \{ x : x \in x, x \notin y' \}$$
$$\Rightarrow \{ x : x \in X \cap Y' \}$$
$$\Rightarrow \{ X - Y \}' = (X \cap Y')'$$
$$= X' \cup (Y')' = X' \cup Y$$

3. (b) Use $n(A \cup B) = n(A) + n(B) - n(A \cap B)$

4. (a) Given, $A = P(\{1, 2\})$, then
$$A = [\phi, \{1\}, \{2\}, \{1, 2\}]$$
$$\Rightarrow \{1, 2\} \in A$$

5. (b) Here, $P(A)$ denotes the power set of A.
Hence, A is a subset of $P(A)$.
$$\therefore \ A \cap P(A) = A$$

6. (a) Given set $= P \cap \{(P - \xi) \cup (\xi - P)\}$
$$= P \cap \{\phi \cup P'\}$$
$$= P \cap P' = \phi$$

7. (c) $F(20) \cap F(16) \subseteq F(y)$
$$\Rightarrow [\{2, 4, 5, 10, 20\} \cap \{2, 4, 8, 16\}] \subseteq F(y)$$
$$\Rightarrow [\{2, 4\}] \subseteq F(y)$$
$$\therefore \text{ The least value of } y \text{ is } 4.$$

8. (b) 1. $A - (B \cup C) = (A - B) \cup (A - C)$
$(A - B) \cup (A - C)$
$\Rightarrow$ For all $(x \in A$ and $x \notin B)$ or $(x \in A$ and $x \notin C)$
$\Rightarrow$ For all $(x \in A$ and $x \notin B$ and $x \notin C)$
$\Rightarrow x \in A - (B \cap C)$
Hence not correct.

2. $x \in (A - (A \cap B)) \Rightarrow x \in A$ and $x \notin A \cap B$
$$\Rightarrow x \in A \text{ and } x \notin B$$
$$\Rightarrow x \in (A - B)$$

3. $x \in [(A \cap B) \cup (A - B)]$
$$\Rightarrow (x \in A \text{ and } x \notin B) \text{ or } (x \in A \text{ and } x \in B')$$
$$\Rightarrow x \in A \text{ and } x \in B \text{ and } x \in B'$$
$$\Rightarrow x \in A \text{ and } x \in B \cap B'$$
$$\Rightarrow x \in A \text{ and } x \in \phi$$
$$\Rightarrow x \in A \cup \phi \Rightarrow x \in A$$
Hence (1) is incorrect.

10. (b) A. $A \cup (B - C) = A \cup (B \cap C')$
B. $A - (B \cup C) = A \cap (B \cup C)'$
$$= A \cap (B' \cup C')$$
C. $A - (B \cap C) = A \cap (B \cap C)'$
$$= A \cap (B' \cup C')$$

D. $A \cap (B - C) = A \cap (B \cap C')$
Hence B is the correct option.

11. (c) $(A \times B) = \{ (1,1), (1,2), (2,1), (2,2), (3,1), (3,2)\}$
$(B \times A) = \{ (1,1), (1,2), (1,3), (2,1), (2,2), (2,3)\}$
$(B \times C) = \{ (1,2), (1,3), (2,2), (2,3)\}$
$(C \times B) = \{ (2,1), (2,2), (3,1), (3,2)\}$
$(A \times C) = \{ (1,2), (1,3), (2,2), (2,3), (3,2) (3,3)\}$
$(C \times A) = \{ (2,1), (2,2), (2,3), (3,1), (3,2) (3,3)\}$
$\therefore (A \times B) \cup (B \times A) = \{(1,1), (1,2), (2,1), (2,2),$
$$(3,1), (3,2) (1,3), (2,3)\}$$
and $(A \times B) \cup (B \times C) = \{ (1,1), (1,2), (2,1),$
$$(2,2), (3,1), (3,2), (1,3), (2,3)\}$$
$\therefore (A \times B) \cup (B \times A) = (A \times B) \cup (B \times C)$

12. (c) $n(A - B) = $ No. of these elements of A which are not common in A and B
$$= n(A) - n(A \cap B)$$

13. (d) Let the number of children who can swim and play chess both are x

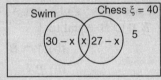

Then $30 - x + x + 27 - x = 40 - 5 = 35$
$$\Rightarrow \quad 57 - x = 35 \Rightarrow x = 22$$
$\therefore$ The number of children who can swim only are $30 - 22 = \mathbf{8}$.

14. (b) We known that the cartesian product of two sets is defined as
$$X \times Y = \{ (x, y) : x \in x \text{ and } x \in y \}$$
$$\therefore \quad A \times (B - C) = (A \times B) - (A \times C)$$

15. (b) 1. $A - (B - C) = A - (B \cap C')$
$$= A \cap (B \cap C')'$$
$$= A \cap (B' \cup (C')')$$
$$= A \cap (B' \cup C)$$
$(A - B) \cup C = (A \cap B') \cup C$
Thus, $A - (B - C) \neq (A - B) \cup C$

2. $A - (B \cup C) = A \cap (B \cap C)'$
$$= A \cap (B' \cap C')$$
$(A - B) - C = (A \cap B') - C$
$$= A \cap B' \cap C'$$
$$\Rightarrow \quad A - (B \cup C) = (A - B) - C$$
Associative property.

16. (c) Here, $n(c) = 125, n(F) = 220, n(H) = 300$

$n(H \cap F) = 28, n(C \cap F) = 70, n(C \cap H) = 32$

and $n(C \cap F \cap H) = 26$

∴ Number of students who did not play any game

$= n(C' \cap F' \cap H')$

$= n((C \cup F \cup H)')$

$= n(\xi) - n(C \cup F \cup H)$

$= n(\xi) - [n(C) + n(F) + n(H) - n(C \cap F)$

$\qquad - n(H \cap F) - n(C \cap H)$

$\qquad + n(C \cap F \cap H)]$

$= 800 - [125 + 220 + 300 - 70 - 28 - 32 - 26]$

$= 800 - 541 = 259$

17. (d) $n(A \Delta B) = n(A - B) + n(B - A)$

∴ $A \Delta B = (A - B) \cup (B - A)$

$= n(A) - n(A \cap B) + n(B) - n(A \cap B)$

$= n(A) + n(B) - 2n(A \cap B)$

$= 65 + 32 - 2 \times 14 = \mathbf{69}.$

18. (b) ∵ The number of subsets of a set containing p elements $= 2^p$,

Here, $2^m - 2^n = 56$

The values of m and n satisfying the given equations from the given options are 6, 3.

$2^6 - 2^3 = 64 - 8 = \mathbf{56}.$

19. (c) $A \cup B \cup C = U$

and $A \cap (B \cup C) = \phi$

$\Rightarrow (B \cup C) = A' = U - A$

$\Rightarrow n(B \cup C) = n(U) - n(A)$

$\qquad = 160 - 50 = 110$

Now, $n(B \cup C) = n(B) + n(C) - n(B \cap C)$

$\Rightarrow 110 = 70 + n(C) - 15$

$\Rightarrow n(C) = 110 + 15 - 70 = \mathbf{55}.$

20. (a) (A) $(E - A) \cup (E - A') = A' \cup A = E$

(B) $E - \{(A \cup A') - (A \cap A')\}$

$= E - \{E - \phi\}$

$= E - E + \phi = \phi$

(C) $\{E \cap (A - A')\} \cup A$

$= \{E \cap A\} \cup A = A \cup A = A$

(D) $\{(E - \phi) \cup (\phi - E)\} - A$

$= \{E \cup \phi\} - A = E - A = A'$

Self Assessment Sheet–29

1. If A and B are subsets of a set X, then what is $(A \cap (X - B)) \cup B$ equal to ?

  (a) $A \cup B$ (b) $A \cap B$

  (c) A (b) B

2. A set contains n elements. The power set of this set contains:

  (a) n^2 elements   (b) $2^{\lambda/2}$ elements

  (c) 2^n elements   (b) n elements

3. For non empty subsets A, B and C of a set X such that $A \cup B = B \cap C$, which one of the following is the strongest inference that can be derived?

  (a) $A = B = C$ (b) $A \subseteq B = C$

  (c) $A = B \subseteq C$ (b) $A \subseteq B \subseteq C$

4. Let ξ = the set of all triangles , P = the set of all isosceles triangles, Q = the set of all equilateral triangles, R = the set of all right angled triangles. What do the sets $P \cap Q$ and $R - P$ represent respectively?

  (a) The set of isosceles triangles; the set of non-isosceles right angled triangles.

  (b) The set of isosceles triangles; the set of right-angled triangles.

  (c) The set of equilateral triangles; the set of right-angled triangles.

  (b) The set of isosceles triangles; the set of equilateral triangles.

5. What does the shaded region represent in the figure given below?

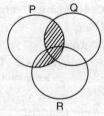

  (a) $(P \cup Q) - (P \cap Q)$

  (b) $P \cap (Q \cup R)$

  (c) $(P \cap Q) \cap (P \cap R)$

  (d) $(P \cap Q) \cup (P \cap R)$

6. Let $A = \{(n, 2n) : n \in z\}$ and

$\qquad B = \{(2n, 3n) : n \in z)\}$, what is $A \cap B$ equal to

  (a) $\{(n, 6n) : n \in z\}$

  (b) $\{(2n, 6n) : n \in z\}$

  (c) $\{(n, 3n) : n \in z\}$

  (d) ϕ

7. While preparing the progress reports of the students, the class teacher found that 70% of the students passed in Hindi, 80% passed in English and only 65% passed in both the subjects. Find out the percentage of students who failed in both the subjects.

   (a) 15%   (b) 20%

   (c) 30%   (d) 35%

8. One hundred twenty five (125) aliens descended on a set of film as Extra Terrestrial beings. 40 had two noses, 30 had three legs, 20 had four ears, 10 had two noses, and three legs, 12 had 3 legs and four ears, 5 had two noses and four ears and 3 had all the unusual features. How many were there without any of these unusual features?

   (a) 5   (b) 35
   (c) 80   (d) None of these

9. If A and B are non empty sets and A' and B' represents their compliments respectively, then

   (a) $A - B = A' - B'$ (b) $A - A' = B - B'$

   (c) $A - B = B' - A'$ (d) $A - B' = A' - B$

10. Let Z_N be the set of non-negative integers, Z_p be the set of non-positive intergers, Z the set of integers, E the set of even integers and P the set of prime numbers. Then,

   (a) $E \cap P = \phi$ (b) $Z_N \cap Z_P = \phi$

   (c) $Z - Z_N = Z_P$ (d) $Z_N \Delta Z_P = Z - \{0\}$

Answers

1. (a)	2. (c)	3. (d)	4. (c)	5. (d)	6. (d)	7. (a)	8. (d)	9. (c)	10. (d)

CO-ORDINATE GEOMETRY

- *Co-ordinate Geometry*

Chapter
31
CO-ORDINATE GEOMETRY

KEY FACTS

1. CO- ORDINATES OF A POINT

(i) The **horizontal line** is labelled as XOX' and called the **x-axis**.

(ii) The **vertical line** is labelled as YOY' and called the **y-axis**.

(iii) The point of intersection of the two axis O is the **origin**.

This set of axis forms a **cartesian co-ordinate system**.

(iv) On the x-axis, to the right of O is positive, to the left is negative. On the y-axis, up is positive and down is negative.

(v) To locate a point (x, y), we go x units along the right on x-axis and then y units up in the vertical direction.

Ex. To locate $(-2, 1)$, we go 2 units horizontally in the negative direction along the x-axis and then 1 unit up parallel to y-axis.

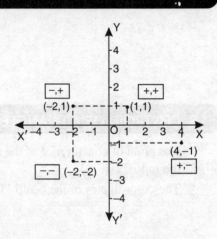

Note : The co-ordinates of the origin are $(0, 0)$.

2. DISTANCE FORMULA

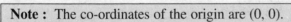

(i) The distance between two points $P(x_1, y_2)$ and $Q(x_2, y_2)$ is given by

$$PQ = \sqrt{(x_2 - x_1)^2 + (y_2 - y_1)^2}$$

Ex. Distance between the points $A(3, 4)$ and $B(6, -3)$ is

$$AB = \sqrt{(6-3)^2 + (-3-4)^2}$$

$$= \sqrt{3^2 + 7^2} = \sqrt{9 + 49} = \sqrt{58}$$

(ii) Distance of a point $P(x, y)$ from the origin is

$$\sqrt{(x-0)^2 + (y-0)^2} = \sqrt{x^2 + y^2}$$

3. IMPORTANT POINTS TO REMEMBER

(i) The co-ordinates of any point on the x-axis is $(x, 0)$.

(ii) The co-ordinates of any point on the y-axis is $(0, y)$.

(iii) To prove that any quadrilateral is a

(a) **Square**, show all sides are equal and diagonals are also equal.

(b) **Rhombus,** show that all sides are equal.

(c) **Parallelogram,** show that both pairs of opposite sides are equal (or) diagonals bisect each other.

(d) **Rectangle,** show that both the pairs of opposite sides are equal and diagonals are also equal.

(iv) To prove that a triangle is

(a) **an isosceles triangle,** show two sides are equal.

(b) **an equilateral triangle,** show all sides are equal.

(c) **a right angled triangle,** show the sum of squares of two sides is equal to the square of the third side.

(v) For three points to be **collinear,** show that the sum of the distance between any pairs of points is equal to distances between remaining pairs of points.

4. MID-POINT FORMULA

If R is the mid-point of $P(x_1, y_1)$ and $Q(x_2, y_2)$, then the co-ordinates of R are $\left(\dfrac{x_1 + x_2}{2}, \dfrac{y_1 + y_2}{2} \right)$

5. CENTROID OF A TRIANGLE

The point of concurrence of the medians of a triangle is called the centroid of the triangle. It divides the median in the ratio 2 : 1.

The co-ordinates of the centroid of a triangle whose vertices are (x_1, y_1), (x_2, y_2) and (x_3, y_3) are given by

$$\left(\frac{x_1 + x_2 + x_3}{3}, \frac{y_1 + y_2 + y_3}{3} \right)$$

6. AREA OF A TRIANGLE

Area of ΔABC, whose vertices are $A(x_1, y_1)$, $B(x_2, y_2)$ and C is (x_3, y_3)

$$\frac{1}{2} \mid x_1 (y_2 - y_3) + x_2 (y_3 - y_1) + x_3 (y_1 - y_2) \mid$$

If $\dfrac{1}{2} \mid x_1 (y_2 - y_3) + x_2 (y_3 - y_1) + x_3 (y_1 - y_2) \mid = 0$, then the points A, B and C are collinear.

Solved Examples

Ex. 1. *Find the distance between the points (a cos 60°, 0) and (0, a sin 60°).*

Sol. Required distance $= \sqrt{(0 - a\cos 60°)^2 + (a\sin 60° - 0)^2}$

$$= \sqrt{\left(\frac{-a}{2} \right)^2 + \left(\frac{\sqrt{3}a}{2} \right)^2}$$

$$= \sqrt{\frac{a^2}{4} + \frac{3a^2}{4}} = \sqrt{\frac{4a^2}{4}} = a \text{ units.}$$

Ex. 2. *Find the value of x if the distance between the points (2, –11) and (x, –3) is 10 units.*

Sol. Given, $\sqrt{(x-2)^2 + (-3-(-11))^2} = 10$

$\Rightarrow \sqrt{(x-2)^2 + 64} = 10$

$\Rightarrow \sqrt{x^2 - 4x + 4 + 64} = 10$

$\Rightarrow x^2 - 4x + 68 = 100$

$\Rightarrow x^2 - 4x - 32 = 0$

$\Rightarrow (x-8)(x+4) = 0 \Rightarrow x = 8 \text{ or} - 4$

Ex. 3. *Show that the points (1, –1), (5, 2) and (9, 5) are collinear.*

Sol. Let $P(1, -1)$, $Q(5, 2)$ and $R(9, 5)$ be the given points. Then,

$$PQ = \sqrt{(5-1)^2 + (2+1)^2} = \sqrt{16+9} = \sqrt{25} = 5$$

$$QR = \sqrt{(9-5)^2 + (5-2)^2} = \sqrt{16+9} = \sqrt{25} = 5$$

$$PR = \sqrt{(9-1)^2 + (5+1)^2} = \sqrt{64+36} = \sqrt{100} = 10$$

$\Rightarrow PQ + QR = PR$

$\Rightarrow P, Q, R$ are collinear points.

Ex. 4. *If P(2, –1), Q (3, 4), R (–2, 3) and S (–3, –2) are four points in a plane, show that PQRS is a rhombus but not a square. Also find its area.*

Sol. $PQ = \sqrt{(3-2)^2 + (4+1)^2} = \sqrt{1+25} = \sqrt{26}$ units

$QR = \sqrt{(-2-3)^2 + (3-4)^2} = \sqrt{25+1} = \sqrt{26}$ units

$RS = \sqrt{(-3+2)^2 + (-2-3)^2} = \sqrt{1+25} = \sqrt{26}$ units

$PS = \sqrt{(-3-2)^2 + (-2+1)^2} = \sqrt{25+1} = \sqrt{26}$ units

$PR = \sqrt{(-2-2)^2 + (3+1)^2} = \sqrt{16+16} = \sqrt{32} = 4\sqrt{2}$ units

$QS = \sqrt{(-3-3)^2 + (-2-4)^2} = \sqrt{36+36} = \sqrt{72} = 6\sqrt{2}$ units

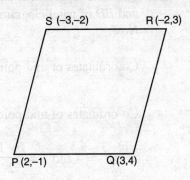

Now, we can see that

$PQ = QR = RS = PS = \sqrt{26}$ units, but
$PR \neq QS$

$\Rightarrow$ PQRS is a quadrilateral whose all sides are equal, but diagonals are not equal.

$\Rightarrow$ PQRS is a rhombus, not a square.

$\therefore$ Area of rhombus $PQRS = \dfrac{1}{2} \times$ (Product of diagonals)

$$= \frac{1}{2} \times PR \times QS$$

$$= \frac{1}{2} \times 4\sqrt{2} \times 6\sqrt{2} = \textbf{24 sq. units}$$

Ex. 5. *Find a point on the x-axis, which is equidistant from the points (7, 6) and (–3, 4).*

Sol. Any point on the x-axis is $(x, 0)$

According to the given condition,

Distance between $(x, 0)$ and $(7, 6)$ = Distance between $(x, 0)$ and $(–3, 4)$

$\Rightarrow \quad \sqrt{(x-7)^2 + (0-6)^2} = \sqrt{(x+3)^2 + (0-4)^2}$

$\Rightarrow \quad \sqrt{x^2 - 14x + 49 + 36} = \sqrt{x^2 + 6x + 9 + 16}$

$\Rightarrow \qquad\qquad x^2 - 14x + 85 = x^2 + 6x + 25 \Rightarrow 60 = 20x \Rightarrow x = 3$

$\therefore$ The required point is **(3, 0)**.

Ex. 6. *If A (–1, 3), B (1, –1) and C (5, 1) are the vertices of triangle ABC, find the length of the median through A.*

Sol. Let AD be the median though A of ΔABC.

Then, D is the mid-point of BC.

Co-ordinates of D are $\left(\dfrac{1+5}{2}, \dfrac{-1+1}{2}\right)$, *i.e.*, $(3, 0)$

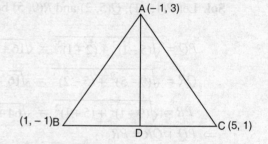

$\therefore \qquad AD = \sqrt{(3+1)^2 + (0-3)^2}$

$\qquad\qquad = \sqrt{16+9} = \sqrt{25} =$ **5 units.**

Ex. 7. *If the points A (a, –10), B (6, b), C (3, 16), D (2, –1) are the vertices of a parallelogram ABCD, find the values of a and b.*

Sol. Since the diagonals of a parallelogram bisect each other, the co-ordinates of the mid-points of the diagonals AC and BD of parallelogram $ABCD$ will be equal.

Now,

Co-ordinates of mid-point of AC are $\left(\dfrac{a+3}{2}, \dfrac{-10+16}{2}\right)$, *i.e.*, $\left(\dfrac{a+3}{2}, 3\right)$

Co-ordinates of mid-point of BD are $\left(\dfrac{6+2}{2}, \dfrac{b-1}{2}\right)$, *i.e.*, $\left(4, \dfrac{b-1}{2}\right)$

$\Rightarrow \quad \dfrac{a+3}{2} = 4$ and $3 = \dfrac{b-1}{2} \Rightarrow a + 3 = 8$ and $b - 1 = 6$

$\Rightarrow \quad$ **a = 5 and b = 7.**

Ex. 8. *Two vertices of a triangle are A (1, 1), B (2, –3). If its centroid is (2, 1) find the third vertex.*

Sol. Let the co-ordinates of the third vertex be (x, y). Then,

Co-ordinates of the centroid are $\left(\dfrac{1+2+x}{3}, \dfrac{1-3+y}{3}\right) = \left(\dfrac{3+x}{3}, \dfrac{-2+y}{3}\right)$

Given, $\left(\dfrac{3+x}{3}, \dfrac{-2+y}{3}\right) = (2, 1)$

$\Rightarrow \quad \dfrac{3+x}{3} = 2$ and $\dfrac{-2+y}{3} = 1$

$\Rightarrow \quad 3 + x = 6$ and $-2 + y = 3$

$\Rightarrow \quad$ **x = 3 and y = 5**

$\therefore$ Co-ordinates of the third vertex are **(3, 5)**.

Question Bank–31

1. The distance between the points $(\cos \theta, \sin \theta)$ and $(\sin \theta, -\cos \theta)$ is

(a) $\sqrt{3}$ (b) $\sqrt{2}$

(c) 1   (d) 0

2. If y is a positive integer such that the distance between the points $(-6, -1)$ and $(-6, y)$ is 12 units, then $y =$

(a) 5   (b) 8

(c) 11   (d) 1

3. If $A(x, y)$ is equidistant from $P(-3, 2)$ and $Q(2, -3)$, then

(a) $2x = y$ (b) $x = -y$

(c) $x = 2y$ (d) $x = y$

4. The nearest point from the origin is

(a) $(2, -3)$ (b) $(6, 0)$

(c) $(-2, -1)$ (d) $(3, 5)$

5. The vertices of a triangle are $A(3, -2)$, $B(-2, 1)$ and $C(5, 2)$. Then the length of the median through B is

(a) $\sqrt{67}$ units   (b) $\sqrt{37}$ units

(c) $\sqrt{35}$ units   (d) 6 units

6. The points (a, a), $(-a, -a)$ and $(-\sqrt{3}a, \sqrt{3}a)$ are the vertices of

(a) right triangle   (b) scalene triangle

(c) equilateral triangle   (d) isosceles triangle

7. The co-ordinates of the vertices of a side of square are $(4, -3)$ and $(-1, -5)$. Its area is

(a) $2\sqrt{29}$ sq. units   (b) $\dfrac{\sqrt{89}}{2}$ sq. units

(c) 89 sq. units   (d) 29 sq. units

8. The quadrilateral $P(-3, 2)$, $Q(-5, -5)$, $R(2, -3)$ and $S(4, 4)$ is a

(a) rectangle   (b) square

(c) rhombus   (d) kite

9. The value of p for which the points $(-1, 3)$, $(2, p)$ and $(5, -1)$ are collinear is

(a) -1 (b) 2

(c) $\dfrac{1}{3}$ (d) 1

10. The centre of a circle is $(x - 2, x + 1)$ and it passes through the points $(4, 4)$. Find the value (or values) of x, if the diameter of the circle is of length $2\sqrt{5}$ units .

(a) 1 or 3   (b) -1 or 4

(c) 5 or 4   (d) 3 or -2

11. Find the area of a rectangle whose vertices are $A(-2, 6)$, $B(5, 3)$, $C(-1, -11)$ and $D(-8, -8)$

(a) $4\sqrt{29}$ sq. units   (b) 116 sq. units

(c) $29\sqrt{5}$ sq. units   (d) $58\sqrt{2}$ sq. units

12. If the points $P(12, 8)$, $Q(-2, a)$ and $R(6, 0)$ are the vertices of a right angled triangle PQR, where $\angle R = 90°$, the value of a is

(a) 6   (b) -2

(c) -4 (d) -6

13. If $A(-2, -1)$, $B(a, 0)$, $C(4, b)$ and $D(1, 2)$ are the vertices of a parallelogram the values of a and b respectively are

(a) 3, 1   (b) -3, 1

(c) 1, 3   (d) -1, -3

14. The co-ordinates of one end-point of a circle are $(-3, 1)$ and the co-ordinates of the centre of the circle are $(2, -4)$. The co-ordinates of the other end-point of the diameter are

(a) $\left(\dfrac{-1}{2}, \dfrac{-3}{2}\right)$ (b) $(-7, 9)$

(c) $(7, -9)$ (d) $\left(\dfrac{1}{2}, \dfrac{3}{2}\right)$

15. If three consecutive vertices of a parallelogram are $(1, -2)$, $(3, 6)$ and $(5, 10)$, find its fourth vertex.

(a) $(2, -3)$ (b) $(-2, -3)$

(c) $(3, 2)$ (d) $(3, -2)$

16. The point on the x-axis which is equidistant from the points $(7, 6)$ and $(-3, 4)$ is

(a) $(0, 3)$ (b) $(3, 0)$

(c) $(-3, 0)$ (d) $(0, -3)$

17. What is the perimeter of the parallelogram $JKLM$, whose co-ordinates are $J(-5, 2)$, $K(-2, 6)$, $L(5, 6)$, $M(2, 2)$.

(a) 30 units   (b) 24 units

(c) 28 units   (d) 21 units

18. Find the area of the right angled triangle whose vertices are $(2, -2)$, $(-2, 1)$ and $(5, 2)$.

(a) $5\sqrt{2}$ sq. units   (b) $\dfrac{25}{2}$ sq. units

(c) $15\sqrt{2}$ sq. units   (d) 10 sq. units

19. The points (3, 4), (11, 10) and (5, 11/2) are

(a) collinear

(b) vertices of an equilateral triangle

(c) vertices of isosceles triangle

(d) vertices of scalene triangle.

20. The co-ordinates of vertices P and Q of an equilateral ΔPQR are $(1, \sqrt{3})$ and $(0, 0)$. Which of the following could be co-ordinates of R?

(a) (1, 2) (b) (2, 0)

(c) $\left(1, \dfrac{\sqrt{3}}{2}\right)$ (d) $(\sqrt{3}, 1)$

Answers

1. (b)	2. (c)	3. (d)	4. (c)	5. (b)	6. (c)	7. (d)	8. (c)	9. (d)	10. (c)
11. (b)	12. (a)	13. (c)	14. (c)	15. (c)	16. (b)	17. (c)	18. (b)	19. (a)	20. (b)

Hints and Solutions

1. (b) Reqd. dist.

$$= \sqrt{(\sin\theta - \cos\theta)^2 + (-\cos\theta - \sin\theta)^2}$$

$$= \sqrt{\begin{array}{c}(\sin^2\theta - 2\sin\theta\cos\theta + \cos^2\theta) \\ + (\cos^2\theta + 2\cos\theta\sin\theta + \sin^2\theta)\end{array}}$$

$$= \sqrt{2(\sin^2\theta + \cos^2\theta)} = \sqrt{2} \ (\because \cos^2\theta + \sin^2\theta = 1)$$

2. (c) Given, $\sqrt{(-6+6)^2 + (y+1)^2} = 12$

$\Rightarrow \sqrt{y^2 + 2y + 1} = 12$

$\Rightarrow y^2 + 2y + 1 = 144 \Rightarrow y^2 + 2y - 143 = 0$

$\Rightarrow y^2 + 13y - 11y - 143 = 0$

$\Rightarrow y(y+13) - 11(y+13) = 0$

$\Rightarrow (y-11)(y+13) = 0 \Rightarrow y = 11$ or -13

The required positive integer is **11**.

3. (d) Given, $AP = AQ$

$\Rightarrow AP^2 = AQ^2$

$\Rightarrow (x+3)^2 + (y-2)^2 = (x-2)^2 + (y+3)^2$

$\Rightarrow x^2 + 6x + 9 + y^2 - 4y + 4$
$\qquad = x^2 - 4x + 4 + y^2 + 6y + 9$

$\Rightarrow 10x = 10y \Rightarrow \boldsymbol{x = y}$

4. (c) Calculating the distance of each point from the origin, we have

(a) $\sqrt{(0-2)^2 + (0+3)^2} = \sqrt{4+9} = \sqrt{13}$

(b) $\sqrt{(0-6)^2 + 0} = \sqrt{36} = 6$

(c) $\sqrt{(0+2)^2 + (0+1)^2} = \sqrt{4+1} = \sqrt{5}$

(d) $\sqrt{(0-3)^2 + (0-5)^2} = \sqrt{9+25} = \sqrt{34}$

Clearly, $(-2, -1)$ is the nearest point from the origin.

5. (b) Let BD be the median through B. Then D is the mid-point of AC. Co-ordinate of D are

$\left(\dfrac{5+3}{2}, \dfrac{2+(-2)}{0}\right)$, *i.e.*, (4, 0).

Now calculate BD.

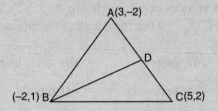

6. (c) Let the vertices be $A\,(a, a)$, $B\,(-a, -a)$ and $C\,(-\sqrt{3}a, \sqrt{3}a)$

$AB = \sqrt{(-a-a)^2 + (-a-a)^2} = \sqrt{4a^2 + 4a^2}$

$\qquad = \sqrt{8a^2} = 2\sqrt{2}a$

$BC = \sqrt{(-\sqrt{3}a + a)^2 + (\sqrt{3}a + a)^2}$

$\qquad = \sqrt{3a^2 + a^2 - 2\sqrt{3}a^2 + 3a^2 + a^2 + 2\sqrt{3}a^2}$

$\qquad = \sqrt{8a^2} = 2\sqrt{2}a$

$AC = \sqrt{(-\sqrt{3}a - a)^2 + (\sqrt{3}a - a)^2}$

$\qquad = \sqrt{8a^2} = 2\sqrt{2}\,a$

Hence, $AB = BC = AC \Rightarrow \Delta ABC$ is equilateral.

7. (d) Length of a side of the square

$\qquad = \sqrt{(-1-4)^2 + (-5+3)^2}$

$\qquad = \sqrt{25+4} = \sqrt{29}$ units.

$\therefore$ Area $= (\text{Side})^2 = (\sqrt{29})^2$ sq.units

$\qquad = \boldsymbol{29 \text{ sq.units}}$

8. (c) Sides $PQ = \sqrt{(-5+3)^2 + (-5-2)^2}$

$\qquad = \sqrt{(-2)^2 + (-7)^2} = \sqrt{4+49} = \sqrt{53}$

$\qquad = \sqrt{(-2)^2 + (-7)^2} = \sqrt{4+49} = \sqrt{53}$

$\quad QR = \sqrt{(2+5)^2 + (-3+5)^2}$

$\qquad = \sqrt{7^2 + 2^2} = \sqrt{49+4} = \sqrt{53}$

$\quad RS = \sqrt{(4-2)^2 + (4+3)^2}$

$\qquad = \sqrt{2^2 + 7^2} = \sqrt{4+49} = \sqrt{53}$

$\quad PS = \sqrt{(4+3)^2 + (4-2)^2}$

$\qquad = \sqrt{7^2 + 2^2} = \sqrt{49+4} = \sqrt{53}$

Diagonals $PR = \sqrt{(2+3)^2 + (-3-2)^2}$

$\qquad = \sqrt{5^2 + 5^2} = \sqrt{25+25} = 5\sqrt{2}$

$\quad QS = \sqrt{(4+5)^2 + (4+5)^2}$

$\qquad = \sqrt{9^2 + 9^2} = \sqrt{81+81} = 9\sqrt{2}$

This shows, sides $PQ = QR = RS = PS$ and diagonals $PR \neq QS$

∴ The quadrilateral is a rhombus.

9. (d) For collinearity of points $A\ (-1, 3)$, $B\ (2, p)$, $C\ (5, -1)$ area of $\triangle ABC$ should be zero, *i.e.*,

$\dfrac{1}{2}\,|-1(p+1) + 2(-1-3) + 5(3-p)| = 0$

$\Rightarrow -p - 1 - 8 + 15 - 5p = 0$

$\Rightarrow -6p + 6 = 0 \Rightarrow -6p = -6 \Rightarrow p = 1.$

10. (c) Radius of the circle = Dist. between the centre and given pt. on the circle

$= \sqrt{(x-2-4)^2 + (x+1-4)^2}$

$= \sqrt{(x-6)^2 + (x-3)^2}$

$= \sqrt{x^2 - 12x + 36 + x^2 - 6x + 9}$

$= \sqrt{2x^2 - 18x + 45}$

Given, diameter $= 2\sqrt{5}$

$\Rightarrow$ radius $= \dfrac{1}{2} \times 2\sqrt{5} = \sqrt{5}$

$\therefore \sqrt{2x^2 - 18x + 45} = \sqrt{5}$

$\Rightarrow 2x^2 - 18x + 45 = 5$

$\Rightarrow 2x^2 - 18x + 40 = 0$

$\Rightarrow x^2 - 9x + 20 = 0$

$\Rightarrow (x-5)(x-4) = 0 \Rightarrow x = 5 \text{ or } 4$

11. (b) $AB = \sqrt{(5+2)^2 + (3-6)^2}$

$\qquad = \sqrt{7^2 + (-3)^2} = \sqrt{49+9} = \sqrt{58}$

$\quad BC = \sqrt{(-1-5)^2 + (-11-3)^2}$

$\qquad = \sqrt{(-6)^2 + (-14)^2} = \sqrt{36+196} = \sqrt{232}$

A (–2,6)   (5,3) B

D (–8,–8)   (–1,–11) C

∴ Area of the rectangle $ABCD = AB \times BC$

$\qquad = \sqrt{58} \times \sqrt{232} = \sqrt{29 \times 2 \times 29 \times 8}$

$\qquad = 29 \times 4 = 116$ Sq. units

12. (a) $\triangle PQR$ is rt. $\angle d$ at R

$\Rightarrow PR^2 + QR^2 = PQ^2$

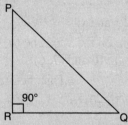

$\Rightarrow \{(6-12)^2 + (0-8)^2\} + \{(6+2)^2 + (0-a)^2\}$

$\qquad\qquad = \{(-2-12)^2 + (a-8)^2\}$

$\Rightarrow (36+64) + (64+a^2) = (196 + a^2 - 16a + 64)$

$\Rightarrow a^2 + 164 = a^2 - 16a + 260$

$\Rightarrow 16a = 260 - 164 = 96 \Rightarrow a = 6.$

13. (c) The diagonals of a parallelogram bisect each other, therefore co-ordinates of mid-points of both the diagonals are the same co-ordinates of mid-point of diagonal AC

$\qquad = \left(\dfrac{4-2}{2}, \dfrac{b-1}{2}\right) = \left(1, \dfrac{b-1}{2}\right)$

Co-ordinate of mid-point of diagonal BD

$\qquad = \left(\dfrac{1+a}{2}, \dfrac{2-0}{2}\right) = \left(\dfrac{1+a}{2}, 1\right)$

$\Rightarrow \dfrac{1+a}{2} = 1$ and $\dfrac{b-1}{2} = 1$

$\Rightarrow 1 + a = 2$ and $b - 1 = 2$

$\Rightarrow a = +1$ and $b = 3$

14. (c) Let the co-ordinates of the other end point of the diameter be (x, y).

Then, $\dfrac{-3+x}{2}=2$ and $\dfrac{1+y}{2}=-4$

$\Rightarrow -3+x=4$ and $1+y=-8$

$\Rightarrow$ **$x=7$ and $y=-9$**

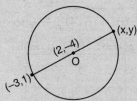

15. (c) Let the vertices of the parallelogram be $A\,(1,-2)$, $B\,(3,6)$, $C\,(5,10)$ and $D\,(x,y)$ since diagonals of a parallelogram bisect each other, mid-point of AC = mid-point of BD

$\Rightarrow \left(\dfrac{5+1}{2},\dfrac{10-2}{2}\right)=\left(\dfrac{x+3}{2},\dfrac{y+6}{2}\right)$

$\Rightarrow (3,4)=\left(\dfrac{x+3}{2},\dfrac{y+6}{2}\right)$

Now find x and y.

16. (b) Let a point on the x-axis be $(a,0)$. Then,
Dist. bet. $(a,0)$ and $(7,6)$
$\qquad\qquad$ = Dist. bet. $(a,0)$ and $(-3,4)$
$(7-a)^2+(6-0)^2=(-3-a)^2+(4-0)^2$
Now solve.

17. (b) Perimeter = $2\,(JK+KL)$.

18. (b) Let the vertices of the Δ be $A\,(2,-2)$, $B\,(-2,1)$ and $C\,(5,2)$.
Then $AB^2=(-2-2)^2+(1+2)^2=16+9=25$
$BC^2=(5+2)^2+(2-1)^2=49+1=50$
$AC^2=(5-2)^2+(2+2)^2=9+16=25$
$\Rightarrow AB^2+AC^2=BC^2$
$\Rightarrow$ The Δ is rt $\angle d$ at A.

$\Rightarrow$ Area of $\Delta ABC=\dfrac{1}{2}\times AC\times AB$

$\qquad =\dfrac{1}{2}\times5\times5=\dfrac{25}{2}$ sq. units

19. (a) Let the points be $P\,(3,4)$, $Q\,(11,10)$ and $R\left(5,\dfrac{11}{2}\right)$

$PQ=\sqrt{(11-3)^2+(10-4)^2}=\sqrt{8^2+6^2}$

$\qquad =\sqrt{64+36}=\sqrt{100}=10$ units

$QR=\sqrt{(5-11)^2+\left(\dfrac{11}{2}-10\right)^2}=\sqrt{(-6)^2+\left(\dfrac{-9}{2}\right)^2}$

$\qquad =\sqrt{36+\dfrac{81}{4}}=\sqrt{\dfrac{225}{4}}=\dfrac{15}{2}$ units

$PR=\sqrt{(5-3)^2+\left(\dfrac{11}{2}-4\right)^2}=\sqrt{2^2+\left(\dfrac{3}{2}\right)^2}$

$\qquad =\sqrt{4+\dfrac{9}{4}}=\sqrt{\dfrac{25}{4}}=\dfrac{5}{2}$ units

Clearly,

$PR+QR=\dfrac{5}{2}+\dfrac{15}{2}=10$ units $=PQ$

Hence, the points P, R and Q are collinear.

20. (b) The co-ordinates of point R which satisfy the equation $PQ=QR=PR$ is the reqd. point.

Self Assessment Sheet–30

1. The distance between the points (a,b) and $(-a,-b)$ is
(a) 0 (b) 1
(c) $\sqrt{ab}$ (d) $2\sqrt{a^2+b^2}$

2. If the points $(a,0)$, $(0,b)$ and $(1,1)$ are collinear, then $\dfrac{1}{a}+\dfrac{1}{b}=$
(a) –1 (b) 1
(c) 0 (d) 2

3. If $Q\,(0,1)$ is equidistant from $P\,(5,-3)$ and $R\,(x,6)$, the positive value of x is

(a) 5 (b) 4
(c) 2 (d) 8

4. The quadrilateral formed by the points $(-1,-2)$, $(1,0)$, $(-1,2)$ and $(-3,0)$ is a
(a) rectangle (b) square
(c) rhombus (d) none of these

5. The third vertex of a triangle, whose two vertices are $(-4,1)$ and $(0,-3)$ and centroid is at the origin is
(a) $(3,1)$ (b) $(-4,1)$
(c) $(4,2)$ (d) $(-1,2)$

6. $(3,2)$, $(-3,2)$ and $(0,2-\sqrt{3})$ are the vertices of _____ triangle of area _____.

(a) isosceles; 81 sq. units

(b) scalene; $9\sqrt{3}$ sq. units

(c) equilateral; $9\sqrt{3}$ sq. units

(d) right angled; 81 sq. units

7. The mid-point of the line segment joining $(2a, 4)$ and $(-2, 2b)$ is $(1, 2a + 1)$. The values of a and b are

(a) $a = 3, b = -1$ (b) $a = 2, b = -3$

(c) $a = 3, b = -2$ (d) $a = 2, b = 3$

8. If $A(1, 0)$, $B(5, 3)$, $C(2, 7)$ and $D(x, y)$ are vertices of a parallelogram $ABCD$, the co-ordinates of D are

(a) $(-2, -3)$ (b) $(-2, 4)$

(c) $(2, -3)$ (d) $(3, 5)$

9. If $M(x, y)$ is equidistant from $A(a + b, b - a)$ and $B(a - b, a + b)$, then

(a) $bx + ay = 0$ (b) $bx - ay = 0$

(c) $ax + by = 0$ (d) $ax - by = 0$

10. The centre of the circle is at the origin and its radius is 10. Which of the following points lies inside the circle?

(a) $(6, 8)$ (b) $(0, 11)$

(c) $(-10, 0)$ (d) $(7, 7)$

Answers

1. (d)   2. (b)   3. (b)   4. (b) [**Hint:** Show all sides equal and diagonals equal.]

5. (c) [**Hint:** Here $\frac{-4+0+x}{3} = 0$ and $\frac{1-3+y}{3} = 0$]   6. (c)   7. (d)   8. (b)   9. (b)   10. (d)

UNIT–9

TRIGONOMETRY

- *Trigonometrical Ratios*
- *Trigonometrical Ratios of Standard Angles*
- *Some Applications of Trigonometry*

Chapter 32

TRIGONOMETRICAL RATIOS

KEY FACTS

1. In a right-angled triangle three sides give the following six ratios.
   Which are called trigonometrical ratios (*t*- ratios)

   (i) The ratio of the *perpendicular to the hypotenuse* is called **sine** of the angle θ and is written as **sin θ.**

   $$\sin\theta = \frac{\text{Perpendicular}}{\text{Hypotenuse}} = \frac{AB}{AC}$$

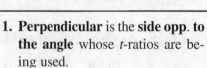

   (ii) The ratio of the *base to the hypotenuse* is called the **cosine** of the angle θ and is written as **cos θ**

   $$\cos\theta = \frac{\text{Base}}{\text{Hypotenuse}} = \frac{BC}{AC}$$

   (iii) The ratio of the *perpendicular to the base* is called the **tangent** of angle θ and is written as **tan θ.**

   $$\tan\theta = \frac{\text{Perpendicular}}{\text{Base}} = \frac{AB}{BC}$$

> 1. **Perpendicular** is the **side opp. to the angle** whose *t*-ratios are being used.
> 2. **Base** is the side **adjacent to the angle** whose *t*-ratios are being used.
> 3. **Hypotenuse** is the side opposite the right angle.

RECIPROCAL RATIOS

The following three are reciprocals of the above ratios.

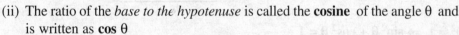

   (iv) The ratio of the *hypotenuse to the perpendicular* is called the **cosecant** of angle θ and is written as **cosec θ.**

   $$\operatorname{cosec}\theta = \frac{\text{Hypotenuse}}{\text{Perpendicular}} = \frac{AC}{AB}$$

   (v) The ratio of the *hypotenuse to the base* is called the **secant** of the angle θ and is written as **sec θ.**

   $$\sec\theta = \frac{\text{Hypotenuse}}{\text{Base}} = \frac{AC}{BC}$$

   (vi) The ratio of the *base to the perpendicular* is called the **cotangent** of angle θ is written as **cot θ.**

   $$\cot\theta = \frac{\text{Base}}{\text{Perpendicular}} = \frac{BC}{AB}$$

Therefore, $\sin\theta = \dfrac{1}{\operatorname{cosec}\theta} \Rightarrow \sin\theta\cdot\operatorname{cosec}\theta = 1$

$\cos\theta = \dfrac{1}{\sec\theta} \Rightarrow \cos\theta\cdot\sec\theta = 1$

$\cot\theta = \dfrac{1}{\tan\theta} \Rightarrow \cot\theta\cdot\tan\theta = 1$

> **Note.** 1. $\sin\theta$ refers to sin of angle θ not $\sin\times\theta$
> 2. $(\sin\theta)^2 = \sin^2\theta$ and is read as "sin square θ"
> 3. $(\cos\theta)^3 = \cos^3\theta$ and is read as "cos cube θ", etc.

2. $\tan\theta = \dfrac{\sin\theta}{\cos\theta}$ and $\cot\theta = \dfrac{\cos\theta}{\sin\theta}$

$\sin\theta = \dfrac{\text{Perp.}}{\text{Hyp.}}$, $\cos\theta = \dfrac{\text{Base}}{\text{Hyp.}} \Rightarrow \dfrac{\sin\theta}{\cos\theta} = \dfrac{\text{Perp./Hyp.}}{\text{Base/Hyp.}} = \dfrac{\text{Perp.}}{\text{Base}} = \mathbf{\tan\theta}$

$\therefore \cot\theta = \dfrac{1}{\tan\theta} = \dfrac{1}{\sin\theta/\cos\theta} = \dfrac{\cos\theta}{\sin\theta}$.

3. Simple Trigonometric Identities

(i) For any rt. $\triangle ABC$,

$(\text{Perp.})^2 + (\text{base})^2 = (\text{hyp.})^2$

$\Rightarrow \quad p^2 + b^2 = h^2$

$\Rightarrow \quad \dfrac{p^2}{h^2} + \dfrac{b^2}{h^2} = 1$

$\Rightarrow \quad (\sin\theta)^2 + (\cos\theta)^2 = 1 \Rightarrow \mathbf{\sin^2\theta + \cos^2\theta = 1}$

$\Rightarrow \sin^2\theta = 1 - \cos^2\theta$ or $\cos^2\theta = 1 - \sin^2\theta$

(ii) Dividing both sides of $\sin^2\theta + \cos^2\theta = 1$ by $\cos^2\theta$, we get

$\dfrac{\sin^2\theta}{\cos^2\theta} + \dfrac{\cos^2\theta}{\cos^2\theta} = \dfrac{1}{\cos^2\theta} \Rightarrow \tan^2\theta + 1 = \sec^2\theta$

$\Rightarrow \quad \mathbf{\sec^2\theta - \tan^2\theta = 1}$

(iii) Similarly, on dividing both sides of Identity (i) by $\sin^2\theta$, we get

$\dfrac{\sin^2\theta}{\sin^2\theta} + \dfrac{\cos^2\theta}{\sin^2\theta} = \dfrac{1}{\sin^2\theta} \Rightarrow 1 + \cot^2\theta = \operatorname{cosec}^2\theta$

$\Rightarrow \quad \mathbf{\operatorname{cosec}^2\theta - \cot^2\theta = 1}$

Solved Examples

Ex. 1. *If* $\cos\theta = \dfrac{12}{13}$, *find the value of* $2\sin\theta - 4\tan\theta$, *where* θ *is a acute.*

Sol. Draw a right angled $\triangle ABC$, such that

$\angle B = 90°$ and $\angle C = \theta$ (acute angle)

$\therefore \qquad \cos\theta = \dfrac{12}{13}$ and $\cos\theta = \dfrac{\text{Base}}{\text{Hyp.}} = \dfrac{BC}{AC}$

$\Rightarrow \quad \dfrac{BC}{AC} = \dfrac{12}{13} \Rightarrow \dfrac{BC}{12} = \dfrac{AC}{13} = k$

$\Rightarrow \quad BC = 12k, \qquad AC = 13k$

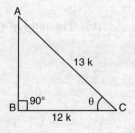

∴   By Pythagoras' Theorem

$$AB^2 + BC^2 = AC^2$$

⇒ $AB^2 = AC^2 - BC^2 = (13\,k)^2 - (12\,k)^2 = 169\,k^2 - 144\,k^2 = 25\,k^2$

⇒ $AB = 5\,k$

∴ $2\sin\theta - 4\tan\theta = 2 \times \dfrac{AB}{AC} - 4 \times \dfrac{AB}{BC}$

$$= 2 \times \frac{5k}{13k} - 4 \times \frac{5k}{12k}$$

$$= \frac{10}{13} - \frac{5}{3} = \frac{30 - 65}{39} = -\frac{35}{39}.$$

Ex. 2. *If* $\tan\theta = \dfrac{1}{\sqrt{3}}$*, prove that* $7\sin^2\theta + 3\cos^2\theta = 4$.

Sol. $\tan\theta = \dfrac{1}{\sqrt{3}} = \dfrac{AB}{BC}$

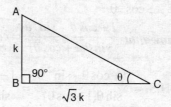

⇒ $\dfrac{AB}{1} = \dfrac{BC}{\sqrt{3}} = k \Rightarrow AB = k$ and $BC = \sqrt{3}k$

∴   By Pythagoras' Theorem

$$AC^2 = AB^2 + BC^2$$

$$= k^2 + (\sqrt{3}\,k)^2 = k^2 + 3k^2 = 4k^2 \Rightarrow AC = 2k$$

∴ $\sin\theta = \dfrac{AB}{AC} = \dfrac{k}{2k} = \dfrac{1}{2}$ and $\cos\theta = \dfrac{BC}{AC} = \dfrac{\sqrt{3}\,k}{2k} = \dfrac{\sqrt{3}}{2}$

∴   L.H.S $= 7\sin^2\theta + 3\cos^2\theta = 7 \times \left(\dfrac{1}{2}\right)^2 + 3 \times \left(\dfrac{\sqrt{3}}{2}\right)^2$

$$= \frac{7}{4} + \frac{9}{4} = \frac{16}{4} = \mathbf{4} = \text{R.H.S}.$$

Ex. 3. *If* $b\tan\theta = a$*, find the value of* $\dfrac{\cos\theta + \sin\theta}{\cos\theta - \sin\theta}$.

Sol. $b\tan\theta = a \Rightarrow \tan\theta = \dfrac{a}{b}$

$$\frac{\cos\theta + \sin\theta}{\cos\theta - \sin\theta} = \frac{\dfrac{\cos\theta}{\cos\theta} + \dfrac{\sin\theta}{\cos\theta}}{\dfrac{\cos\theta}{\cos\theta} - \dfrac{\sin\theta}{\cos\theta}}$$ (Dividing both numerator and denominator by $\cos\theta$)

$$= \frac{1 + \tan\theta}{1 - \tan\theta}$$

$$= \frac{1 + \dfrac{a}{b}}{1 - \dfrac{a}{b}} = \frac{\dfrac{b+a}{b}}{\dfrac{b-a}{b}} = \frac{\mathbf{b+a}}{\mathbf{b-a}}.$$

Ex. 4. *Prove that* $(1 + \cot^2\theta)(\sin^2\theta) = 1$.

Sol. L.H.S. $= (1 + \cot^2\theta) . \sin^2\theta = \text{cosec}^2\theta . \sin^2\theta$ (∵ $\text{cosec}^2\theta - \cot^2\theta = 1$)

$$= \frac{1}{\sin^2\theta} . \sin^2\theta = \mathbf{1}.$$

Ex. 5. *Prove that $sec^2\theta + cosec^2\theta = sec^2\theta . cosec^2\theta$*

Sol. L.H.S. $= \sec^2\theta + \csc^2\theta = \dfrac{1}{\cos^2\theta} + \dfrac{1}{\sin^2\theta}$

$= \dfrac{\sin^2\theta + \cos^2\theta}{\cos^2\theta . \sin^2\theta} = \dfrac{1}{\cos^2\theta . \sin^2\theta}$ $(\because \sin^2\theta + \cos^2\theta = 1)$

$= \dfrac{1}{\cos^2\theta} \dfrac{1}{\sin^2\theta} = \sec^2\theta \times \csc^2\theta$

Ex. 6. *Show that $(1 + tan^2\ \theta)\ (1 - sin\theta)\ (1 + sin\theta)\ = 1$*

Sol. $(1 + \tan^2\theta)\ (1 - \sin\theta)\ (1 + \sin\theta)$

$= \sec^2 (1 - \sin^2\theta)$

$= \sec^2\theta . \cos^2\theta$

$= \dfrac{1}{\cos^2\theta} . \cos^2\theta = \mathbf{1}.$

$\begin{pmatrix} \because \sec^2\theta - \tan^2\theta = 1 \\ (a+b)(a-b) = a^2 - b^2 \\ \sin^2\theta + \cos^2\theta = 1 \end{pmatrix}$

Ex. 7. *Show that $\dfrac{1 + cos\ \theta - sin^2\theta}{sin\ \theta(1 + cos\ \theta)} = cot\ \theta$*

Sol. L.H.S. $= \dfrac{1 + \cos\theta - \sin^2\theta}{\sin\theta(1 + \cos\theta)} = \dfrac{1 - \sin^2\theta + \cos\theta}{\sin\theta(1 + \cos\theta)}$

$= \dfrac{\cos^2\theta + \cos\theta}{\sin\theta(1 + \cos\theta)} = \dfrac{\cos\theta(\cos\theta + 1)}{\sin\theta(1 + \cos\theta)}$ $(\because \sin^2\theta + \cos^2\theta = 1)$

$= \cot\theta = $ R.H.S

Ex. 8. *Prove that $\dfrac{sin\theta}{1 + cos\theta} + \dfrac{1 + cos\theta}{sin\theta} = 2\ cosec\ \theta$*

Sol. L.H.S. $= \dfrac{\sin\theta}{1 + \cos\theta} + \dfrac{1 + \cos\theta}{\sin\theta} = \dfrac{\sin^2\theta + (1 + \cos\theta)^2}{\sin\theta(1 + \cos\theta)}$

$= \dfrac{\sin^2\theta + 1 + 2\cos\theta + \cos^2\theta}{\sin\theta(1 + \cos\theta)}$ $(\because (a + b)^2 = a^2 + 2ab + b^2)$

$= \dfrac{\sin^2\theta + \cos^2\theta + 1 + 2\cos\theta}{\sin\theta(1 + \cos\theta)} = \dfrac{1 + 1 + 2\cos\theta}{\sin\theta(1 + \cos\theta)} = \dfrac{2(1 + \cos\theta)}{\sin\theta(1 + \cos\theta)}$

$= \dfrac{2}{\sin\theta} = 2\cos ec\ \theta = $ R.H.S

Question Bank–32

1. If $\cos\theta = \dfrac{5}{13}$, θ being an acute angle, then the value

of $\dfrac{\cos\theta + 5\cot\theta}{\cos ec\theta - \cos\theta}$ will be

(a) $\dfrac{169}{109}$ (b) $\dfrac{155}{109}$

(c) $\dfrac{385}{109}$ (d) $\dfrac{395}{109}$

2. If $p \sin x = q$ and x is acute then $\sqrt{p^2 - q^2}\ \tan x$ is equal to

(a) p (b) q

(c) pq (d) $p + q$

3. If $8 \tan A = 15$, then the value of $\dfrac{\sin A - \cos A}{\sin A + \cos A}$ is

(a) $\dfrac{7}{23}$ (b) $\dfrac{11}{23}$

(c) $\dfrac{13}{23}$ (d) $\dfrac{17}{23}$

4. If $\sin \theta : \cos \theta :: a : b$, then the value of $\sec \theta$ is

(a) $\dfrac{\sqrt{a^2 + b^2}}{a}$

(b) $\dfrac{b}{\sqrt{a^2 + b^2}}$

(c) $\dfrac{\sqrt{a^2 + b^2}}{b}$

(d) $\dfrac{a}{\sqrt{a^2 + b^2}}$

5. If $\tan x = \dfrac{4}{3}$, then the value of

$\sqrt{\dfrac{(1 - \sin x)(1 + \sin x)}{(1 + \cos x)(1 - \cos x)}}$ is

(a) $\dfrac{9}{16}$

(b) $\dfrac{3}{4}$

(c) $\dfrac{4}{3}$

(d) $\dfrac{16}{9}$

6. If $\cos A = 0.6$, then $5 \sin A - 3 \tan A$ is equal to

(a) 0

(b) 1

(c) 2

(d) 8

7. If $\cos \theta = \dfrac{3}{5}$, then the value of $\dfrac{\sin \theta . \tan \theta + 1}{2 \tan^2 \theta}$ is

(a) $\dfrac{88}{160}$

(b) $\dfrac{91}{160}$

(c) $\dfrac{92}{160}$

(d) $\dfrac{93}{160}$

8. If $\tan \theta = \dfrac{x}{y}$, then $\dfrac{x \sin \theta + y \cos \theta}{x \sin \theta - y \cos \theta}$ is equal to

(a) $\dfrac{x^2 + y^2}{x^2 - y^2}$

(b) $\dfrac{x^2 - y^2}{x^2 + y^2}$

(c) $\dfrac{x}{\sqrt{x^2 + y^2}}$

(d) $\dfrac{y}{\sqrt{x^2 + y^2}}$

9. In a $\triangle ABC$ right angled at B, it $\tan A = \dfrac{1}{\sqrt{3}}$, find the value of $\cos A \cos C - \sin A \sin C$.

(a) 1

(b) 0

(c) –1

(d) 2

10. If $7 \sin A = 24 \cos A$, then $14 \tan A + 25 \cos A - 7 \sec A$ equals.

(a) 0

(b) 1

(c) 30

(d) 32

11. The value of $\sin^2\theta \cos^2\theta (\sec^2\theta + \text{cosec}^2\theta)$ is

(a) 2

(b) 4

(c) 1

(d) 3

12. Which of the following is not an identity?

(a) $(1 - \sin^2 A) . \sec^2 A = 1$

(b) $(\sec^2 \theta - 1) (1 - \text{cosec}^2 \theta) = 1$

(c) $\dfrac{\sin \theta}{1 + \cos \theta} + \dfrac{1 + \cos \theta}{\sin \theta} = \dfrac{2}{\sin \theta}$

(d) $\sin^4 \theta - \cos^4 \theta = \sin^2 \theta - \cos^2 \theta$

13. If $x = a \sec \theta + b \tan \theta$

$y = b \sec \theta + a \tan \theta$

then $x^2 - y^2$ is equal to

(a) $4ab \ \sec \theta \tan \theta$

(b) $a^2 - b^2$

(c) $b^2 - a^2$

(d) $a^2 + b^2$

14. If $x = a \cos^3 \theta$ and $y = b \sin^3 \theta$, then the value of $\left(\dfrac{x}{a}\right)^{2/3} + \left(\dfrac{y}{b}\right)^{2/3}$ is

(a) 1

(b) –2

(c) 2

(d) –1

15. What is the value of $(\text{cosec } A - \sin A) (\sec A - \cos A) (\tan A + \cot A)$?

(a) 0

(b) 1

(c) 2

(d) 3

16. What is the value of $\dfrac{\sin^3 x + \cos^3 x}{\sin x + \cos x} + \sin x \cos x$?

(a) 0

(b) $\sin x$

(c) $\cos x$

(d) 1

17. $\dfrac{\cos \theta}{1 - \sin \theta} + \dfrac{\cos \theta}{1 + \sin \theta}$ equals

(a) $2 \tan \theta$

(b) $2 \text{ cosec } \theta$

(c) $2 \cot \theta$

(d) $2 \sec \theta$

18. Consider the following:

1. $\tan^2 \theta - \sin^2 \theta = \tan^2 \theta \sin^2 \theta$

2. $(1 + \cot^2 \theta) (1 - \cos \theta) (1 + \cos \theta) = 1$

Which of the statements given below is correct?

(a) 1 only is the identity

(b) 2 only is the identity

(c) Both 1 and 2 are identities

(d) Neither 1 nor 2 is the identity

19. $\sin^4 \theta + 2\cos^2 \theta \left(1 - \dfrac{1}{\sec^2 \theta}\right) + \cos^4 \theta =$

(a) 1

(b) 2

(c) $\sqrt{2}$

(d) 0

20. If $\sin \theta + \sin^2 \theta = 1$, then $\cos^2 \theta + \cos^4 \theta =$

(a) 1

(b) $\sqrt{2}$

(c) 0

(d) 2

Answers

1. (c)	**2.** (b)	**3.** (a)	**4.** (c)	**5.** (b)	**6.** (a)	**7.** (d)	**8.** (a)	**9.** (b)	**10.** (c)
11. (c)	**12.** (b)	**13.** (b)	**14.** (a)	**15.** (b)	**16.** (d)	**17.** (d)	**18.** (c)	**19.** (a)	**20.** (a)

Hints and Solutions

1. (c) $\cos\theta = \dfrac{5}{13} \Rightarrow \sin\theta = \sqrt{1 - \cos^2\theta} = \sqrt{1 - \left(\dfrac{5}{13}\right)^2}$

$\qquad = \sqrt{1 - \dfrac{25}{169}} = \sqrt{\dfrac{144}{169}} = \dfrac{12}{13}$

$\therefore \ \operatorname{cosec}\theta = \dfrac{1}{\sin\theta} = \dfrac{13}{12}; \ \cot\theta = \dfrac{\cos\theta}{\sin\theta} = \dfrac{5/13}{12/13} = \dfrac{5}{12}$

Now substitute the values in the exp.

$\dfrac{\cos\theta + 5\cot\theta}{\operatorname{cosec}\theta - \cos\theta}$

2. (b) $\sqrt{p^2 - q^2}\,\tan x = \sqrt{p^2 - p^2\sin^2 x}\,.\,\tan x$

$\qquad = \sqrt{p^2(1 - \sin^2 x)}\,.\,\tan x = p\sqrt{(1 - \sin^2 x)}\,.\,\tan x$

$\qquad\qquad\qquad (\because \cos^2\theta + \sin^2\theta = 1)$

$\qquad = p\sqrt{\cos^2 x}\,.\,\tan x = p\cos x\,.\,\tan x = p\sin x = q$

3. (a) $8\tan A = 15 \Rightarrow \tan A = \dfrac{15}{8}$

Now, $\dfrac{\sin A - \cos A}{\sin A + \cos A} = \dfrac{\dfrac{\sin A}{\cos A} - \dfrac{\cos A}{\cos A}}{\dfrac{\sin A}{\cos A} + \dfrac{\cos A}{\cos A}}$

$\qquad = \dfrac{\tan A - 1}{\tan A + 1}$

Now, substitute the value of $\tan A$.

4. (c) Given, $\dfrac{\sin\theta}{\cos\theta} = \dfrac{a}{b} \Rightarrow \tan\theta = \dfrac{a}{b}$

Also, $\sec^2\theta = 1 + \tan^2\theta \Rightarrow \sec\theta = \sqrt{1 + \tan^2\theta}$

$\Rightarrow \sec\theta = \sqrt{1 + \dfrac{a^2}{b^2}} = \sqrt{\dfrac{b^2 + a^2}{b^2}} = \dfrac{\sqrt{a^2 + b^2}}{b}$

5. (b) $\sqrt{\dfrac{(1 - \sin x)(1 + \sin x)}{(1 + \cos x)(1 - \cos x)}} = \sqrt{\dfrac{1 - \sin^2 x}{1 - \cos^2 x}}$

$\qquad\qquad (\because (a - b)(a + b) = a^2 - b^2)$

$\qquad = \sqrt{\dfrac{\cos^2 x}{\sin^2 x}} \quad (\because \sin^2 x + \cos^2 x = 1)$

$\qquad = \dfrac{\cos x}{\sin x} = \cot\theta = \dfrac{1}{\tan\theta} = \dfrac{1}{\dfrac{4}{3}} = \dfrac{3}{4}.$

6. (a) $\cos A = 0.6 = \dfrac{6}{10} = \dfrac{3}{5}$

$\Rightarrow \cos A = \dfrac{BA}{AC} = \dfrac{3}{5} = k$

$\Rightarrow BA = 3k$ and $AC = 5k$

Using Pythagoras' Theorem

$AC^2 = BC^2 + AB^2 \Rightarrow (5k)^2 = BC^2 + (3k)^2$

$\Rightarrow BC^2 = 25k^2 - 9k^2 = 16k^2 \Rightarrow BC = 4k$

Now, $\sin A = \dfrac{BC}{AC} = \dfrac{4k}{5k} = \dfrac{4}{5},$

$\tan A = \dfrac{BC}{AB} = \dfrac{4k}{3k} = \dfrac{4}{3}.$ Now solve.

7. (d) Similar to Q. 6.

8. (a) $\dfrac{x\sin\theta + y\cos\theta}{x\sin\theta - y\cos\theta} = \dfrac{x\dfrac{\sin\theta}{\cos\theta} + y\dfrac{\cos\theta}{\cos\theta}}{x\dfrac{\sin\theta}{\cos\theta} - y\dfrac{\cos\theta}{\cos\theta}}$

$\qquad = \dfrac{x\tan\theta + y}{x\tan\theta - y} = \dfrac{x \times \dfrac{x}{y} + y}{x \times \dfrac{x}{y} - y}$

$\qquad = \dfrac{\dfrac{x^2 + y^2}{y}}{\dfrac{x^2 - y^2}{y}} = \dfrac{x^2 + y^2}{x^2 - y^2}$

9. (b) Consider the $\triangle ABC$ in which $\angle B = 90°$

Since, $\tan A = \dfrac{1}{\sqrt{3}}$

$\therefore \ \dfrac{1}{\sqrt{3}} = \dfrac{BC}{AB}$

$\Rightarrow \dfrac{AB}{\sqrt{3}} = \dfrac{BC}{1} = k$

$\Rightarrow AB = \sqrt{3}k$ and $BC = k$

∴ By Pythagoras' Theorem,

$$AC^2 = AB^2 + BC^2$$

$$= (\sqrt{3}\,k)^2 + k^2 + 4k^2$$

$$\Rightarrow AC = 2k$$

Now **for** $\angle A$, Base $= AB$, Perp. $= BC$, Hyp.$= AC$

$$\therefore \quad \sin A = \frac{\text{Perp.}}{\text{Hyp.}} = \frac{BC}{AC} = \frac{k}{2k} = \frac{1}{2}$$

$$\cos A = \frac{\text{Base}}{\text{Hyp.}} = \frac{AB}{AC} = \frac{\sqrt{3}k}{2k} = \frac{\sqrt{3}}{2}$$

For $\angle C$, Base $= BC$, Perp. $= AB$, Hyp. $= AC$

$$\therefore \quad \sin C = \frac{\text{Perp.}}{\text{Hyp.}} = \frac{AB}{AC} = \frac{\sqrt{3}k}{2k} = \frac{\sqrt{3}}{2}$$

$$\cos C = \frac{\text{Base}}{\text{Hyp.}} = \frac{BC}{AC} = \frac{k}{2k} = \frac{1}{2}$$

$$\therefore \ \cos A \cos C - \sin A \sin C = \frac{\sqrt{3}}{2} \times \frac{1}{2} - \frac{1}{2} \times \frac{\sqrt{3}}{2} = 0$$

10. (c) $7 \sin A = 24 \cos A$

$$\Rightarrow \frac{\sin A}{\cos A} = \frac{24}{7} \Rightarrow \tan A = \frac{24}{7}$$

Now make the figure and find all the other t-ratios and solve.

11. (c) $\sin^2 \theta \cos^2 \theta (\sec^2 \theta + \operatorname{cosec}^2 \theta)$

$$= \sin^2 \theta \cos^2 \theta \left(\frac{1}{\cos^2 \theta} + \frac{1}{\sin^2 \theta} \right)$$

$$= \sin^2 \theta \cos^2 \theta \left(\frac{\sin^2 \theta + \cos^2 \theta}{\cos^2 \theta \sin^2 \theta} \right)$$

$$= \sin^2 \theta + \cos^2 \theta = 1$$

12. (a) L.H.S. $= (1 - \sin^2 A)\sec^2 A = \cos^2 A \cdot \sec^2 A$

$$= \cos^2 A \cdot \frac{1}{\cos^2 A} = 1 = \text{R.H.S.}$$

(b) $(\sec^2 \theta - 1)(1 - \operatorname{cosec}^2 \theta) = \tan^2 \theta \times \cot^2 \theta$

$$= -1 \neq \text{R.H.S.}$$

$$\left(\begin{array}{l} \because \sec^2 \theta - \tan^2 \theta = 1 \\ \operatorname{cosec}^2 \theta - \cot^2 \theta = 1 \end{array} \right)$$

(c) $\dfrac{\sin \theta}{1 + \cos \theta} + \dfrac{1 + \cos \theta}{\sin \theta} = \dfrac{\sin^2 \theta + (1 + \cos \theta)^2}{\sin \theta (1 + \cos \theta)}$

$$= \frac{\overline{\sin^2 \theta + 1 + 2\cos \theta + \cos^2 \theta}}{\sin \theta (1 + \cos \theta)}$$

$$= \frac{2 + 2\cos \theta}{\sin \theta (1 + \cos \theta)} = \frac{2(1 + \cos \theta)}{\sin \theta (1 + \cos \theta)} = \frac{2}{\sin \theta} = \text{R.H.S.}$$

(d) $\sin^4 \theta - \cos^4 \theta = (\sin^2 \theta)^2 - (\cos^2 \theta)^2$

$$= (\sin^2 \theta + \cos^2 \theta)(\sin^2 \theta - \cos^2 \theta)$$

$$= \sin^2 \theta - \cos^2 \theta$$

$$(\because \sin^2 \theta + \cos^2 \theta = 1)$$

$$= \text{R.H.S.}$$

Hence, (b) is not an identity.

13. (b) $x^2 - y^2 = (a \sec \theta + b \tan \theta)^2 - (b \sec \theta + a \tan \theta)^2$

$$= a^2 \sec^2 \theta + 2ab \sec \theta \tan \theta + b^2 \tan^2 \theta$$

$$- (b^2 \sec^2 \theta + 2ab \sec \theta \tan \theta + a^2 \tan^2 \theta)$$

$$= (a^2 - b^2) \sec^2 \theta + (b^2 - a^2) \tan^2 \theta$$

$$= (a^2 - b^2)(\sec^2 \theta - \tan^2 \theta)$$

$$(\because a^2 - b^2 = -(b^2 - a^2))$$

$$= a^2 - b^2 \qquad (\because \sec^2 \theta - \tan^2 \theta = 1)$$

14. (a) $\left(\dfrac{x}{a} \right)^{\frac{2}{3}} + \left(\dfrac{y}{b} \right)^{\frac{2}{3}}$

$$= \left(\sqrt[3]{\frac{a \cos^3 \theta}{a}} \right)^2 + \left(\sqrt[3]{\frac{b \sin^3 \theta}{b}} \right)^2$$

$$= (\sin \theta)^2 + (\cos \theta)^2 = \sin^2 \theta + \cos^2 \theta = 1$$

15. (b) $(\operatorname{cosec} A - \sin A)(\sec A - \cos A)(\tan A + \cot A)$

$$= \left(\frac{1}{\sin A} - \sin A \right)\left(\frac{1}{\cos A} - \cos A \right)\left(\frac{\sin A}{\cos A} + \frac{\cos A}{\sin A} \right)$$

$$= \left(\frac{1 - \sin^2 A}{\sin A} \right)\left(\frac{1 - \cos^2 A}{\cos A} \right)\left(\frac{\sin^2 A + \cos^2 A}{\cos A \sin A} \right)$$

$$= \frac{\cos^2 A . \sin^2 A . 1}{\sin A . \cos A . \cos A \sin A} = 1$$

16. (d) $\dfrac{\sin^3 x + \cos^3 x}{\sin x + \cos x} + \sin x \cos x$

$$= \frac{(\sin x + \cos x)(\sin^2 x - \sin x \cos x + \cos^2 x)}{\sin x + \cos x} + \sin x \cos x$$

$$(\because a^3 + b^3 = (a + b)(a^2 - ab + b^2))$$

$$= 1 - \sin x \cos x + \sin x \cos x = 1$$

$$(\because \sin^2 x + \cos^2 x = 1)$$

17. (d) $\dfrac{\cos\theta}{1-\sin\theta}+\dfrac{\cos\theta}{1+\sin\theta}=\dfrac{\cos\theta(1+\sin\theta)+\cos\theta(1-\sin\theta)}{(1-\sin\theta)\,(1+\sin\theta)}$

$$=\dfrac{\cos\theta+\cos\theta\,\sin\theta+\cos\theta-\cos\theta\,\sin\theta}{1-\sin^2\theta}$$

$$=\dfrac{2\cos\theta}{\cos^2\theta}=\dfrac{2}{\cos\theta}=2\sec\theta$$

18. 1. $\tan^2\theta-\sin^2\theta=\dfrac{\sin^2\theta}{\cos^2\theta}-\sin^2\theta$

$$=\dfrac{\sin^2\theta-\sin^2\theta\cos^2\theta}{\cos^2\theta}$$

$$=\dfrac{\sin^2\theta(1-\cos^2\theta)}{\cos^2\theta}=\dfrac{\sin^2\theta}{\cos^2\theta}\times\sin^2\theta=\tan^2\theta\,\sin^2\theta$$

2. $(1+\cot^2\theta)\,(1-\cos\theta)\,(1+\cos\theta)$

$$=(1+\cot^2\theta)\,(1-\cos^2\theta)$$

$$=\left(1+\dfrac{\cos^2\theta}{\sin^2\theta}\right)\!\left(1-\cos^2\theta\right)$$

$$=\left(\dfrac{\sin^2\theta+\cos^2\theta}{\sin^2\theta}\right)\times\sin^2\theta$$

$$(\because\,1-\cos^2\theta=\sin^2\theta)$$

$$=1.\,(\cos^2\theta+\sin^2\theta=1)$$

$\therefore$ (c) is the correct option.

19. (a) $\sin^4\theta+2\cos^2\theta\left(1-\dfrac{1}{\sec^2\theta}\right)+\cos^4\theta$

$$=\sin^4\theta+2\cos^2\theta(1-\cos^2\theta)+\cos^4\theta$$

$$=\sin^4\theta+2\cos^2\theta.\sin^2\theta+\cos^4\theta$$

$$=(\sin^2\theta+\cos^2\theta)^2=1^2=1$$

20. (a) Given, $\sin\theta=1-\sin^2\theta=\cos^2\theta$

$\therefore\ \cos^2\theta+\cos^4\theta=\sin\theta+\sin^2\theta=1$

Self Assessment Sheet–31

1. If $\sec\theta=\dfrac{13}{5}$, then what is the value of $\dfrac{2\sin\theta-3\cos\theta}{4\sin\theta-9\cos\theta}$?

(a) 1   (b) 2

(c) 3   (d) 4

2. If $\tan x=\dfrac{x}{y}$, where x and y are whole numbers, then $\sin x$ is:

(a) $\dfrac{y}{\sqrt{y^2-x^2}}$ (b) $\dfrac{x}{\sqrt{x^2+y^2}}$

(c) $\dfrac{y}{\sqrt{x^2+y^2}}$ (d) $\dfrac{x}{\sqrt{y^2-x^2}}$

3. If $4\sin\theta=3\cos\theta$, then $\dfrac{\sec^2\theta}{4(1-\tan^2\theta)}$ is

(a) $\dfrac{25}{16}$ (b) $\dfrac{25}{28}$

(c) $\dfrac{1}{4}$ (d) $\dfrac{16}{25}$

4. If $\cos x=\dfrac{12}{13}$ and x is an acute angle, then $\sqrt{\left(1+\dfrac{\sin x}{\cos x}\right)(1-\tan x)}$ is

(a) $\dfrac{\sqrt{115}}{12}$ (b) $\dfrac{\sqrt{116}}{12}$

(c) $\dfrac{\sqrt{119}}{12}$ (d) $\dfrac{\sqrt{117}}{12}$

5. If $\dfrac{1}{\cos\theta}=a+\dfrac{1}{4a}$, then the value of $\left(\tan\theta+\dfrac{1}{\cos\theta}\right)$ is

(a) a (b) $2a$

(c) $3a$ (d) $4a$

6. If $p=\cos x-\sin x,\ q=\dfrac{1-\sin^3 x}{1-\sin x},\ r=\dfrac{1+\cos^3 x}{1+\cos x}$, what is the value of $p+q+r$?

(a) 0   (b) 1

(c) 2   (d) 3

7. $\dfrac{\cos\alpha+\cos\beta}{\sin\alpha+\sin\beta}+\dfrac{\sin\alpha-\sin\beta}{\cos\alpha-\cos\beta}=$

(a) 1   (b) 2

(c) $\sqrt{2}$ (d) 0

8. $\dfrac{1+\tan^2 A}{1+\cot^2 A}$ equals

(a) $\sec^2 A$

(b) -1

(c) $\cot^2 A$

(d) $\tan^2 A$

9. ABC is a right angled triangle, right angled at c. D is the midpoint of BC. Then, $\dfrac{\tan \theta}{\tan \phi} =$

(a) 1

(b) $\dfrac{1}{\sqrt{2}}$

(c) $\dfrac{1}{2}$

(d) 2

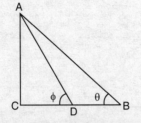

10. Which of the following is not an identity?

(a) $\dfrac{1}{\sec A - 1} + \dfrac{1}{\sec A + 1} = 2 \operatorname{cosec} A \, \cot A$

(b) $\cos^2 A + \dfrac{1}{1 + \cot^2 A} = 1$

(c) $\tan \theta - \cot \theta = \dfrac{2 \sin^2 \theta - 1}{\sin \theta \cos \theta}$

(d) $\cot \theta + \tan \theta = \operatorname{cosec} \theta \cdot \cos \theta$

Answers

1. (c)   2. (b)   3. (b)   4. (c)   5. (b)   6. (d) $\left[\text{**Hint.** Use } a^3 - b^3 = (a - b)(a^2 + ab + b^2) \text{ and } \right.$
$\left. a^3 + b^3 = (a + b)(a^2 - ab - b^2). \right]$

7. (d)   8. (d)   9. (c)   10. (d)

Chapter
33 TRIGONOMETRICAL RATIOS OF STANDARD ANGLES

KEY
FACTS

1. T-ratios of 0°

Consider the right angled $\triangle OMP$, rt.$\angle d$ at M. Here, for acute $\angle O$, OP = hyp.= h, PM = perp. = p, OM = base = b.

When $\theta = 0°$, P coincides with M, so that
$$p = 0 \text{ and } b = h$$

Then,
$$\sin\theta = \frac{p}{h} = \frac{0}{h} = 0$$

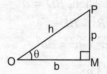

$$\cos\theta = \frac{b}{h} = \frac{h}{h} = 1$$

$$\tan\theta = \frac{\sin\theta}{\cos\theta} = \frac{0}{1} = 0$$

Hence, **sin 0° = 0, cos 0°= 1, tan 0° = 0.**

2. T-ratios of 90°

Again in the rt $\angle d$ $\triangle OMP$, $\angle O$ is increased so that it becomes 90°. Hence $\angle P$ reduces, thus making side OM reduce such that the point O coincides with point M and $\angle P = 0°$.

When $\theta = 90°$, base = 0 , perp. = hyp.

$\therefore$
$$\sin\theta = \frac{\text{perp.}}{\text{hyp.}} = \frac{\text{hyp.}}{\text{hyp.}} = 1$$

$$\cos\theta = \frac{\text{base}}{\text{hyp.}} = \frac{0}{\text{hyp.}} = 0$$

$$\tan\theta = \frac{\sin\theta}{\cos\theta} = \frac{1}{0} = \text{not defined (approaches } \infty)$$

Hence, **sin 90° = 1, cos 90° = 0, tan 90° is not defined.**

> **Note .** sin θ increases from 0 to 1
> cos θ decrease from 1 to 0
> tan θ increases from 0 to ∞

3. T-ratios of 45°

Draw $\triangle OPM$ right angled at M such that $\angle POM = 45°$. Then $\angle MPO = 45°$.

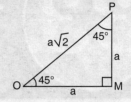

$\therefore$ $\triangle OPM$ is an isosceles triangle with $MP = MO = a$ (say)

Then, by Pythagoras' Theorem,

$$OP^2 = OM^2 + PM^2$$
$$= a^2 + a^2 = 2a^2$$
$$\Rightarrow \qquad OP = \sqrt{2}a$$

Hence, $\sin 45° = \dfrac{PM}{OP} = \dfrac{a}{a\sqrt{2}} = \dfrac{1}{\sqrt{2}}$; $\cos 45° = \dfrac{OM}{OP} = \dfrac{a}{a\sqrt{2}} = \dfrac{1}{\sqrt{2}}$

$$\tan 45° = \dfrac{PM}{OM} = \dfrac{a}{a} = 1$$

4. T-ratios of 30° and 60°

Let ABC be an equilateral triangle. Draw $AD \perp$ to BC. Then, D is the mid-point of BC. (*The altitude of an equilateral $\triangle$ co-incides with the median and bisects the vertical angle*)

Let $AB = AC = BC = 2a$. Then, $BD = a$

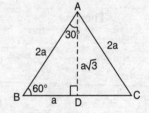

$\therefore$ In $\triangle ADB$, $AD^2 = AB^2 - BD^2$
$$= 4a^2 - a^2 = 3a^2$$
$$AD = \sqrt{3}\, a$$

Here, $\angle A = 30°$, $\angle B = 60°$

$\therefore$ In $\triangle ABD$,

$$\sin 30° = \dfrac{BD}{AB} = \dfrac{a}{2a} = \dfrac{1}{2}; \qquad \cos 30° = \dfrac{AD}{AB} = \dfrac{a\sqrt{3}}{2a} = \dfrac{\sqrt{3}}{2}; \qquad \tan 30° = \dfrac{BD}{AD} = \dfrac{a}{a\sqrt{3}} = \dfrac{1}{\sqrt{3}};$$

$$\sin 60° = \dfrac{AD}{AB} = \dfrac{a\sqrt{3}}{2a} = \dfrac{\sqrt{3}}{2}; \qquad \cos 60° = \dfrac{BD}{AB} = \dfrac{a}{2a} = \dfrac{1}{2}; \qquad \tan 60° = \dfrac{AD}{BD} = \dfrac{a\sqrt{3}}{a} = \sqrt{3}$$

Thus, the consolidated table of the *t*-ratios of some standard angles is

θ	$\sin \theta$	$\cos \theta$	$\tan \theta$
0°	0	1	0
30°	$\dfrac{1}{2}$	$\dfrac{\sqrt{3}}{2}$	$\dfrac{1}{\sqrt{3}}$
45°	$\dfrac{1}{\sqrt{2}}$	$\dfrac{1}{\sqrt{2}}$	1
60°	$\dfrac{\sqrt{3}}{2}$	$\dfrac{1}{2}$	$\sqrt{3}$
90°	1	0	not defined

5. Trigonometric ratios of complementary angles

Theorem. If θ is an acute angle

(i) $\sin (90° - \theta) = \cos \theta$ (ii) $\cos (90° - \theta) = \sin \theta$

(iii) $\tan (90° - \theta) = \cot \theta$ (iv) $\cot (90° - \theta) = \tan \theta$

(v) $\sec (90° - \theta) = \operatorname{cosec} \theta$ (vi) $\operatorname{cosec} (90° - \theta) = \sec \theta$

Ex. $\sin(90° - 18°) = \cos 18° \Rightarrow \sin 72° = \cos 18°$

$\cos(90° - 36°) = \sin 36° \Rightarrow \cos 54° = \sin 36°$

$\tan(90° - 30°) = \cot 30° \Rightarrow \tan 60° = \cot 30°$

$\cot(90° - 25°) = \tan 25° \Rightarrow \cot 65° = \tan 25°$

$\sec(90° - 15°) = \operatorname{cosec} 15° \Rightarrow \sec 75° = \operatorname{cosec} 15°$

$\operatorname{cosec}(90° - 42°) = \sec 42° \Rightarrow \operatorname{cosec} 48° = \sec 42°$

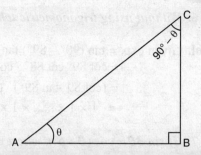

Solved Examples

Ex. 1. *Evaluate : 2 tan² 45° + cos² 30° – sin² 60°*

Sol. $2 \tan^2 45° + \cos^2 30° - \sin^2 60° = 2 \times (1)^2 + \left(\dfrac{\sqrt{3}}{2}\right)^2 - \left(\dfrac{\sqrt{3}}{2}\right)^2$

$$= 2 + \frac{3}{4} - \frac{3}{4} = \mathbf{2}.$$

Ex. 2. *Prove that :* $\dfrac{4}{3}$ *tan² 30° + sin² 60° – 3 cos² 60° +* $\dfrac{3}{4}$ *tan² 60° – 2 tan² 45° =* $\dfrac{25}{36}$

Sol. Given exp $= \dfrac{4}{3} \times \left(\dfrac{1}{\sqrt{3}}\right)^2 + \left(\dfrac{\sqrt{3}}{2}\right)^2 - 3 \times \left(\dfrac{1}{2}\right)^2 + \dfrac{3}{4} \times \left(\sqrt{3}\right)^2 - 2 \times (1)^2$

$= \dfrac{4}{3} \times \dfrac{1}{3} + \dfrac{3}{4} - \dfrac{3}{4} + \dfrac{9}{4} - 2$

$= \dfrac{4}{9} + \dfrac{9}{4} - 2 = \dfrac{16 + 81 - 72}{36} = \dfrac{\mathbf{25}}{\mathbf{36}}.$

Ex. 3. *Find the value of x, if tan 3x = sin 45° cos 45° + sin 30°*

Sol. $\tan 3x = \sin 45° \cos 45° + \sin 30°$

$= \dfrac{1}{\sqrt{2}} \times \dfrac{1}{\sqrt{2}} + \dfrac{1}{2} = \dfrac{1}{2} + \dfrac{1}{2} = 1$

$\Rightarrow \quad \tan 3x = \tan 45° \Rightarrow 3x = 45° \Rightarrow x = \mathbf{15°}$

Ex. 4. *Without using trigonometric tables, show that* $\dfrac{cos 70°}{sin 20°}$ *+ cos 49° sin 41° = 2*

Sol. $\dfrac{\cos 70°}{\sin 20°} + \cos 49° \operatorname{cosec} 41° = \dfrac{\cos(90° - 20°)}{\sin 20°} + \cos(90° - 41°) \operatorname{cosec} 41°$

$= \dfrac{\sin 20°}{\sin 20°} + \sin 41° \times \dfrac{1}{\sin 41°} = 1 + 1 = \mathbf{2}.$

Ex. 5. *Evaluate: tan 7° tan 23° tan 60° tan 67° tan 83°*

Sol. $\tan 7° \tan 23° \tan 60° \tan 67° \tan 83°$

$= \tan 7° \cdot \tan 23° \cdot \sqrt{3} \cdot \tan(90° - 23°) \cdot \tan(90° - 7°)$ ($\because \tan(90° - \theta) = \cot \theta$)

$= \tan 7° \cdot \tan 23° \cdot \sqrt{3} \cdot \cot 23° \cdot \cot 7°$

$= \tan 7° \cdot \cot 7° \cdot \tan 23° \cdot \cot 23° \cdot \sqrt{3}$ ($\because \tan \theta \cdot \cot \theta = 1$)

$= 1 \times 1 \times \sqrt{3} = \mathbf{\sqrt{3}}.$

Ex. 6. *Without using trigonometric tables, prove that: tan 1° · tan 2° · tan 3° · .... tan 89° = 1*

Sol. Given exp. = tan (90° − 89°) tan (90° − 88°) tan (90° − 87°) ... tan 87° · tan 88° · tan 89°

= cot 89°·cot 88° · cot 87° ..... tan 87° · tan 88° · tan 89°

= (cot 89°·tan 89°) (cot 88°·tan 88°) .... (cot 44°·tan 44°).tan45°

= 1 × 1 × ...... × 1 × 1 = **1**   (∵ tan 45° = 1, cot θ. tan θ = 1)

Ex. 7. *If sin 3θ = cos (θ − 6°), where 3θ and θ − 6° are acute angles, find the value of θ.*

Sol. sin 3θ = cos (θ − 6°)

⇒   cos (90° − 3θ) = cos (θ − 6°)

⇒   90° − 3θ = θ − 6° ⇒ 4θ = 96° ⇒ θ = **24°**.

Ex. 8. *Show that* $\dfrac{1}{1+\cos(90°-\theta)} + \dfrac{1}{1-\cos(90°-\theta)} = 2\,cosec^2(90°-\theta)$

Sol. L.H.S. = $\dfrac{1}{1+\sin\theta} + \dfrac{1}{1-\sin\theta} = \dfrac{1-\sin\theta+1+\sin\theta}{(1+\sin\theta)(1-\sin\theta)}$

$= \dfrac{2}{1-\sin^2\theta} = \dfrac{2}{\cos^2\theta} = 2\sec^2\theta$

R.H.S.= 2 cosec² (90° − θ) = 2 sec² θ.

∴ L.H.S. = R.H.S

Question Bank–33

1. $\dfrac{1-\tan^2 45°}{1+\tan^2 45°} =$

(a) tan 90°   (b) 1

(c) sin 45°   (d) 0

2. If $x \tan 30° = \dfrac{\sin 30° + \cos 60°}{\tan 60° + \sin 60°}$, then the value of x is :

(a) $\dfrac{2}{3}$ (b) $\dfrac{2}{\sqrt{3}}$

(c) $\dfrac{2}{3\sqrt{3}}$ (d) $\dfrac{3}{2}$

3. The value of sin 0° + cos 30° − tan 45° + cosec 60° + cot 90° is equal to

(a) $\dfrac{5\sqrt{3}-6}{6}$ (b) $\dfrac{-6+7\sqrt{3}}{6}$

(c) 0   (d) 2

4. If $2 \sin^2 x + \cos^2 45° = \tan 45°$ and x is an acute angle, then the value of tan x is:

(a) 1   (b) $\sqrt{3}$

(c) $\dfrac{1}{\sqrt{3}}$ (d) 3

5. The value of $a \sin 0° + b \cos 90° + c \tan 45°$ is:

(a) $a + b + c$ (b) $b + c$

(c) $a + c$ (d) c

6. The value of $\dfrac{\sin 30° - \cos 60° + \tan 45°}{\cos 90° - \tan 45° + \sin 90°}$ is:

(a) $\dfrac{1}{2}$ (b) 1

(c) $\sqrt{3}$ (d) ∞

7. The value of $\dfrac{1}{2} \sin^2 90° \sin^2 30° \cos^2 45° +$ $4 \tan^2 30° + \dfrac{1}{2} \sin^2 90° - 2 \cos^2 90°$ is:

(a) $\dfrac{45}{24}$ (b) $\dfrac{46}{24}$

(c) $\dfrac{47}{24}$ (d) $\dfrac{49}{24}$

8. The value of $(\cos 0° + \sin 45° + \sin 30°)$ $(\sin 90° - \cos 45° + \cos 60°)$ is:

(a) 0   (b) 1

(c) $\dfrac{7}{4}$ (d) $\dfrac{1}{2}$

9. $\dfrac{\tan 60° - \tan 30°}{1 + \tan 60° \tan 30°}$ equals

   (a) tan 60°   (b) tan 0°

   (c) tan 30°   (d) $\dfrac{1}{3}$

10. Find the value of x, if
   $\sin 2x = \sin 60° \cos 30° - \cos 60° \sin 30°$

   (a) 20°   (b) 15°

   (c) 30°   (d) 45°

All the questions from Q.No. 11 – Q.No. 18. Should be attempted without using the trigonometric tables.

11. tan 26° – cot 64° equals

   (a) –1   (b) 1

   (c) 0   (d) 2

12. The value of $\dfrac{\sin 19°}{\cos 71°} + \dfrac{\cos 73°}{\sin 17°}$ is equal to :

   (a) 0   (b) 1

   (c) 2   (d) $\dfrac{1}{2}$

13. Consider the following equations:

   1. $\dfrac{\cos 75°}{\sin 15°} + \dfrac{\sin 12°}{\cos 78°} - \dfrac{\text{cosec} 18°}{\sec 72°} = 1$

   2. $\dfrac{\tan 50° + \sec 50°}{\cot 40° + \text{cosec} 40°} + \cos 40° \text{cosec} 50° = 2$

   3. $\dfrac{\sin 80°}{\cos 10°} - \sin 59° \sec 31° = 0$

   Which of these statements given below is correct?

   (a) 1 only is correct

   (b) 3 only is correct

   (c) All 1, 2 and 3 are correct

   (d) 2 and 3 are correct

14. $\sin^2 25° + \sin^2 65°$ is equal to

   (a) 0   (b) $2 \sin^2 25°$

   (c) $2 \cos^2 65°$ (d) 1

15. If $\sin (30° - \theta) = \cos (60° + \phi)$, then

   (a) $\phi - \theta = 30°$ (b) $\phi - \theta = 0°$

   (c) $\phi + \theta = 60°$ (d) $\phi - \theta = 60°$

16. The value of cot 15° cot 16° cot 17° ....cot 73° cot 74° cot 75° is:

17. If $\sin \theta = \cos \theta$, then value of θ is:

   (a) $\dfrac{1}{2}$ (b) 0

   (c) 1   (d) –1

17. If $\sin \theta = \cos \theta$, then value of θ is:

   (a) 60°   (b) 0°

   (c) 45°   (d) 90°

18. Value of $\cos^2 5° + \cos^2 10° + \cos^2 80° + \cos^2 85°$ is:

   (a) 1   (b) 0

   (c) 2   (d) 3

19. If $\sin 3\theta = \cos (\theta - 2°)$ where 3θ and $(\theta - 2°)$ are acute angles, what is the value of θ?

   (a) 22°   (b) 23°

   (c) 24°   (d) 25°

20. If $\tan \theta = 1$ and $\sin \theta = \dfrac{1}{\sqrt{2}}$, then the value of $\cos (\theta + \phi)$ is:

   (a) –1   (b) 0

   (c) 1   (d) $\dfrac{\sqrt{3}}{2}$

21. If $x \cos 60° + y \cos 0° = 3$ and $4x \sin 30° - y \cot 45° = 2$, then what is the value of x?

   (a) –1   (b) 0

   (c) 1   (d) 2

22. Which one of the following is true?

   (a) $\tan x > 1$; $45° < x < 90°$

   (b) $\sin x > \dfrac{1}{2}$; $0° < x < 30°$

   (c) $\cos x > \dfrac{1}{2}$; $60° < x < 90°$

   (d) $\sin x = \cos x$ for some value of x; $30° < x < 45°$

23. If $x + y = 90°$, then what is $\sqrt{\cos x \, \text{cosec} \, y - \cos x \sin y}$ equal to

   (a) cos x (b) sin x

   (c) $\sqrt{\cos x}$ (d) $\sqrt{\sin x}$

24. If $0° < \theta < 90°$ and $\cos^2 \theta - \sin^2 \theta = \dfrac{1}{2}$, then what is the value of θ?

   (a) 30°   (b) 45°

   (c) 60°   (d) 90°

25. The value of $\sin^2 (90° - \theta) [1 + \cot^2 (90° - \theta)]$ is

   (a) –1   (b) 0

   (c) $\dfrac{1}{2}$ (d) 1

Answers

1. (d)	**2.** (a)	**3.** (b)	**4.** (c)	**5.** (d)	**6.** (d)	**7.** (c)	**8.** (c)	**9.** (c)	**10.** (b)
11. (c)	**12.** (c)	**13.** (c)	**14.** (d)	**15.** (b)	**16.** (c)	**17.** (c)	**18.** (c)	**19.** (b)	**20.** (b)
21. (d)	**22.** (a)	**23.** (b)	**24.** (a)	**25.** (d)					

Hints and Solutions

1. (d) $\dfrac{1-\tan^2 45°}{1+\tan^2 45°} = \dfrac{1-1}{1+1} = \dfrac{0}{2} = 0$ $(\because \tan 45° = 1)$

2. (a) $x\tan 30° = \dfrac{\sin 30° + \cos 60°}{\tan 60° + \sin 60°}$

$= \dfrac{\dfrac{1}{2}+\dfrac{1}{2}}{\sqrt{3}+\dfrac{\sqrt{3}}{2}} = \dfrac{1}{\dfrac{2\sqrt{3}+\sqrt{3}}{2}} = \dfrac{2}{3\sqrt{3}}$

$x \times \dfrac{1}{\sqrt{3}} = \dfrac{2}{3\sqrt{3}} \Rightarrow x = \dfrac{2\sqrt{3}}{3\sqrt{3}} = \dfrac{2}{3}.$

3. (b) Given exp. $= 0 + \dfrac{\sqrt{3}}{2} - 1 + \dfrac{2}{\sqrt{3}} + 0$

$= \dfrac{\sqrt{3}}{2} - 1 + \dfrac{2}{\sqrt{3}} = \dfrac{3-2\sqrt{3}+4}{2\sqrt{3}} = \dfrac{7-2\sqrt{3}}{2\sqrt{3}} \times \dfrac{\sqrt{3}}{\sqrt{3}}$

$= \dfrac{7\sqrt{3}-6}{6}.$

4. (c) $2\sin^2 x + \cos^2 45° = \tan 45°$

$\Rightarrow 2\sin^2 x + \left(\dfrac{1}{\sqrt{2}}\right)^2 = 1$

$\Rightarrow 2\sin^2 x = 1 - \dfrac{1}{2} = \dfrac{1}{2}$

$\Rightarrow \sin^2 x = \dfrac{1}{4} \Rightarrow \sin x = \dfrac{1}{2} = \sin 30° \Rightarrow x = 30°$

$\therefore \tan x = \tan 30° = \dfrac{1}{\sqrt{3}}.$

5. (d) $a\sin 0° + b\cos 90° + c\tan 45°$

$= a \times 0 + b \times 0 + c \times 1 = c.$

6. (d) Given exp. $= \dfrac{\dfrac{1}{2}-\dfrac{1}{2}+1}{0-1+1} = \dfrac{1}{0} = \infty.$

7. (c) Given exp.

$= \left(\dfrac{1}{2}\right)^2 \times \left(\dfrac{1}{\sqrt{2}}\right)^2 + 4 \times \left(\dfrac{1}{\sqrt{3}}\right)^2 + \dfrac{1}{2} \times (1)^2 - 2 \times 0$

$= \dfrac{1}{4} \times \dfrac{1}{2} + 4 \times \dfrac{1}{3} + \dfrac{1}{2} = \dfrac{1}{8} + \dfrac{4}{3} + \dfrac{1}{2} = \dfrac{3+32+12}{24} = \dfrac{47}{24}$

8. Do yourself.

9. Do yourself.

10. (b) $\sin 2x = \dfrac{\sqrt{3}}{2} \times \dfrac{\sqrt{3}}{2} - \dfrac{1}{2} \times \dfrac{1}{2}$

$= \dfrac{3}{4} - \dfrac{1}{4} = \dfrac{1}{2}$

$\therefore \sin 2x = \dfrac{1}{2} = \sin 30° \Rightarrow 2x = 30° \Rightarrow x = 15°.$

11. (c) $\tan 26° - \cot 64°$
$= \tan(90° - 64°) - \cot 64°$
$= \cot 64° - \cot 64° = 0 \ (\because \tan(90° - \theta) = \cot\theta)$

12. (c) Given exp. $= \dfrac{\sin(90°-71°)}{\cos 71°} + \dfrac{\cos 73°}{\sin(90°-73°)}$

$= \dfrac{\cos 71°}{\cos 71°} + \dfrac{\cos 73°}{\cos 73°} (\because \sin(90°-\theta) = \cos\theta)$
$= 1 + 1 = 2.$

13. (c) 1. Do yourself.

2. Given exp. $= \dfrac{\tan(90°-40°) + \sec(90°-40°)}{\cot 40° + \mathrm{cosec}\, 40°}$

$+ \cos(90°-50°)\,\mathrm{cosec}\, 50°$

$= \dfrac{\cot 40° + \mathrm{cosec}\, 40°}{\cot 40° + \mathrm{cosec}\, 40°} + \sin 50°\,\mathrm{cosec}\, 50°$

$= 1 + 1 = 2.$

3. Do yourself.

14. (d) $\sin^2 25° + \sin^2 65° = \sin^2(90°-65°) + \sin^2 65°$
$= \cos^2 65° + \sin^2 65° = 1.$

15. (b) $\sin(30°-\theta) = \cos(60°+\phi)$
$\Rightarrow \cos(90°-(30°-\theta)) = \cos(60°+\phi)$
$(\because \sin\theta = \cos(90°-\theta))$
$\Rightarrow \cos(60°+\theta) = \cos(60°+\phi)$
$\Rightarrow 60°+\theta = 60°+\phi \Rightarrow \theta-\phi = 0.$

16. Type Solved Example 5.

17. (c) $\sin\theta = \cos\theta \Rightarrow \cos(90°-\theta) = \cos\theta$
Now do yourself.

18. Do yourself.

19. (b) $\sin 3\theta = \cos(\theta-2°)$
$\Rightarrow \cos(90°-3\theta) = \cos(\theta-2°)$
$\Rightarrow 90°-3\theta = \theta-2°$
$\Rightarrow 4\theta = 92° \Rightarrow \theta = 23°.$

20. (b) $\tan\theta = 1 \Rightarrow \tan\theta = \tan 45° \Rightarrow \theta = 45°$

$\sin\phi = \dfrac{1}{\sqrt{2}} \Rightarrow \sin\phi = \sin 45° \Rightarrow \phi = 45°$

$\therefore \cos(\theta+\phi) = \cos(45°+45°)$
$= \cos 90° = 0$

21. (d) $x\cos 60° + y\cos\theta° = 3$

$\Rightarrow x \times \dfrac{1}{2} + y \times 1 = 3$

$\Rightarrow x + 2y = 6$ ...(i)
and $4x\sin 30° - y\cot 45° = 2$

$\Rightarrow 4x \times \dfrac{1}{2} - y \times 1 = 2$

$\Rightarrow$ $2x - y = 2$ ...(ii)
Now solve for x and y.

22. (a) $\tan 45° = 1$ and $\tan 90° =$ undefined (∞)
$\therefore$ when x lies between $45°$ and $90°$, $\tan x > 1$.

23. (b) $x + y = 90° \Rightarrow y = 90° - x$
$\therefore \sqrt{\cos x \operatorname{cosec} y - \cos x \sin y}$
$= \sqrt{\cos x \operatorname{cosec} (90° - x) - \cos x \sin(90° - x)}$
$= \sqrt{\cos x \sec x - \cos x \cos x}$
$= \sqrt{1 - \cos^2 x} = \sqrt{\sin^2 x} = \sin x$

24. (a) $\cos^2 \theta - \sin^2 \theta = \dfrac{1}{2}$

$\Rightarrow (1 - \sin^2 \theta) - \sin^2 \theta = \dfrac{1}{2}$
$\Rightarrow 1 - 2 \sin^2 \theta = \dfrac{1}{2}$
$\Rightarrow 2 \sin^2 \theta = \dfrac{1}{2} \Rightarrow \sin^2 \theta = \dfrac{1}{4}$
$\Rightarrow \sin \theta = \dfrac{1}{2} = \sin 30° \Rightarrow \theta = 30°$

25. (d) Given exp. $= \cos^2 \theta [1 + \tan^2 \theta]$
$= \cos^2 \theta \cdot \sec^2 \theta = 1$

$\left[\begin{array}{l} \because \sin (90° - \theta) = \cos \theta \\ \cot (90° - \theta) = \tan \theta \end{array} \right]$

Self Assessment Sheet–32

1. Which of the following is not defined?
(a) $\sin 90°$
(b) $\tan 0°$
(c) $\cot 90°$
(d) $\operatorname{cosec} 0°$

2. What is the value of $\dfrac{\sin 60°}{\cos^2 45°} - 3 \tan 30° + 5 \cos 90°$
(a) 1
(b) –1
(c) $\dfrac{2}{5}$
(d) 0

3. The value of $2\sqrt{2} \cos 45° \cdot \cos 60° + 2\sqrt{3} \sin 30°$. $\tan 60° - \cos 0°$ is
(a) $\dfrac{1}{3}$
(b) 3
(c) –3
(d) 0

4. If $2 \cos \theta = \sqrt{3}$, evaluate $3 \sin \theta - 4 \sin^3 \theta$
(a) 3
(b) $\dfrac{4}{3}$
(c) 1
(d) 2

5. What is the value of
$\dfrac{\sin 2° \sin 4° \sin 6° ..... \sin 88°}{\cos 88° \cos 86° \cos 84° .... \cos 2°}$
(Do not use trigonometric tables)
(a) 0
(b) 1
(c) 2
(d) 4

6. If $x + y = 90°$, then what is the value of $\left(1 + \dfrac{\tan x}{\tan y}\right)$ $\sin^2 y$?
(a) 0
(b) $\dfrac{1}{2}$
(c) 1
(d) 2

7. If $\sin A = \cos A$ and A is acute, $\tan A - \cot A$ is equal to :
(a) 2
(b) 1
(c) $\dfrac{1}{2}$
(d) 0

8. Without using trigonometric tables, find the value
of: $\dfrac{2}{3}\left(\dfrac{\sec 56°}{\operatorname{cosec} 34°}\right) - 2\cos^2 20° + \dfrac{1}{2} \cot 28° \cot 35°$
$\cot 45° \cot 62° \cot 55° - 2 \cos^2 70°$
(a) $\dfrac{4}{5}$
(b) $-\dfrac{3}{4}$
(c) $-\dfrac{5}{6}$
(d) 1

9. If $11x$ is an acute angle and $\tan 11x = \cot 7x$, then what is the value of x?
(a) 5°
(b) 6°
(c) 7°
(d) 8°

10. Evaluate $\sin (50° + \theta) - \cos (40° - \theta) + \tan 1°$ $\tan 15° \tan 20° \tan 70° \tan 65° \tan 89° + \sec (90° - \theta).$ $\operatorname{cosec} \theta - \tan (90° - \theta) \cot \theta.$
(a) 0
(b) 1
(c) 2
(d) 3

Answers

1. (d)   **2.** (d)   **3.** (b)   **4.** (c)   **5.** (b)   **6.** (c)   **7.** (d)   **8.** (c)   **9.** (a)   **10.** (c)

Chapter 34

SOME APPLICATIONS OF TRIGONOMETRY

KEY FACTS

1. **Line of Sight**: It is the line drawn from the eyes of the observer to a point on the object, which the observer is viewing.

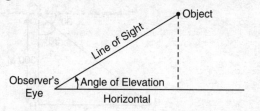

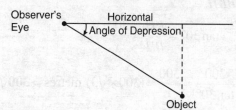

2. **Angle of Elevation:** It is the angle formed by the line of sight with the horizontal through the observer's eye. when the object is above the horizontal level.

3. **Angle of Depression:** It is the angle formed by the line of sight with the horizontal through the observer's eye, when the object is below the horizontal level.

Solved Examples

Ex. 1. *The angle of elevation of the top of a tower from a point at a distance of 100 metres from its foot on a horizontal plane is found to be 60°. Find the height of the tower.*

Sol. Let the height of the tower AC be h metres. Given, distance $BC = 100$ m, $\angle ABC = 60°$.

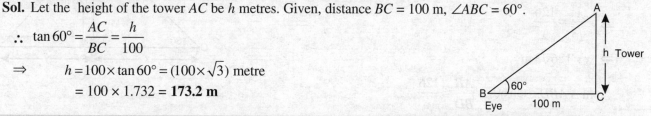

$$\therefore \quad \tan 60° = \frac{AC}{BC} = \frac{h}{100}$$

$$\Rightarrow \quad h = 100 \times \tan 60° = (100 \times \sqrt{3}) \text{ metre}$$

$$= 100 \times 1.732 = \textbf{173.2 m}$$

Ex. 2. *A circus artist is climbing a 20 m long rope, which is tightly stretched and tied from the top of a vertical pole to the ground. Find the height of the pole, if the angle made by the rope with the ground level is 30°.*

Sol. Let $AB = h$ metres be the height of the pole. Given,

$AC = 20$ m be the rope which the circus artist is climbing,

$$\angle ACB = 30°$$

In $\triangle ABC$, $\sin 30° = \dfrac{AB}{AC} = \dfrac{h}{20}$

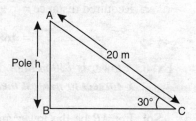

$$\Rightarrow \quad \frac{h}{20} = \frac{1}{2} \Rightarrow h = \frac{1}{2} \times 20 \text{ m} = \textbf{10 m.}$$

Ex. 3. *A kite is flying at a height of 60 m above the ground. The string attached to the kite is temporarily tied to a point on the ground. If the length of the string is $40\sqrt{3}$ m, find the inclination of the string with the ground.*

Sol. Let θ be the inclination of the kite with the ground.

Height of kite above horizontal = KP = 60 m

Length of string of kite = OK = $40\sqrt{3}$ m

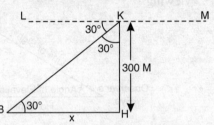

∴ $\sin\theta = \dfrac{KP}{OK} = \dfrac{60}{40\sqrt{3}} = \dfrac{3}{2\sqrt{3}} = \dfrac{\sqrt{3}}{2} = \sin 60°$

⇒ **θ = 60°.**

Ex. 4. *The angle of depression of a boat B from the top K of a cliff HK, 300 metres high is 30°. Find the distance of the boat from the foot H of the cliff.*

Sol. Let the required distance $BH = x$ metres. HK is the cliff and $\angle LKB$ is the angle of depression of B from K. Then, $\angle KBH = \angle LKB = 30°$ (∵ $LM \parallel BH$, alt. $\angle s$)

∴ In $\triangle BKH$,

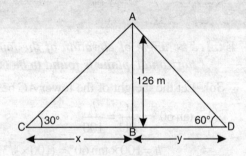

$$\tan 30° = \dfrac{KH}{BH} = \dfrac{300}{x}$$

⇒ $x = \dfrac{300}{\tan 30°} = \dfrac{300}{\frac{1}{\sqrt{3}}} = (300\times\sqrt{3})$ metres = **$300\sqrt{3}$ m.**

Ex. 5. *Two men on either side of a temple 126 m high observe the angle of elevation of the top of the temple to be 30° and 60° respectively. Find the distance between the two men?*

Sol. Let the height of the temple be AB = 126 m

Let the distances of two men from the base of the temple be
$BC = x$ m and $BD = y$ m respectively.

Then, in $\triangle ABC$, $\tan 30° = \dfrac{AB}{BC} = \dfrac{126}{x}$

⇒ $\dfrac{1}{\sqrt{3}} = \dfrac{126}{x}$

⇒ $x = 126\sqrt{3}$ m

In $\triangle ABD$, $\tan 60° = \dfrac{AB}{BD} = \dfrac{126}{y}$

⇒ $\sqrt{3} = \dfrac{126}{y} \Rightarrow y = \dfrac{126}{\sqrt{3}} \times \dfrac{\sqrt{3}}{\sqrt{3}} = \dfrac{126\sqrt{3}}{3}$ m $= 42\sqrt{3}$ m

∴ Required distance = $x + y = 126\sqrt{3}$ m $+ 42\sqrt{3}$ m

$= \mathbf{168\sqrt{3}}$ **m**

Ex. 6. *A tower is 120 m high. Its shadow is x m shorter, when the sun's altitude is 60° than when it was 45°. Find x correct to nearest metre.*

Sol. Let AB be the tower and its shadows be BD and BC corresponding to the angle of elevations 60° and 45° respectively.

In $\triangle ABC$, $\tan 45° = \dfrac{AB}{BC}$

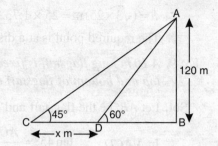

$$\Rightarrow 1 = \frac{120}{BC} \Rightarrow BC = 120 \text{ m}$$

In $\triangle ABD$, $\tan 60° = \frac{AB}{BD} \Rightarrow \sqrt{3} = \frac{120}{BD}$

$$\Rightarrow BD = \frac{120}{\sqrt{3}} = \frac{120}{\sqrt{3}} \times \frac{\sqrt{3}}{\sqrt{3}} = \frac{120\sqrt{3}}{3} = 40\sqrt{3} \text{ metres.}$$

$$\therefore \quad x = BC - BD = 120 - 40\sqrt{3} = 120 \text{ m} - 69.28 \text{ m}$$

$$= 50.72 \text{ m} = \mathbf{51 \ m} \ (\text{ to the nearest m}).$$

Ex. 7. *The angular elevation of a tower from a point is 30°. As we more 100 m nearer to the base of the tower, the angle of elevation becomes 60°. Find the height of the tower and the distance of the first point from the base of the tower.*

Sol. Let AB represent the tower, P, Q the points of observation. It is given $PQ = 100$ m. Angles of elevation at P and Q are respectively 30° and 60° respectively.

Let the required height of the tower $AB = h$ metres and $BQ = x$ metres.

Then, in $\triangle BAQ$, $\frac{h}{x} = \tan 60°$

$$\Rightarrow h = x \tan 60° = x\sqrt{3} \qquad \qquad ...(i)$$

In $\triangle ABP$, $\frac{h}{100 + x} = \tan 30°$

$$\Rightarrow \frac{h}{100 + x} = \frac{1}{\sqrt{3}} \Rightarrow h\sqrt{3} = 100 + x$$

$$\Rightarrow x\sqrt{3} \times \sqrt{3} = 100 + x \Rightarrow 3x = 100 + x \Rightarrow 2x = 100 \Rightarrow x = 50$$

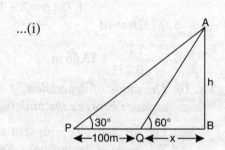

$$\therefore \text{ From (i), } h = 50\sqrt{3} \text{ m} = 50 \times 1.732 \text{ m} = \mathbf{86.6 \ m}$$

Also, $AP = (x + 100) \text{ m} = (50 + 100) \text{ m} = \mathbf{150 \ m.}$

Ex. 8. *Two pillars are of equal height on either sides of a road, which is 100 m wide. The angles of elevation of the top of the pillars are 60° and 30° at a point on the road between the pillars. Find the position of the point between the pillars and height of each pillar.*

Sol. Let AB and ED be two pillars each of height h metres.

Let C be a point on the road BD such that
$BC = x$ metres. Then, $CD = (100 - x)$ metres
Given, $\angle ACB = 60°$ and $\angle ECD = 30°$

In $\triangle ABC$, $\tan 60° = \frac{AB}{BC}$

$$\Rightarrow \sqrt{3} = \frac{h}{x} \Rightarrow h = \sqrt{3}x \qquad \qquad ...(i)$$

In $\triangle ECD$, $\tan 30° = \frac{ED}{CD}$

$$\Rightarrow \frac{1}{\sqrt{3}} = \frac{h}{100 - x} \Rightarrow h\sqrt{3} = 100 - x \qquad \qquad ...(ii)$$

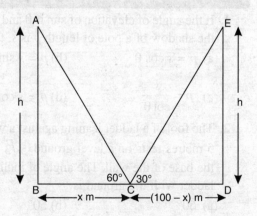

$\therefore$ Subst. the value of h from (i) in (ii), we get

$$\sqrt{3}x \cdot x = 100 - x \Rightarrow 3x = 100 - x \Rightarrow 4x = 100 \Rightarrow x = 25 \text{ m}$$

$\therefore \ h = (\sqrt{3} \times 25)\,m = 25 \times 1.732\,m = 43.3\,m$

$\therefore$ The required point is at a distance of **25 m** from the pillar B and the height of each pillar is **43.3 m.**

Ex. 9. *A 10 m long flagstaff is fixed on the top of a tower from a point on the ground, the angles of elevation of the top and bottom of flagstaff are 45° and 30° respectively. Find the height of the tower.*

Sol. Let AB be the flagstaff and BC be the tower. Let $CD = x$ m and $BC = h$ m

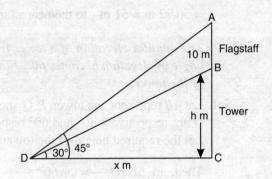

In $\triangle ACD,$ $\tan 45° = \dfrac{AC}{DC} = \dfrac{h+10}{x}$

$\Rightarrow 1 = \dfrac{h+10}{x} \Rightarrow x = h + 10$ ... (i)

In $\triangle BCD,$ $\tan 30° = \dfrac{BC}{DC} = \dfrac{h}{x}$

$\Rightarrow \dfrac{1}{\sqrt{3}} = \dfrac{h}{x} \Rightarrow x = h\sqrt{3} = 1.732h$ ... (ii)

$\therefore$ From (i) and (ii)

$$1.732\,h = h + 10$$

$\Rightarrow 0.732h = 10$

$\Rightarrow h = \dfrac{10}{0.732} = \textbf{13.66 m}.$

Ex. 10. *The angle of depression of 47 m high building from the top of a tower 137 m high is 30°. Calculate the distance between the building and the tower.*

Sol. Let AB and CD represent the tower and building respectively.

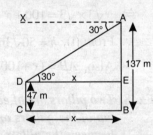

The angle of depression $\angle XAD = 30°.$

In $\triangle ADE,$ $DE = CB = x$

$\angle ADE = \angle XAD = 30°$ (alt. $\angle$s)

$AE = AB - EB = AB - DC = (137 - 47)\,m = 90\,m$

$\therefore \ \tan 30° = \dfrac{AE}{DE} \Rightarrow \dfrac{1}{\sqrt{3}} = \dfrac{90}{x} \Rightarrow x = \textbf{90}\sqrt{\textbf{3}}\,\textbf{m}.$

Question Bank–34

1. If the angle of elevation of sun is θ and the length of the shadow of a pole of length p is s, then
   (a) $p = s \cos \theta$ (b) $p = s \sin \theta$
   (c) $p = \dfrac{s}{\cot \theta}$ (d) $p = s \cot \theta$

2. The foot of a ladder leaning against a wall of length 5 metres rests on a level ground $5\sqrt{3}$ metres from the base of the wall. The angle of inclination of the ladder with the ground is
   (a) 60°   (b) 50°
   (c) 40°   (d) 30°

3. From the top of a light house 60 metre high, with its base at the sea level, the angle of depression of a boat is 30°. The distance of the boat from the foot of the light-house is
   (a) $60\sqrt{3}$ metres   (b) $\dfrac{60}{\sqrt{3}}$ metres
   (c) 60 metres   (d) $30\sqrt{2}$ metres

4. A pole is standing erect on the ground which is horizontal. The tip of the pole is tied tight with a rope of length $\sqrt{12}$ m to a point on the ground. If the rope is making 30° angle with the horizontal, then the height of the pole is
   (a) $2\sqrt{3}$ m   (b) $3\sqrt{2}$ m
   (c) 3 m   (d) $\sqrt{3}$ m

5. Two observers are stationed due north of a tower at a distance of 20 m from each other. If the elevations of the tower observed by them are 30° and 45° respectively, then the height of the tower is :
   - (a) 10 m
   - (b) 16.32
   - (c) $10(\sqrt{3}+1)\,m$
   - (d) 30 m

6. Two ships are sailing in the sea on either side of a light-house. The angle of depression of the two ships are 45° each. If the height of the light-house is 300 metres, then the distance between the ships is :
   - (a) 600 m
   - (b) $600/\sqrt{3}\,m$
   - (c) $300\sqrt{3}\,m$
   - (d) 300 m

7. Two posts are k metres apart. If from the middle point of the line joining their feet, an observer finds the angles of elevations of their tops to be 60° and 30° respectively, then the ratio of heights of the posts respectively is:
   - (a) $\sqrt{3}$
   - (b) $\dfrac{k}{\sqrt{3}}$
   - (c) $k\sqrt{3}$
   - (d) 3

8. A person standing on the bank of a river observed that the angle subtended by the top of a tree on the opposite bank is 60°. If he moves 40 m away from the bank, he finds the angle to be 30°. The breadth of the river is :
   - (a) 20 m
   - (b) 40 m
   - (c) $20\sqrt{3}\,m$
   - (d) $40\sqrt{3}\,m$

9. The horizontal distance between two tree of different heights is 60 m. The angle of depression of the top of the first tree as seen from the top of the second tree is 45°. If the height of the second tree is 80 m, then the height of the first tree is
   - (a) 20 m
   - (b) 24 m
   - (c) 40 m
   - (d) 64 m

10. 2 posts are k metres apart and the height of one is double that of the other. If from the middle point of the line joining their feet, an observer find the angular elevations of their tops to be complementary, then the height (in metres) of the shorter post is :
   - (a) $\dfrac{k}{2\sqrt{2}}$
   - (b) $\dfrac{k}{4}$
   - (c) $k\sqrt{2}$
   - (d) $\dfrac{k}{\sqrt{2}}$

11. A tree AC is broken over by wind from B. D is the point where the top of the broken tree touches the ground and BD makes an angle of 45° with the ground. If the distance between the base of the tree and the point $D = 10$ m. What is the height of the tree?
   - (a) 20 m
   - (b) $10(1+\sqrt{2})\,m$
   - (c) $10\sqrt{2}\,m$
   - (d) $20\sqrt{2}\,m$

12. From a point on the ground the angles of elevation of the bottom and top of a transmission tower fixed at the top of 20 m high building are 45° and 60° respectively. Find the height of the tower?
   - (a) $20\sqrt{3}\,m$
   - (b) 60 m
   - (c) $20(\sqrt{3}-1)\,m$
   - (d) $40\sqrt{3}\,m$

13. A straight highway leads to the foot of a tower of height 50 m. From the top of the tower, angles of depressions of two cars standing on the highway are 30° and 60°. What is the distance between the cars?
   - (a) 43.3 m
   - (b) 57.66 m
   - (c) 86.6 m
   - (d) 100 m

14. The angle of elevation of the top of a hill at the foot of a tower is 60° and the angle of elevation of the top of the tower from the foot of the hill is 30°. If the tower is 50 m high, what is the height of the hill?
   - (a) 100 m
   - (b) 125 m
   - (c) $50\sqrt{3}\,m$
   - (d) 150 m

15. An aeroplane when 3000 m high passes vertically above another aeroplane at an instance when their angles of elevation at the same observation point are 60° and 45°. How many metres higher is this one than the other.
   - (a) 1350 m
   - (b) 1268 m
   - (c) 1000 m
   - (d) 1160 m

Answers

1. (c)	2. (d)	3. (a)	4. (d)	5. (c)	6. (a)	7. (d)	8. (a)	9. (a)	10. (a)
11. (b)	12. (c)	13. (b)	14. (d)	15. (b)					

Hints and Solutions

1. (c) $\dfrac{p}{s} = \tan\theta \Rightarrow p = s\tan\theta = \dfrac{s}{\cot\theta}$

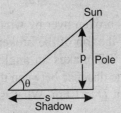

2. (d) $\tan\theta = \dfrac{AB}{BC} = \dfrac{5}{5\sqrt{3}} = \dfrac{1}{\sqrt{3}}$

$\therefore \quad \theta = 30°$

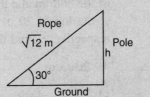

3. (a) Type Solved Ex. 4.

4. (d) Type Solved Ex. 2.

$\therefore \quad \dfrac{h}{\sqrt{12}} = \sin 30°$

Now solve.

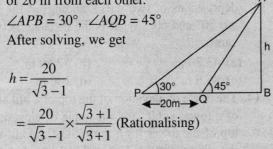

5. (c) Type Solved Ex. 7.

Let P and Q be the two observers at a distance of 20 m from each other.

$\angle APB = 30°$, $\angle AQB = 45°$

After solving, we get

$h = \dfrac{20}{\sqrt{3}-1}$

$= \dfrac{20}{\sqrt{3}-1} \times \dfrac{\sqrt{3}+1}{\sqrt{3}+1}$ (Rationalising)

$= \dfrac{20\sqrt{3}+20}{3-1} = 10(\sqrt{3}+1)\,\text{m}.$

6. (a) Type Solved Ex. 5.

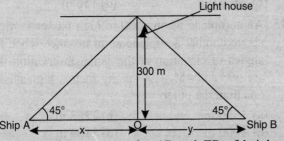

7. (d) Let the two posts be AB and ED of heights h_1 and h_2 respectively. Let the angle of elevation for the top of AB be 60° and the top of ED be 30° respectively. Given, $BC = CD = k/2$.

In $\triangle ABC$, $\dfrac{h_1}{k/2} = \tan 60° \Rightarrow h_1 = \dfrac{k}{2} \times \sqrt{3} = \dfrac{k\sqrt{3}}{2}$

In $\triangle EDC$, $\dfrac{h_2}{k/2} = \tan 30° \Rightarrow h_2 = \dfrac{k}{2} \times \dfrac{1}{\sqrt{3}} = \dfrac{k}{2\sqrt{3}}$

$\therefore \dfrac{h_1}{h_2} = \dfrac{k\sqrt{3}}{2} \Big/ \dfrac{k}{2\sqrt{3}} = \sqrt{3} \times \sqrt{3} = 3.$

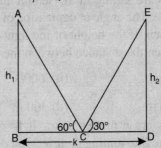

8. (a) Type Solved Example 7.

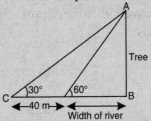

9. (a) Let BF be the first tree of height x m. Then, AS is the second tree of height 80 metres.

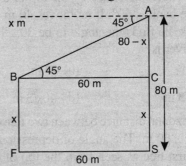

Now, $BC = FS = 60$ m

$CS = BF = x$ m

$\therefore AC = AS - CS = (80 - x)$ m

$\angle ABC = 45°$

In $\triangle ABC$, $\tan 45° = \dfrac{AC}{BC} = \dfrac{80-x}{60}$

$\Rightarrow 1 = \dfrac{80-x}{60} \Rightarrow 60 = 80 - x \Rightarrow x = \mathbf{20\,m.}$

10. (a) Let the two posts be

$AB = 2$ h metres and

$CD = $ h metres

$BE = EC = k/2$

Let $\angle DEC = \theta$

Then, $\angle AEB = 90° - \theta$

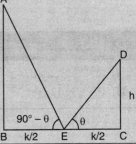

∴ In Δ *ABE*,

$$\frac{2h}{k/2} = \tan(90° - \theta) = \cot\theta$$

$$\Rightarrow \frac{4h}{k} = \cot\theta \qquad \ldots(i)$$

In Δ *DEC*, $\frac{h}{k/2} = \frac{2h}{k} = \tan\theta \qquad \ldots(ii)$

Multiplying (i) and (ii), we get,

$$\frac{4h}{k} \times \frac{2h}{k} = \cot\theta \times \tan\theta \Rightarrow \frac{8h^2}{k^2} = 1 \Rightarrow h^2 = \frac{k^2}{8}$$

$$\Rightarrow h = \frac{k}{2\sqrt{2}}$$

11. (b) Let the tree *AC* be broken from point *B*.

Let *BC* = *x* metres and *AB* = *y* metres

Then *AB* = *BD* (broken part)

∠*BDC* = 45°

In Δ *BDC*,

$$\tan 45° = \frac{BC}{DC} = \frac{x}{10} \Rightarrow 1 = \frac{x}{10} \Rightarrow x = 10\,\text{m}$$

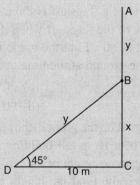

Also,

$$\cos 45° = \frac{DC}{BD} = \frac{10}{y} \Rightarrow \frac{1}{\sqrt{2}} = \frac{10}{y} \Rightarrow y = 10\sqrt{2}\,\text{m}$$

∴ Height of the tree = *x* + *y* = $(10 + 10\sqrt{2})$ m

$$= 10(1 + \sqrt{2})\,\text{m}$$

12. (c) Let *PQ* be the building of height 20 metres,

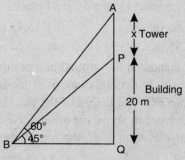

AP be the transmission tower of height *x* metres.

∠*PBQ* = 45°, ∠*ABQ* = 60°

In Δ*PBQ*, $\frac{PQ}{BQ} = \tan 45°$

$$\Rightarrow BQ = PQ = 20\,\text{m} \because \tan 45° = 1$$

In Δ*ABQ*, $\frac{AQ}{BQ} = \tan 60° \Rightarrow \frac{x+20}{20} = \sqrt{3}$

$$\Rightarrow x + 20 = 20\sqrt{3}$$

$$\Rightarrow x = 20\sqrt{3} - 20 = 20(\sqrt{3} - 1)\,\text{metres}.$$

13. (b) Type Solved Ex. 6:

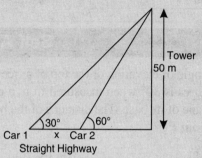

14. (d) Let the tower be *AB* = 50 m and hill be *CD* of height *h* meters.

Given, ∠*DBC* = 60°, ∠*ACB* = 30°

In Δ*ABC*,

$$\frac{AB}{BC} = \tan 30° \Rightarrow \frac{50}{BC} = \frac{1}{\sqrt{3}}$$

$$\Rightarrow BC = 50\sqrt{3}\,\text{metres}.$$

In Δ*DCB*,

$$\frac{DC}{CB} = \tan 60° \Rightarrow \frac{h}{50\sqrt{3}} = \sqrt{3}$$

$$\Rightarrow h = 50\sqrt{3} \times \sqrt{3} = 150\,\text{metres}.$$

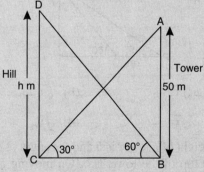

15. (b) Let the aeroplane at a height of 3000 m at point *A* passes over another aeroplane at *B* at a height *x* metres.

∠*ADC* = 60°

(Angle of elevation of higher aeroplane)

∠*BDC* = 45°

(Angle of elevation of lower aeroplane)

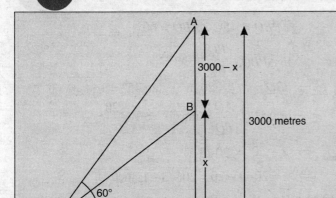

∴   In △ ADC,

$$\frac{AC}{DC} = \tan 60° \Rightarrow \frac{3000}{DC} = \sqrt{3}$$

$$\Rightarrow DC = \frac{3000}{\sqrt{3}} \text{ metres} = 1732 \text{ metres (approx)}$$

∴ In △ BDC,

$$\frac{BC}{DC} = \tan 45° \Rightarrow \frac{x}{1732} = 1 \Rightarrow x = 1732 \text{ m}$$

∴ Difference in heights = (3000 – 1732) metres
= **1268 metres**.

Self Assessment Sheet–33

1. The angle of elevation of the top of a tree of height 18 metres is 30° when measured from a point P in the plane of its base. The distance of the base of the tree from P is

   (a) 6 m   (b) $6\sqrt{3}$ m

   (c) 19 m   (d) $18\sqrt{3}$ m

2. If the shadow of a pole 3 metre high is $3\sqrt{3}$ metre long, then the angle of elevation of the sun is
   (a) 30°   (b) 45°
   (c) 60°   (d) 75°

3. The angle of depression of two posts P and Q at a distance of 2 metres on the same side of a road from a balloon (B) vertically over the road are observed to be 45° and 60°. What is the height of the balloon?

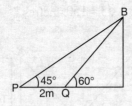

   (a) $3 - \sqrt{3}$ (b) $\sqrt{3} - 1$

   (c) $3 + \sqrt{3}$ (d) $\sqrt{3} + 1$

4. The height of a tower is h and the angle of elevation of the top of the tower is α. On moving a distance $h/2$ towards the tower, the angle of elevation becomes β. What is the value of cot α – cot β.

   (a) $\frac{1}{2}$ (b) $\frac{2}{3}$

   (c) 1   (d) 2

5. Two houses are collinear with the base of a tower and are at distances 3 m and 12 m (on the same

side) from the base of the tower. The angles of elevation from these two houses of the top of the tower are complementary. What is the height of the tower?

   (a) 4 m   (b) 6 m
   (c) 7.5 m   (d) 36 m

6. The shadow of a flagstaff is three times as long as the shadow of the flagstaff when the sun rays meet the ground at 60°. Find the angle between the sun rays and the ground at the time of longer shadow.

   (a) 45°   (b) 30°
   (c) 15°   (d) 90°

7. From a point A on the ground, the angles of elevation of the top of a 10 m tall building and a helicopter hovering at some height of the building are 30° and 60° respectively. Find the height of the helicopter above the building.

   (a) $10\sqrt{3}$ (b) $20(3 + \sqrt{3})$ m

   (c) 20 m   (d) 30 m

8. From two points A and B on the same side of a building the angles of elevation of the top of the building are 30° and 60° respectively. If the height of the building is 10 m, find the distances between A and B correct to two decimal places

   (a) 10.66 m   (b) 13.43 m
   (c) 11.55 m   (d) 12.26 m

9. A man stands on the ground at a point A, which is on the same horizontal plane as B, the foot of a vertical pole BC. The height of the pole is 10 m. The man's eye is 2 m above the ground. He observes the angle of elevation at C, the top of the pole as $x°$, where $\tan x° = \frac{2}{5}$. The distance AB (in metres) is

(a) 15 m   (b) 18 m
(c) 20 m   (d) 16 m

10. The angle of elevation of the top of an unfinished pillar at a point 150 m from its base is 30°. If the angle of elevation at the same point is to be 45°,

then the pillar has to be raised to a height by how many metres?

(a) 59.4 m   (b) 61.4 m
(c) 62.4 m   (d) 63.4 m

Answers

1. (d) **2.** (a) **3.** (c) **4.** (a) **5.** (b) **6.** (b) **7.** (c) **8.** (c) **9.** (c) **10.** (d)

Unit Test – 6

1. If $\tan x = \dfrac{3}{4}$, $0 < x < 90°$, then what is the value of $\sin x \cos x$?

   (a) $\dfrac{3}{5}$ (b) $\dfrac{4}{5}$

   (c) $\dfrac{12}{25}$ (d) $\dfrac{13}{25}$

2. What is the expression $\dfrac{\tan x}{1+\sec x} - \dfrac{\tan x}{1-\sec x}$ equal to?

   (a) $\operatorname{cosec} x$ (b) $2\operatorname{cosec} x$
   (c) $2\sin x$ (d) $2\cos x$

3. If $\tan \theta = 1$ and $\sin \phi = \dfrac{1}{\sqrt{2}}$, then the value of $\cos(\theta + \phi)$ is

   (a) -1 (b) 0

   (c) 1 (d) $\dfrac{\sqrt{3}}{2}$

4. If $\cos \theta = \dfrac{3}{5}$, then the value of $\dfrac{\sin\theta - \tan\theta + 1}{2\tan^2\theta}$ is

   (a) $\dfrac{13}{15}$ (b) $\dfrac{91}{160}$

   (c) $\dfrac{14}{15}$ (d) $\dfrac{92}{160}$

5. Given $x\cos\theta + y\sin\theta = 2$ and $x\cos\theta - y\sin\theta = 0$, then which of the following is correct ?

   (a) $x^2 + y^2 = 1$ (b) $\dfrac{1}{x^2} + \dfrac{1}{y^2} = 1$

   (c) $xy = 1$ (d) $x^2 - y^2 = 1$

6. Which of the following is / are the value(s) of the expression?

   $\sin A (1 + \tan A) + \cos A (1 + \cot A)$?

   1. $\sec A + \operatorname{cosec} A$

2. $2 \operatorname{cosec} A (\sin A + \cos A)$

3. $\tan A + \cot A$

Select the correct answer using the code given below:

   (a) 1 only   (b) 1 and 2 only
   (c) 2 only   (d) 1 and 3 only

7. If $\sin A = \dfrac{2mn}{m^2+n^2}$, what is the value of $\tan A$?

   (a) $\dfrac{2mn}{m^2+n^2}$ (b) $\dfrac{2mn}{m^2-n^2}$

   (c) $\dfrac{m^2-n^2}{2mn}$ (d) $\dfrac{m^2+n^2}{m^2-n^2}$

8. If $\sec^2\theta + \tan^2\theta = \dfrac{5}{3}$ and $0 \le \theta \le \dfrac{\pi}{2}$, then the value of θ is equal to

   (a) $15°$ (b) $30°$
   (c) $45°$ (d) $60°$

9. Evaluate: $\dfrac{5\sin^2 30° + \cos^2 45° + 4\tan^2 60°}{2\sin 30°\cos 60° + \tan 45°}$

   (a) 1 (b) $9\dfrac{1}{6}$

   (c) $7\dfrac{3}{7}$ (d) $\dfrac{47}{12}$

10. Evaluate: $\dfrac{5\cos^2 60° + 4\sec^2 30° - \tan^2 45°}{\sin^2 30° + \cos^2 30°}$

   (a) $2\dfrac{5}{16}$ (b) $\dfrac{67}{12}$

   (c) 0 (d) 1

11. The value of $\sin^2 1° + \sin^2 2° + \sin^2 3° + ... + \sin^2 89° + \sin^2 90°$ is

   (a) 1 (b) 0
   (c) 45.5 (d) 44

12. If $\tan 2A = \cot(A - 60°)$, where $2A$ is an acute angle, then the value of A is
    (a) 30°　　　　　　　(b) 60°
    (c) 50°　　　　　　　(d) 24°

13. Evaluate : $\dfrac{3\cos 53° \operatorname{cosec} 37°}{(\cos^2 29° + \cos^2 61°)} - 3\tan^2 45°$
    (a) 1　　　　　　　　(b) 3
    (c) 6　　　　　　　　(d) 0

14. Evaluate: $\sin\theta\cos\theta - \dfrac{\sin\theta\cos(90° - \theta)\cos\theta}{\sec(90° - \theta)}$
    $- \dfrac{\cos\theta\sin(90° - \theta)\sin\theta}{\operatorname{cosec}(90° - \theta)}$
    (a) –1　　　　　　　(b) 2
    (c) 0　　　　　　　　(d) 1

15. Using trignometric identities, $5\operatorname{cosec}^2\theta - 5\cot^2\theta - 3$ expressed as an integer is
    (a) 5　　　　　　　　(b) 3
    (c) 2　　　　　　　　(d) 0

16. The angle of elevation of the top of a tower at a horizontal distance equal to the height of the tower from the base of the tower is
    (a) 30°　　　　　　　(b) 45°
    (c) 60°　　　　　　　(d) any acute angle

17. A person aims at a bird on top of a 5 metre high pole with an elevation of 30°. If the bullet is fired, it will travel k metre before reaching the bird. The value of k (in meters) is
    (a) $5\sqrt{3}/2$　　　　　(b) 10
    (c) $5\sqrt{3}$　　　　　　(d) $10\sqrt{3}$

18. Horizontal distance between two pillars of different heights is 60 m. It was observed that the angular elevation from the top of the shorter pillar to the top of the taller pillar is 45°. If the height of taller pillar is 130 m, the height of the shorter pillar is
    (a) 45 m　　　　　　(b) 70 m
    (c) 80 m　　　　　　(d) 60 m

19. The angles of elevation of the top of a tower h metre tall from two different points on the same horizontal line are x and y, $(x > y)$. What is the distance between the points ?
    (a) $h(\tan x - \tan y)$　　(b) $\dfrac{h}{\tan x \tan y}$
    (c) $\dfrac{h(\tan x - \tan y)}{\tan x \tan y}$　(d) $\dfrac{h(\tan x \tan y)}{\tan x - \tan y}$

20. A radio transmitter antenna of height 100 m stands at the top of a tall building. At a point on the ground, the angle of elevation of the bottom of the antenna is 45° and that of the top of the antenna is 60°. What is the height of the building?
    (a) 100 m　　　　　(b) 50 m
    (c) $50(\sqrt{3} + 1)\,$m　　(d) $50(\sqrt{3} - 1)\,$m

Answers

1. (c)	2. (b)	3. (b)	4. (c)	5. (b)	6. (a)		

7. (b) $\left[\textbf{Hint.}\ \cos A = \sqrt{1 - \sin^2 A} = \sqrt{1 - \left(\dfrac{2mn}{m^2 + n^2}\right)^2}\ \right]$　　　8. (b)　　9. (b)　　10. (b)

11. (c)	12. (c)	13. (d)	14. (c)	15. (c)	16. (b)	17. (b)	18. (b)	19. (c)	20. (d)